NEC® POCKET GUIDE TO COMMERCIAL AND INDUSTRIAL ELECTRICAL INSTALLATIONS

2017 Edition

TIM McCLINTOCK
JEFF SARGENT

NATIONAL FIRE PROTECTION ASSOCIATION
The leading information and knowledge resource on fire, electrical and related hazards

Product Manager: Debra Rose
Project Editor: Kenneth Ritchie
Composition: Shepherd, Inc.
Printer: Dickinson Press, LLC.

National Fire Protection Association
One Batterymarch Park
Quincy, Massachusetts 02169-7471

NFPA No.: PGNECC17
ISBN: 978-1-4559-1381-7

Library of Congress Control Number: 2016946637

Printed in the United States of America

16 17 18 19 20 5 4 3 2 1

IMPORTANT NOTICES AND DISCLAIMERS OF LIABILITY

This *Pocket Guide* is a compilation of extracts from the 2017 edition of the *National Electrical Code*® (*NEC*®). These extracts have been selected and arranged by the Authors of the *Pocket Guide*, who have also supplied introductory material located at the beginning of each chapter.

The *NEC*, like all NFPA® codes and standards, is a document that is developed through a consensus standards development process approved by the American National Standards Institute. This process brings together volunteers representing varied viewpoints and interests to achieve consensus on fire, electrical, and other safety issues. While the NFPA administers the process and establishes rules to promote fairness in the development of consensus, it does not independently test, evaluate, or verify the accuracy of any information or the soundness of any judgments contained in its codes and standards.

The NFPA and the Authors disclaim liability for any personal injury, property or other damages of any nature whatsoever, whether special, indirect, consequential, or compensatory, directly or indirectly resulting from the publication, use of, or reliance on the *NEC* or this *Pocket Guide*. The NFPA and the Authors also make no guaranty or warranty as to the accuracy or completeness of any information published herein.

In issuing and making the *NEC* and this *Pocket Guide* available, the NFPA and the Author are not undertaking to render professional or other services for or on behalf of any person or entity. Nor is the NFPA or the Authors undertaking to perform any duty owed by any person or entity to someone else. Anyone using these materials should rely on his or her own independent judgment or, as appropriate, seek the advice of a competent professional in determining the exercise of reasonable care in any given circumstances.

This *Pocket Guide* is not intended to be a substitute for the *NEC* itself, which should always be consulted to ensure full *Code* compliance.

Additional Important Notices and Legal Disclaimers concerning the use of the *NEC* can be viewed at www.nfpa.org/disclaimers.

NOTICE CONCERNING CODE INTERPRETATIONS

The introductory materials and the selection and arrangement of the *NEC* extracts contained in this *Pocket Guide* reflect the personal opinions and judgments of the Authors and do not necessarily represent the official position of the NFPA (which can only be obtained through Formal Interpretations processed in accordance with NFPA rules).

REMINDER: UPDATING OF NFPA DOCUMENTS

NFPA 70®, National Electrical Code®, like all NFPA codes, standards, recommended practices, and guides ("NFPA Standards"), may be amended from time to time through the issuance of Tentative Interim Amendments or corrected by Errata. An official NFPA Standard at any point in time consists of the current edition of the document together with any Tentative Interim Amendment and any Errata then in effect.

In order to determine whether an NFPA Standard has been amended through the issuance of Tentative Interim Amendments or corrected by Errata, visit the "Codes & Standards" section on NFPA's website. There, the document information pages located at the "List of NFPA Codes & Standards" provide up-to-date, document-specific information including any issued Tentative Interim Amendments and Errata.

To view the document information page for a specific NFPA Standard, go to http://www.nfpa.org/docinfo to choose from the list of NFPA Standards or use the search feature to select the NFPA Standard number (e.g., *NFPA 70*). The document information page includes postings of all existing Tentative Interim Amendments and Errata. It also includes the option to register for an "Alert" feature to receive an automatic email notification when new updates and other information are posted regarding the document.

CONTENTS/TABS

CONTENTS/TABS

CHAPTER/ARTICLE CROSS-REFERENCE TABLE

Note: *NEC* article number refers to selected requirements from that article.

Chapter No.	*NEC* Article No.
1	90
2	110
3	590
4	220
5	230
6	215
7	408, 404
8	240
9	200
10	250
11	314, 312
12	330, 332, 336
13	342, 344, 348, 350, 352, 356, 358, 362, 368, 376, 378, 386, 388
14	300, 310
15	210
16	210, 406
17	404
18	210, 410, 411
19	430
20	440
21	445
22	450
23	680
24	690, 705
25	700
26	702
27	Chapter 9 Tables, Annex C

ARTICLE/CHAPTER CROSS-REFERENCE TABLE

Note: *NEC* article number refers to selected requirements from that article.

NEC Article No.	Chapter No.
90	1
110	2
200	9
210	15, 16, 18
215	6
220	4
230	5
240	8
250	10
300	14
310	14
312	11
314	11
330	12
332	12
336	12
342	13
344	13
348	13
350	13
352	13
356	13
358	13
362	13
368	13
376	13
378	13
386	13
388	13
404	7, 17
406	16
408	7
410	18
411	18
430	19
440	20
445	21
590	3
680	23
690	24
700	25
702	26
705	24
Annex C	27
Chapter 9 Tables	27

INTRODUCTION

The *Pocket Guide to Commercial and Industrial Electrical Installations* has been developed to provide a portable and practical comprehensive field reference source for electricians, contractors, inspectors, and others responsible for applying the rules of the *National Electrical Code*®. Together with its companion volume, *Pocket Guide to Residential Electrical Installations*, the most widely applied general requirements of the *NEC*® are compiled in these two extremely useful on-the-job, on-the-go documents.

The compiled requirements follow an order similar to the progression of an electrical system installation on the construction site, rather than the normal sequence of requirements in the *NEC*. Introductory chapters are followed by the requirements pertaining to temporary electrical installations. The *Pocket Guide* continues with branch circuit, feeder, and service load calculation requirements from Article 220 and then proceeds into other general requirements extracted from Chapters 2, 3, and 4 of the *NEC.*

Due to its compact size, the *Pocket Guide* does not contain all of the *Code* requirements — it is a complement to, not a replacement of, the 2017 *NEC.* When paired with the entire volume of electrical safety requirements contained in the 2017 *Code,* the *Pocket Guide* is an effective tool that everyone concerned with safe electrical installations will want to have. In-depth explanations of *NEC* requirements are available in NFPA's *National Electrical Code Handbook*, the only *NEC* handbook with the entire text of the *NEC*, plus useful explanation and commentary developed by NFPA staff.

Users of the *Pocket Guide to Commercial and Industrial Electrical Installations* must also understand that some adopting jurisdictions enact requirements to address unique local environmental, geographical, or other considerations. Always consult with the local authority having jurisdiction for information on locally enacted requirements.

The highly specialized requirements contained in Chapters 5 through 8 of the *Code* are not included in this compilation, as this material is outside of the intended purpose of this document. The requirements included in the *Pocket Guide* are those that apply to virtually all commercial and industrial electrical installations and are based on a practical working knowledge of how a commercial and industrial electrical construction project evolves from beginning to completion. As indicated above, however, the *Pocket Guide* is not intended to be a substitute for the *NEC* itself, which should always be consulted to ensure full *Code* compliance.

It is hoped that this 2017 edition of the *Pocket Guide to Commercial and Industrial Electrical Installations* proves to be one of the most valuable tools that you use on a daily basis in performing safe, *NEC*-compliant electrical installations.

New Features of the 2017 Edition

The 2017 edition of this *Pocket Guide* includes features to aid the user in identifying changes in the requirements between the 2014 and 2017 editions of the *NEC*. These usability features parallel those used in the 2017 *NEC*.

Changes other than editorial are indicated with gray shading within sections. An entire figure caption with gray shading indicates a change to an existing figure. New sections, tables, and figures are indicated by a bold, italic ***N*** in a gray box to the left of the new material. An ***N*** next to an Article title indicates that the entire Article is new. Where one or more complete paragraphs have been deleted, the deletion is indicated by a bullet (•) between the paragraphs that remain.

CHAPTER 1

INTRODUCTION TO THE NATIONAL ELECTRICAL CODE

INTRODUCTION

This chapter contains requirements from Article 90—Introduction, which contains foundational elements necessary for proper application and use of the *National Electrical Code®*. Besides providing the purpose of the *Code,* Article 90 specifies the document scope, describes how the *Code* is arranged, provides information on enforcement, and contains other global requirements such as equipment evaluation and how metric conversions are applied. Section 90.1(A) states that "the purpose of this *Code* is the practical safeguarding of persons and property from hazards arising from the use of electricity." Safety — for persons *and* property — has always been the primary objective of the *Code.* Section 90.1(B) specifies that compliance with the *Code* and proper maintenance thereafter results in an installation essentially free from electrical hazards, but not necessarily efficient, convenient, or adequate for good service or future expansion of electrical use. The *Code* is a compilation of installation requirements focused on electrical safety, and its primary focus is not effective or efficient electrical system design. Although state and local jurisdictions typically adopt the *Code* exactly as written, there are some that amend the *Code* or incorporate requirements to address specific local conditions. Therefore, obtain a copy of any amendments or additional rules and regulations for the area(s) where you work.

ARTICLE 90
Introduction

90.1 Purpose.

(A) Practical Safeguarding. The purpose of this *Code* is the practical safeguarding of persons and property from hazards arising from the use of electricity. This *Code* is not intended as a design specification or an instruction manual for untrained persons.

(B) Adequacy. This *Code* contains provisions that are considered necessary for safety. Compliance therewith and proper maintenance result in an installation that is essentially free from hazard but not necessarily efficient, convenient, or adequate for good service or future expansion of electrical use.

90.2 Scope.

(A) Covered. This *Code* covers the installation and removal of electrical conductors, equipment, and raceways; signaling and communications conductors, equipment, and raceways; and optical fiber cables and raceways for the following:

(1) Public and private premises, including buildings, structures, mobile homes, recreational vehicles, and floating buildings
(2) Yards, lots, parking lots, carnivals, and industrial substations
(3) Installations of conductors and equipment that connect to the supply of electricity
(4) Installations used by the electric utility, such as office buildings, warehouses, garages, machine shops, and recreational buildings, that are not an integral part of a generating plant, substation, or control center

(B) Not Covered. This *Code* does not cover the following:

(5) Installations under the exclusive control of an electric utility where such installations

a. Consist of service drops or service laterals, and associated metering, or

b. Are on property owned or leased by the electric utility for the purpose of communications, metering, generation, control, transformation, transmission, energy storage, or distribution of electric energy, or
c. Are located in legally established easements or rights-of-way, or
d. Are located by other written agreements either designated by or recognized by public service commissions, utility commissions, or other regulatory agencies having jurisdiction for such installations. These written agreements shall be limited to installations for the purpose of communications, metering, generation, control, transformation, transmission, energy storage, or distribution of electric energy where legally established easements or rights-of-way cannot be obtained. These installations shall be limited to federal lands, Native American reservations through the U.S. Department of the Interior Bureau of Indian Affairs, military bases, lands controlled by port authorities and state agencies and departments, and lands owned by railroads.

90.3 Code Arrangement. This *Code* is divided into the introduction and nine chapters, as shown in Figure 90.3. Chapters 1, 2, 3, and 4 apply generally. Chapters 5, 6, and 7 apply to special occupancies, special equipment, or other special conditions and may supplement or modify the requirements in Chapters 1 through 7.

Chapter 8 covers communications systems and is not subject to the requirements of Chapters 1 through 7 except where the requirements are specifically referenced in Chapter 8.

Chapter 9 consists of tables that are applicable as referenced.

Informative annexes are not part of the requirements of this *Code* but are included for informational purposes only.

90.4 Enforcement. This *Code* is intended to be suitable for mandatory application by governmental bodies that exercise legal jurisdiction over electrical installations, including signaling and communications systems, and for use by insurance inspectors. The authority having jurisdiction for enforcement of the *Code* has the responsibility for making interpretations of the rules, for deciding on the approval of equipment and materials, and for granting the special permission contemplated in a number of the rules.

By special permission, the authority having jurisdiction may waive specific requirements in this *Code* or permit alternative methods where it is assured that equivalent objectives can be achieved by establishing and maintaining effective safety.

This *Code* may require new products, constructions, or materials that may not yet be available at the time the *Code* is adopted. In such event, the authority having jurisdiction may permit the use of the products, constructions, or materials that comply with the most recent previous edition of this *Code* adopted by the jurisdiction.

90.5 Mandatory Rules, Permissive Rules, and Explanatory Material.

(A) Mandatory Rules. Mandatory rules of this *Code* are those that identify actions that are specifically required or prohibited and are characterized by the use of the terms *shall* or *shall not*.

(B) Permissive Rules. Permissive rules of this *Code* are those that identify actions that are allowed but not required, are normally used to describe options or alternative methods, and are characterized by the use of the terms *shall be permitted* or *shall not be required*.

(C) Explanatory Material. Explanatory material, such as references to other standards, references to related sections of this *Code*, or information related to a *Code* rule, is included in this *Code* in the form of informational notes. Such notes are informational only and are not enforceable as requirements of this *Code*.

Brackets containing section references to another NFPA document are for informational purposes only and are provided as a guide to indicate the source of the extracted text. These bracketed references immediately follow the extracted text.

(D) Informative Annexes. Nonmandatory information relative to the use of the *NEC* is provided in informative annexes. Informative annexes are not part of the enforceable requirements of the *NEC*, but are included for information purposes only.

90.7 Examination of Equipment for Safety. For specific items of equipment and materials referred to in this *Code*, examinations for safety made under standard conditions provide a basis for approval where the record is made generally available through promulgation by organizations properly equipped and qualified for experimental testing, inspections of the run of goods at factories, and service-value determination through field inspections. This avoids the necessity for repetition of examinations by different examiners, frequently with inadequate facilities for such work, and the confusion that would result from conflicting reports on the suitability of devices and materials examined for a given purpose.

It is the intent of this *Code* that factory-installed internal wiring or the construction of equipment need not be inspected at the time of installation of the equipment, except to detect alterations or damage, if the equipment has been listed by a qualified electrical testing laboratory that is recognized as having the facilities described in the preceding paragraph and that requires suitability for installation in accordance with this *Code*. Suitability shall be determined by application of requirements that are compatible with this *Code*.

90.9 Units of Measurement.

(C) Permitted Uses of Soft Conversion. The cases given in 90.9(C)(1) through (C)(4) shall not be required to use hard conversion and shall be permitted to use soft conversion.

(1) Trade Sizes. Where the actual measured size of a product is not the same as the nominal size, trade size designators shall be used rather than dimensions. Trade practices shall be followed in all cases.

(2) Extracted Material. Where material is extracted from another standard, the context of the original material shall not be compromised or violated. Any editing of the extracted text shall be confined to making the style consistent with that of the *NEC*.

(3) Industry Practice. Where industry practice is to express units in inch-pound units, the inclusion of SI units shall not be required.

(4) Safety. Where a negative impact on safety would result, soft conversion shall be used.

(D) Compliance. Conversion from inch-pound units to SI units shall be permitted to be an approximate conversion. Compliance with the numbers shown in either the SI system or the inch-pound system shall constitute compliance with this *Code*.

Informational Note No. 1: Hard conversion is considered a change in dimensions or properties of an item into new sizes that might or might not be interchangeable with the sizes used in the original measurement. Soft conversion is considered a direct mathematical conversion and involves a change in the description of an existing measurement but not in the actual dimension.

CHAPTER 2

GENERAL INSTALLATION REQUIREMENTS

INTRODUCTION

This chapter contains requirements from Article 110—Requirements for Electrical Installations. These important general requirements address fundamental safety concepts that are essential to the installation of conductors and equipment. The requirements of Article 110 apply to all electrical installations covered within the scope of the *Code.* Included in Article 110 are requirements for the approval, examination, identification, installation, use, and access to and spaces about electrical equipment (and conductors). The general installation requirements are segregated into two major voltages ranges: 0–1000 volts, and 1000 volts and above. Compliance with the requirements covering the working space around electric equipment is essential to the safety of workers. Sufficient access and working space must be provided and maintained around all electrical equipment to permit ready and safe operation and maintenance of such equipment. Section 110.26 provides specific dimensions for the height, width, and depth of the working space for equipment rated up to 1000 volts. The work space requirements of 110.26(A) apply to all equipment that is likely to require examination, adjustment, servicing, or maintenance while energized; whereas the dedicated space requirements of 110.26(E) are applicable only to switchboards, panelboards, distribution boards, and motor control centers. For equipment rated over 1000 volts, work space requirements are contained in 110.34.

ARTICLE 110
Requirements for Electrical Installations

PART I. GENERAL

110.2 Approval. The conductors and equipment required or permitted by this *Code* shall be acceptable only if approved.

110.3 Examination, Identification, Installation, Use, and Listing (Product Certification) of Equipment.

(A) Examination. In judging equipment, considerations such as the following shall be evaluated:

(1) Suitability for installation and use in conformity with the provisions of this *Code*
(2) Mechanical strength and durability, including, for parts designed to enclose and protect other equipment, the adequacy of the protection thus provided
(3) Wire-bending and connection space
(4) Electrical insulation
(5) Heating effects under normal conditions of use and also under abnormal conditions likely to arise in service
(6) Arcing effects
(7) Classification by type, size, voltage, current capacity, and specific use
(8) Other factors that contribute to the practical safeguarding of persons using or likely to come in contact with the equipment

(B) Installation and Use. Listed or labeled equipment shall be installed and used in accordance with any instructions included in the listing or labeling.

110.8 Wiring Methods. Only wiring methods recognized as suitable are included in this *Code*. The recognized methods of wiring shall be permitted to be installed in any type of building or occupancy, except as otherwise provided in this *Code*.

110.9 Interrupting Rating. Equipment intended to interrupt current at fault levels shall have an interrupting rating at nominal circuit voltage at least equal to the current that is available at the line terminals of the equipment.

Equipment intended to interrupt current at other than fault levels shall have an interrupting rating at nominal circuit voltage at least equal to the current that must be interrupted.

110.10 Circuit Impedance, Short-Circuit Current Ratings, and Other Characteristics. The overcurrent protective devices, the total impedance, the equipment short-circuit current ratings, and other characteristics of the circuit to be protected shall be selected and coordinated to permit the circuit protective devices used to clear a fault to do so without extensive damage to the electrical equipment of the circuit. This fault shall be assumed to be either between two or more of the circuit conductors or between any circuit conductor and the equipment grounding conductor(s) permitted in 250.118. Listed equipment applied in accordance with their listing shall be considered to meet the requirements of this section.

110.11 Deteriorating Agents. Unless identified for use in the operating environment, no conductors or equipment shall be located in damp or wet locations; where exposed to gases, fumes, vapors, liquids, or other agents that have a deteriorating effect on the conductors or equipment; or where exposed to excessive temperatures.

Equipment not identified for outdoor use and equipment identified only for indoor use, such as "dry locations," "indoor use only," "damp locations," or enclosure Types 1, 2, 5, 12, 12K, and/or 13, shall be protected against damage from the weather during construction.

110.12 Mechanical Execution of Work. Electrical equipment shall be installed in a neat and workmanlike manner.

(A) Unused Openings. Unused openings, other than those intended for the operation of equipment, those intended for mounting purposes, or those permitted as part of the design for listed equipment, shall be closed to afford protection substantially equivalent to the wall of the equipment. Where metallic plugs or plates are used with nonmetallic enclosures, they shall be recessed at least 6 mm (¼ in.) from the outer surface of the enclosure.

(B) Integrity of Electrical Equipment and Connections. Internal parts of electrical equipment, including busbars, wiring terminals, insulators, and other surfaces, shall not be damaged or contaminated by foreign materials such as paint, plaster, cleaners, abrasives, or corrosive residues. There shall be no damaged parts that may adversely affect safe operation or mechanical strength of the equipment such as parts that are broken; bent; cut; or deteriorated by corrosion, chemical action, or overheating.

110.13 Mounting and Cooling of Equipment.

(A) Mounting. Electrical equipment shall be firmly secured to the surface on which it is mounted. Wooden plugs driven into holes in masonry, concrete, plaster, or similar materials shall not be used.

(B) Cooling. Electrical equipment that depends on the natural circulation of air and convection principles for cooling of exposed surfaces shall be installed so that room airflow over such surfaces is not prevented by walls or by adjacent installed equipment. For equipment designed for floor mounting, clearance between top surfaces and adjacent surfaces shall be provided to dissipate rising warm air.

Electrical equipment provided with ventilating openings shall be installed so that walls or other obstructions do not prevent the free circulation of air through the equipment.

110.14 Electrical Connections. Because of different characteristics of dissimilar metals, devices such as pressure terminal or pressure splicing connectors and soldering lugs shall be identified for the material of the conductor and shall be properly installed and used. Conductors of dissimilar metals shall not be intermixed in a terminal or splicing connector where physical contact occurs between dissimilar conductors (such as copper and aluminum, copper and copper-clad aluminum, or aluminum and copper-clad aluminum), unless the device is identified for the purpose and conditions of use. Materials such as solder, fluxes, inhibitors, and compounds, where employed, shall be suitable for the use and shall be of a type that will not adversely affect the conductors, installation, or equipment.

Connectors and terminals for conductors more finely stranded than Class B and Class C stranding as shown in Chapter 9, Table 10, shall be identified for the specific conductor class or classes.

(A) Terminals. Connection of conductors to terminal parts shall ensure a thoroughly good connection without damaging the conductors and shall be made by means of pressure connectors (including set-screw type), solder lugs, or splices to flexible leads. Connection by means of wire-binding screws or studs and nuts that have upturned lugs or the equivalent shall be permitted for 10 AWG or smaller conductors.

Terminals for more than one conductor and terminals used to connect aluminum shall be so identified.

(B) Splices. Conductors shall be spliced or joined with splicing devices identified for the use or by brazing, welding, or soldering with a fusible metal or alloy. Soldered splices shall first be spliced or joined so as to be mechanically and electrically secure without solder and then be soldered. All splices and joints and the free ends of conductors shall be covered with an insulation equivalent to that of the conductors or with an identified insulating device.

Wire connectors or splicing means installed on conductors for direct burial shall be listed for such use.

(C) Temperature Limitations. The temperature rating associated with the ampacity of a conductor shall be selected and coordinated so as not to exceed the lowest temperature rating of any connected termination, conductor, or device. Conductors with temperature ratings higher than specified for terminations shall be permitted to be used for ampacity adjustment, correction, or both.

(1) Equipment Provisions. The determination of termination provisions of equipment shall be based on 110.14(C)(1)(a) or (C)(1)(b). Unless the equipment is listed and marked otherwise, conductor ampacities used in determining equipment termination provisions shall be based on Table 310.15(B)(16) as appropriately modified by 310.15(B)(7).

(a) Termination provisions of equipment for circuits rated 100 amperes or less, or marked for 14 AWG through 1 AWG conductors, shall be used only for one of the following:

(1) Conductors rated 60°C (140°F).
(2) Conductors with higher temperature ratings, provided the ampacity of such conductors is determined based on the 60°C (140°F) ampacity of the conductor size used.

(3) Conductors with higher temperature ratings if the equipment is listed and identified for use with such conductors.
(4) For motors marked with design letters B, C, or D, conductors having an insulation rating of 75°C (167°F) or higher shall be permitted to be used, provided the ampacity of such conductors does not exceed the 75°C (167°F) ampacity.

(b) Termination provisions of equipment for circuits rated over 100 amperes, or marked for conductors larger than 1 AWG, shall be used only for one of the following:

(1) Conductors rated 75°C (167°F)
(2) Conductors with higher temperature ratings, provided the ampacity of such conductors does not exceed the 75°C (167°F) ampacity of the conductor size used, or up to their ampacity if the equipment is listed and identified for use with such conductors

(2) Separate Connector Provisions. Separately installed pressure connectors shall be used with conductors at the ampacities not exceeding the ampacity at the listed and identified temperature rating of the connector.

N **(D) Installation.** Where a tightening torque is indicated as a numeric value on equipment or in installation instructions provided by the manufacturer, a calibrated torque tool shall be used to achieve the indicated torque value, unless the equipment manufacturer has provided installation instructions for an alternative method of achieving the required torque.

110.15 High-Leg Marking. On a 4-wire, delta-connected system where the midpoint of one phase winding is grounded, only the conductor or busbar having the higher phase voltage to ground shall be durably and permanently marked by an outer finish that is orange in color or by other effective means. Such identification shall be placed at each point on the system where a connection is made if the grounded conductor is also present.

110.16 Arc-Flash Hazard Warning.

(A) General. Electrical equipment, such as switchboards, switchgear, panelboards, industrial control panels, meter socket enclosures, and motor control centers, that is in other than dwelling units, and is likely to require examination, adjustment, servicing, or maintenance while energized, shall be field or factory marked to warn qualified persons of potential electric arc flash hazards. The marking shall meet the requirements in 110.21(B) and shall be located so as to be clearly visible to qualified persons before examination, adjustment, servicing, or maintenance of the equipment.

(B) Service Equipment. In other than dwelling units, in addition to the requirements in (A), a permanent label shall be field or factory applied to service equipment rated 1200 amps or more. The label shall meet the requirements of 110.21(B) and contain the following information:

(1) Nominal system voltage
(2) Available fault current at the service overcurrent protective devices
(3) The clearing time of service overcurrent protective devices based on the available fault current at the service equipment
(4) The date the label was applied

Exception: Service equipment labeling shall not be required if an arc flash label is applied in accordance with acceptable industry practice.

110.21 Marking.

(A) Equipment Markings.

(1) General. The manufacturer's name, trademark, or other descriptive marking by which the organization responsible for the product can be identified shall be placed on all electrical equipment. Other markings that indicate voltage, current, wattage, or other ratings shall be provided as specified elsewhere in this *Code*. The marking or label shall be of sufficient durability to withstand the environment involved.

N **(2) Reconditioned Equipment.** Reconditioned equipment shall be marked with the name, trademark, or other descriptive marking by which the organization responsible for reconditioning the electrical equipment can be identified, along with the date of the reconditioning.

Reconditioned equipment shall be identified as "reconditioned" and approval of the reconditioned equipment shall not be based solely on the equipment's original listing.

Exception: In industrial occupancies, where conditions of maintenance and supervision ensure that only qualified persons service the equipment, the markings indicated in 110.21(A)(2) shall not be required.

(B) Field-Applied Hazard Markings. Where caution, warning, or danger signs or labels are required by this *Code*, the labels shall meet the following requirements:

(1) The marking shall warn of the hazards using effective words, colors, symbols, or any combination thereof.
(2) The label shall be permanently affixed to the equipment or wiring method and shall not be handwritten.
(3) The label shall be of sufficient durability to withstand the environment involved.

110.22 Identification of Disconnecting Means.

(A) General. Each disconnecting means shall be legibly marked to indicate its purpose unless located and arranged so the purpose is evident. The marking shall be of sufficient durability to withstand the environment involved.

110.24 Available Fault Current.

(A) Field Marking. Service equipment at other than dwelling units shall be legibly marked in the field with the maximum available fault current. The field marking(s) shall include the date the fault-current calculation was performed and be of sufficient durability to withstand the environment involved. The calculation shall be documented and made available to those authorized to design, install, inspect, maintain, or operate the system.

(B) Modifications. When modifications to the electrical installation occur that affect the maximum available fault current at the service, the maximum available fault current shall be verified or recalculated as necessary to ensure the service equipment ratings are sufficient for the maximum available fault current at the line terminals of the equipment.

The required field marking(s) in 110.24(A) shall be adjusted to reflect the new level of maximum available fault current.

Exception: The field marking requirements in 110.24(A) and 110.24(B) shall not be required in industrial installations where conditions of maintenance and supervision ensure that only qualified persons service the equipment.

110.25 Lockable Disconnecting Means. If a disconnecting means is required to be lockable open elsewhere in this *Code*, it shall be capable of being locked in the open position. The provisions for locking shall remain in place with or without the lock installed.

Exception: Locking provisions for a cord-and-plug connection shall not be required to remain in place without the lock installed.

PART II. 1000 VOLTS, NOMINAL, OR LESS

110.26 Spaces About Electrical Equipment. Access and working space shall be provided and maintained about all electrical equipment to permit ready and safe operation and maintenance of such equipment.

(A) Working Space. Working space for equipment operating at 1000 volts, nominal, or less to ground and likely to require examination, adjustment, servicing, or maintenance while energized shall comply with the dimensions of 110.26(A)(1), (A)(2), (A)(3), and (A)(4) or as required or permitted elsewhere in this *Code*.

(1) Depth of Working Space. The depth of the working space in the direction of live parts shall not be less than that specified in Table 110.26(A)(1) unless the requirements of 110.26(A)(1)(a), (A)(1)(b), or (A)(1)(c) are met. Distances shall be measured from the exposed live parts or from the enclosure or opening if the live parts are enclosed.

(a) *Dead-Front Assemblies.* Working space shall not be required in the back or sides of assemblies, such as dead-front switchboards, switchgear, or motor control centers, where all connections and all renewable or adjustable parts,

Table 110.26(A)(1) Working Spaces

Nominal Voltage to Ground	Minimum Clear Distance		
	Condition 1	Condition 2	Condition 3
0–150	900 mm (3 ft)	900 mm (3 ft)	900 mm (3 ft)
151–600	900 mm (3 ft)	1.0 m (3 ft 6 in.)	1.2 m (4 ft)
601–1000	900 mm (3 ft)	1.2 m (4 ft)	1.5 m (5 ft)

Note: Where the conditions are as follows:

Condition 1 — Exposed live parts on one side of the working space and no live or grounded parts on the other side of the working space, or exposed live parts on both sides of the working space that are effectively guarded by insulating materials.

Condition 2 — Exposed live parts on one side of the working space and grounded parts on the other side of the working space. Concrete, brick, or tile walls shall be considered as grounded.

Condition 3 — Exposed live parts on both sides of the working space.

such as fuses or switches, are accessible from locations other than the back or sides. Where rear access is required to work on nonelectrical parts on the back of enclosed equipment, a minimum horizontal working space of 762 mm (30 in.) shall be provided.

(b) *Low Voltage.* By special permission, smaller working spaces shall be permitted where all exposed live parts operate at not greater than 30 volts rms, 42 volts peak, or 60 volts dc.

(c) *Existing Buildings.* In existing buildings where electrical equipment is being replaced, Condition 2 working clearance shall be permitted between dead-front switchboards, switchgear, panelboards, or motor control centers located across the aisle from each other where conditions of maintenance and supervision ensure that written procedures have been adopted to prohibit equipment on both sides of the aisle from being open at the same time and qualified persons who are authorized will service the installation.

(2) Width of Working Space. The width of the working space in front of the electrical equipment shall be the width of the equipment or 762 mm (30 in.), whichever is greater. In all cases, the work space shall permit at least a 90 degree opening of equipment doors or hinged panels.

(3) Height of Working Space. The work space shall be clear and extend from the grade, floor, or platform to a height of 2.0 m (6½ ft) or the height of the equipment, whichever is greater. Within the height requirements of this section, other equipment that is associated with the electrical installation and is located above or below the electrical equipment shall be permitted to extend not more than 150 mm (6 in.) beyond the front of the electrical equipment.

Exception No. 2: Meters that are installed in meter sockets shall be permitted to extend beyond the other equipment. The meter socket shall be required to follow the rules of this section.

(B) Clear Spaces. Working space required by this section shall not be used for storage. When normally enclosed live parts are exposed for inspection or servicing, the working space, if in a passageway or general open space, shall be suitably guarded.

(C) Entrance to and Egress from Working Space.

(1) Minimum Required. At least one entrance of sufficient area shall be provided to give access to and egress from working space about electrical equipment.

(2) Large Equipment. For equipment rated 1200 amperes or more and over 1.8 m (6 ft) wide that contains overcurrent devices, switching devices, or control devices, there shall be one entrance to and egress from the required working space not less than 610 mm (24 in.) wide and 2.0 m (6½ ft) high at each end of the working space.

A single entrance to and egress from the required working space shall be permitted where either of the conditions in 110.26(C)(2)(a) or (C)(2)(b) is met.

(a) *Unobstructed Egress.* Where the location permits a continuous and unobstructed way of egress travel, a single entrance to the working space shall be permitted.

(b) *Extra Working Space.* Where the depth of the working space is twice that required by 110.26(A)(1), a single entrance shall be permitted. It shall be located such that the distance from the equipment to the nearest edge of the entrance is not less than the minimum clear distance specified in Table 110.26(A)(1) for equipment operating at that voltage and in that condition.

(3) Personnel Doors. Where equipment rated 800 A or more that contains overcurrent devices, switching devices, or control devices is installed and there is a personnel door(s) intended for entrance to and egress from the working space less than 7.6 m (25 ft) from the nearest edge of the working space, the door(s) shall open in the direction of egress and be equipped with listed panic hardware.

(D) Illumination. Illumination shall be provided for all working spaces about service equipment, switchboards, switchgear, panelboards, or motor control centers installed indoors. Control by automatic means only shall not be permitted. Additional lighting outlets shall not be required where the work space is illuminated by an adjacent light source or as permitted by 210.70(A)(1), Exception No. 1, for switched receptacles.

(E) Dedicated Equipment Space. All switchboards, switchgear, panelboards, and motor control centers shall be located in dedicated spaces and protected from damage.

Exception: Control equipment that by its very nature or because of other rules of the Code must be adjacent to or within sight of its operating machinery shall be permitted in those locations.

(1) Indoor. Indoor installations shall comply with 110.26(E)(1)(a) through (E)(1)(d).

(a) *Dedicated Electrical Space.* The space equal to the width and depth of the equipment and extending from the floor to a height of 1.8 m (6 ft) above the equipment or to the structural ceiling, whichever is lower, shall be dedicated to the electrical installation. No piping, ducts, leak protection apparatus, or other equipment foreign to the electrical installation shall be located in this zone.

Exception: Suspended ceilings with removable panels shall be permitted within the 1.8-m (6-ft) zone.

(b) *Foreign Systems.* The area above the dedicated space required by 110.26(E)(1)(a) shall be permitted to contain foreign systems, provided protection is installed to avoid damage to the electrical equipment from condensation, leaks, or breaks in such foreign systems.

(c) *Sprinkler Protection.* Sprinkler protection shall be permitted for the dedicated space where the piping complies with this section.

(d) *Suspended Ceilings.* A dropped, suspended, or similar ceiling that does not add strength to the building structure shall not be considered a structural ceiling.

(2) Outdoor. Outdoor installations shall comply with 110.26(E)(2)(a) through (c).

(a) *Installation Requirements.* Outdoor electrical equipment shall be the following:

(1) Installed in identified enclosures
(2) Protected from accidental contact by unauthorized personnel or by vehicular traffic
(3) Protected from accidental spillage or leakage from piping systems

(b) *Work Space.* The working clearance space shall include the zone described in 110.26(A). No architectural appurtenance or other equipment shall be located in this zone.

Exception: Structural overhangs or roof extensions shall be permitted in this zone.

(c) *Dedicated Equipment Space.* The space equal to the width and depth of the equipment, and extending from grade to a height of 1.8 m (6 ft) above the equipment, shall be dedicated to the electrical installation. No piping or other equipment foreign to the electrical installation shall be located in this zone.

(F) Locked Electrical Equipment Rooms or Enclosures. Electrical equipment rooms or enclosures housing electrical apparatus that are controlled by a lock(s) shall be considered accessible to qualified persons.

110.27 Guarding of Live Parts.

(A) Live Parts Guarded Against Accidental Contact. Except as elsewhere required or permitted by this *Code,* live parts of electrical equipment operating at 50 to 1000 volts, nominal shall be guarded against accidental contact by approved enclosures or by any of the following means:

(1) By location in a room, vault, or similar enclosure that is accessible only to qualified persons.
(2) By permanent, substantial partitions or screens arranged so that only qualified persons have access to the space within reach of the live parts. Any openings in such partitions or screens shall be sized and located so that persons are not likely to come into accidental contact with the live parts or to bring conducting objects into contact with them.
(3) By location on a balcony, gallery, or platform elevated and arranged so as to exclude unqualified persons.
(4) By elevation above the floor or other working surface as follows:

    a. A minimum of 2.5 m (8 ft) for 50 volts to 300 volts between ungrounded conductors
    b. A minimum of 2.6 m (8 ft 6 in.) for 301 volts to 600 volts between ungrounded conductors
    c. A minimum of 2.62 m (8 ft 7 in.) for 601 volts to 1000 volts between ungrounded conductors

(B) Prevent Physical Damage. In locations where electrical equipment is likely to be exposed to physical damage, enclosures or guards shall be so arranged and of such strength as to prevent such damage.

(C) Warning Signs. Entrances to rooms and other guarded locations that contain exposed live parts shall be marked with conspicuous warning signs forbidding unqualified persons to enter. The marking shall meet the requirements in 110.21(B).

110.28 Enclosure Types. Enclosures (other than surrounding fences or walls covered in 110.31) of switchboards, switchgear, panelboards, industrial control panels, motor control centers, meter sockets, enclosed switches, transfer switches, power outlets, circuit breakers, adjustable-speed drive systems, pullout switches, portable power distribution equipment, termination boxes, general-purpose transformers, fire pump controllers, fire pump motors, and motor controllers, rated not over 1000 volts nominal and intended for such locations, shall be marked with an enclosure-type number as shown in Table 110.28.

Table 110.28 shall be used for selecting these enclosures for use in specific locations other than hazardous (classified) locations. The enclosures are not intended to protect against conditions such as condensation, icing, corrosion, or contamination that may occur within the enclosure or enter via the conduit or unsealed openings.

PART III. OVER 1000 VOLTS, NOMINAL

110.31 Enclosure for Electrical Installations. Electrical installations in a vault, room, or closet or in an area surrounded by a wall, screen, or fence, access to which is controlled by a lock(s) or other approved means, shall be considered to be accessible to qualified persons only. The type of enclosure used in a given case shall be designed and constructed according to the nature and degree of the hazard(s) associated with the installation.

For installations other than equipment as described in 110.31(D), a wall, screen, or fence shall be used to enclose an outdoor electrical installation to deter access by persons who are not qualified. A fence shall not be less than 2.1 m (7 ft) in height or a combination of 1.8 m (6 ft) or more of fence fabric and a 300 mm (1 ft) or more extension utilizing three or more strands of barbed wire or equivalent. The distance from the fence to live parts shall be not less than given in Table 110.31.

Table 110.28 Enclosure Selection

Provides a Degree of Protection Against the Following Environmental Conditions	For Outdoor Use — Enclosure-Type Number									
	3	3R	3S	3X	3RX	3SX	4	4X	6	6P
Incidental contact with the enclosed equipment	X	X	X	X	X	X	X	X	X	X
Rain, snow, and sleet	X	X	X	X	X	X	X	X	X	X
Sleet*	—	—	X	—	—	X	—	—	—	—
Windblown dust	X	—	X	X	—	X	X	X	X	X
Hosedown	—	—	—	—	—	—	X	X	X	X
Corrosive agents	—	—	—	X	X	X	—	X	—	X
Temporary submersion	—	—	—	—	—	—	—	—	X	X
Prolonged submersion	—	—	—	—	—	—	—	—	—	X

Provides a Degree of Protection Against the Following Environmental Conditions	For Indoor Use — Enclosure-Type Number									
	1	2	4	4X	5	6	6P	12	12K	13
Incidental contact with the enclosed equipment	X	X	X	X	X	X	X	X	X	X
Falling dirt	X	X	X	X	X	X	X	X	X	X
Falling liquids and light splashing	—	X	X	X	X	X	X	X	X	X

Circulating dust, lint, fibers, and flyings	—	—	X	X	—	X	X	X	X	X
Settling airborne dust, lint, fibers, and flyings	—	—	X	X	X	X	X	X	X	X
Hosedown and splashing water	—	—	X	X	—	X	X	—	—	—
Oil and coolant seepage	—	—	—	—	—	—	—	X	X	X
Oil or coolant spraying and splashing	—	—	—	—	—	—	—	—	—	X
Corrosive agents	—	—	—	X	—	—	X	—	—	—
Temporary submersion	—	—	—	—	—	X	X	—	—	—
Prolonged submersion	—	—	—	—	—	—	X	—	—	—

*Mechanism shall be operable when ice covered.

Informational Note No. 1: The term *raintight* is typically used in conjunction with Enclosure Types 3, 3S, 3SX, 3X, 4, 4X, 6, and 6P. The term *rainproof* is typically used in conjunction with Enclosure Types 3R and 3RX. The term *watertight* is typically used in conjunction with Enclosure Types 4, 4X, 6, and 6P. The term *driptight* is typically used in conjunction with Enclosure Types 2, 5, 12, 12K, and 13. The term *dusttight* is typically used in conjunction with Enclosure Types 3, 3S, 3SX, 3X, 5, 12, 12K, and 13.

Informational Note No. 2: Ingress protection (IP) ratings may be found in ANSI/IEC 60529, *Degrees of Protection Provided by Enclosures*. IP ratings are not a substitute for Enclosure Type ratings.

Table 110.31 Minimum Distance from Fence to Live Parts

Nominal Voltage	Minimum Distance to Live Parts	
	m	ft
1001–13,799	3.05	10
13,800–230,000	4.57	15
Over 230,000	5.49	18

Note: For clearances of conductors for specific system voltages and typical BIL ratings, see ANSI/IEEE C2-2012, *National Electrical Safety Code*.

110.32 Work Space About Equipment. Sufficient space shall be provided and maintained about electrical equipment to permit ready and safe operation and maintenance of such equipment. Where energized parts are exposed, the minimum clear work space shall be not less than 2.0 m (6½ ft) high (measured vertically from the floor or platform) or not less than 914 mm (3 ft) wide (measured parallel to the equipment). The depth shall be as required in 110.34(A). In all cases, the work space shall permit at least a 90 degree opening of doors or hinged panels.

110.33 Entrance to Enclosures and Access to Working Space.

(A) Entrance. At least one entrance to enclosures for electrical installations as described in 110.31 not less than 610 mm (24 in.) wide and 2.0 m (6½ ft) high shall be provided to give access to the working space about electrical equipment.

(1) Large Equipment. On switchgear and control panels exceeding 1.8 m (6 ft) in width, there shall be one entrance at each end of the equipment. A single entrance to the required working space shall be permitted where either of the conditions in 110.33(A)(1)(a) or (A)(1)(b) is met.

(a) *Unobstructed Exit.* Where the location permits a continuous and unobstructed way of exit travel, a single entrance to the working space shall be permitted.

(b) *Extra Working Space.* Where the depth of the working space is twice that required by 110.34(A), a single entrance shall be permitted. It shall be located so that the distance from the equipment to the nearest edge of the entrance is not less than the minimum clear distance specified in Table 110.34(A) for equipment operating at that voltage and in that condition.

(2) Guarding. Where bare energized parts at any voltage or insulated energized parts above 1000 volts, nominal, are located adjacent to such entrance, they shall be suitably guarded.

(3) Personnel Doors. Where there is a personnel door(s) intended for entrance to and egress from the working space less than 7.6 m (25 ft) from the nearest edge of the working space, the door(s) shall open in the direction of egress and be equipped with listed panic hardware.

(B) Access. Permanent ladders or stairways shall be provided to give safe access to the working space around electrical equipment installed on platforms, balconies, or mezzanine floors or in attic or roof rooms or spaces.

110.34 Work Space and Guarding.

(A) Working Space. Except as elsewhere required or permitted in this *Code*, equipment likely to require examination, adjustment, servicing, or maintenance while energized shall have clear working space in the direction of access to live parts of the electrical equipment and shall be not less than specified in Table 110.34(A). Distances shall be measured from the live parts, if such are exposed, or from the enclosure front or opening if such are enclosed.

Exception: Working space shall not be required in back of equipment such as switchgear or control assemblies where there are no renewable or adjustable parts (such as fuses or switches) on the back and where all connections are accessible from locations other than the back. Where rear access is required to work on nonelectrical parts on the back of enclosed equipment, a minimum working space of 762 mm (30 in.) horizontally shall be provided.

Table 110.34(A) Minimum Depth of Clear Working Space at Electrical Equipment

Nominal Voltage to Ground	Minimum Clear Distance		
	Condition 1	Condition 2	Condition 3
1001–2500 V	900 mm (3 ft)	1.2 m (4 ft)	1.5 m (5 ft)
2501–9000 V	1.2 m (4 ft)	1.5 m (5 ft)	1.8 m (6 ft)
9001–25,000 V	1.5 m (5 ft)	1.8 m (6 ft)	2.8 m (9 ft)
25,001 V–75 kV	1.8 m (6 ft)	2.5 m (8 ft)	3.0 m (10 ft)
Above 75 kV	2.5 m (8 ft)	3.0 m (10 ft)	3.7 m (12 ft)

Note: Where the conditions are as follows:

(1) **Condition 1** — Exposed live parts on one side of the working space and no live or grounded parts on the other side of the working space, or exposed live parts on both sides of the working space that are effectively guarded by insulating materials.

(2) **Condition 2** — Exposed live parts on one side of the working space and grounded parts on the other side of the working space. Concrete, brick, or tile walls shall be considered as grounded.

(3) **Condition 3** — Exposed live parts on both sides of the working space.

(B) Separation from Low-Voltage Equipment. Where switches, cutouts, or other equipment operating at 1000 volts, nominal, or less are installed in a vault, room, or enclosure where there are exposed live parts or exposed wiring operating at over 1000 volts, nominal, the high-voltage equipment shall be effectively separated from the space occupied by the low-voltage equipment by a suitable partition, fence, or screen.

Exception: Switches or other equipment operating at 1000 volts, nominal, or less and serving only equipment within the high-voltage vault, room, or enclosure shall be permitted to be installed in the high-voltage vault, room, or enclosure without a partition, fence, or screen if accessible to qualified persons only.

(C) Locked Rooms or Enclosures. The entrance to all buildings, vaults, rooms, or enclosures containing exposed live parts or exposed conductors operating at over 1000 volts, nominal, shall be kept locked unless such entrances are under the observation of a qualified person at all times.

Permanent and conspicuous danger signs shall be provided. The danger sign shall meet the requirements in 110.21(B) and shall read as follows:

DANGER — HIGH VOLTAGE — KEEP OUT

(D) Illumination. Illumination shall be provided for all working spaces about electrical equipment. Control by automatic means only shall not be permitted. The lighting outlets shall be arranged so that persons changing lamps or making repairs on the lighting system are not endangered by live parts or other equipment.

The points of control shall be located so that persons are not likely to come in contact with any live part or moving part of the equipment while turning on the lights.

(E) Elevation of Unguarded Live Parts. Unguarded live parts above working space shall be maintained at elevations not less than required by Table 110.34(E).

Table 110.34(E) Elevation of Unguarded Live Parts Above Working Space

Nominal Voltage Between Phases	Elevation	
	m	ft
1001–7500 V	2.7	9
7501–35,000 V	2.9	9 ft 6 in.
Over 35 kV	Add 9.5 mm per kV above 35 kV	Add 0.37 in. per kV above 35 kV

(F) Protection of Service Equipment, Switchgear, and Industrial Control Assemblies. Pipes or ducts foreign to the electrical installation and requiring periodic maintenance or whose malfunction would endanger the operation of the electrical system shall not be located in the vicinity of the service equipment, switchgear, or industrial control assemblies. Protection shall be provided where necessary to avoid damage from condensation leaks and breaks in such foreign systems. Piping and other facilities shall not be considered foreign if provided for fire protection of the electrical installation.

110.36 Circuit Conductors. Circuit conductors shall be permitted to be installed in raceways; in cable trays; as metal-clad cable Type MC; as bare wire, cable, and busbars; or as Type MV cables or conductors as provided in 300.37, 300.39, 300.40, and 300.50. Bare live conductors shall comply with 490.24.

110.40 Temperature Limitations at Terminations. Conductors shall be permitted to be terminated based on the 90°C (194°F) temperature rating and ampacity as given in Table 310.60(C)(67) through Table 310.60(C)(86), unless otherwise identified.

CHAPTER 3

TEMPORARY WIRING

INTRODUCTION

This chapter contains requirements from Article 590—Temporary Installations. Article 590 applies to temporary electrical power and lighting installations. New construction and many renovation or addition projects generally require the installation of temporary electrical equipment, which, particularly in the case of new construction, often involves setting a pole and building a temporary service. Although the feeders, branch circuits, and receptacles have specific requirements in Article 590, the service must be installed in accordance with Article 230 (see Chapter 5 of this *Pocket Guide*). For example, the minimum clearance of overhead service-drop conductors must comply with 230.24. Of course, not all electrical projects will require the installation of a temporary service. It is important to note that all of the general requirements in Chapters 1 through 4 of the *Code* apply to temporary installations unless there is a specific modification to a general rule specified in Article 590.

Because of the inherent hazards associated with construction sites, ground-fault protection of receptacles with certain ratings that are in use by personnel is required for all temporary wiring installations, not just for new construction sites with temporary services. If a receptacle(s) is installed or exists as part of the permanent wiring or the building or structure and is used for temporary electric power, ground fault circuit-interrupter (GFCI) protection for personnel must be provided. This may require replacing the existing receptacle with a GFCI receptacle or the use of a portable GFCI device. All other outlets (besides 125-volt, single-phase, 15-, 20-, and 30-ampere receptacle outlets) must have GFCI protection for personnel, special purpose ground-fault protection, or an assured equipment grounding conductor program in accordance with 590.6(B)(2).

ARTICLE 590
Temporary Installations

590.2 All Wiring Installations.

(A) Other Articles. Except as specifically modified in this article, all other requirements of this *Code* for permanent wiring shall apply to temporary wiring installations.

(B) Approval. Temporary wiring methods shall be acceptable only if approved based on the conditions of use and any special requirements of the temporary installation.

590.3 Time Constraints.

(A) During the Period of Construction. Temporary electric power and lighting installations shall be permitted during the period of construction, remodeling, maintenance, repair, or demolition of buildings, structures, equipment, or similar activities.

(B) 90 Days. Temporary electric power and lighting installations shall be permitted for a period not to exceed 90 days for holiday decorative lighting and similar purposes.

(D) Removal. Temporary wiring shall be removed immediately upon completion of construction or purpose for which the wiring was installed.

590.4 General.

(A) Services. Services shall be installed in conformance with Parts I through VIII of Article 230, as applicable.

(B) Feeders. Overcurrent protection shall be provided in accordance with 240.4, 240.5, 240.100, and 240.101. Conductors shall be permitted within cable assemblies or within multiconductor cords or cables of a type identified in Table 400.4 for hard usage or extra-hard usage. For the purpose of this section, the following wiring methods shall be permitted:

(1) Type NM, Type NMC, and Type SE cables shall be permitted to be used in any dwelling, building, or structure without any height limitation or limitation by building construction type and without concealment within walls, floors, or ceilings.

(2) Type SE cable shall be permitted to be installed in a raceway in an underground installation.

(C) Branch Circuits. All branch circuits shall originate in an approved power outlet, switchgear, switchboard or panelboard, motor control center, or fused switch enclosure. Conductors shall be permitted within cable assemblies or within multiconductor cord or cable of a type identified in Table 400.4 for hard usage or extra-hard usage. Conductors shall be protected from overcurrent as provided in 240.4, 240.5, and 240.100. For the purposes of this section, the following wiring methods shall be permitted:

(1) Type NM, Type NMC, and Type SE cables shall be permitted to be used in any dwelling, building, or structure without any height limitation or limitation by building construction type and without concealment within walls, floors, or ceilings.
(2) Type SE cable shall be permitted to be installed in a raceway in an underground installation.

Exception: Branch circuits installed for the purposes specified in 590.3(B) or 590.3(C) shall be permitted to be run as single insulated conductors. Where the wiring is installed in accordance with 590.3(B), the voltage to ground shall not exceed 150 volts, the wiring shall not be subject to physical damage, and the conductors shall be supported on insulators at intervals of not more than 3.0 m (10 ft); or, for festoon lighting, the conductors shall be so arranged that excessive strain is not transmitted to the lampholders.

(D) Receptacles.

(1) All Receptacles. All receptacles shall be of the grounding type. Unless installed in a continuous metal raceway that qualifies as an equipment grounding conductor in accordance with 250.118 or a continuous metal-covered cable that qualifies as an equipment grounding conductor in accordance with 250.118, all branch circuits shall include a separate equipment grounding conductor, and all receptacles shall be electrically connected to the equipment grounding conductor(s). Receptacles on construction sites shall not be installed on any branch circuit that supplies temporary lighting.

(2) Receptacles in Wet Locations. All 15- and 20-ampere, 125- and 250-volt receptacles installed in a wet location shall comply with 406.9(B)(1).

(E) Disconnecting Means. Suitable disconnecting switches or plug connectors shall be installed to permit the disconnection of all ungrounded conductors of each temporary circuit. Multiwire branch circuits shall be provided with a means to disconnect simultaneously all ungrounded conductors at the power outlet or panelboard where the branch circuit originated. Identified handle ties shall be permitted.

(F) Lamp Protection. All lamps for general illumination shall be protected from accidental contact or breakage by a suitable luminaire or lampholder with a guard.

Brass shell, paper-lined sockets, or other metal-cased sockets shall not be used unless the shell is grounded.

(G) Splices. A box, conduit body, or other enclosure, with a cover installed, shall be required for all splices except where:

(1) The circuit conductors being spliced are all from nonmetallic multiconductor cord or cable assemblies, provided that the equipment grounding continuity is maintained with or without the box.
(2) The circuit conductors being spliced are all from metal sheathed cable assemblies terminated in listed fittings that mechanically secure the cable sheath to maintain effective electrical continuity.

(H) Protection from Accidental Damage. Flexible cords and cables shall be protected from accidental damage. Sharp corners and projections shall be avoided. Where passing through doorways or other pinch points, protection shall be provided to avoid damage.

(I) Termination(s) at Devices. Flexible cords and cables entering enclosures containing devices requiring termination shall be secured to the box with fittings listed for connecting flexible cords and cables to boxes designed for the purpose.

(J) Support. Cable assemblies and flexible cords and cables shall be supported in place at intervals that ensure that they will be protected from physical damage. Support shall be in the form of staples, cable ties, straps, or similar type fittings installed so as not to cause damage. Cable assemblies and flexible cords and cables installed as branch circuits or feeders shall not be installed on the floor or on the ground. Extension cords shall not be required to comply with 590.4(J). Vegetation shall not be used for support of overhead spans of branch circuits or feeders.

590.5 Listing of Decorative Lighting. Decorative lighting used for holiday lighting and similar purposes, in accordance with 590.3(B), shall be listed and shall be labeled on the product.

590.6 Ground-Fault Protection for Personnel. Ground-fault protection for personnel for all temporary wiring installations shall be provided to comply with 590.6(A) and (B). This section shall apply only to temporary wiring installations used to supply temporary power to equipment used by personnel during construction, remodeling, maintenance, repair, or demolition of buildings, structures, equipment, or similar activities. This section shall apply to power derived from an electric utility company or from an on-site-generated power source.

(A) Receptacle Outlets. Temporary receptacle installations used to supply temporary power to equipment used by personnel during construction, remodeling, maintenance, repair, or demolition of buildings, structures, equipment, or similar activities shall comply with the requirements of 590.6(A)(1) through (A)(3), as applicable.

Exception: In industrial establishments only, where conditions of maintenance and supervision ensure that only qualified personnel are involved, an assured equipment grounding conductor program as specified in 590.6(B)(3) shall be permitted for only those receptacle outlets used to supply equipment that would create a greater hazard if power were interrupted or having a design that is not compatible with GFCI protection.

(1) Receptacle Outlets Not Part of Permanent Wiring. All 125-volt, single-phase, 15-, 20-, and 30-ampere receptacle outlets that are not a part of the permanent wiring of the building or structure and that are in use by personnel shall have ground-fault circuit-interrupter protection for personnel.

In addition to this required ground-fault circuit-interrupter protection for personnel, listed cord sets or devices incorporating listed ground-fault circuit-interrupter protection for personnel identified for portable use shall be permitted.

(2) Receptacle Outlets Existing or Installed as Permanent Wiring. Ground-fault circuit-interrupter protection for personnel shall be provided for all 125-volt, single-phase, 15-, 20-, and 30-ampere receptacle outlets installed or existing as part of the permanent wiring of the building or structure and used for temporary electric power. Listed cord sets or devices incorporating listed ground-fault circuit-interrupter protection for personnel identified for portable use shall be permitted.

(3) Receptacles on 15-kW or less Portable Generators. All 125-volt and 125/250-volt, single-phase, 15-, 20-, and 30-ampere receptacle outlets that are a part of a 15-kW or smaller portable generator shall have listed ground-fault circuit-interrupter protection for personnel. All 15- and 20-ampere, 125- and 250-volt receptacles, including those that are part of a portable generator, used in a damp or wet location shall comply with 406.9(A) and (B). Listed cord sets or devices incorporating listed ground-fault circuit-interrupter protection for personnel identified for portable use shall be permitted for use with 15-kW or less portable generators manufactured or remanufactured prior to January 1, 2011.

(B) Use of Other Outlets. For temporary wiring installations, receptacles, other than those covered by 590.6(A)(1) through (A)(3) used to supply temporary power to equipment used by personnel during construction, remodeling, maintenance, repair, or demolition of buildings, structures, or equipment, or similar activities, shall have protection in accordance with (B)(1), (B)(2), or the assured equipment grounding conductor program in accordance with (B)(3).

(1) GFCI Protection. Ground-fault circuit-interrupter protection for personnel.

N **(2) SPGFCI Protection.** Special purpose ground-fault circuit-interrupter protection for personnel.

(3) Assured Equipment Grounding Conductor Program. A written assured equipment grounding conductor program continuously enforced at the site by one or more designated persons to ensure that equipment grounding conductors for all cord sets, receptacles that are not a part of the permanent wiring of the building or structure, and equipment connected by cord and plug are installed and maintained in accordance with the applicable requirements of 250.114, 250.138, 406.4(C), and 590.4(D).

(a) The following tests shall be performed on all cord sets, receptacles that are not part of the permanent wiring of the building or structure, and cord-and-plug-connected equipment required to be connected to an equipment grounding conductor:

(1) All equipment grounding conductors shall be tested for continuity and shall be electrically continuous.
(2) Each receptacle and attachment plug shall be tested for correct attachment of the equipment grounding conductor. The equipment grounding conductor shall be connected to its proper terminal.
(3) All required tests shall be performed as follows:
    a. Before first use on site
    b. When there is evidence of damage
    c. Before equipment is returned to service following any repairs
    d. At intervals not exceeding 3 months

(b) The tests required in item (3)(a) shall be recorded and made available to the authority having jurisdiction.

CHAPTER 4

BRANCH-CIRCUIT, FEEDER, AND SERVICE CALCULATIONS

INTRODUCTION

This chapter contains requirements from Article 220—Branch-Circuit, Feeder, and Service Calculations. Article 220 provides requirements for calculating branch-circuit, feeder, and service loads. This *Pocket Guide* includes requirements from four of the five parts of Article 220. The four parts are Part I General, Part II Branch Circuit Load Calculations, Part III Feeders and Services, and Part IV Optional Feeder and Service Calculations. These requirements cover sizing branch-circuits, feeders, and services. The calculation provisions apply to all commercial and industrial occupancy types. Load calculation requirements for specific types of equipment are located elsewhere in the *Code* and are referenced in Table 220.3. For instance, Article 220 specifies that motor loads must be calculated in accordance with Article 430 and that hermetic refrigerant motor compressors must be calculated in accordance with Article 440.

ARTICLE 220

Branch-Circuit, Feeder, and Service Load Calculations

PART I. GENERAL

220.3 Other Articles for Specific-Purpose Calculations. Table 220.3 shall provide references for specific-purpose calculation requirements not located in Chapters 5, 6, or 7 that amend or supplement the requirements of this article.

Table 220.3 Specific-Purpose Calculation References

Calculation	Article	Section (or Part)
Air-conditioning and refrigerating equipment, branch-circuit conductor sizing	440	Part IV
Fixed electric heating equipment for pipelines and vessels, branch-circuit sizing	427	427.4
Fixed electric space-heating equipment, branch-circuit sizing	424	424.3
Fixed outdoor electric deicing and snow-melting equipment, branch-circuit sizing	426	426.4
Motors, feeder demand factor	430	430.26
Motors, multimotor and combination-load equipment	430	430.25
Motors, several motors or a motor(s) and other load(s)	430	430.24
Over 600-volt branch-circuit calculations	210	210.19(B)
Over 600-volt feeder calculations	215	215.2(B)
Phase converters, conductors	455	455.6
Storage-type water heaters	422	422.11(E)

220.5 Calculations.

(A) Voltages. Unless other voltages are specified, for purposes of calculating branch-circuit and feeder loads, nominal system voltages of 120, 120/240, 208Y/120, 240, 347, 480Y/277, 480, 600Y/347, and 600 volts shall be used.

(B) Fractions of an Ampere. Calculations shall be permitted to be rounded to the nearest whole ampere, with decimal fractions smaller than 0.5 dropped.

PART II. BRANCH-CIRCUIT LOAD CALCULATIONS

220.12 Lighting Load for Specified Occupancies. A unit load of not less than that specified in Table 220.12 for occupancies specified shall constitute the minimum lighting load. The floor area for each floor shall be calculated from the outside dimensions of the building, dwelling unit, or other area involved. For dwelling units, the calculated floor area shall not include open porches, garages, or unused or unfinished spaces not adaptable for future use.

Exception No. 1: Where the building is designed and constructed to comply with an energy code adopted by the local authority, the lighting load shall be permitted to be calculated at the values specified in the energy code where the following conditions are met:

(1) A power monitoring system is installed that will provide continuous information regarding the total general lighting load of the building.

(2) The power monitoring system will be set with alarm values to alert the building owner or manager if the lighting load exceeds the values set by the energy code.

(3) The demand factors specified in 220.42 are not applied to the general lighting load.

Exception No. 2: Where a building is designed and constructed to comply with an energy code adopted by the local authority and specifying an overall lighting density of less than 13.5 volt-amperes/13.5 m² (1.2 volt-amperes/1.2 ft²), the unit lighting loads in Table 220.12 for office and bank areas within the building shall be permitted to be reduced by 11 volt-amperes/11 m² (1 volt-amperes/1 ft²).

Table 220.12 General Lighting Loads by Occupancy

Type of Occupancy	Unit Load	
	Volt-amperes/ m^2	Volt-amperes/ ft^2
Armories and auditoriums	11	1
Banks	39[b]	3½[b]
Barber shops and beauty parlors	33	3
Churches	11	1
Clubs	22	2
Courtrooms	22	2
Dwelling units[a]	33	3
Garages — commercial (storage)	6	½
Hospitals	22	2
Hotels and motels, including apartment houses without provision for cooking by tenants[a]	22	2
Industrial commercial (loft) buildings	22	2
Lodge rooms	17	1½
Office buildings	39[b]	3½[b]
Restaurants	22	2
Schools	33	3
Stores	33	3
Warehouses (storage)	3	¼
In any of the preceding occupancies except one-family dwellings and individual dwelling units of two-family and multifamily dwellings:		
Assembly halls and auditoriums	11	1
Halls, corridors, closets, stairways	6	½
Storage spaces	3	¼

[a]See 220.14(J).

[b]See 220.14(K).

220.14 Other Loads — All Occupancies. In all occupancies, the minimum load for each outlet for general-use receptacles and outlets not used for general illumination shall not be less than that calculated in 220.14(A) through (L), the loads shown being based on nominal branch-circuit voltages.

(A) Specific Appliances or Loads. An outlet for a specific appliance or other load not covered in 220.14(B) through (L) shall be calculated based on the ampere rating of the appliance or load served.

(B) Electric Dryers and Electric Cooking Appliances in Dwellings and Household Cooking Appliances Used in Instructional Programs. Load calculations shall be permitted as specified in 220.54 for electric dryers and in 220.55 for electric ranges and other cooking appliances.

(C) Motor Outlets. Loads for motor outlets shall be calculated in accordance with the requirements in 430.22, 430.24, and 440.6.

(D) Luminaires. An outlet supplying luminaire(s) shall be calculated based on the maximum volt-ampere rating of the equipment and lamps for which the luminaire(s) is rated.

(E) Heavy-Duty Lampholders. Outlets for heavy-duty lampholders shall be calculated at a minimum of 600 volt-amperes.

(F) Sign and Outline Lighting. Sign and outline lighting outlets shall be calculated at a minimum of 1200 volt-amperes for each required branch circuit specified in 600.5(A).

> **[600.5 Branch Circuits.**
>
> **(A) Required Branch Circuit.** Each commercial building and each commercial occupancy accessible to pedestrians shall be provided with at least one outlet in an accessible location at each entrance to each tenant space for sign or outline lighting system use. The outlet(s) shall be supplied by a branch circuit rated at least 20 amperes that supplies no other load. Service hallways or corridors shall not be considered accessible to pedestrians.]

(G) Show Windows. Show windows shall be calculated in accordance with either of the following:

(1) The unit load per outlet as required in other provisions of this section
(2) At 200 volt-amperes per linear 300 mm (1 ft) of show window

(H) Fixed Multioutlet Assemblies. Fixed multioutlet assemblies used in other than dwelling units or the guest rooms or guest suites of hotels or motels shall be calculated in accordance with (H)(1) or (H)(2). For the purposes of this section, the calculation shall be permitted to be based on the portion that contains receptacle outlets.

(1) Where appliances are unlikely to be used simultaneously, each 1.5 m (5 ft) or fraction thereof of each separate and continuous length shall be considered as one outlet of not less than 180 volt-amperes.
(2) Where appliances are likely to be used simultaneously, each 300 mm (1 ft) or fraction thereof shall be considered as an outlet of not less than 180 volt-amperes.

(I) Receptacle Outlets. Except as covered in 220.14(J) and (K), receptacle outlets shall be calculated at not less than 180 volt-amperes for each single or for each multiple receptacle on one yoke. A single piece of equipment consisting of a multiple receptacle comprised of four or more receptacles shall be calculated at not less than 90 volt-amperes per receptacle.

(K) Banks and Office Buildings. In banks or office buildings, the receptacle loads shall be calculated to be the larger of (1) or (2):

(1) The calculated load from 220.14(I)
(2) 11 volt-amperes/m^2 or 1 volt-ampere/ft^2

(L) Other Outlets. Other outlets not covered in 220.14(A) through (K) shall be calculated based on 180 volt-amperes per outlet.

220.18 Maximum Loads. The total load shall not exceed the rating of the branch circuit, and it shall not exceed the maximum loads specified in 220.18(A) through (C) under the conditions specified therein.

(A) Motor-Operated and Combination Loads. Where a circuit supplies only motor-operated loads, Article 430 shall apply. Where a circuit supplies only air-conditioning equipment, refrigerating equipment, or both, Article 440 shall apply. For circuits supplying loads consisting of motor-operated utilization equipment that is fastened in place and has a motor larger than ⅛ hp in combination with other loads, the total calculated load shall be based on 125 percent of the largest motor load plus the sum of the other loads.

(B) Inductive and LED Lighting Loads. For circuits supplying lighting units that have ballasts, transformers, autotransformers, or LED drivers, the calculated load shall be based on the total ampere ratings of such units and not on the total watts of the lamps.

PART III. FEEDER AND SERVICE LOAD CALCULATIONS

220.40 General. The calculated load of a feeder or service shall not be less than the sum of the loads on the branch circuits supplied, as determined by Part II of this article, after any applicable demand factors permitted by Part III or IV or required by Part V have been applied.

220.42 General Lighting. The demand factors specified in Table 220.42 shall apply to that portion of the total branch-circuit load calculated for general illumination. They shall not be applied in determining the number of branch circuits for general illumination.

Table 220.42 Lighting Load Demand Factors

Type of Occupancy	Portion of Lighting Load to Which Demand Factor Applies (Volt-Amperes)	Demand Factor (%)
Hospitals*	First 50,000 or less at	40
	Remainder over 50,000 at	20
Hotels and motels, including apartment houses without provision for cooking by tenants*	First 20,000 or less at	50
	From 20,001 to 100,000 at	40
	Remainder over 100,000 at	30
Warehouses (storage)	First 12,500 or less at	100
	Remainder over 12,500 at	50
All others	Total volt-amperes	100

*The demand factors of this table shall not apply to the calculated load of feeders or services supplying areas in hospitals, hotels, and motels where the entire lighting is likely to be used at one time, as in operating rooms, ballrooms, or dining rooms.

220.43 Show-Window and Track Lighting.

(A) Show Windows. For show-window lighting, a load of not less than 660 volt-amperes/linear meter or 200 volt-amperes/linear foot shall be included for a show window, measured horizontally along its base.

Informational Note: See 220.14(G) for branch circuits supplying show windows.

(B) Track Lighting. For track lighting in other than dwelling units or guest rooms or guest suites of hotels or motels, an additional load of 150 volt-amperes shall be included for every 600 mm (2 ft) of lighting track or fraction thereof. Where multicircuit track is installed, the load shall be considered to be divided equally between the track circuits.

Exception: If the track lighting is supplied through a device that limits the current to the track, the load shall be permitted to be calculated based on the rating of the device used to limit the current.

220.44 Receptacle Loads — Other Than Dwelling Units. Receptacle loads calculated in accordance with 220.14(H) and (I) shall be permitted to be made subject to the demand factors given in Table 220.42 or Table 220.44.

Table 220.44 Demand Factors for Non-Dwelling Receptacle Loads

Portion of Receptacle Load to Which Demand Factor Applies (Volt-Amperes)	Demand Factor (%)
First 10 kVA or less at	100
Remainder over 10 kVA at	50

220.50 Motors. Motor loads shall be calculated in accordance with 430.24, 430.25, and 430.26 and with 440.6 for hermetic refrigerant motor-compressors.

220.51 Fixed Electric Space Heating. Fixed electric space-heating loads shall be calculated at 100 percent of the total connected load. However, in no case shall a feeder or service load current rating be less than the rating of the largest branch circuit supplied.

Exception: Where reduced loading of the conductors results from units operating on duty-cycle, intermittently, or from all units not operating at the same time, the authority having jurisdiction may grant permission for feeder and service conductors to have an ampacity less than 100 percent, provided the conductors have an ampacity for the load so determined.

220.56 Kitchen Equipment — Other Than Dwelling Unit(s). It shall be permissible to calculate the load for commercial electric cooking equipment, dishwasher booster heaters, water heaters, and other kitchen equipment in accordance with Table 220.56. These demand factors shall be applied to all equipment that has either thermostatic control or intermittent use as kitchen equipment. These demand factors shall not apply to space-heating, ventilating, or air-conditioning equipment.

However, in no case shall the feeder or service calculated load be less than the sum of the largest two kitchen equipment loads.

Table 220.56 Demand Factors for Kitchen Equipment — Other Than Dwelling Unit(s)

Number of Units of Equipment	Demand Factor (%)
1	100
2	100
3	90
4	80
5	70
6 and over	65

220.60 Noncoincident Loads. Where it is unlikely that two or more noncoincident loads will be in use simultaneously, it shall be permissible to use only the largest load(s) that will be used at one time for calculating the total load of a feeder or service.

220.61 Feeder or Service Neutral Load.

(A) Basic Calculation. The feeder or service neutral load shall be the maximum unbalance of the load determined by this article. The maximum unbalanced load shall be the maximum net calculated load between the neutral conductor and any one ungrounded conductor.

(B) Permitted Reductions. A service or feeder supplying the following loads shall be permitted to have an additional demand factor of 70 percent applied to the amount in 220.61(B)(1) or portion of the amount in 220.61(B)(2) determined by the following basic calculations:

(2) That portion of the unbalanced load in excess of 200 amperes where the feeder or service is supplied from a 3-wire dc or single-phase ac system; or a 4-wire, 3-phase system; or a 3-wire, 2-phase system; or a 5-wire, 2-phase system

(C) Prohibited Reductions. There shall be no reduction of the neutral or grounded conductor capacity applied to the amount in 220.61(C)(1), or portion of the amount in (C)(2), from that determined by the basic calculation:

(1) Any portion of a 3-wire circuit consisting of 2 ungrounded conductors and the neutral conductor of a 4-wire, 3-phase, wye-connected system
(2) That portion consisting of nonlinear loads supplied from a 4-wire, wye-connected, 3-phase system

Informational Note: A 3-phase, 4-wire, wye-connected power system used to supply power to nonlinear loads may necessitate that the power system design allow for the possibility of high harmonic neutral conductor currents.

PART IV. OPTIONAL FEEDER AND SERVICE LOAD CALCULATIONS

220.86 Schools. The calculation of a feeder or service load for schools shall be permitted in accordance with Table 220.86 in lieu of Part III of this article where equipped with electric space heating, air conditioning, or both. The connected load to which the demand factors of Table 220.86 apply shall include all of the interior and exterior lighting, power, water heating, cooking, other loads, and the larger of the air-conditioning load or space-heating load within the building or structure.

Feeders and service conductors whose calculated load is determined by this optional calculation shall be permitted to have the neutral load determined by 220.61. Where the building or structure load is calculated by this optional method, feeders within the building or structure shall have ampacity as permitted in Part III of this article; however, the ampacity of an individual feeder shall not be required to be larger than the ampacity for the entire building.

This section shall not apply to portable classroom buildings.

Table 220.86 Optional Method — Demand Factors for Feeders and Service Conductors for Schools

Connected Load		Demand Factor (Percent)
First 33 VA/m^2	(3 VA/ft^2) at	100
Plus,		
Over 33 through 220 VA/m^2	(3 through 20 VA/ft^2) at	75
Plus,		
Remainder over 220 VA/m^2	(20 VA/ft^2) at	25

220.87 Determining Existing Loads. The calculation of a feeder or service load for existing installations shall be permitted to use actual maximum demand to determine the existing load under all of the following conditions:

(1) The maximum demand data is available for a 1-year period.

Exception: If the maximum demand data for a 1-year period is not available, the calculated load shall be permitted to be based on the maximum demand (the highest average kilowatts reached and maintained for a 15-minute interval) continuously recorded over a minimum 30-day period using a recording ammeter or power meter connected to the highest loaded phase of the feeder or service, based on the initial loading at the start of the recording. The recording shall reflect the maximum demand of the feeder or service by being taken when the building or space is occupied and shall include by measurement or calculation the larger of the heating or cooling equipment load, and other loads that may be periodic in nature due to seasonal or similar conditions.

(2) The maximum demand at 125 percent plus the new load does not exceed the ampacity of the feeder or rating of the service.
(3) The feeder has overcurrent protection in accordance with 240.4, and the service has overload protection in accordance with 230.90.

220.88 New Restaurants. Calculation of a service or feeder load, where the feeder serves the total load, for a new restaurant shall be permitted in accordance with Table 220.88 in lieu of Part III of this article.

The overload protection of the service conductors shall be in accordance with 230.90 and 240.4.

Feeder conductors shall not be required to be of greater ampacity than the service conductors.

Service or feeder conductors whose calculated load is determined by this optional calculation shall be permitted to have the neutral load determined by 220.61.

Table 220.88 Optional Method — Permitted Load Calculations for Service and Feeder Conductors for New Restaurants

Total Connected Load (kVA)	All Electric Restaurant Calculated Loads (kVA)	Not All Electric Restaurant Calculated Loads (kVA)
0–200	80%	100%
201–325	10% (amount over 200) + 160.0	50% (amount over 200) + 200.0
326–800	50% (amount over 325) + 172.5	45% (amount over 325) + 262.5
Over 800	50% (amount over 800) + 410.0	20% (amount over 800) + 476.3

Note: Add all electrical loads, including both heating and cooling loads, to calculate the total connected load. Select the one demand factor that applies from the table, then multiply the total connected load by this single demand factor.

CHAPTER 5

SERVICES

INTRODUCTION

This chapter contains requirements from Article 230—Services. Article 230 provides installation requirements for conductors and equipment that supply, control, and protect services. Contained in this *Pocket Guide* are requirements from the eight parts that comprise Article 230: Part I General; Part II Overhead Service Conductors; Part III Underground Service Conductors; Part IV Service-Entrance Conductors; Part V Service Equipment—General; Part VI Service Equipment—Disconnecting Means; Part VII Service Equipment—Overcurrent Protection; and Part VIII Services Exceeding 1000 Volts, Nominal.

While all service installations must comply with Parts I, IV, V, VI, and VII, compliance with Parts II and III depends on the type of service. If the service is brought to the premises using overhead conductors, Part II applies; if the service is brought to the premises using underground conductors, Part III applies. All requirements in Parts I through VII for services rated 1000 volts apply to services exceeding 1000 volts, unless there is a unique requirement in Part VIII that amends or modifies the requirements for the "lower" voltage services.

ARTICLE 230
Services

PART I. GENERAL

230.2 Number of Services. A building or other structure served shall be supplied by only one service unless permitted in 230.2(A) through (D). For the purpose of 230.40, Exception No. 2 only, underground sets of conductors, 1/0 AWG and larger, running to the same location and connected together at their supply end but not connected together at their load end shall be considered to be supplying one service.

(A) Special Conditions. Additional services shall be permitted to supply the following:

(1) Fire pumps
(2) Emergency systems
(3) Legally required standby systems
(4) Optional standby systems
(5) Parallel power production systems
(6) Systems designed for connection to multiple sources of supply for the purpose of enhanced reliability

(B) Special Occupancies. By special permission, additional services shall be permitted for either of the following:

(1) Multiple-occupancy buildings where there is no available space for service equipment accessible to all occupants
(2) A single building or other structure sufficiently large to make two or more services necessary

(C) Capacity Requirements. Additional services shall be permitted under any of the following:

(1) Where the capacity requirements are in excess of 2000 amperes at a supply voltage of 1000 volts or less
(2) Where the load requirements of a single-phase installation are greater than the serving agency normally supplies through one service
(3) By special permission

(D) Different Characteristics. Additional services shall be permitted for different voltages, frequencies, or phases, or for different uses, such as for different rate schedules.

(E) Identification. Where a building or structure is supplied by more than one service, or any combination of branch circuits, feeders, and services, a permanent plaque or directory shall be installed at each service disconnect location denoting all other services, feeders, and branch circuits supplying that building or structure and the area served by each.

230.3 One Building or Other Structure Not to Be Supplied Through Another. Service conductors supplying a building or other structure shall not pass through the interior of another building or other structure.

230.6 Conductors Considered Outside the Building. Conductors shall be considered outside of a building or other structure under any of the following conditions:

(1) Where installed under not less than 50 mm (2 in.) of concrete beneath a building or other structure
(2) Where installed within a building or other structure in a raceway that is encased in concrete or brick not less than 50 mm (2 in.) thick
(3) Where installed in any vault that meets the construction requirements of Article 450, Part III
(4) Where installed in conduit and under not less than 450 mm (18 in.) of earth beneath a building or other structure
(5) Where installed within rigid metal conduit (Type RMC) or intermediate metal conduit (Type IMC) used to accommodate the clearance requirements in 230.24 and routed directly through an eave but not a wall of a building

230.7 Other Conductors in Raceway or Cable. Conductors other than service conductors shall not be installed in the same service raceway or service cable in which the service conductors are installed.

Exception No. 1: Grounding electrode conductors or supply side bonding jumpers or conductors shall be permitted within service raceways.

Exception No. 2: Load management control conductors having overcurrent protection shall be permitted within service raceways.

230.8 Raceway Seal. Where a service raceway enters a building or structure from an underground distribution system, it shall be sealed in accordance with 300.5(G). Spare or unused raceways shall also be sealed. Sealants shall be identified for use with the cable insulation, shield, or other components.

230.9 Clearances on Buildings. Service conductors and final spans shall comply with 230.9(A), (B), and (C).

(A) Clearances. Service conductors installed as open conductors or multiconductor cable without an overall outer jacket shall have a clearance of not less than 900 mm (3 ft) from windows that are designed to be opened, doors, porches, balconies, ladders, stairs, fire escapes, or similar locations.

Exception: Conductors run above the top level of a window shall be permitted to be less than the 900 mm (3 ft) requirement.

(B) Vertical Clearance. The vertical clearance of final spans above, or within 900 mm (3 ft) measured horizontally of platforms, projections, or surfaces that will permit personal contact shall be maintained in accordance with 230.24(B).

(C) Building Openings. Overhead service conductors shall not be installed beneath openings through which materials may be moved, such as openings in farm and commercial buildings, and shall not be installed where they obstruct entrance to these building openings.

230.10 Vegetation as Support. Vegetation such as trees shall not be used for support of overhead service conductors or service equipment.

PART II. OVERHEAD SERVICE CONDUCTORS

230.22 Insulation or Covering. Individual conductors shall be insulated or covered.

230.23 Size and Rating.

(A) General. Conductors shall have sufficient ampacity to carry the current for the load as calculated in accordance with Article 220 and shall have adequate mechanical strength.

(B) Minimum Size. The conductors shall not be smaller than 8 AWG copper or 6 AWG aluminum or copper-clad aluminum.

(C) Grounded Conductors. The grounded conductor shall not be less than the minimum size as required by 250.24(C).

230.24 Clearances. Overhead service conductors shall not be readily accessible and shall comply with 230.24(A) through (E) for services not over 1000 volts, nominal.

(A) Above Roofs. Conductors shall have a vertical clearance of not less than 2.5 m (8 ft) above the roof surface. The vertical clearance above the roof level shall be maintained for a distance of not less than 900 mm (3 ft) in all directions from the edge of the roof.

Exception No. 1: The area above a roof surface subject to pedestrian or vehicular traffic shall have a vertical clearance from the roof surface in accordance with the clearance requirements of 230.24(B).

Exception No. 2: Where the voltage between conductors does not exceed 300 and the roof has a slope of 100 mm in 300 mm (4 in. in 12 in.) or greater, a reduction in clearance to 900 mm (3 ft) shall be permitted.

Exception No. 3: Where the voltage between conductors does not exceed 300, a reduction in clearance above only the overhanging portion of the roof to not less than 450 mm (18 in.) shall be permitted if (1) not more than 1.8 m (6 ft) of overhead service conductors, 1.2 m (4 ft) horizontally, pass above the roof overhang, and (2) they are terminated at a through-the-roof raceway or approved support.

Exception No. 4: The requirement for maintaining the vertical clearance 900 mm (3 ft) from the edge of the roof shall not apply to the final conductor span where the service drop or overhead service conductors are attached to the side of a building.

Exception No. 5: Where the voltage between conductors does not exceed 300 and the roof area is guarded or isolated, a reduction in clearance to 900 mm (3 ft) shall be permitted.

(B) Vertical Clearance for Overhead Service Conductors. Overhead service conductors, where not in excess of 600 volts, nominal, shall have the following minimum clearance from final grade:

(1) 3.0 m (10 ft) — at the electrical service entrance to buildings, also at the lowest point of the drip loop of the building electrical entrance, and above areas or sidewalks accessible only to pedestrians, measured from final grade or other accessible surface only for overhead service conductors supported on and cabled together with a grounded bare messenger where the voltage does not exceed 150 volts to ground
(2) 3.7 m (12 ft) — over residential property and driveways, and those commercial areas not subject to truck traffic where the voltage does not exceed 300 volts to ground
(3) 4.5 m (15 ft) — for those areas listed in the 3.7 m (12 ft) classification where the voltage exceeds 300 volts to ground
(4) 5.5 m (18 ft) — over public streets, alleys, roads, parking areas subject to truck traffic, driveways on other than residential property, and other land such as cultivated, grazing, forest, and orchard
(5) 7.5 m (24½) over tracks of railroads

230.26 Point of Attachment. The point of attachment of the overhead service conductors to a building or other structure shall provide the minimum clearances as specified in 230.9 and 230.24. In no case shall this point of attachment be less than 3.0 m (10 ft) above finished grade.

230.27 Means of Attachment. Multiconductor cables used for overhead service conductors shall be attached to buildings or other structures by fittings identified for use with service conductors. Open conductors shall be attached to fittings identified for use with service conductors or to noncombustible, nonabsorbent insulators securely attached to the building or other structure.

230.28 Service Masts as Supports. Only power service-drop or overhead service conductors shall be permitted to be attached to a service mast. Service masts used for the support of service-drop or overhead service conductors shall be installed in accordance with 230.28(A) and (B).

(A) Strength. The service mast shall be of adequate strength or be supported by braces or guys to withstand safely the strain imposed by the service-drop or overhead service conductors. Hubs intended for use with a conduit that serves as a service mast shall be identified for use with service-entrance equipment.

(B) Attachment. Service-drop or overhead service conductors shall not be attached to a service mast between a weatherhead or the end of the conduit and a coupling, where the coupling is located above the last point of securement to the building or other structure or is located above the building or other structure.

230.29 Supports over Buildings. Service conductors passing over a roof shall be securely supported by substantial structures. For a grounded system, where the substantial structure is metal, it shall be bonded by means of a bonding jumper and listed connector to the grounded overhead service conductor. Where practicable, such supports shall be independent of the building.

PART III. UNDERGROUND SERVICE CONDUCTORS

230.30 Installation.

(A) Insulation. Underground service conductors shall be insulated for the applied voltage.

Exception: A grounded conductor shall be permitted to be uninsulated as follows:

(1) Bare copper used in a raceway
(2) Bare copper for direct burial where bare copper is approved for the soil conditions
(3) Bare copper for direct burial without regard to soil conditions where part of a cable assembly identified for underground use
(4) Aluminum or copper-clad aluminum without individual insulation or covering where part of a cable assembly identified for underground use in a raceway or for direct burial

(B) Wiring Methods. Underground service conductors shall be installed in accordance with the applicable requirements of this *Code* covering the type of wiring method used and shall be limited to the following methods:

(1) Type RMC conduit
(2) Type IMC conduit
(3) Type NUCC conduit
(4) Type HDPE conduit
(5) Type PVC conduit
(6) Type RTRC conduit
(7) Type IGS cable
(8) Type USE conductors or cables
(9) Type MV or Type MC cable identified for direct burial applications
(10) Type MI cable, where suitably protected against physical damage and corrosive conditions

230.31 Size and Rating.

(A) General. Underground service conductors shall have sufficient ampacity to carry the current for the load as calculated in accordance with Article 220 and shall have adequate mechanical strength.

(B) Minimum Size. The conductors shall not be smaller than 8 AWG copper or 6 AWG aluminum or copper-clad aluminum.

Exception: Conductors supplying only limited loads of a single branch circuit — such as small polyphase power, controlled water heaters, and similar loads — shall not be smaller than 12 AWG copper or 10 AWG aluminum or copper-clad aluminum.

(C) Grounded Conductors. The grounded conductor shall not be less than the minimum size required by 250.24(C).

230.32 Protection Against Damage. Underground service conductors shall be protected against damage in accordance with 300.5. Service conductors entering a building or other structure shall be installed in accordance with 230.6 or protected by a raceway wiring method identified in 230.43.

230.33 Spliced Conductors. Service conductors shall be permitted to be spliced or tapped in accordance with 110.14, 300.5(E), 300.13, and 300.15.

PART IV. SERVICE-ENTRANCE CONDUCTORS

230.40 Number of Service-Entrance Conductor Sets. Each service drop, set of overhead service conductors, set of underground service conductors, or service lateral shall supply only one set of service-entrance conductors.

Exception No. 1: A building with more than one occupancy shall be permitted to have one set of service-entrance conductors for each service, as defined in 230.2, run to each occupancy or group of occupancies. If the number of service disconnect locations for any given classification of service does not exceed six, the requirements of 230.2(E) shall apply at each location. If the number of service disconnect locations exceeds six for any given supply classification, all service disconnect locations for all supply characteristics, together with any branch circuit or feeder supply sources, if applicable, shall be clearly described using suitable graphics or text, or both, on one or more plaques located in an approved, readily accessible location(s) on the building or structure served and as near as practicable to the point(s) of attachment or entry(ies) for each service drop or service lateral, and for each set of overhead or underground service conductors.

Exception No. 2: Where two to six service disconnecting means in separate enclosures are grouped at one location and supply separate loads from one service drop, set of overhead service conductors, set of underground service conductors, or service lateral, one set of service-entrance conductors shall be permitted to supply each or several such service equipment enclosures.

Exception No. 4: Two-family dwellings, multifamily dwellings, and multiple occupancy buildings shall be permitted to have one set of service-entrance conductors installed to supply the circuits covered in 210.25.

Exception No. 5: One set of service-entrance conductors connected to the supply side of the normal service disconnecting means shall be permitted to supply each or several systems covered by 230.82(5) or 230.82(6).

230.41 Insulation of Service-Entrance Conductors. Service-entrance conductors entering or on the exterior of buildings or other structures shall be insulated.

Exception: A grounded conductor shall be permitted to be uninsulated as follows:

(1) Bare copper used in a raceway or part of a service cable assembly
(2) Bare copper for direct burial where bare copper is approved for the soil conditions
(3) Bare copper for direct burial without regard to soil conditions where part of a cable assembly identified for underground use
(4) Aluminum or copper-clad aluminum without individual insulation or covering where part of a cable assembly or identified for underground use in a raceway, or for direct burial
(5) Bare conductors used in an auxiliary gutter

230.42 Minimum Size and Rating.

(A) General. Service-entrance conductors shall have an ampacity of not less than the maximum load to be served. Conductors shall be sized to carry not less than the largest of 230.42(A)(1) or (A)(2). Loads shall be determined in accordance with Part III, IV, or V of Article 220, as applicable. Ampacity shall be determined from 310.15. The maximum allowable current of busways shall be that value for which the busway has been listed or labeled.

(1) Where the service-entrance conductors supply continuous loads or any combination of noncontinuous and continuous loads, the minimum service-entrance conductor size shall have an allowable ampacity not less than the sum of the noncontinuous loads plus 125 percent of continuous loads.
(2) The minimum service-entrance conductor size shall have an ampacity not less than the maximum load to be served after the application of any adjustment or correction factors.

(B) Specific Installations. In addition to the requirements of 230.42(A), the minimum ampacity for ungrounded conductors for specific installations shall not be less than the rating of the service disconnecting means specified in 230.79(A) through (D).

(C) Grounded Conductors. The grounded conductor shall not be smaller than the minimum size as required by 250.24(C).

230.43 Wiring Methods for 1000 Volts, Nominal, or Less. Service-entrance conductors shall be installed in accordance with the applicable requirements of this *Code* covering the type of wiring method used and shall be limited to the following methods:

(1) Open wiring on insulators
(2) Type IGS cable
(3) Rigid metal conduit (RMC)
(4) Intermediate metal conduit (IMC)
(5) Electrical metallic tubing (EMT)
(6) Electrical nonmetallic tubing
(7) Service-entrance cables
(8) Wireways
(9) Busways
(10) Auxiliary gutters
(11) Rigid polyvinyl chloride conduit (PVC)
(12) Cablebus
(13) Type MC cable
(14) Mineral-insulated, metal-sheathed cable, Type MI
(15) Flexible metal conduit (FMC) not over 1.8 m (6 ft) long or liquidtight flexible metal conduit (LFMC) not over 1.8 m (6 ft) long between a raceway, or between a raceway and service equipment, with a supply-side bonding jumper routed with the flexible metal conduit (FMC) or the liquidtight flexible metal conduit (LFMC) according to the provisions of 250.102(A), (B), (C), and (E)
(16) Liquidtight flexible nonmetallic conduit (LFNC)
(17) High density polyethylene conduit (HDPE)
(18) Nonmetallic underground conduit with conductors (NUCC)
(19) Reinforced thermosetting resin conduit (RTRC)

230.44 Cable Trays. Cable tray systems shall be permitted to support service-entrance conductors. Cable trays used to support service-entrance conductors shall contain only service-entrance conductors and shall be limited to the following methods:

(1) Type SE cable
(2) Type MC cable
(3) Type MI cable
(4) Type IGS cable
(5) Single conductors 1/0 and larger that are listed for use in cable tray

Such cable trays shall be identified with permanently affixed labels with the wording "Service-Entrance Conductors." The labels shall be located so as to be visible after installation with a spacing not to exceed 3 m (10 ft) so that the service-entrance conductors are able to be readily traced through the entire length of the cable tray.

Exception: Conductors, other than service-entrance conductors, shall be permitted to be installed in a cable tray with service-entrance conductors, provided a solid fixed barrier of a material compatible with the cable tray is installed to separate the service-entrance conductors from other conductors installed in the cable tray.

230.46 Spliced Conductors. Service-entrance conductors shall be permitted to be spliced or tapped in accordance with 110.14, 300.5(E), 300.13, and 300.15.

230.50 Protection Against Physical Damage.

(A) Underground Service-Entrance Conductors. Underground service-entrance conductors shall be protected against physical damage in accordance with 300.5.

(B) All Other Service-Entrance Conductors. All other service-entrance conductors, other than underground service entrance conductors, shall be protected against physical damage as specified in 230.50(B)(1) or (B)(2).

(1) Service-Entrance Cables. Service-entrance cables, where subject to physical damage, shall be protected by any of the following:

(1) Rigid metal conduit (RMC)
(2) Intermediate metal conduit (IMC)

(3) Schedule 80 PVC conduit
(4) Electrical metallic tubing (EMT)
(5) Reinforced thermosetting resin conduit (RTRC)
(6) Other approved means

(2) Other Than Service-Entrance Cables. Individual open conductors and cables, other than service-entrance cables, shall not be installed within 3.0 m (10 ft) of grade level or where exposed to physical damage.

Exception: Type MI and Type MC cable shall be permitted within 3.0 m (10 ft) of grade level where not exposed to physical damage or where protected in accordance with 300.5(D).

230.51 Mounting Supports. Service-entrance cables or individual open service-entrance conductors shall be supported as specified in 230.51(A), (B), or (C).

(A) Service-Entrance Cables. Service-entrance cables shall be supported by straps or other approved means within 300 mm (12 in.) of every service head, gooseneck, or connection to a raceway or enclosure and at intervals not exceeding 750 mm (30 in.).

(B) Other Cables. Cables that are not approved for mounting in contact with a building or other structure shall be mounted on insulating supports installed at intervals not exceeding 4.5 m (15 ft) and in a manner that maintains a clearance of not less than 50 mm (2 in.) from the surface over which they pass.

230.53 Raceways to Drain. Where exposed to the weather, raceways enclosing service-entrance conductors shall be listed or approved for use in wet locations and arranged to drain. Where embedded in masonry, raceways shall be arranged to drain.

230.54 Overhead Service Locations.

(A) Service Head. Service raceways shall be equipped with a service head at the point of connection to service-drop or overhead service conductors. The service head shall be listed for use in wet locations.

(B) Service-Entrance Cables Equipped with Service Head or Gooseneck. Service-entrance cables shall be equipped with a service head. The service head shall be listed for use in wet locations.

(C) Service Heads and Goosenecks Above Service-Drop or Overhead Service Attachment. Service heads on raceways or service-entrance cables and goosenecks in service-entrance cables shall be located above the point of attachment of the service-drop or overhead service conductors to the building or other structure.

Exception: Where it is impracticable to locate the service head or gooseneck above the point of attachment, the service head or gooseneck location shall be permitted not farther than 600 mm (24 in.) from the point of attachment.

(D) Secured. Service-entrance cables shall be held securely in place.

(E) Separately Bushed Openings. Service heads shall have conductors of different potential brought out through separately bushed openings.

(F) Drip Loops. Drip loops shall be formed on individual conductors. To prevent the entrance of moisture, service-entrance conductors shall be connected to the service-drop or overhead service conductors either (1) below the level of the service head or (2) below the level of the termination of the service-entrance cable sheath.

(G) Arranged That Water Will Not Enter Service Raceway or Equipment. Service-entrance and overhead service conductors shall be arranged so that water will not enter service raceway or equipment.

230.56 Service Conductor with the Higher Voltage to Ground. On a 4-wire, delta-connected service where the midpoint of one phase winding is grounded, the service conductor having the higher phase voltage to ground shall be durably and permanently marked by an outer finish that is orange in color, or by other effective means, at each termination or junction point.

PART V. SERVICE EQUIPMENT — GENERAL

230.62 Service Equipment — Enclosed or Guarded. Energized parts of service equipment shall be enclosed as specified in 230.62(A) or guarded as specified in 230.62(B).

(A) Enclosed. Energized parts shall be enclosed so that they will not be exposed to accidental contact or shall be guarded as in 230.62(B).

(B) Guarded. Energized parts that are not enclosed shall be installed on a switchboard, panelboard, or control board and guarded in accordance with 110.18 and 110.27. Where energized parts are guarded as provided in 110.27(A)(1) and (A)(2), a means for locking or sealing doors providing access to energized parts shall be provided.

230.66 Marking. Service equipment rated at 1000 volts or less shall be marked to identify it as being suitable for use as service equipment. All service equipment shall be listed or field labeled. Individual meter socket enclosures shall not be considered service equipment but shall be listed and rated for the voltage and ampacity of the service.

Exception: Meter sockets supplied by and under the exclusive control of an electric utility shall not be required to be listed.

PART VI. SERVICE EQUIPMENT — DISCONNECTING MEANS

230.70 General. Means shall be provided to disconnect all conductors in a building or other structure from the service-entrance conductors.

(A) Location. The service disconnecting means shall be installed in accordance with 230.70(A)(1), (A)(2), and (A)(3).

(1) Readily Accessible Location. The service disconnecting means shall be installed at a readily accessible location either outside of a building or structure or inside nearest the point of entrance of the service conductors.

(2) Bathrooms. Service disconnecting means shall not be installed in bathrooms.

(3) Remote Control. Where a remote control device(s) is used to actuate the service disconnecting means, the service disconnecting means shall be located in accordance with 230.70(A)(1).

(B) Marking. Each service disconnect shall be permanently marked to identify it as a service disconnect.

(C) Suitable for Use. Each service disconnecting means shall be suitable for the prevailing conditions. Service equipment installed in hazardous (classified) locations shall comply with the requirements of Articles 500 through 517.

230.71 Maximum Number of Disconnects.

(A) General. The service disconnecting means for each service permitted by 230.2, or for each set of service-entrance conductors permitted by 230.40, Exception No. 1, 3, 4, or 5, shall consist of not more than six switches or sets of circuit breakers, or a combination of not more than six switches and sets of circuit breakers, mounted in a single enclosure, in a group of separate enclosures, or in or on a switchboard or in switchgear. There shall be not more than six sets of disconnects per service grouped in any one location.

For the purpose of this section, disconnecting means installed as part of listed equipment and used solely for the following shall not be considered a service disconnecting means:

(1) Power monitoring equipment
(2) Surge-protective device(s)
(3) Control circuit of the ground-fault protection system
(4) Power-operable service disconnecting means

(B) Single-Pole Units. Two or three single-pole switches or breakers, capable of individual operation, shall be permitted on multiwire circuits, one pole for each ungrounded conductor, as one multipole disconnect, provided they are equipped with identified handle ties or a master handle to disconnect all conductors of the service with no more than six operations of the hand.

230.72 Grouping of Disconnects.

(A) General. The two to six disconnects as permitted in 230.71 shall be grouped. Each disconnect shall be marked to indicate the load served.

Exception: One of the two to six service disconnecting means permitted in 230.71, where used only for a water pump also intended to provide fire protection, shall be permitted to be located remote from the other disconnecting means. If remotely installed in accordance with this exception, a plaque shall be posted at the location of the remaining grouped disconnects denoting its location.

(B) Additional Service Disconnecting Means. The one or more additional service disconnecting means for fire pumps, emergency systems, legally required standby, or optional standby services permitted by 230.2 shall be installed remote from the one to six service disconnecting means for normal service to minimize the possibility of simultaneous interruption of supply.

(C) Access to Occupants. In a multiple-occupancy building, each occupant shall have access to the occupant's service disconnecting means.

Exception: In a multiple-occupancy building where electric service and electrical maintenance are provided by the building management and where these are under continuous building management supervision, the service disconnecting means supplying more than one occupancy shall be permitted to be accessible to authorized management personnel only.

230.74 Simultaneous Opening of Poles. Each service disconnect shall simultaneously disconnect all ungrounded service conductors that it controls from the premises wiring system.

230.75 Disconnection of Grounded Conductor. Where the service disconnecting means does not disconnect the grounded conductor from the premises wiring, other means shall be provided for this purpose in the service equipment. A terminal or bus to which all grounded conductors can be attached by means of pressure connectors shall be permitted for this purpose. In a multisection switchboard or switchgear, disconnects for the grounded conductor shall be permitted to be in any section of the switchboard or switchgear, if the switchboard or switchgear section is marked to indicate a grounded conductor disconnect is located within.

230.76 Manually or Power Operable. The service disconnecting means for ungrounded service conductors shall consist of one of the following:

(1) A manually operable switch or circuit breaker equipped with a handle or other suitable operating means
(2) A power-operated switch or circuit breaker, provided the switch or circuit breaker can be opened by hand in the event of a power supply failure

230.77 Indicating. The service disconnecting means shall plainly indicate whether it is in the open (off) or closed (on) position.

230.79 Rating of Service Disconnecting Means. The service disconnecting means shall have a rating not less than the calculated load to be carried, determined in accordance with Part III, IV, or V of Article 220, as applicable. In no case shall the rating be lower than specified in 230.79(A), (B), (C), or (D).

(A) One-Circuit Installations. For installations to supply only limited loads of a single branch circuit, the service disconnecting means shall have a rating of not less than 15 amperes.

(B) Two-Circuit Installations. For installations consisting of not more than two 2-wire branch circuits, the service disconnecting means shall have a rating of not less than 30 amperes.

(D) All Others. For all other installations, the service disconnecting means shall have a rating of not less than 60 amperes.

230.80 Combined Rating of Disconnects. Where the service disconnecting means consists of more than one switch or circuit breaker, as permitted by 230.71, the combined ratings of all the switches or circuit breakers used shall not be less than the rating required by 230.79.

230.82 Equipment Connected to the Supply Side of Service Disconnect. Only the following equipment shall be permitted to be connected to the supply side of the service disconnecting means:

(1) Cable limiters or other current-limiting devices.
(2) Meters and meter sockets nominally rated not in excess of 1000 volts, if all metal housings and service enclosures are grounded in accordance with Part VII and bonded in accordance with Part V of Article 250.
(3) Meter disconnect switches nominally rated not in excess of 1000 V that have a short-circuit current rating equal to or greater than the available short-circuit current, if all metal housings and service enclosures are grounded in accordance with Part VII and bonded in accordance with Part V of Article 250. A meter disconnect switch shall be capable of interrupting the load served. A meter disconnect shall be legibly field marked on its exterior in a manner suitable for the environment as follows:

METER DISCONNECT
NOT SERVICE EQUIPMENT

(4) Instrument transformers (current and voltage), impedance shunts, load management devices, surge arresters, and Type 1 surge-protective devices.
(5) Taps used only to supply load management devices, circuits for standby power systems, fire pump equipment, and fire and sprinkler alarms, if provided with service equipment and installed in accordance with requirements for service-entrance conductors.
(6) Solar photovoltaic systems, fuel cell systems, wind electric systems, energy storage systems, or interconnected electric power production sources.
(7) Control circuits for power-operable service disconnecting means, if suitable overcurrent protection and disconnecting means are provided.
(8) Ground-fault protection systems or Type 2 surge-protective devices, where installed as part of listed equipment,

if suitable overcurrent protection and disconnecting means are provided.

(9) Connections used only to supply listed communications equipment under the exclusive control of the serving electric utility, if suitable overcurrent protection and disconnecting means are provided. For installations of equipment by the serving electric utility, a disconnecting means is not required if the supply is installed as part of a meter socket, such that access can only be gained with the meter removed..

PART VII. SERVICE EQUIPMENT — OVERCURRENT PROTECTION

230.90 Where Required. Each ungrounded service conductor shall have overload protection.

(A) Ungrounded Conductor. Such protection shall be provided by an overcurrent device in series with each ungrounded service conductor that has a rating or setting not higher than the allowable ampacity of the conductor. A set of fuses shall be considered all the fuses required to protect all the ungrounded conductors of a circuit. Single-pole circuit breakers, grouped in accordance with 230.71(B), shall be considered as one protective device.

Exception No. 1: For motor-starting currents, ratings that comply with 430.52, 430.62, and 430.63 shall be permitted.

Exception No. 2: Fuses and circuit breakers with a rating or setting that complies with 240.4(B) or (C) and 240.6 shall be permitted.

Exception No. 3: Two to six circuit breakers or sets of fuses shall be permitted as the overcurrent device to provide the overload protection. The sum of the ratings of the circuit breakers or fuses shall be permitted to exceed the ampacity of the service conductors, provided the calculated load does not exceed the ampacity of the service conductors.

Exception No. 4: Overload protection for fire pump supply conductors shall comply with 695.4(B)(2)(a).

Exception No. 5: Overload protection for 120/240-volt, 3-wire, single-phase dwelling services shall be permitted in accordance with the requirements of 310.15(B)(7).

(B) Not in Grounded Conductor. No overcurrent device shall be inserted in a grounded service conductor except a circuit breaker that simultaneously opens all conductors of the circuit.

230.91 Location. The service overcurrent device shall be an integral part of the service disconnecting means or shall be located immediately adjacent thereto. Where fuses are used as the service overcurrent device, the disconnecting means shall be located ahead of the supply side of the fuses.

230.92 Locked Service Overcurrent Devices. Where the service overcurrent devices are locked or sealed or are not readily accessible to the occupant, branch-circuit or feeder overcurrent devices shall be installed on the load side, shall be mounted in a readily accessible location, and shall be of lower ampere rating than the service overcurrent device.

230.94 Relative Location of Overcurrent Device and Other Service Equipment. The overcurrent device shall protect all circuits and devices.

Exception No. 1: The service switch shall be permitted on the supply side.

Exception No. 2: High-impedance shunt circuits, surge arresters, Type 1 surge-protective devices, surge-protective capacitors, and instrument transformers (current and voltage) shall be permitted to be connected and installed on the supply side of the service disconnecting means as permitted by 230.82.

Exception No. 3: Circuits for load management devices shall be permitted to be connected on the supply side of the service overcurrent device where separately provided with overcurrent protection.

Exception No. 4: Circuits used only for the operation of fire alarm, other protective signaling systems, or the supply to fire pump equipment shall be permitted to be connected on the supply side of the service overcurrent device where separately provided with overcurrent protection.

Exception No. 5: Meters nominally rated not in excess of 600 volts shall be permitted, provided all metal housings and service enclosures are grounded.

Exception No. 6: Where service equipment is power operable, the control circuit shall be permitted to be connected ahead of the service equipment if suitable overcurrent protection and disconnecting means are provided.

230.95 Ground-Fault Protection of Equipment. Ground-fault protection of equipment shall be provided for solidly grounded wye electric services of more than 150 volts to ground but not exceeding 1000 volts phase-to-phase for each service disconnect rated 1000 amperes or more. The grounded conductor for the solidly grounded wye system shall be connected directly to ground through a grounding electrode system, as specified in 250.50, without inserting any resistor or impedance device.

The rating of the service disconnect shall be considered to be the rating of the largest fuse that can be installed or the highest continuous current trip setting for which the actual overcurrent device installed in a circuit breaker is rated or can be adjusted.

Exception: The ground-fault protection provisions of this section shall not apply to a service disconnect for a continuous industrial process where a nonorderly shutdown will introduce additional or increased hazards.

(A) Setting. The ground-fault protection system shall operate to cause the service disconnect to open all ungrounded conductors of the faulted circuit. The maximum setting of the ground-fault protection shall be 1200 amperes, and the maximum time delay shall be one second for ground-fault currents equal to or greater than 3000 amperes.

(B) Fuses. If a switch and fuse combination is used, the fuses employed shall be capable of interrupting any current higher than the interrupting capacity of the switch during a time that the ground-fault protective system will not cause the switch to open.

(C) Performance Testing. The ground-fault protection system shall be performance tested when first installed on site. This testing shall be conducted by a qualified person(s) using a test process of primary current injection, in accordance with instructions that shall be provided with the equipment. A written record of this testing shall be made and shall be available to the authority having jurisdiction.

PART VIII. SERVICES EXCEEDING 1000 VOLTS, NOMINAL

230.200 General. Service conductors and equipment used on circuits exceeding 1000 volts, nominal, shall comply with all the applicable provisions of the preceding sections of this article and with the following sections that supplement or modify the preceding sections. In no case shall the provisions of Part VIII apply to equipment on the supply side of the service point.

CHAPTER 6

FEEDERS

INTRODUCTION

This chapter contains requirements from Article 215—Feeders. This article covers installation, overcurrent protection, minimum size, and ampacity of conductors for feeders that supply other feeders and/or branch-circuit loads. Feeder loads must be calculated in accordance with Article 220. A feeder consists of all circuit conductors located between the service equipment, the source of a separately derived system, or other power supply source and the final branch-circuit overcurrent device. Feeder conductors supply power to feeder panelboards or other distribution equipment from which branch circuits are supplied. Feeder panelboards are often referred to in the industry as subpanels and are sometimes misidentified as service equipment. Service equipment is supplied only by service conductors, never by a feeder. Feeder conductors typically are provided with the full range of overcurrent protection meeting the applicable requirements of Article 240 at the point they receive their supply. There are allowances within Article 240 to provide one level of overcurrent protection at the point the feeder receives its supply, with the overload protection being provided at the point the feeder conductor terminates. The conductors used in this type of application are referred to in the industry as "tap" conductors.

ARTICLE 215

Feeders

215.2 Minimum Rating and Size.

(A) Feeders Not More Than 600 Volts.

(1) General. Feeder conductors shall have an ampacity not less than required to supply the load as calculated in Parts III, IV, and V of Article 220. Conductors shall be sized to carry not less than the larger of 215.2(A)(1)(a) or (b).

(a) Where a feeder supplies continuous loads or any combination of continuous and noncontinuous loads, the minimum feeder conductor size shall have an allowable ampacity not less than the noncontinuous load plus 125 percent of the continuous load.

Exception No. 1: If the assembly, including the overcurrent devices protecting the feeder(s), is listed for operation at 100 percent of its rating, the allowable ampacity of the feeder conductors shall be permitted to be not less than the sum of the continuous load plus the noncontinuous load.

Exception No. 2: Where a portion of a feeder is connected at both its supply and load ends to separately installed pressure connections as covered in 110.14(C)(2), it shall be permitted to have an allowable ampacity not less than the sum of the continuous load plus the noncontinuous load. No portion of a feeder installed under the provisions of this exception shall extend into an enclosure containing either the feeder supply or the feeder load terminations, as covered in 110.14(C)(1).

Exception No. 3: Grounded conductors that are not connected to an overcurrent device shall be permitted to be sized at 100 percent of the continuous and noncontinuous load.

(b) The minimum feeder conductor size shall have an allowable ampacity not less than the maximum load to be served after the application of any adjustment or correction factors.

(2) Grounded Conductor. The size of the feeder circuit grounded conductor shall not be smaller than that required by 250.122, except that 250.122(F) shall not apply where grounded conductors are run in parallel.

Additional minimum sizes shall be as specified in 215.2(A)(3) under the conditions stipulated.

(3) Ampacity Relative to Service Conductors. The feeder conductor ampacity shall not be less than that of the service conductors where the feeder conductors carry the total load supplied by service conductors with an ampacity of 55 amperes or less.

(B) Feeders over 600 Volts. The ampacity of conductors shall be in accordance with 310.15 and 310.60 as applicable. Where installed, the size of the feeder-circuit grounded conductor shall not be smaller than that required by 250.122, except that 250.122(F) shall not apply where grounded conductors are run in parallel. Feeder conductors over 600 volts shall be sized in accordance with 215.2(B)(1), (B)(2), or (B)(3).

(1) Feeders Supplying Transformers. The ampacity of feeder conductors shall not be less than the sum of the nameplate ratings of the transformers supplied when only transformers are supplied.

(2) Feeders Supplying Transformers and Utilization Equipment. The ampacity of feeders supplying a combination of transformers and utilization equipment shall not be less than the sum of the nameplate ratings of the transformers and 125 percent of the designed potential load of the utilization equipment that will be operated simultaneously.

(3) Supervised Installations. For supervised installations, feeder conductor sizing shall be permitted to be determined by qualified persons under engineering supervision. Supervised installations are defined as those portions of a facility where all of the following conditions are met:

(1) Conditions of design and installation are provided under engineering supervision.
(2) Qualified persons with documented training and experience in over 600-volt systems provide maintenance, monitoring, and servicing of the system.

215.3 Overcurrent Protection. Feeders shall be protected against overcurrent in accordance with the provisions of Part I of Article 240. Where a feeder supplies continuous loads or any combination of continuous and noncontinuous loads, the rating of the overcurrent device shall not be less than the noncontinuous load plus 125 percent of the continuous load.

Exception No. 1: Where the assembly, including the overcurrent devices protecting the feeder(s), is listed for operation at 100 percent of its rating, the ampere rating of the overcurrent device shall be permitted to be not less than the sum of the continuous load plus the noncontinuous load.

215.4 Feeders with Common Neutral Conductor.

(A) Feeders with Common Neutral. Up to three sets of 3-wire feeders or two sets of 4-wire or 5-wire feeders shall be permitted to utilize a common neutral.

(B) In Metal Raceway or Enclosure. Where installed in a metal raceway or other metal enclosure, all conductors of all feeders using a common neutral conductor shall be enclosed within the same raceway or other enclosure as required in 300.20.

215.5 Diagrams of Feeders. If required by the authority having jurisdiction, a diagram showing feeder details shall be provided prior to the installation of the feeders. Such a diagram shall show the area in square feet of the building or other structure supplied by each feeder, the total calculated load before applying demand factors, the demand factors used, the calculated load after applying demand factors, and the size and type of conductors to be used.

215.6 Feeder Equipment Grounding Conductor. Where a feeder supplies branch circuits in which equipment grounding conductors are required, the feeder shall include or provide an equipment grounding conductor in accordance with the provisions of 250.134, to which the equipment grounding conductors of the branch circuits shall be connected. Where the feeder supplies a separate building or structure, the requirements of 250.32(B) shall apply.

215.9 Ground-Fault Circuit-Interrupter Protection for Personnel. Feeders supplying 15- and 20-ampere receptacle branch circuits shall be permitted to be protected by a ground-fault circuit interrupter installed in a readily accessible location in lieu of the provisions for such interrupters as specified in 210.8 and 590.6(A).

215.10 Ground-Fault Protection of Equipment. Each feeder disconnect rated 1000 amperes or more and installed on solidly grounded wye electrical systems of more than 150 volts to ground, but not exceeding 600 volts phase-to-phase, shall be provided with ground-fault protection of equipment in accordance with the provisions of 230.95.

Exception No. 1: The provisions of this section shall not apply to a disconnecting means for a continuous industrial process where a nonorderly shutdown will introduce additional or increased hazards.

Exception No. 2: The provisions of this section shall not apply if ground-fault protection of equipment is provided on the supply side of the feeder and on the load side of any transformer supplying the feeder.

215.11 Circuits Derived from Autotransformers. Feeders shall not be derived from autotransformers unless the system supplied has a grounded conductor that is electrically connected to a grounded conductor of the system supplying the autotransformer.

Exception No. 1: An autotransformer shall be permitted without the connection to a grounded conductor where transforming from a nominal 208 volts to a nominal 240-volt supply or similarly from 240 volts to 208 volts.

Exception No. 2: In industrial occupancies, where conditions of maintenance and supervision ensure that only qualified persons service the installation, autotransformers shall be permitted to supply nominal 600-volt loads from nominal 480-volt systems, and 480-volt loads from nominal 600-volt systems, without the connection to a similar grounded conductor.

215.12 Identification for Feeders.

(A) Grounded Conductor. The grounded conductor of a feeder, if insulated, shall be identified in accordance with 200.6.

(B) Equipment Grounding Conductor. The equipment grounding conductor shall be identified in accordance with 250.119.

(C) Identification of Ungrounded Conductors. Ungrounded conductors shall be identified in accordance with 215.12(C)(1) or (C)(2), as applicable.

(1) Feeders Supplied from More Than One Nominal Voltage System. Where the premises wiring system has feeders supplied from more than one nominal voltage system, each ungrounded conductor of a feeder shall be identified by phase or line and system at all termination, connection, and splice points in compliance with 215.12(C)(1)(a) and (b).

(a) *Means of Identification.* The means of identification shall be permitted to be by separate color coding, marking tape, tagging, or other approved means.

(b) *Posting of Identification Means.* The method utilized for conductors originating within each feeder panelboard or similar feeder distribution equipment shall be documented in a manner that is readily available or shall be permanently posted at each feeder panelboard or similar feeder distribution equipment.

(2) Feeders Supplied from Direct-Current Systems. Where a feeder is supplied from a dc system operating at more than 60 volts, each ungrounded conductor of 4 AWG or larger shall be identified by polarity at all termination, connection, and splice points by marking tape, tagging, or other approved means; each ungrounded conductor of 6 AWG or smaller shall be identified by polarity at all termination, connection, and splice points in compliance with 215.12(C)(2)(a) and (b). The identification methods utilized for conductors originating within each feeder panelboard or similar feeder distribution equipment shall be documented in a manner that is readily available or shall be permanently posted at each feeder panelboard or similar feeder distribution equipment.

(a) *Positive Polarity, Sizes 6 AWG or Smaller.* Where the positive polarity of a dc system does not serve as the connection for the grounded conductor, each positive ungrounded conductor shall be identified by one of the following means:

(1) A continuous red outer finish
(2) A continuous red stripe durably marked along the conductor's entire length on insulation of a color other than green, white, gray, or black
(3) Imprinted plus signs (+) or the word POSITIVE or POS durably marked on insulation of a color other than green, white, gray, or black, and repeated at intervals not exceeding 610 mm (24 in.) in accordance with 310.120(B)
(4) An approved permanent marking means such as sleeving or shrink-tubing that is suitable for the conductor size, at all termination, connection, and splice points, with imprinted plus signs (+) or the word POSITIVE or POS durably marked on insulation of a color other than green, white, gray, or black

(b) *Negative Polarity, Sizes 6 AWG or Smaller.* Where the negative polarity of a dc system does not serve as the connection for the grounded conductor, each negative ungrounded conductor shall be identified by one of the following means:

(1) A continuous black outer finish
(2) A continuous black stripe durably marked along the conductor's entire length on insulation of a color other than green, white, gray, or red
(3) Imprinted minus signs (–) or the word NEGATIVE or NEG durably marked on insulation of a color other than green, white, gray, or red, and repeated at intervals not exceeding 610 mm (24 in.) in accordance with 310.120(B)
(4) An approved permanent marking means such as sleeving or shrink-tubing that is suitable for the conductor size, at all termination, connection, and splice points, with imprinted minus signs (–) or the word NEGATIVE or NEG durably marked on insulation of a color other than green, white, gray, or red

CHAPTER 7

SWITCHBOARDS, PANELBOARDS, AND DISCONNECTS

INTRODUCTION

This chapter contains requirements from Article 408—Switchboards, Switchgear, and Panelboards, and Article 404—Switches. Article 408 covers switchboards, switchgear, and panelboards, installed for the control of light and power circuits. A switchboard is a large single panel, frame, or assembly of panels on which are mounted, on the front or back (or both), switches, overcurrent and other protective devices, buses, and usually instruments. Switchboards are generally accessible from both the front and the back and are not intended to be installed in cabinets. Switchboards and switchgear are similar in construction in that both types of equipment are free-standing and fully enclosed (other than ventilating openings) and can be an assembly of sections. Internal construction differences such as full compartmentalization for overcurrent protective devices and for bus and terminal connections and the ability to withdraw circuit breakers with enclosure doors closed distinguish low voltage (600 volts and below) switchgear from switchboards. Additionally, switchgear is constructed for installation in systems rated over 600 volts and is commonly used in medium voltage systems.

A panelboard consists of one (or more) panel units designed for assembly in the form of a single panel. It includes buses and automatic overcurrent devices and may contain switches for the control of light, heat, or power circuits. Panelboards are accessible only from the front and are designed to be enclosed within cabinets or cutout boxes. Switchboards and panelboards that are used as service equipment must be specifically identified for such applications per 230.66.

Requirements from Article 404 are located in two chapters of this *Pocket Guide,* Chapters 7 and 17. This chapter covers switches and circuit breakers used as a switch or disconnecting means. The switches discussed here are more commonly called disconnect switches or safety switches. Disconnects may contain overcurrent protective devices (fuses or circuit breakers), or they may be nonfusible or molded-case switches.

ARTICLE 408
Switchboards, Switchgear, and Panelboards

PART I. GENERAL

408.3 Support and Arrangement of Busbars and Conductors.

(A) Conductors and Busbars on a Switchboard, Switchgear, or Panelboard. Conductors and busbars on a switchboard, switchgear, or panelboard shall comply with 408.3(A)(1), (A)(2), and (A)(3) as applicable.

(1) Location. Conductors and busbars shall be located so as to be free from physical damage and shall be held firmly in place.

(2) Service Panelboards, Switchboards, and Switchgear. Barriers shall be placed in all service panelboards, switchboards, and switchgear such that no uninsulated, ungrounded service busbar or service terminal is exposed to inadvertent contact by persons or maintenance equipment while servicing load terminations.

Exception: This requirement shall not apply to service panelboards with provisions for more than one service disconnect within a single enclosure as permitted in 408.36, Exceptions 1, 2, and 3.

(3) Same Vertical Section. Other than the required interconnections and control wiring, only those conductors that are intended for termination in a vertical section of a switchboard or switchgear shall be located in that section.

Exception: Conductors shall be permitted to travel horizontally through vertical sections of switchboards and switchgear where such conductors are isolated from busbars by a barrier.

(B) Overheating and Inductive Effects. The arrangement of busbars and conductors shall be such as to avoid overheating due to inductive effects.

(C) Used as Service Equipment. Each switchboard, switchgear, or panelboard, if used as service equipment, shall be provided with a main bonding jumper sized in accordance with 250.28(D) or the equivalent placed within the panelboard or one of the sections of the switchboard or switchgear for connecting the grounded service conductor on its supply side to the switchboard, switchgear, or panelboard frame. All sections of a switchboard or switchgear shall be bonded together using an equipment-bonding jumper or a supply-side bonding jumper sized in accordance with 250.122 or 250.102(C)(1) as applicable.

Exception: Switchboards, switchgear, and panelboards used as service equipment on high-impedance grounded neutral systems in accordance with 250.36 shall not be required to be provided with a main bonding jumper.

(E) Bus Arrangement.

(1) AC Phase Arrangement. Alternating-current phase arrangement on 3-phase buses shall be A, B, C from front to back, top to bottom, or left to right, as viewed from the front of the switchboard, switchgear, or panelboard. The B phase shall be that phase having the higher voltage to ground on 3-phase, 4-wire, delta-connected systems. Other busbar arrangements shall be permitted for additions to existing installations and shall be marked.

(2) DC Bus Arrangement. Direct-current ungrounded buses shall be permitted to be in any order. Arrangement of dc buses shall be field marked as to polarity, grounding system, and nominal voltage.

(F) Switchboard, Switchgear, or Panelboard Identification. A caution sign(s) or a label(s) provided in accordance with 408.3(F)(1) through (F)(5) shall comply with 110.21(B).

(1) High-Leg Identification. A switchboard, switchgear, or panelboard containing a 4-wire, delta-connected system where the midpoint of one phase winding is grounded shall be legibly and permanently field marked as follows:

"Caution _____ Phase Has _____ Volts to Ground"

(2) Ungrounded AC Systems. A switchboard, switchgear, or panelboard containing an ungrounded ac electrical system as permitted in 250.21 shall be legibly and permanently field marked as follows:

"Caution Ungrounded System Operating — _____ Volts Between Conductors"

(3) High-Impedance Grounded Neutral AC System. A switchboard, switchgear, or panelboard containing a high-impedance grounded neutral ac system in accordance with 250.36 shall be legibly and permanently field marked as follows:

CAUTION: HIGH-IMPEDANCE GROUNDED NEUTRAL AC SYSTEM OPERATING — _____ VOLTS BETWEEN CONDUCTORS AND MAY OPERATE — _____ VOLTS TO GROUND FOR INDEFINITE PERIODS UNDER FAULT CONDITIONS

(4) Ungrounded DC Systems. A switchboard, switchgear, or panelboard containing an ungrounded dc electrical system in accordance with 250.169 shall be legibly and permanently field marked as follows:

CAUTION: UNGROUNDED DC SYSTEM OPERATING — _____ VOLTS BETWEEN CONDUCTORS

(5) Resistively Grounded DC Systems. A switchboard, switchgear, or panelboard containing a resistive connection between current-carrying conductors and the grounding system to stabilize voltage to ground shall be legibly and permanently field marked as follows:

CAUTION: DC SYSTEM OPERATING — _____ VOLTS BETWEEN CONDUCTORS AND MAY OPERATE — _____ VOLTS TO GROUND FOR INDEFINITE PERIODS UNDER FAULT CONDITIONS

(G) Minimum Wire-Bending Space. The minimum wire-bending space at terminals and minimum gutter space provided in switchboards, switchgear, and panelboards shall be as required in 312.6.

408.4 Field Identification Required.

(A) Circuit Directory or Circuit Identification. Every circuit and circuit modification shall be legibly identified as to its clear, evident, and specific purpose or use. The identification shall include an approved degree of detail that allows each circuit to be distinguished from all others. Spare positions that contain unused overcurrent devices or switches shall be described accordingly. The identification shall be included in a circuit directory that is located on the face or inside of the panel door in the case of a panelboard and at each switch or circuit breaker in a switchboard or switchgear. No circuit shall be described in a manner that depends on transient conditions of occupancy.

(B) Source of Supply. All switchboards, switchgear, and panelboards supplied by a feeder(s) in other than one- or two-family dwellings shall be permanently marked to indicate each device or equipment where the power originates. The label shall be permanently affixed, of sufficient durability to withstand the environment involved, and not handwritten.

408.5 Clearance for Conductor Entering Bus Enclosures.

Where conduits or other raceways enter a switchboard, switchgear, floor-standing panelboard, or similar enclosure at the bottom, approved space shall be provided to permit installation of conductors in the enclosure. The wiring space shall not be less than shown in Table 408.5 where the conduit or raceways enter or leave the enclosure below the busbars, their supports, or other obstructions. The conduit or

Table 408.5 Clearance for Conductors Entering Bus Enclosures

Conductor	Minimum Spacing Between Bottom of Enclosure and Busbars, Their Supports, or Other Obstructions	
	mm	in.
Insulated busbars, their supports, or other obstructions	200	8
Noninsulated busbars	250	10

raceways, including their end fittings, shall not rise more than 75 mm (3 in.) above the bottom of the enclosure.

408.7 Unused Openings. Unused openings for circuit breakers and switches shall be closed using identified closures, or other approved means that provide protection substantially equivalent to the wall of the enclosure.

PART II. SWITCHBOARDS AND SWITCHGEAR

408.16 Switchboards and Switchgear in Damp or Wet Locations. Switchboards and switchgear in damp or wet locations shall be installed in accordance with 312.2.

408.17 Location Relative to Easily Ignitible Material. Switchboards and switchgear shall be placed so as to reduce to a minimum the probability of communicating fire to adjacent combustible materials. Where installed over a combustible floor, suitable protection thereto shall be provided.

408.18 Clearances.

(A) From Ceiling. For other than a totally enclosed switchboard or switchgear, a space not less than 900 mm (3 ft) shall be provided between the top of the switchboard or switchgear and any combustible ceiling, unless a noncombustible shield is provided between the switchboard or switchgear and the ceiling.

(B) Around Switchboards and Switchgear. Clearances around switchboards and switchgear shall comply with the provisions of 110.26.

408.20 Location of Switchboards and Switchgear. Switchboards and switchgear that have any exposed live parts shall be located in permanently dry locations and then only where under competent supervision and accessible only to qualified persons. Switchboards and switchgear shall be located such that the probability of damage from equipment or processes is reduced to a minimum.

PART III. PANELBOARDS

408.30 General. All panelboards shall have a rating not less than the minimum feeder capacity required for the load calculated in accordance with Part III, IV, or V of Article 220, as applicable.

408.36 Overcurrent Protection. In addition to the requirement of 408.30, a panelboard shall be protected by an overcurrent protective device having a rating not greater than that of the panelboard. This overcurrent protective device shall be located within or at any point on the supply side of the panelboard.

Exception No. 1: Individual protection shall not be required for a panelboard used as service equipment with multiple disconnecting means in accordance with 230.71. In panelboards protected by three or more main circuit breakers or sets of fuses, the circuit breakers or sets of fuses shall not supply a second bus structure within the same panelboard assembly.

Exception No. 2: Individual protection shall not be required for a panelboard protected on its supply side by two main circuit breakers or two sets of fuses having a combined rating not greater than that of the panelboard. A panelboard constructed or wired under this exception shall not contain more than 42 overcurrent devices. For the purposes of determining the maximum of 42 overcurrent devices, a 2-pole or a 3-pole circuit breaker shall be considered as two or three overcurrent devices, respectively.

Exception No. 3: For existing panelboards, individual protection shall not be required for a panelboard used as service equipment for an individual residential occupancy.

(A) Snap Switches Rated at 30 Amperes or Less. Panelboards equipped with snap switches rated at 30 amperes or less shall have overcurrent protection of 200 amperes or less.

(B) Supplied Through a Transformer. Where a panelboard is supplied through a transformer, the overcurrent protection required by 408.36 shall be located on the secondary side of the transformer.

Exception: A panelboard supplied by the secondary side of a transformer shall be considered as protected by the overcurrent protection provided on the primary side of the transformer where that protection is in accordance with 240.21(C)(1).

(C) Delta Breakers. A 3-phase disconnect or overcurrent device shall not be connected to the bus of any panelboard that has less than 3-phase buses. Delta breakers shall not be installed in panelboards.

(D) Back-Fed Devices. Plug-in-type overcurrent protection devices or plug-in type main lug assemblies that are backfed and used to terminate field-installed ungrounded supply conductors shall be secured in place by an additional fastener that requires other than a pull to release the device from the mounting means on the panel.

408.37 Panelboards in Damp or Wet Locations. Panelboards in damp or wet locations shall be installed to comply with 312.2.

408.40 Grounding of Panelboards. Panelboard cabinets and panelboard frames, if of metal, shall be in physical contact with each other and shall be connected to an equipment grounding conductor. Where the panelboard is used with nonmetallic raceway or cable or where separate equipment grounding conductors are provided, a terminal bar for the equipment grounding conductors shall be secured inside the cabinet. The terminal bar shall be bonded to the cabinet and panelboard frame, if of metal; otherwise it shall be connected to the equipment grounding conductor that is run with the conductors feeding the panelboard.

Exception: Where an isolated equipment grounding conductor is provided as permitted by 250.146(D), the insulated equipment grounding conductor that is run with the circuit conductors shall be permitted to pass through the panelboard without being connected to the panelboard's equipment grounding terminal bar.

Equipment grounding conductors shall not be connected to a terminal bar provided for grounded conductors or neutral conductors unless the bar is identified for the purpose and is located where interconnection between equipment grounding conductors and grounded circuit conductors is permitted or required by Article 250.

408.41 Grounded Conductor Terminations. Each grounded conductor shall terminate within the panelboard in an individual terminal that is not also used for another conductor.

Exception: Grounded conductors of circuits with parallel conductors shall be permitted to terminate in a single terminal if the terminal is identified for connection of more than one conductor.

ARTICLE 404
Switches

404.2 Switch Connections.

(B) Grounded Conductors. Switches or circuit breakers shall not disconnect the grounded conductor of a circuit.

Exception: A switch or circuit breaker shall be permitted to disconnect a grounded circuit conductor where all circuit conductors are disconnected simultaneously, or where the device is arranged so that the grounded conductor cannot be disconnected until all the ungrounded conductors of the circuit have been disconnected.

404.3 Enclosure.

(A) General. Switches and circuit breakers shall be of the externally operable type mounted in an enclosure listed for the intended use. The minimum wire-bending space at terminals and minimum gutter space provided in switch enclosures shall be as required in 312.6.

Exception No. 2:, Switches and circuit breakers installed in accordance with 110.27(A)(1), (A)(2), (A)(3), or (A)(4) shall be permitted without enclosures.

(B) Used as a Raceway. Enclosures shall not be used as junction boxes, auxiliary gutters, or raceways for conductors feeding through or tapping off to other switches or overcurrent devices, unless the enclosure complies with 312.8.

404.4 Damp or Wet Locations.

(A) Surface-Mounted Switch or Circuit Breaker. Asurface-mounted switch or circuit breaker shall be enclosed in a weatherproof enclosure or cabinet that complies with 312.2.

(B) Flush-Mounted Switch or Circuit Breaker. A flush-mounted switch or circuit breaker shall be equipped with a weatherproof cover.

404.5 Time Switches, Flashers, and Similar Devices. Time switches, flashers, and similar devices shall be of the enclosed type or shall be mounted in cabinets or boxes or equipment enclosures. Energized parts shall be barriered to prevent operator exposure when making manual adjustments or switching.

Exception: Devices mounted so they are accessible only to qualified persons shall be permitted without barriers, provided they are located within an enclosure such that any energized parts within 152 mm (6.0 in.) of the manual adjustment or switch are covered by suitable barriers.

404.6 Position and Connection of Switches.

(A) Single-Throw Knife Switches. Single-throw knife switches shall be placed so that gravity will not tend to close them. Single-throw knife switches, approved for use in the inverted position, shall be provided with an integral mechanical means that ensures that the blades remain in the open position when so set.

404.7 Indicating. General-use and motor-circuit switches, circuit breakers, and molded case switches, where mounted in an enclosure as described in 404.3, shall clearly indicate whether they are in the open (off) or closed (on) position.

Where these switch or circuit breaker handles are operated vertically rather than rotationally or horizontally, the up position of the handle shall be the closed (on) position.

Exception No. 1: Vertically operated double-throw switches shall be permitted to be in the closed (on) position with the handle in either the up or down position.

Exception No. 2: On busway installations, tap switches employing a center-pivoting handle shall be permitted to be open or closed with either end of the handle in the up or down position. The switch position shall be clearly indicating and shall be visible from the floor or from the usual point of operation.

404.8 Accessibility and Grouping.

(A) Location. All switches and circuit breakers used as switches shall be located so that they may be operated from a readily accessible place. They shall be installed such that the center of the grip of the operating handle of the switch or circuit breaker, when in its highest position, is not more than 2.0 m (6 ft 7 in.) above the floor or working platform.

Exception No. 1: On busway installations, fused switches and circuit breakers shall be permitted to be located at the same level as the busway. Suitable means shall be provided to operate the handle of the device from the floor.

Exception No. 2: Switches and circuit breakers installed adjacent to motors, appliances, or other equipment that they supply shall be permitted to be located higher than 2.0 m (6 ft 7 in.) and to be accessible by portable means.

Exception No. 3: Hookstick operable isolating switches shall be permitted at greater heights.

404.11 Circuit Breakers as Switches. A hand-operable circuit breaker equipped with a lever or handle, or a power-operated circuit breaker capable of being opened by hand in the event of a power failure, shall be permitted to serve as a switch if it has the required number of poles.

404.12 Grounding of Enclosures. Metal enclosures for switches or circuit breakers shall be connected to an equipment grounding conductor as specified in Part IV of Article 250. Metal enclosures for switches or circuit breakers used as service equipment shall comply with the provisions of Part V of Article 250. Where nonmetallic enclosures are used with metal raceways or metal-armored cables, provision shall be made for connecting the equipment grounding conductor(s).

Except as covered in 404.9(B), Exception No. 1, nonmetallic boxes for switches shall be installed with a wiring method that provides or includes an equipment grounding conductor.

CHAPTER 8

OVERCURRENT PROTECTION

INTRODUCTION

This chapter contains requirements from Article 240—Overcurrent Protection. Article 240 provides general requirements for overcurrent protection of conductors and equipment and for overcurrent protective devices. Generally for equipment rated 1000 volts and less, overcurrent protective devices are plug fuses, cartridge fuses, and circuit breakers. Overcurrent protection for conductors and equipment is provided to open the circuit if the current reaches a value that will cause an excessive or dangerous temperature in conductors or conductor insulation.

As a general rule, overcurrent protection of conductors is required to be provided at the point a conductor receives its supply. Using this approach, overload, short-circuit, and line to ground fault protection is provided using a single overcurrent protective device. However, this approach is not always attainable, as in the case of conductors connected to a transformer secondary or to the output terminals of a generator. Alternative protection arrangements are provided by 240.21 that meet the overall objective to provide the full range of overcurrent protection for conductors.

Contained in this *Pocket Guide* are eight of the nine parts included in Article 240: Part I General; Part II Location; Part III Enclosures; Part IV Disconnecting and Guarding; Part VI Cartridge Fuses and Fuseholders; Part VII Circuit Breakers; Part VIII Supervised Industrial Installations; and Part IX Overcurrent Protection over 1000 Volts, Nominal.

ARTICLE 240
Overcurrent Protection

PART I. GENERAL

240.1 Scope. Parts I through VII of this article provide the general requirements for overcurrent protection and overcurrent protective devices not more than 1000 volts, nominal. Part VIII covers overcurrent protection for those portions of supervised industrial installations operating at voltages of not more than 1000 volts, nominal. Part IX covers overcurrent protection over 1000 volts, nominal.

240.2 Definitions.

Supervised Industrial Installation. For the purposes of Part VIII, the industrial portions of a facility where all of the following conditions are met:

(1) Conditions of maintenance and engineering supervision ensure that only qualified persons monitor and service the system.
(2) The premises wiring system has 2500 kVA or greater of load used in industrial process(es), manufacturing activities, or both, as calculated in accordance with Article 220.
(3) The premises has at least one service or feeder that is more than 150 volts to ground and more than 300 volts phase-to-phase.

This definition excludes installations in buildings used by the industrial facility for offices, warehouses, garages, machine shops, and recreational facilities that are not an integral part of the industrial plant, substation, or control center.

Tap Conductor. A conductor, other than a service conductor, that has overcurrent protection ahead of its point of supply that exceeds the value permitted for similar conductors that are protected as described elsewhere in 240.4.

240.4 Protection of Conductors. Conductors, other than flexible cords, flexible cables, and fixture wires, shall be protected against overcurrent in accordance with their ampacities specified in 310.15, unless otherwise permitted or required in 240.4(A) through (G).

(A) Power Loss Hazard. Conductor overload protection shall not be required where the interruption of the circuit would create a hazard, such as in a material-handling magnet circuit or fire pump circuit. Short-circuit protection shall be provided.

(B) Overcurrent Devices Rated 800 Amperes or Less. The next higher standard overcurrent device rating (above the ampacity of the conductors being protected) shall be permitted to be used, provided all of the following conditions are met:

(1) The conductors being protected are not part of a branch circuit supplying more than one receptacle for cord-and-plug-connected portable loads.
(2) The ampacity of the conductors does not correspond with the standard ampere rating of a fuse or a circuit breaker without overload trip adjustments above its rating (but that shall be permitted to have other trip or rating adjustments).
(3) The next higher standard rating selected does not exceed 800 amperes.

(C) Overcurrent Devices Rated over 800 Amperes. Where the overcurrent device is rated over 800 amperes, the ampacity of the conductors it protects shall be equal to or greater than the rating of the overcurrent device defined in 240.6.

(D) Small Conductors. Unless specifically permitted in 240.4(E) or (G), the overcurrent protection shall not exceed that required by (D)(1) through (D)(7) after any correction factors for ambient temperature and number of conductors have been applied.

(1) 18 AWG Copper. 7 amperes, provided all the following conditions are met:

(1) Continuous loads do not exceed 5.6 amperes.
(2) Overcurrent protection is provided by one of the following:
    a. Branch-circuit-rated circuit breakers listed and marked for use with 18 AWG copper wire

b. Branch-circuit-rated fuses listed and marked for use with 18 AWG copper wire
c. Class CC, Class J, or Class T fuses

(2) 16 AWG Copper. 10 amperes, provided all the following conditions are met:

(1) Continuous loads do not exceed 8 amperes.
(2) Overcurrent protection is provided by one of the following:
 a. Branch-circuit-rated circuit breakers listed and marked for use with 16 AWG copper wire
 b. Branch-circuit-rated fuses listed and marked for use with 16 AWG copper wire
 c. Class CC, Class J, or Class T fuses

(3) 14 AWG Copper. 15 amperes

(4) 12 AWG Aluminum and Copper-Clad Aluminum. 15 amperes

(5) 12 AWG Copper. 20 amperes

(6) 10 AWG Aluminum and Copper-Clad Aluminum. 25 amperes

(7) 10 AWG Copper. 30 amperes

(E) Tap Conductors. Tap conductors shall be permitted to be protected against overcurrent in accordance with the following:

(1) 210.19(A)(3)and (A)(4), Household Ranges and Cooking Appliances and Other Loads
(2) 240.5(B)(2), Fixture Wire
(3) 240.21, Location in Circuit
(4) 368.17(B), Reduction in Ampacity Size of Busway
(5) 368.17(C), Feeder or Branch Circuits (busway taps)
(6) 430.53(D), Single Motor Taps

(F) Transformer Secondary Conductors. Single-phase (other than 2-wire) and multiphase (other than delta-delta, 3-wire) transformer secondary conductors shall not be considered to be protected by the primary overcurrent protective device. Conductors supplied by the secondary side of a single-phase transformer having a 2-wire (single-voltage) secondary, or a three-phase, delta-delta connected transformer having a 3-wire (single-voltage) secondary, shall be permitted to be protected by overcurrent protection provided on the primary (supply) side of the transformer, provided this protection is in accordance with 450.3 and does not exceed the value determined by multiplying the secondary conductor ampacity by the secondary-to-primary transformer voltage ratio.

(G) Overcurrent Protection for Specific Conductor Applications. Overcurrent protection for the specific conductors shall be permitted to be provided as referenced in Table 240.4(G).

Table 240.4(G) Specific Conductor Applications

Conductor	Article	Section
Air-conditioning and refrigeration equipment circuit conductors	440, Parts III, VI	
Capacitor circuit conductors	460	460.8(B) and 460.25(A)–(D)
Control and instrumentation circuit conductors (Type ITC)	727	727.9
Electric welder circuit conductors	630	630.12 and 630.32
Fire alarm system circuit conductors	760	760.43, 760.45, 760.121, and Chapter 9, Tables 12(A) and 12(B)
Motor-operated appliance circuit conductors	422, Part II	
Motor and motor-control circuit conductors	430, Parts II, III, IV, V, VI, VII	
Phase converter supply conductors	455	455.7
Remote-control, signaling, and power-limited circuit conductors	725	725.43, 725.45, 725.121, and Chapter 9, Tables 11(A) and 11(B)
Secondary tie conductors	450	450.6

240.5 Protection of Flexible Cords, Flexible Cables, and Fixture Wires. Flexible cord and flexible cable, including tinsel cord and extension cords, and fixture wires shall be protected against overcurrent by either 240.5(A) or (B).

(A) Ampacities. Flexible cord and flexible cable shall be protected by an overcurrent device in accordance with their ampacity as specified in Table 400.5(A)(1) and Table 400.5(A)(2). Fixture wire shall be protected against overcurrent in accordance with its ampacity as specified in Table 402.5. Supplementary overcurrent protection, as covered in 240.10, shall be permitted to be an acceptable means for providing this protection.

(B) Branch-Circuit Overcurrent Device. Flexible cord shall be protected, where supplied by a branch circuit, in accordance with one of the methods described in 240.5(B)(1), (B)(3), or (B)(4). Fixture wire shall be protected, where supplied by a branch circuit, in accordance with 240.5(B)(2).

(2) Fixture Wire. Fixture wire shall be permitted to be tapped to the branch-circuit conductor of a branch circuit in accordance with the following:

(1) 20-ampere circuits — 18 AWG, up to 15 m (50 ft) of run length
(2) 20-ampere circuits — 16 AWG, up to 30 m (100 ft) of run length
(3) 20-ampere circuits — 14 AWG and larger
(4) 30-ampere circuits — 14 AWG and larger
(5) 40-ampere circuits — 12 AWG and larger
(6) 50-ampere circuits — 12 AWG and larger

240.6 Standard Ampere Ratings.

(A) Fuses and Fixed-Trip Circuit Breakers. The standard ampere ratings for fuses and inverse time circuit breakers shall be considered as shown in Table 240.6(A). Additional standard ampere ratings for fuses shall be 1, 3, 6, 10, and 601. The use of fuses and inverse time circuit breakers with nonstandard ampere ratings shall be permitted.

(B) Adjustable-Trip Circuit Breakers. The rating of adjustable-trip circuit breakers having external means for adjusting the current setting (long-time pickup setting), not meeting

N **Table 240.6(A) Standard Ampere Ratings for Fuses and Inverse Time Circuit Breakers**

Standard Ampere Ratings				
15	20	25	30	35
40	45	50	60	70
80	90	100	110	125
150	175	200	225	250
300	350	400	450	500
600	700	800	1000	1200
1600	2000	2500	3000	4000
5000	6000	—	—	—

the requirements of 240.6(C), shall be the maximum setting possible.

(C) Restricted Access Adjustable-Trip Circuit Breakers. A circuit breaker(s) that has restricted access to the adjusting means shall be permitted to have an ampere rating(s) that is equal to the adjusted current setting (long-time pickup setting). Restricted access shall be defined as located behind one of the following:

(1) Removable and sealable covers over the adjusting means
(2) Bolted equipment enclosure doors
(3) Locked doors accessible only to qualified personnel

240.8 Fuses or Circuit Breakers in Parallel. Fuses and circuit breakers shall be permitted to be connected in parallel where they are factory assembled in parallel and listed as a unit. Individual fuses, circuit breakers, or combinations thereof shall not otherwise be connected in parallel.

240.9 Thermal Devices. Thermal relays and other devices not designed to open short circuits or ground faults shall not be used for the protection of conductors against overcurrent due to short circuits or ground faults, but the use of such devices shall be permitted to protect motor branch-circuit conductors from overload if protected in accordance with 430.40.

240.10 Supplementary Overcurrent Protection. Where supplementary overcurrent protection is used for luminaires, appliances, and other equipment or for internal circuits and components of equipment, it shall not be used as a substitute for required branch-circuit overcurrent devices or in place of the required branch-circuit protection. Supplementary overcurrent devices shall not be required to be readily accessible.

240.12 Electrical System Coordination. Where an orderly shutdown is required to minimize the hazard(s) to personnel and equipment, a system of coordination based on the following two conditions shall be permitted:

(1) Coordinated short-circuit protection
(2) Overload indication based on monitoring systems or devices

240.13 Ground-Fault Protection of Equipment. Ground-fault protection of equipment shall be provided in accordance with the provisions of 230.95 for solidly grounded wye electrical systems of more than 150 volts to ground but not exceeding 1000 volts phase-to-phase for each individual device used as a building or structure main disconnecting means rated 1000 amperes or more.

The provisions of this section shall not apply to the disconnecting means for the following:

(1) Continuous industrial processes where a nonorderly shutdown will introduce additional or increased hazards
(2) Installations where ground-fault protection is provided by other requirements for services or feeders
(3) Fire pumps

240.15 Ungrounded Conductors.

(A) Overcurrent Device Required. A fuse or an overcurrent trip unit of a circuit breaker shall be connected in series with each ungrounded conductor. A combination of a current transformer and overcurrent relay shall be considered equivalent to an overcurrent trip unit.

(B) Circuit Breaker as Overcurrent Device. Circuit breakers shall open all ungrounded conductors of the circuit both manually and automatically unless otherwise permitted in 240.15(B)(1), (B)(2), (B)(3), and (B)(4).

(1) Multiwire Branch Circuits. Individual single-pole circuit breakers, with identified handle ties, shall be permitted as the protection for each ungrounded conductor of multiwire branch circuits that serve only single-phase line-to-neutral loads.

(2) Grounded Single-Phase Alternating-Current Circuits. In grounded systems, individual single-pole circuit breakers rated 120/240 volts ac, with identified handle ties, shall be permitted as the protection for each ungrounded conductor for line-to-line connected loads for single-phase circuits.

(3) 3-Phase and 2-Phase Systems. For line-to-line loads in 4-wire, 3-phase systems or 5-wire, 2-phase systems, individual single-pole circuit breakers rated 120/240 volts ac with identified handle ties shall be permitted as the protection for each ungrounded conductor, if the systems have a grounded neutral point and the voltage to ground does not exceed 120 volts.

(4) 3-Wire Direct-Current Circuits. Individual single-pole circuit breakers rated 125/250 volts dc with identified handle ties shall be permitted as the protection for each ungrounded conductor for line-to-line connected loads for 3-wire, direct-current circuits supplied from a system with a grounded neutral where the voltage to ground does not exceed 125 volts.

PART II. LOCATION

240.21 Location in Circuit. Overcurrent protection shall be provided in each ungrounded circuit conductor and shall be located at the point where the conductors receive their supply except as specified in 240.21(A) through (H). Conductors supplied under the provisions of 240.21(A) through (H) shall not supply another conductor except through an overcurrent protective device meeting the requirements of 240.4.

(A) Branch-Circuit Conductors. Branch-circuit tap conductors meeting the requirements specified in 210.19 shall be permitted to have overcurrent protection as specified in 210.20.

(B) Feeder Taps. Conductors shall be permitted to be tapped, without overcurrent protection at the tap, to a feeder as specified in 240.21(B)(1) through (B)(5). The provisions of 240.4(B) shall not be permitted for tap conductors.

(1) Taps Not over 3 m (10 ft) Long. If the length of the tap conductors does not exceed 3 m (10 ft) and the tap conductors comply with all of the following:

(1) The ampacity of the tap conductors is

a. Not less than the combined calculated loads on the circuits supplied by the tap conductors, and
b. Not less than the rating of the equipment containing an overcurrent device(s) supplied by the tap conductors or not less than the rating of the overcurrent protective device at the termination of the tap conductors.

Exception to b: Where listed equipment, such as a surge protective device(s) [SPD(s)], is provided with specific instructions on minimum conductor sizing, the ampacity of the tap conductors supplying that equipment shall be permitted to be determined based on the manufacturer's instructions.

(2) The tap conductors do not extend beyond the switchboard, switchgear, panelboard, disconnecting means, or control devices they supply.
(3) Except at the point of connection to the feeder, the tap conductors are enclosed in a raceway, which extends from the tap to the enclosure of an enclosed switchboard, switchgear, a panelboard, or control devices, or to the back of an open switchboard.
(4) For field installations, if the tap conductors leave the enclosure or vault in which the tap is made, the ampacity of the tap conductors is not less than one-tenth of the rating of the overcurrent device protecting the feeder conductors.

(2) Taps Not over 7.5 m (25 ft) Long. Where the length of the tap conductors does not exceed 7.5 m (25 ft) and the tap conductors comply with all the following:

(1) The ampacity of the tap conductors is not less than one-third of the rating of the overcurrent device protecting the feeder conductors.
(2) The tap conductors terminate in a single circuit breaker or a single set of fuses that limit the load to the ampacity of the tap conductors. This device shall be permitted to supply any number of additional overcurrent devices on its load side.
(3) The tap conductors are protected from physical damage by being enclosed in an approved raceway or by other approved means.

(3) Taps Supplying a Transformer [Primary Plus Secondary Not over 7.5 m (25 ft) Long]. Where the tap conductors supply a transformer and comply with all the following conditions:

(1) The conductors supplying the primary of a transformer have an ampacity at least one-third the rating of the overcurrent device protecting the feeder conductors.
(2) The conductors supplied by the secondary of the transformer shall have an ampacity that is not less than the value of the primary-to-secondary voltage ratio multiplied by one-third of the rating of the overcurrent device protecting the feeder conductors.
(3) The total length of one primary plus one secondary conductor, excluding any portion of the primary conductor that is protected at its ampacity, is not over 7.5 m (25 ft).
(4) The primary and secondary conductors are protected from physical damage by being enclosed in an approved raceway or by other approved means.
(5) The secondary conductors terminate in a single circuit breaker or set of fuses that limit the load current to not more than the conductor ampacity that is permitted by 310.15.

(4) Taps over 7.5 m (25 ft) Long. Where the feeder is in a high bay manufacturing building over 11 m (35 ft) high at walls and the installation complies with all the following conditions:

(1) Conditions of maintenance and supervision ensure that only qualified persons service the systems.

(2) The tap conductors are not over 7.5 m (25 ft) long horizontally and not over 30 m (100 ft) total length.
(3) The ampacity of the tap conductors is not less than one-third the rating of the overcurrent device protecting the feeder conductors.
(4) The tap conductors terminate at a single circuit breaker or a single set of fuses that limit the load to the ampacity of the tap conductors. This single overcurrent device shall be permitted to supply any number of additional overcurrent devices on its load side.
(5) The tap conductors are protected from physical damage by being enclosed in an approved raceway or by other approved means.
(6) The tap conductors are continuous from end-to-end and contain no splices.
(7) The tap conductors are sized 6 AWG copper or 4 AWG aluminum or larger.
(8) The tap conductors do not penetrate walls, floors, or ceilings.
(9) The tap is made no less than 9 m (30 ft) from the floor.

(5) Outside Taps of Unlimited Length. Where the conductors are located outside of a building or structure, except at the point of load termination, and comply with all of the following conditions:

(1) The tap conductors are protected from physical damage in an approved manner.
(2) The tap conductors terminate at a single circuit breaker or a single set of fuses that limits the load to the ampacity of the tap conductors. This single overcurrent device shall be permitted to supply any number of additional overcurrent devices on its load side.
(3) The overcurrent device for the tap conductors is an integral part of a disconnecting means or shall be located immediately adjacent thereto.
(4) The disconnecting means for the tap conductors is installed at a readily accessible location complying with one of the following:

a. Outside of a building or structure
b. Inside, nearest the point of entrance of the tap conductors

c. Where installed in accordance with 230.6, nearest the point of entrance of the tap conductors

(C) Transformer Secondary Conductors. A set of conductors feeding a single load, or each set of conductors feeding separate loads, shall be permitted to be connected to a transformer secondary, without overcurrent protection at the secondary, as specified in 240.21(C)(1) through (C)(6). The provisions of 240.4(B) shall not be permitted for transformer secondary conductors.

(1) Protection by Primary Overcurrent Device. Conductors supplied by the secondary side of a single-phase transformer having a 2-wire (single-voltage) secondary, or a three-phase, delta-delta connected transformer having a 3-wire (single-voltage) secondary, shall be permitted to be protected by overcurrent protection provided on the primary (supply) side of the transformer, provided this protection is in accordance with 450.3 and does not exceed the value determined by multiplying the secondary conductor ampacity by the secondary-to-primary transformer voltage ratio.

Single-phase (other than 2-wire) and multiphase (other than delta-delta, 3-wire) transformer secondary conductors are not considered to be protected by the primary overcurrent protective device.

(2) Transformer Secondary Conductors Not over 3 m (10 ft) Long. If the length of secondary conductor does not exceed 3 m (10 ft) and complies with all of the following:

(1) The ampacity of the secondary conductors is

  a. Not less than the combined calculated loads on the circuits supplied by the secondary conductors, and
  b. Not less than the rating of the equipment containing an overcurrent device(s) supplied by the secondary conductors or not less than the rating of the overcurrent protective device at the termination of the secondary conductors.

Exception: Where listed equipment, such as a surge protective device(s) [SPD(s)], is provided with specific instructions on minimum conductor sizing, the ampacity of the tap conductors supplying that equipment shall be permitted to be determined based on the manufacturer's instructions.

(2) The secondary conductors do not extend beyond the switchboard, switchgear, panelboard, disconnecting means, or control devices they supply.
(3) The secondary conductors are enclosed in a raceway, which shall extend from the transformer to the enclosure of an enclosed switchboard, switchgear, a panelboard, or control devices or to the back of an open switchboard.
(4) For field installations where the secondary conductors leave the enclosure or vault in which the supply connection is made, the rating of the overcurrent device protecting the primary of the transformer, multiplied by the primary to secondary transformer voltage ratio, shall not exceed 10 times the ampacity of the secondary conductor.

(3) Industrial Installation Secondary Conductors Not over 7.5 m (25 ft) Long. For the supply of switchgear or switchboards in industrial installations only, where the length of the secondary conductors does not exceed 7.5 m (25 ft) and complies with all of the following:

(1) Conditions of maintenance and supervision ensure that only qualified persons service the systems.
(2) The ampacity of the secondary conductors is not less than the secondary current rating of the transformer, and the sum of the ratings of the overcurrent devices does not exceed the ampacity of the secondary conductors.
(3) All overcurrent devices are grouped.
(4) The secondary conductors are protected from physical damage by being enclosed in an approved raceway or by other approved means.

(4) Outside Secondary Conductors. Where the conductors are located outside of a building or structure, except at the point of load termination, and comply with all of the following conditions:

(1) The conductors are protected from physical damage in an approved manner.
(2) The conductors terminate at a single circuit breaker or a single set of fuses that limit the load to the ampacity of the conductors. This single overcurrent device shall be

permitted to supply any number of additional overcurrent devices on its load side.

(3) The overcurrent device for the conductors is an integral part of a disconnecting means or shall be located immediately adjacent thereto.

(4) The disconnecting means for the conductors is installed at a readily accessible location complying with one of the following:

a. Outside of a building or structure
b. Inside, nearest the point of entrance of the conductors
c. Where installed in accordance with 230.6, nearest the point of entrance of the conductors

(5) Secondary Conductors from a Feeder Tapped Transformer. Transformer secondary conductors installed in accordance with 240.21(B)(3) shall be permitted to have overcurrent protection as specified in that section.

(6) Secondary Conductors Not over 7.5 m (25 ft) Long. Where the length of secondary conductor does not exceed 7.5 m (25 ft) and complies with all of the following:

(1) The secondary conductors shall have an ampacity that is not less than the value of the primary-to-secondary voltage ratio multiplied by one-third of the rating of the overcurrent device protecting the primary of the transformer.

(2) The secondary conductors terminate in a single circuit breaker or set of fuses that limit the load current to not more than the conductor ampacity that is permitted by 310.15.

(3) The secondary conductors are protected from physical damage by being enclosed in an approved raceway or by other approved means.

(D) Service Conductors. Service conductors shall be permitted to be protected by overcurrent devices in accordance with 230.91.

(E) Busway Taps. Busways and busway taps shall be permitted to be protected against overcurrent in accordance with 368.17.

(F) Motor Circuit Taps. Motor-feeder and branch-circuit conductors shall be permitted to be protected against overcurrent in accordance with 430.28 and 430.53, respectively.

(G) Conductors from Generator Terminals. Conductors from generator terminals that meet the size requirement in 445.13 shall be permitted to be protected against overload by the generator overload protective device(s) required by 445.12.

(H) Battery Conductors. Overcurrent protection shall be permitted to be installed as close as practicable to the storage battery terminals in an unclassified location. Installation of the overcurrent protection within a hazardous (classified) location shall also be permitted.

240.23 Change in Size of Grounded Conductor. Where a change occurs in the size of the ungrounded conductor, a similar change shall be permitted to be made in the size of the grounded conductor.

240.24 Location in or on Premises.

(A) Accessibility. Switches containing fuses and circuit breakers shall be readily accessible and installed so that the center of the grip of the operating handle of the switch or circuit breaker, when in its highest position, is not more than 2.0 m (6 ft 7 in.) above the floor or working platform, unless one of the following applies:

(1) For busways, as provided in 368.17(C).
(2) For supplementary overcurrent protection, as described in 240.10.
(3) For overcurrent devices, as described in 225.40 and 230.92.
(4) For overcurrent devices adjacent to utilization equipment that they supply, access shall be permitted to be by portable means.

Exception: The use of a tool shall be permitted to access overcurrent devices located within listed industrial control panels or similar enclosures.

(B) Occupancy. Each occupant shall have ready access to all overcurrent devices protecting the conductors supplying that occupancy, unless otherwise permitted in 240.24(B)(1) and (B)(2).

(1) Service and Feeder Overcurrent Devices. Where electric service and electrical maintenance are provided by the building management and where these are under continuous building management supervision, the service overcurrent devices and feeder overcurrent devices supplying more than one occupancy shall be permitted to be accessible only to authorized management personnel in the following:

(1) Multiple-occupancy buildings
(2) Guest rooms or guest suites

(2) Branch-Circuit Overcurrent Devices. Where electric service and electrical maintenance are provided by the building management and where these are under continuous building management supervision, the branch-circuit overcurrent devices supplying any guest rooms or guest suites without permanent provisions for cooking shall be permitted to be accessible only to authorized management personnel.

(C) Not Exposed to Physical Damage. Overcurrent devices shall be located where they will not be exposed to physical damage.

(D) Not in Vicinity of Easily Ignitible Material. Overcurrent devices shall not be located in the vicinity of easily ignitible material, such as in clothes closets.

(E) Not Located in Bathrooms. In dwelling units, dormitories, and guest rooms or guest suites, overcurrent devices, other than supplementary overcurrent protection, shall not be located in bathrooms.

(F) Not Located over Steps. Overcurrent devices shall not be located over steps of a stairway.

PART III. ENCLOSURES

240.30 General.

(A) Protection from Physical Damage. Overcurrent devices shall be protected from physical damage by one of the following:

(1) Installation in enclosures, cabinets, cutout boxes, or equipment assemblies

(2) Mounting on open-type switchboards, panelboards, or control boards that are in rooms or enclosures free from dampness and easily ignitible material and are accessible only to qualified personnel

(B) Operating Handle. The operating handle of a circuit breaker shall be permitted to be accessible without opening a door or cover.

240.32 Damp or Wet Locations. Enclosures for overcurrent devices in damp or wet locations shall comply with 312.2.

240.33 Vertical Position. Enclosures for overcurrent devices shall be mounted in a vertical position unless that is shown to be impracticable. Circuit breaker enclosures shall be permitted to be installed horizontally where the circuit breaker is installed in accordance with 240.81. Listed busway plug-in units shall be permitted to be mounted in orientations corresponding to the busway mounting position.

PART IV. DISCONNECTING AND GUARDING

240.40 Disconnecting Means for Fuses. Cartridge fuses in circuits of any voltage where accessible to other than qualified persons, and all fuses in circuits over 150 volts to ground, shall be provided with a disconnecting means on their supply side so that each circuit containing fuses can be independently disconnected from the source of power. A current-limiting device without a disconnecting means shall be permitted on the supply side of the service disconnecting means as permitted by 230.82. A single disconnecting means shall be permitted on the supply side of more than one set of fuses as permitted by 430.112, Exception, for group operation of motors and 424.22(C) for fixed electric space-heating equipment.

240.41 Arcing or Suddenly Moving Parts. Arcing or suddenly moving parts shall comply with 240.41(A) and (B).

(A) Location. Fuses and circuit breakers shall be located or shielded so that persons will not be burned or otherwise injured by their operation.

(B) Suddenly Moving Parts. Handles or levers of circuit breakers, and similar parts that may move suddenly in such a way that persons in the vicinity are likely to be injured by being struck by them, shall be guarded or isolated.

PART VI. CARTRIDGE FUSES AND FUSEHOLDERS

240.60 General.

(A) Maximum Voltage — 300-Volt Type. Cartridge fuses and fuseholders of the 300-volt type shall be permitted to be used in the following circuits:

(1) Circuits not exceeding 300 volts between conductors
(2) Single-phase line-to-neutral circuits supplied from a 3-phase, 4-wire, solidly grounded neutral source where the line-to-neutral voltage does not exceed 300 volts

(B) Noninterchangeable — 0–6000-Ampere Cartridge Fuseholders. Fuseholders shall be designed so that it will be difficult to put a fuse of any given class into a fuseholder that is designed for a current lower, or voltage higher, than that of the class to which the fuse belongs. Fuseholders for current-limiting fuses shall not permit insertion of fuses that are not current-limiting.

(C) Marking. Fuses shall be plainly marked, either by printing on the fuse barrel or by a label attached to the barrel showing the following:

(1) Ampere rating
(2) Voltage rating
(3) Interrupting rating where other than 10,000 amperes
(4) Current limiting where applicable
(5) The name or trademark of the manufacturer

The interrupting rating shall not be required to be marked on fuses used for supplementary protection.

(D) Renewable Fuses. Class H cartridge fuses of the renewable type shall be permitted to be used only for replacement in existing installations where there is no evidence of overfusing or tampering.

N **240.67 Arc Energy Reduction.** Where fuses rated 1200 A or higher are installed, 240.67(A) and (B) shall apply. This requirement shall become effective January 1, 2020.

(A) Documentation. Documentation shall be available to those authorized to design, install, operate, or inspect the installation as to the location of the fuses.

(B) Method to Reduce Clearing Time. A fuse shall have a clearing time of 0.07 seconds or less at the available arcing current, or one of the following shall be provided:

(1) Differential relaying
(2) Energy-reducing maintenance switching with local status indicator
(3) Energy-reducing active arc flash mitigation system
(4) An approved equivalent means

PART VII. CIRCUIT BREAKERS

240.80 Method of Operation. Circuit breakers shall be trip free and capable of being closed and opened by manual operation. Their normal method of operation by other than manual means, such as electrical or pneumatic, shall be permitted if means for manual operation are also provided.

240.81 Indicating. Circuit breakers shall clearly indicate whether they are in the open "off" or closed "on" position.

Where circuit breaker handles are operated vertically rather than rotationally or horizontally, the "up" position of the handle shall be the "on" position.

240.83 Marking.

(D) Used as Switches. Circuit breakers used as switches in 120-volt and 277-volt fluorescent lighting circuits shall be listed and shall be marked SWD or HID. Circuit breakers used as switches in high-intensity discharge lighting circuits shall be listed and shall be marked as HID.

(E) Voltage Marking. Circuit breakers shall be marked with a voltage rating not less than the nominal system voltage that is indicative of their capability to interrupt fault currents between phases or phase to ground.

240.85 Applications. A circuit breaker with a straight voltage rating, such as 240V or 480V, shall be permitted to be applied in a circuit in which the nominal voltage between any two conductors does not exceed the circuit breaker's voltage rating. A two-pole circuit breaker shall not be used for protecting a 3-phase, corner-grounded delta circuit unless the circuit breaker is marked 1ϕ–3ϕ to indicate such suitability.

A circuit breaker with a slash rating, such as 120/240V or 480Y/277V, shall be permitted to be applied in a solidly grounded circuit where the nominal voltage of any conductor to ground does not exceed the lower of the two values of the circuit breaker's voltage rating and the nominal voltage between any two conductors does not exceed the higher value of the circuit breaker's voltage rating.

240.86 Series Ratings. Where a circuit breaker is used on a circuit having an available fault current higher than the marked interrupting rating by being connected on the load side of an acceptable overcurrent protective device having a higher rating, the circuit breaker shall meet the requirements specified in (A) or (B), and (C).

(A) Selected Under Engineering Supervision in Existing Installations. The series rated combination devices shall be selected by a licensed professional engineer engaged primarily in the design or maintenance of electrical installations. The selection shall be documented and stamped by the professional engineer. This documentation shall be available to those authorized to design, install, inspect, maintain, and operate the system. This series combination rating, including identification of the upstream device, shall be field marked on the end use equipment.

For calculated applications, the engineer shall ensure that the downstream circuit breaker(s) that are part of the series combination remain passive during the interruption period of the line side fully rated, current-limiting device.

(B) Tested Combinations. The combination of line-side overcurrent device and load-side circuit breaker(s) is tested and marked on the end use equipment, such as switchboards and panelboards.

(C) Motor Contribution. Series ratings shall not be used where

(1) Motors are connected on the load side of the higher-rated overcurrent device and on the line side of the lower-rated overcurrent device, and
(2) The sum of the motor full-load currents exceeds 1 percent of the interrupting rating of the lower-rated circuit breaker.

240.87 Arc Energy Reduction. Where the highest continuous current trip setting for which the actual overcurrent device installed in a circuit breaker is rated or can be adjusted is 1200 A or higher, 240.87(A) and (B) shall apply.

(A) Documentation. Documentation shall be available to those authorized to design, install, operate, or inspect the installation as to the location of the circuit breaker(s).

(B) Method to Reduce Clearing Time. One of the following means shall be provided:

(1) Zone-selective interlocking
(2) Differential relaying
(3) Energy-reducing maintenance switching with local status indicator
(4) Energy-reducing active arc flash mitigation system
(5) An instantaneous trip setting that is less than the available arcing current
(6) An instantaneous override that is less than the available arcing current
(7) An approved equivalent means

PART VIII. SUPERVISED INDUSTRIAL INSTALLATIONS

240.90 General. Overcurrent protection in areas of supervised industrial installations shall comply with all of the other applicable provisions of this article, except as provided in Part VIII. The provisions of Part VIII shall be permitted to apply only to those portions of the electrical system in the supervised industrial installation used exclusively for manufacturing or process control activities.

240.91 Protection of Conductors. Conductors shall be protected in accordance with 240.91(A) or 240.91(B).

(A) General. Conductors shall be protected in accordance with 240.4.

(B) Devices Rated Over 800 Amperes. Where the overcurrent device is rated over 800 amperes, the ampacity of the conductors it protects shall be equal to or greater than 95 percent of the rating of the overcurrent device specified in 240.6 in accordance with (B)(1) and (2).

(1) The conductors are protected within recognized time vs. current limits for short-circuit currents
(2) All equipment in which the conductors terminate is listed and marked for the application

240.92 Location in Circuit. An overcurrent device shall be connected in each ungrounded circuit conductor as required in 240.92(A) through (E).

(A) Feeder and Branch-Circuit Conductors. Feeder and branch-circuit conductors shall be protected at the point the conductors receive their supply as permitted in 240.21 or as otherwise permitted in 240.92(B), (C), (D), or (E).

(B) Feeder Taps. For feeder taps specified in 240.21(B)(2), (B)(3), and (B)(4), the tap conductors shall be permitted to be sized in accordance with Table 240.92(B).

(C) Transformer Secondary Conductors of Separately Derived Systems. Conductors shall be permitted to be connected to a transformer secondary of a separately derived system, without overcurrent protection at the connection, where the conditions of 240.92(C)(1), (C)(2), and (C)(3) are met.

(1) Short-Circuit and Ground-Fault Protection. The conductors shall be protected from short-circuit and ground-fault conditions by complying with one of the following conditions:

(1) The length of the secondary conductors does not exceed 30 m (100 ft), and the transformer primary overcurrent device has a rating or setting that does not exceed

Table 240.92(B) Tap Conductor Short-Circuit Current Ratings

Tap conductors are considered to be protected under short-circuit conditions when their short-circuit temperature limit is not exceeded. Conductor heating under short-circuit conditions is determined by (1) or (2):

(1) *Short-Circuit Formula for Copper Conductors*

$$(I^2/A^2)t = 0.0297 \log_{10} [(T_2 + 234)/(T_1 + 234)]$$

(2) *Short-Circuit Formula for Aluminum Conductors*

$$(I^2/A^2)t = 0.0125 \log_{10} [(T_2 + 228)/(T_1 + 228)]$$

where:

I = short-circuit current in amperes

A = conductor area in circular mils

t = time of short circuit in seconds (for times less than or equal to 10 seconds)

T_1 = initial conductor temperature in degrees Celsius

T_2 = final conductor temperature in degrees Celsius

Copper conductor with paper, rubber, varnished cloth insulation, $T_2 = 200$

Copper conductor with thermoplastic insulation, $T_2 = 150$

Copper conductor with cross-linked polyethylene insulation, $T_2 = 250$

Copper conductor with ethylene propylene rubber insulation, $T_2 = 250$

Aluminum conductor with paper, rubber, varnished cloth insulation, $T_2 = 200$

Aluminum conductor with thermoplastic insulation, $T_2 = 150$

Aluminum conductor with cross-linked polyethylene insulation, $T_2 = 250$

Aluminum conductor with ethylene propylene rubber insulation, $T_2 = 250$

150 percent of the value determined by multiplying the secondary conductor ampacity by the secondary-to-primary transformer voltage ratio.

(2) The conductors are protected by a differential relay with a trip setting equal to or less than the conductor ampacity.

(3) The conductors shall be considered to be protected if calculations, made under engineering supervision, determine that the system overcurrent devices will protect the conductors within recognized time vs. current limits for all short-circuit and ground-fault conditions.

(2) Overload Protection. The conductors shall be protected against overload conditions by complying with one of the following:

(1) The conductors terminate in a single overcurrent device that will limit the load to the conductor ampacity.
(2) The sum of the overcurrent devices at the conductor termination limits the load to the conductor ampacity. The overcurrent devices shall consist of not more than six circuit breakers or sets of fuses mounted in a single enclosure, in a group of separate enclosures, or in or on a switchboard or switchgear. There shall be no more than six overcurrent devices grouped in any one location.
(3) Overcurrent relaying is connected [with a current transformer(s), if needed] to sense all of the secondary conductor current and limit the load to the conductor ampacity by opening upstream or downstream devices.
(4) Conductors shall be considered to be protected if calculations, made under engineering supervision, determine that the system overcurrent devices will protect the conductors from overload conditions.

(3) Physical Protection. The secondary conductors are protected from physical damage by being enclosed in an approved raceway or by other approved means.

(D) Outside Feeder Taps. Outside conductors shall be permitted to be tapped to a feeder or to be connected at a transformer secondary, without overcurrent protection at the tap or connection, where all the following conditions are met:

(1) The conductors are protected from physical damage in an approved manner.
(2) The sum of the overcurrent devices at the conductor termination limits the load to the conductor ampacity. The overcurrent devices shall consist of not more than six circuit breakers or sets of fuses mounted in a single enclosure, in a group of separate enclosures, or in or on a switchboard or switchgear. There shall be no more than six overcurrent devices grouped in any one location.
(3) The tap conductors are installed outdoors of a building or structure except at the point of load termination.

(4) The overcurrent device for the conductors is an integral part of a disconnecting means or is located immediately adjacent thereto.

(5) The disconnecting means for the conductors are installed at a readily accessible location complying with one of the following:

   a. Outside of a building or structure
   b. Inside, nearest the point of entrance of the conductors
   c. Where installed in accordance with 230.6, nearest the point of entrance of the conductors

(E) Protection by Primary Overcurrent Device. Conductors supplied by the secondary side of a transformer shall be permitted to be protected by overcurrent protection provided on the primary (supply) side of the transformer, provided the primary device time–current protection characteristic, multiplied by the maximum effective primary-to-secondary transformer voltage ratio, effectively protects the secondary conductors.

PART IX. OVERCURRENT PROTECTION OVER 1000 VOLTS, NOMINAL

240.100 Feeders and Branch Circuits.

(A) Location and Type of Protection. Feeder and branch-circuit conductors shall have overcurrent protection in each ungrounded conductor located at the point where the conductor receives its supply or at an alternative location in the circuit when designed under engineering supervision that includes but is not limited to considering the appropriate fault studies and time–current coordination analysis of the protective devices and the conductor damage curves. The overcurrent protection shall be permitted to be provided by either 240.100(A)(1) or (A)(2).

(1) Overcurrent Relays and Current Transformers. Circuit breakers used for overcurrent protection of 3-phase circuits shall have a minimum of three overcurrent relay elements operated from three current transformers. The separate overcurrent relay elements (or protective functions) shall be permitted to be part of a single electronic protective relay unit.

On 3-phase, 3-wire circuits, an overcurrent relay element in the residual circuit of the current transformers shall be permitted to replace one of the phase relay elements.

An overcurrent relay element, operated from a current transformer that links all phases of a 3-phase, 3-wire circuit, shall be permitted to replace the residual relay element and one of the phase-conductor current transformers. Where the neutral conductor is not regrounded on the load side of the circuit as permitted in 250.184(B), the current transformer shall be permitted to link all 3-phase conductors and the grounded circuit conductor (neutral).

(2) Fuses. A fuse shall be connected in series with each ungrounded conductor.

(B) Protective Devices. The protective device(s) shall be capable of detecting and interrupting all values of current that can occur at their location in excess of their trip-setting or melting point.

(C) Conductor Protection. The operating time of the protective device, the available short-circuit current, and the conductor used shall be coordinated to prevent damaging or dangerous temperatures in conductors or conductor insulation under short-circuit conditions.

CHAPTER 9

GROUNDED (NEUTRAL) CONDUCTORS

INTRODUCTION

This chapter contains requirements from Article 200—Use and Identification of Grounded Conductors. Article 200 provides the general requirements on the installation of grounded conductors in premises wiring systems and specifies identification methods for these conductors and their terminals. A grounded conductor is a system or circuit conductor that is intentionally grounded (connected to earth). The grounded conductor in many electrical systems meets the Article 100 definition of *neutral conductor*, but not all grounded conductors are neutral conductors. For instance, a 120 volt, two-wire system has a conductor that is intentionally grounded; it is indeed a grounded conductor, but it is not a neutral conductor. The use and identification requirements of Article 200 apply to all electrical systems conductors that are intentionally grounded, regardless of what the conductor is called on the jobsite. Grounded conductors are generally identified by a continuous white or gray outer finish or by three continuous white or gray stripes on other than green insulation along the entire length. Other methods of identification are permitted for certain wiring methods and for flexible cords and cables. Although Article 200 is a short article, the proper use and identification of grounded conductors is extremely important for ensuring the safety of those who work on electrical systems. Other requirements covering the use and installation of grounded circuit conductors are located in Article 250 of the *Code*.

ARTICLE 200

Use and Identification of Grounded Conductors

200.2 General. Grounded conductors shall comply with 200.2(A) and (B).

(A) Insulation. The grounded conductor, if insulated, shall have insulation that is (1) suitable, other than color, for any ungrounded conductor of the same circuit for systems of 1000 volts or less, or impedance grounded neutral systems of over 1000 volts, or (2) rated not less than 600 volts for solidly grounded neutral systems of over 1000 volts as described in 250.184(A).

(B) Continuity. The continuity of a grounded conductor shall not depend on a connection to a metallic enclosure, raceway, or cable armor.

200.3 Connection to Grounded System. Premises wiring shall not be electrically connected to a supply system unless the latter contains, for any grounded conductor of the interior system, a corresponding conductor that is grounded. For the purpose of this section, *electrically connected* shall mean connected so as to be capable of carrying current, as distinguished from connection through electromagnetic induction.

Exception: Listed utility-interactive inverters identified for use in distributed resource generation systems such as photovoltaic and fuel cell power systems shall be permitted to be connected to premises wiring without a grounded conductor where the connected premises wiring or utility system includes a grounded conductor.

200.4 Neutral Conductors. Neutral conductors shall be installed in accordance with 200.4(A) and (B).

(A) Installation. Neutral conductors shall not be used for more than one branch circuit, for more than one multiwire branch circuit, or for more than one set of ungrounded feeder conductors unless specifically permitted elsewhere in this *Code*.

(B) Multiple Circuits. Where more than one neutral conductor associated with different circuits is in an enclosure, grounded circuit conductors of each circuit shall be identified or grouped to correspond with the ungrounded circuit conductor(s) by wire markers, cable ties, or similar means in at least one location within the enclosure.

Exception No. 1: The requirement for grouping or identifying shall not apply if the branch-circuit or feeder conductors enter from a cable or a raceway unique to the circuit that makes the grouping obvious.

Exception No. 2: The requirement for grouping or identifying shall not apply where branch-circuit conductors pass through a box or conduit body without a loop as described in 314.16(B)(1) or without a splice or termination.

200.6 Means of Identifying Grounded Conductors.

(A) Sizes 6 AWG or Smaller. An insulated grounded conductor of 6 AWG or smaller shall be identified by one of the following means:

(1) A continuous white outer finish.
(2) A continuous gray outer finish.
(3) Three continuous white or gray stripes along the conductor's entire length on other than green insulation.
(4) Wires that have their outer covering finished to show a white or gray color but have colored tracer threads in the braid identifying the source of manufacture shall be considered as meeting the provisions of this section.
(5) The grounded conductor of a mineral-insulated, metal-sheathed cable (Type MI) shall be identified at the time of installation by distinctive marking at its terminations.
(6) A single-conductor, sunlight-resistant, outdoor-rated cable used as a grounded conductor in photovoltaic power systems, as permitted by 690.31, shall be identified at the time of installation by distinctive white marking at all terminations.
(7) Fixture wire shall comply with the requirements for grounded conductor identification as specified in 402.8.
(8) For aerial cable, the identification shall be as above, or by means of a ridge located on the exterior of the cable so as to identify it.

(B) Sizes 4 AWG or Larger. An insulated grounded conductor 4 AWG or larger shall be identified by one of the following means:

(1) A continuous white outer finish.
(2) A continuous gray outer finish.
(3) Three continuous white or gray stripes along the conductor's entire length on other than green insulation.
(4) At the time of installation, by a distinctive white or gray marking at its terminations. This marking shall encircle the conductor or insulation.

(D) Grounded Conductors of Different Systems. Where grounded conductors of different systems are installed in the same raceway, cable, box, auxiliary gutter, or other type of enclosure, each grounded conductor shall be identified by system. Identification that distinguishes each system grounded conductor shall be permitted by one of the following means:

(1) One system grounded conductor shall have an outer covering conforming to 200.6(A) or (B).
(2) The grounded conductor(s) of other systems shall have a different outer covering conforming to 200.6(A) or 200.6(B) or by an outer covering of white or gray with a readily distinguishable colored stripe other than green running along the insulation.
(3) Other and different means of identification allowed by 200.6(A) or (B) shall distinguish each system grounded conductor.

The means of identification shall be documented in a manner that is readily available or shall be permanently posted where the conductors of different systems originate.

(E) Grounded Conductors of Multiconductor Cables. The insulated grounded conductors in a multiconductor cable shall be identified by a continuous white or gray outer finish or by three continuous white or gray stripes on other than green insulation along its entire length. Multiconductor flat cable 4 AWG or larger shall be permitted to employ an external ridge on the grounded conductor.

Exception No. 1: Where the conditions of maintenance and supervision ensure that only qualified persons service the installation, grounded conductors in multiconductor cables shall be permitted to be permanently identified at their terminations at the time of installation by a distinctive white marking or other equally effective means.

Exception No. 2: The grounded conductor of a multiconductor varnished-cloth-insulated cable shall be permitted to be identified at its terminations at the time of installation by a distinctive white marking or other equally effective means.

Informational Note: The color gray may have been used in the past as an ungrounded conductor. Care should be taken when working on existing systems.

200.7 Use of Insulation of a White or Gray Color or with Three Continuous White or Gray Stripes.

(A) General. The following shall be used only for the grounded circuit conductor, unless otherwise permitted in 200.7(B) and (C):

(1) A conductor with continuous white or gray covering
(2) A conductor with three continuous white or gray stripes on other than green insulation
(3) A marking of white or gray color at the termination

(B) Circuits of Less Than 50 Volts. A conductor with white or gray color insulation or three continuous white stripes or having a marking of white or gray at the termination for circuits of less than 50 volts shall be required to be grounded only as required by 250.20(A).

(C) Circuits of 50 Volts or More. The use of insulation that is white or gray or that has three continuous white or gray stripes for other than a grounded conductor for circuits of 50 volts or more shall be permitted only as in (1) and (2).

(1) If part of a cable assembly that has the insulation permanently reidentified to indicate its use as an ungrounded conductor by marking tape, painting, or other effective means at its termination and at each location where the conductor is visible and accessible. Identification shall

encircle the insulation and shall be a color other than white, gray, or green. If used for single-pole, 3-way or 4-way switch loops, the reidentified conductor with white or gray insulation or three continuous white or gray stripes shall be used only for the supply to the switch, but not as a return conductor from the switch to the outlet.

(2) A flexible cord having one conductor identified by a white or gray outer finish or three continuous white or gray stripes, or by any other means permitted by 400.22, that is used for connecting an appliance or equipment permitted by 400.10. This shall apply to flexible cords connected to outlets whether or not the outlet is supplied by a circuit that has a grounded conductor.

200.10 Identification of Terminals.

(A) Device Terminals. All devices, excluding panelboards, provided with terminals for the attachment of conductors and intended for connection to more than one side of the circuit shall have terminals properly marked for identification, unless the electrical connection of the terminal intended to be connected to the grounded conductor is clearly evident.

Exception: Terminal identification shall not be required for devices that have a normal current rating of over 30 amperes, other than polarized attachment plugs and polarized receptacles for attachment plugs as required in 200.10(B).

(B) Receptacles, Plugs, and Connectors. Receptacles, polarized attachment plugs, and cord connectors for plugs and polarized plugs shall have the terminal intended for connection to the grounded conductor identified as follows:

(1) Identification shall be by a metal or metal coating that is substantially white in color or by the word *white* or the letter *W* located adjacent to the identified terminal.

(2) If the terminal is not visible, the conductor entrance hole for the connection shall be colored white or marked with the word *white* or the letter *W*.

(C) Screw Shells. For devices with screw shells, the terminal for the grounded conductor shall be the one connected to the screw shell.

200.11 Polarity of Connections. No grounded conductor shall be attached to any terminal or lead so as to reverse the designated polarity.

CHAPTER 10

GROUNDING AND BONDING

INTRODUCTION

Chapter 10 contains requirements from Article 250 — Grounding and Bonding. These requirements cover grounding and bonding of systems, circuits, and equipment. *Grounded* (grounding) is defined in Article 100 as connected (connecting) to ground or to a conductive body that extends the ground connection. Connecting to ground simply means connecting to the earth. A system or circuit conductor that is intentionally grounded is a grounded conductor. Bonding is connecting parts of an electrical system together into a conductive path. There are bonding requirements for circuit components that are normally current-carrying (e.g., main and system bonding jumpers) and there are requirements to bond normally noncurrent carrying conductive parts together to form a continuous electrical conductor. Bonded conductors and equipment are not grounded until an intentional connection to earth is completed.

Understanding the terminology in Article 250 enables *Code* users to better understand and apply the requirements in Article 250. Some of these terms are defined in Article 250, but many are used in other articles of the *Code* in addition to Article 250 and are defined in Article 100. While Article 250 contains the majority of requirements on grounding and bonding, there are requirements elsewhere in the *Code* that have a unique application to the equipment or occupancy covered in a particular article. Table 250.3 provides a compilation of those special requirements.

Contained in this *Pocket Guide* are eight of the ten parts that are included in Article 250: Part I General; Part II System Grounding; Part III Grounding Electrode System and Grounding Electrode Conductor; Part IV Enclosure, Raceway, and Service Cable Grounding; Part V Bonding; Part VI Equipment

Grounding and Equipment Grounding Conductors; Part VII Methods of Equipment Grounding; and Part VIII Direct-Current Systems.

See Part X of Article 250 in the *NEC* for grounding of systems and circuits of over 1 kV.

ARTICLE 250

Grounding and Bonding

PART I. GENERAL

250.2 Definition.

Bonding Jumper, Supply-Side. A conductor installed on the supply side of a service or within a service equipment enclosure(s), or for a separately derived system, that ensures the required electrical conductivity between metal parts required to be electrically connected.

250.4 General Requirements for Grounding and Bonding. The following general requirements identify what grounding and bonding of electrical systems are required to accomplish. The prescriptive methods contained in Article 250 shall be followed to comply with the performance requirements of this section.

(A) Grounded Systems.

(1) Electrical System Grounding. Electrical systems that are grounded shall be connected to earth in a manner that will limit the voltage imposed by lightning, line surges, or unintentional contact with higher-voltage lines and that will stabilize the voltage to earth during normal operation.

(2) Grounding of Electrical Equipment. Normally non–current-carrying conductive materials enclosing electrical conductors or equipment, or forming part of such equipment, shall be connected to earth so as to limit the voltage to ground on these materials.

(3) Bonding of Electrical Equipment. Normally non–current-carrying conductive materials enclosing electrical conductors or equipment, or forming part of such equipment, shall be connected together and to the electrical supply source in a manner that establishes an effective ground-fault current path.

Table 250.3 Additional Grounding and Bonding Requirements

Conductor/Equipment	Article	Section
Agricultural buildings		547.9 and 547.10
Audio signal processing, amplification, and reproduction equipment		640.7
Branch circuits		210.5, 210.6, 406.3
Cablebus		370.9
Cable trays	392	392.60
Capacitors		460.10, 460.27
Circuits and equipment operating at less than 50 volts	720	
Communications circuits	800	
Community antenna television and radio distribution systems		820.93, 820.100, 820.103, 820.106
Conductors for general wiring	310	
Cranes and hoists	610	
Electrically driven or controlled irrigation machines		675.11(C), 675.12, 675.13, 675.14, 675.15
Electric signs and outline lighting	600	
Electrolytic cells	668	
Elevators, dumbwaiters, escalators, moving walks, wheelchair lifts, and stairway chairlifts	620	
Fixed electric heating equipment for pipelines and vessels		427.29, 427.48
Fixed outdoor electric deicing and snow-melting equipment		426.27
Flexible cords and cables		400.22, 400.23
Floating buildings		553.8, 553.10, 553.11

Grounding-type receptacles, adapters, cord connectors, and attachment plugs		406.9
Hazardous (classified) locations	500–517	
Health care facilities	517	
Induction and dielectric heating equipment	665	
Industrial machinery	670	
Information technology equipment		645.15
Intrinsically safe systems		504.50
Luminaires and lighting equipment		410.40, 410.42, 410.46, 410.155(B)
Luminaires, lampholders, and lamps	410	
Marinas and boatyards		555.15
Mobile homes and mobile home park	550	
Motion picture and television studios and similar locations		530.20, 530.64(B)
Motors, motor circuits, and controllers	430	
Natural and artificially made bodies of water	682	682.30, 682.31, 682.32, 682.33
Network Powered Broadband Communications Circuits		830.93, 830.100, 830.106
Optical Fiber Cables		770.100
Outlet, device, pull, and junction boxes; conduit bodies; and fittings		314.4, 314.25
Over 600 volts, nominal, underground wiring methods		300.50(C)
Panelboards		408.40
Pipe organs	650	
Radio and television equipment	810	
Receptacles and cord connectors		406.3

(continued)

Table 250.3 Additional Grounding and Bonding Requirements *Continued*

Conductor/Equipment	Article	Section
Recreational vehicles and recreational vehicle parks	551	
Services	230	
Solar photovoltaic systems		690.41, 690.42, 690.43, 690.45, 690.47
Swimming pools, fountains, and similar installations	680	
Switchboards and panelboards		408.3(D)
Switches		404.12
Theaters, audience areas of motion picture and television studios, and similar locations		520.81
Transformers and transformer vaults		450.10
Use and identification of grounded conductors	200	
X-ray equipment	660	517.78

(4) Bonding of Electrically Conductive Materials and Other Equipment. Normally non–current-carrying electrically conductive materials that are likely to become energized shall be connected together and to the electrical supply source in a manner that establishes an effective ground-fault current path.

(5) Effective Ground-Fault Current Path. Electrical equipment and wiring and other electrically conductive material likely to become energized shall be installed in a manner that creates a low-impedance circuit facilitating the operation of the overcurrent device or ground detector for high-impedance grounded systems. It shall be capable of safely carrying the maximum ground-fault current likely to be imposed on it from any point on the wiring system where a ground fault may occur to the electrical supply source. The earth shall not be considered as an effective ground-fault current path.

(B) Ungrounded Systems.

(1) Grounding Electrical Equipment. Non–current-carrying conductive materials enclosing electrical conductors or equipment, or forming part of such equipment, shall be connected to earth in a manner that will limit the voltage imposed by lightning or unintentional contact with higher-voltage lines and limit the voltage to ground on these materials.

(2) Bonding of Electrical Equipment. Non–current-carrying conductive materials enclosing electrical conductors or equipment, or forming part of such equipment, shall be connected together and to the supply system grounded equipment in a manner that creates a low-impedance path for ground-fault current that is capable of carrying the maximum fault current likely to be imposed on it.

(3) Bonding of Electrically Conductive Materials and Other Equipment. Electrically conductive materials that are likely to become energized shall be connected together and to the supply system grounded equipment in a manner that creates a low-impedance path for ground-fault current that is capable of carrying the maximum fault current likely to be imposed on it.

(4) Path for Fault Current. Electrical equipment, wiring, and other electrically conductive material likely to become energized shall be installed in a manner that creates a low-impedance circuit from any point on the wiring system to the electrical supply source to facilitate the operation of overcurrent devices should a second ground fault from a different phase occur on the wiring system. The earth shall not be considered as an effective fault-current path.

250.6 Objectionable Current.

(A) Arrangement to Prevent Objectionable Current. The grounding of electrical systems, circuit conductors, surge arresters, surge-protective devices, and conductive normally non–current-carrying metal parts of equipment shall be installed and arranged in a manner that will prevent objectionable current.

(B) Alterations to Stop Objectionable Current. If the use of multiple grounding connections results in objectionable current and the requirements of 250.4(A)(5) or (B)(4) are met, one or more of the following alterations shall be permitted:

(1) Discontinue one or more but not all of such grounding connections.
(2) Change the locations of the grounding connections.
(3) Interrupt the continuity of the conductor or conductive path causing the objectionable current.
(4) Take other suitable remedial and approved action.

(C) Temporary Currents Not Classified as Objectionable Currents. Temporary currents resulting from abnormal conditions, such as ground faults, shall not be classified as objectionable current for the purposes specified in 250.6(A) and (B).

(D) Limitations to Permissible Alterations. The provisions of this section shall not be considered as permitting electronic equipment from being operated on ac systems or branch circuits that are not connected to an equipment grounding conductor as required by this article. Currents that introduce noise or data errors in electronic equipment shall not be considered the objectionable currents addressed in this section.

(E) Isolation of Objectionable Direct-Current Ground Currents. Where isolation of objectionable dc ground currents from cathodic protection systems is required, a listed ac coupling/dc isolating device shall be permitted in the equipment grounding conductor path to provide an effective return path for ac ground-fault current while blocking dc current.

250.8 Connection of Grounding and Bonding Equipment.

(A) Permitted Methods. Equipment grounding conductors, grounding electrode conductors, and bonding jumpers shall be connected by one or more of the following means:

(1) Listed pressure connectors
(2) Terminal bars
(3) Pressure connectors listed as grounding and bonding equipment
(4) Exothermic welding process
(5) Machine screw-type fasteners that engage not less than two threads or are secured with a nut
(6) Thread-forming machine screws that engage not less than two threads in the enclosure
(7) Connections that are part of a listed assembly
(8) Other listed means

(B) Methods Not Permitted. Connection devices or fittings that depend solely on solder shall not be used.

250.10 Protection of Ground Clamps and Fittings. Ground clamps or other fittings exposed to physical damage shall be enclosed in metal, wood, or equivalent protective covering.

250.12 Clean Surfaces. Nonconductive coatings (such as paint, lacquer, and enamel) on equipment to be grounded shall be removed from threads and other contact surfaces to ensure good electrical continuity or be connected by means of fittings designed so as to make such removal unnecessary.

PART II. SYSTEM GROUNDING

250.20 Alternating-Current Systems to Be Grounded. Alternating-current systems shall be grounded as provided for in 250.20(A), (B), (C), or (D). Other systems shall be permitted to be grounded. If such systems are grounded, they shall comply with the applicable provisions of this article.

(A) Alternating-Current Systems of Less Than 50 Volts. Alternating-current systems of less than 50 volts shall be grounded under any of the following conditions:

(1) Where supplied by transformers, if the transformer supply system exceeds 150 volts to ground
(2) Where supplied by transformers, if the transformer supply system is ungrounded
(3) Where installed outside as overhead conductors

(B) Alternating-Current Systems of 50 Volts to 1000 Volts. Alternating-current systems of 50 volts to 1000 volts that supply premises wiring and premises wiring systems shall be grounded under any of the following conditions:

(1) Where the system can be grounded so that the maximum voltage to ground on the ungrounded conductors does not exceed 150 volts
(2) Where the system is 3-phase, 4-wire, wye connected in which the neutral conductor is used as a circuit conductor
(3) Where the system is 3-phase, 4-wire, delta connected in which the midpoint of one phase winding is used as a circuit conductor

250.21 Alternating-Current Systems of 50 Volts to 1000 Volts Not Required to Be Grounded.

(B) Ground Detectors. Ground detectors shall be installed in accordance with 250.21(B)(1) and (B)(2).

(1) Ungrounded ac systems as permitted in 250.21(A)(1) through (A)(4) operating at not less than 120 volts and at 1000 volts or less shall have ground detectors installed on the system.
(2) The ground detection sensing equipment shall be connected as close as practicable to where the system receives its supply.

250.24 Grounding Service-Supplied Alternating-Current Systems.

(A) System Grounding Connections. A premises wiring system supplied by a grounded ac service shall have a grounding electrode conductor connected to the grounded service conductor, at each service, in accordance with 250.24(A)(1) through (A)(5).

(1) General. The grounding electrode conductor connection shall be made at any accessible point from the load end of the overhead service conductors, service drop, underground service conductors, or service lateral to, including the terminal or bus to which the grounded service conductor is connected at the service disconnecting means.

(2) Outdoor Transformer. Where the transformer supplying the service is located outside the building, at least one additional grounding connection shall be made from the grounded service conductor to a grounding electrode, either at the transformer or elsewhere outside the building.

(3) Dual-Fed Services. For services that are dual fed (double ended) in a common enclosure or grouped together in separate enclosures and employing a secondary tie, a single grounding electrode conductor connection to the tie point of the grounded conductor(s) from each power source shall be permitted.

(4) Main Bonding Jumper as Wire or Busbar. Where the main bonding jumper specified in 250.28 is a wire or busbar and is installed from the grounded conductor terminal bar or bus to the equipment grounding terminal bar or bus in the service equipment, the grounding electrode conductor shall be permitted to be connected to the equipment grounding terminal, bar, or bus to which the main bonding jumper is connected.

(5) Load-Side Grounding Connections. A grounded conductor shall not be connected to normally non–current-carrying metal parts of equipment, to equipment grounding conductor(s), or be reconnected to ground on the load side of the service disconnecting means except as otherwise permitted in this article.

(B) Main Bonding Jumper. For a grounded system, an unspliced main bonding jumper shall be used to connect the equipment grounding conductor(s) and the service-disconnect enclosure to the grounded conductor within the enclosure for each service disconnect in accordance with 250.28.

Exception No. 1: Where more than one service disconnecting means is located in an assembly listed for use as service equipment, an unspliced main bonding jumper shall bond the grounded conductor(s) to the assembly enclosure.

(C) Grounded Conductor Brought to Service Equipment. Where an ac system operating at 1000 volts or less is grounded at any point, the grounded conductor(s) shall be routed with the ungrounded conductors to each service disconnecting means and shall be connected to each disconnecting means grounded conductor(s) terminal or bus. A main bonding jumper shall connect the grounded conductor(s) to each service disconnecting means enclosure. The grounded conductor(s) shall be installed in accordance with 250.24(C)(1) through 250.24(C)(4).

(1) Sizing for a Single Raceway or Cable. The grounded conductor shall not be smaller than specified in Table 250.102(C)(1).

(2) Parallel Conductors in Two or More Raceways or Cables. If the ungrounded service-entrance conductors are installed in parallel in two or more raceways or cables, the grounded conductor shall also be installed in parallel. The size of the grounded conductor in each raceway or cable shall be based on the total circular mil area of the parallel ungrounded conductors in the raceway or cable, as indicated in 250.24(C)(1), but not smaller than 1/0 AWG.

(D) Grounding Electrode Conductor. A grounding electrode conductor shall be used to connect the equipment grounding conductors, the service-equipment enclosures, and, where the system is grounded, the grounded service conductor to the grounding electrode(s) required by Part III of this article. This conductor shall be sized in accordance with 250.66.

(E) Ungrounded System Grounding Connections. A premises wiring system that is supplied by an ac service that is ungrounded shall have, at each service, a grounding electrode conductor connected to the grounding electrode(s) required by Part III of this article. The grounding electrode conductor shall be connected to a metal enclosure of the service conductors at any accessible point from the load end of the overhead service conductors, service drop, underground service conductors, or service lateral to the service disconnecting means.

250.26 Conductor to Be Grounded — Alternating-Current Systems. For ac premises wiring systems, the conductor to be grounded shall be as specified in the following:

(1) Single-phase, 2-wire — one conductor
(2) Single-phase, 3-wire — the neutral conductor
(3) Multiphase systems having one wire common to all phases — the neutral conductor
(4) Multiphase systems where one phase is grounded — one phase conductor
(5) Multiphase systems in which one phase is used as in (2) — the neutral conductor

250.28 Main Bonding Jumper and System Bonding Jumper. For a grounded system, main bonding jumpers and system bonding jumpers shall be installed as follows:

(A) Material. Main bonding jumpers and system bonding jumpers shall be of copper or other corrosion-resistant material. A main bonding jumper and a system bonding jumper shall be a wire, bus, screw, or similar suitable conductor.

(B) Construction. Where a main bonding jumper or a system bonding jumper is a screw only, the screw shall be identified with a green finish that shall be visible with the screw installed.

(C) Attachment. Main bonding jumpers and system bonding jumpers shall be connected in the manner specified by the applicable provisions of 250.8.

(D) Size. Main bonding jumpers and system bonding jumpers shall be sized in accordance with 250.28(D)(1) through (D)(3).

(1) General. Main bonding jumpers and system bonding jumpers shall not be smaller than specified in Table 250.102(C)(1).

(2) Main Bonding Jumper for Service with More Than One Enclosure. Where a service consists of more than a single enclosure as permitted in 230.71(A), the main bonding jumper for each enclosure shall be sized in accordance with 250.28(D)(1) based on the largest ungrounded service conductor serving that enclosure.

(3) Separately Derived System with More Than One Enclosure. Where a separately derived system supplies more than a single enclosure, the system bonding jumper for each enclosure shall be sized in accordance with 250.28(D)(1) based on the largest ungrounded feeder conductor serving that enclosure, or a single system bonding jumper shall be installed at the source and sized in accordance with 250.28(D)(1) based on the equivalent size of the largest supply conductor determined by the largest sum of the areas of the corresponding conductors of each set.

250.30 Grounding Separately Derived Alternating-Current Systems. In addition to complying with 250.30(A) for grounded systems, or as provided in 250.30(B) for ungrounded systems, separately derived systems shall comply with 250.20, 250.21, 250.22, or 250.26, as applicable. Multiple separately derived systems that are connected in parallel shall be installed in accordance with 250.30.

Informational Note No. 1: An alternate ac power source, such as an on-site generator, is not a separately derived system if the grounded conductor is solidly interconnected to a service-supplied system grounded conductor. An example of such a situation is where alternate source transfer equipment does not include a switching action in the grounded conductor and allows it to remain solidly connected to the service-supplied grounded conductor when the alternate source is operational and supplying the load served.

(A) Grounded Systems. A separately derived ac system that is grounded shall comply with 250.30(A)(1) through (A)(8). Except as otherwise permitted in this article, a grounded conductor shall not be connected to normally non–current-carrying metal parts of equipment, be connected to equipment grounding conductors, or be reconnected to ground on the load side of the system bonding jumper.

(1) System Bonding Jumper. An unspliced system bonding jumper shall comply with 250.28(A) through (D). This connection shall be made at any single point on the separately derived system from the source to the first system disconnecting means or overcurrent device, or it shall be made at the source of a separately derived system that has no disconnecting means or overcurrent devices, in accordance with 250.30(A)(1)(a) or (b). The system bonding jumper shall remain within the enclosure where it originates. If the source is located outside the building or structure supplied, a system bonding jumper shall be installed at the grounding electrode connection in compliance with 250.30(C).

Exception No. 1: For systems installed in accordance with 450.6, a single system bonding jumper connection to the tie point of the grounded circuit conductors from each power source shall be permitted.

Exception No. 2: If a building or structure is supplied by a feeder from an outdoor separately derived system, a system bonding jumper at both the source and the first disconnecting means shall be permitted if doing so does not establish a parallel path for the grounded conductor. If a grounded conductor is used in this manner, it shall not be smaller than the size specified for the system bonding jumper but shall not be required to be larger than the ungrounded conductor(s). For the purposes of this exception, connection through the earth shall not be considered as providing a parallel path.

Exception No. 3: The size of the system bonding jumper for a system that supplies a Class 1, Class 2, or Class 3 circuit, and is derived from a transformer rated not more than 1000 volt-amperes, shall not be smaller than the derived ungrounded conductors and shall not be smaller than 14 AWG copper or 12 AWG aluminum.

(a) *Installed at the Source.* The system bonding jumper shall connect the grounded conductor to the supply-side bonding jumper and the normally non–current-carrying metal enclosure.

(b) *Installed at the First Disconnecting Means.* The system bonding jumper shall connect the grounded conductor to the supply-side bonding jumper, the disconnecting means enclosure, and the equipment grounding conductor(s).

(2) Supply-Side Bonding Jumper. If the source of a separately derived system and the first disconnecting means are located in separate enclosures, a supply-side bonding jumper shall be installed with the circuit conductors from the source enclosure to the first disconnecting means. A supply-side bonding jumper shall not be required to be larger than the derived ungrounded conductors. The supply-side bonding jumper shall be permitted to be of nonflexible metal raceway type or of the wire or bus type as follows:

(a) A supply-side bonding jumper of the wire type shall comply with 250.102(C), based on the size of the derived ungrounded conductors.

(b) A supply-side bonding jumper of the bus type shall have a cross-sectional area not smaller than a supply-side bonding jumper of the wire type as determined in 250.102(C).

(3) Grounded Conductor. If a grounded conductor is installed and the system bonding jumper connection is not located at the source, 250.30(A)(3)(a) through (A)(3)(d) shall apply.

(a) *Sizing for a Single Raceway.* The grounded conductor shall not be smaller than specified in Table 250.102(C)(1).

(b) *Parallel Conductors in Two or More Raceways.* If the ungrounded conductors are installed in parallel in two or more raceways, the grounded conductor shall also be installed in parallel. The size of the grounded conductor in each raceway shall be based on the total circular mil area of the parallel derived ungrounded conductors in the raceway as indicated in 250.30(A)(3)(a), but not smaller than 1/0 AWG.

(4) Grounding Electrode. The building or structure grounding electrode system shall be used as the grounding electrode for the separately derived system. If located outdoors, the grounding electrode shall be in accordance with 250.30(C).

Exception: If a separately derived system originates in equipment that is listed and identified as suitable for use as service equipment, the grounding electrode used for the service or feeder equipment shall be permitted to be used as the grounding electrode for the separately derived system.

(5) Grounding Electrode Conductor, Single Separately Derived System. A grounding electrode conductor for a single separately derived system shall be sized in accordance with 250.66 for the derived ungrounded conductors. It shall be used to connect the grounded conductor of the derived system to the grounding electrode in accordance with 250.30(A)(4), or as permitted in 250.68(C)(1) and (2). This connection shall be made at the same point on the separately derived system where the system bonding jumper is connected.

Exception No. 1: If the system bonding jumper specified in 250.30(A)(1) is a wire or busbar, it shall be permitted to connect the grounding electrode conductor to the equipment grounding terminal, bar, or bus if the equipment grounding terminal, bar, or bus is of sufficient size for the separately derived system.

Exception No. 2: If the source of a separately derived system is located within equipment listed and identified as suitable for use as service equipment, the grounding electrode conductor from the service or feeder equipment to the grounding electrode shall be permitted as the grounding electrode conductor for the separately derived system, if the grounding electrode conductor is of sufficient size for the separately derived system. If the equipment grounding bus internal to the equipment is not smaller than the required grounding electrode conductor for the separately derived system, the grounding electrode connection for the separately derived system shall be permitted to be made to the bus.

Exception No. 3: A grounding electrode conductor shall not be required for a system that supplies a Class 1, Class 2, or

Class 3 circuit and is derived from a transformer rated not more than 1000 volt-amperes, provided the grounded conductor is bonded to the transformer frame or enclosure by a jumper sized in accordance with 250.30(A)(1), Exception No. 3, and the transformer frame or enclosure is grounded by one of the means specified in 250.134.

(6) Grounding Electrode Conductor, Multiple Separately Derived Systems. A common grounding electrode conductor for multiple separately derived systems shall be permitted. If installed, the common grounding electrode conductor shall be used to connect the grounded conductor of the separately derived systems to the grounding electrode as specified in 250.30(A)(4). A grounding electrode conductor tap shall then be installed from each separately derived system to the common grounding electrode conductor. Each tap conductor shall connect the grounded conductor of the separately derived system to the common grounding electrode conductor. This connection shall be made at the same point on the separately derived system where the system bonding jumper is connected.

Exception No. 1: If the system bonding jumper specified in 250.30(A)(1) is a wire or busbar, it shall be permitted to connect the grounding electrode conductor tap to the equipment grounding terminal, bar, or bus, provided the equipment grounding terminal, bar, or bus is of sufficient size for the separately derived system.

Exception No. 2: A grounding electrode conductor shall not be required for a system that supplies a Class 1, Class 2, or Class 3 circuit and is derived from a transformer rated not more than 1000 volt-amperes, provided the system grounded conductor is bonded to the transformer frame or enclosure by a jumper sized in accordance with 250.30(A)(1), Exception No. 3, and the transformer frame or enclosure is grounded by one of the means specified in 250.134.

(a) *Common Grounding Electrode Conductor.* The common grounding electrode conductor shall be permitted to be one of the following:

(1) A conductor of the wire type not smaller than 3/0 AWG copper or 250 kcmil aluminum
(2) A metal water pipe that complies with 250.68(C)(1)
(3) The metal structural frame of the building or structure that complies with 250.68(C)(2) or is connected to the grounding electrode system by a conductor not smaller than 3/0 AWG copper or 250 kcmil aluminum

(b) *Tap Conductor Size.* Each tap conductor shall be sized in accordance with 250.66 based on the derived ungrounded conductors of the separately derived system it serves.

Exception: If the source of a separately derived system is located within equipment listed and identified as suitable for use as service equipment, the grounding electrode conductor from the service or feeder equipment to the grounding electrode shall be permitted as the grounding electrode conductor for the separately derived system, if the grounding electrode conductor is of sufficient size for the separately derived system. If the equipment grounding bus internal to the equipment is not smaller than the required grounding electrode conductor for the separately derived system, the grounding electrode connection for the separately derived system shall be permitted to be made to the bus.

(c) *Connections.* All tap connections to the common grounding electrode conductor shall be made at an accessible location by one of the following methods:

(1) A connector listed as grounding and bonding equipment.
(2) Listed connections to aluminum or copper busbars not smaller than 6 mm thick × 50 mm wide (¼ in. thick × 2 in. wide) and of sufficient length to accommodate the number of terminations necessary for the installation. If aluminum busbars are used, the installation shall also comply with 250.64(A).
(3) The exothermic welding process.

Tap conductors shall be connected to the common grounding electrode conductor in such a manner that the common grounding electrode conductor remains without a splice or joint.

(7) Installation. The installation of all grounding electrode conductors shall comply with 250.64(A), (B), (C), and (E).

(8) Bonding. Structural steel and metal piping shall be connected to the grounded conductor of a separately derived system in accordance with 250.104(D).

(B) Ungrounded Systems. The equipment of an ungrounded separately derived system shall be grounded and bonded as specified in 250.30(B)(1) through (B)(3).

(1) Grounding Electrode Conductor. A grounding electrode conductor, sized in accordance with 250.66 for the largest derived ungrounded conductor (s) or set of derived ungrounded conductors, shall be used to connect the metal enclosures of the derived system to the grounding electrode as specified in 250.30(A)(5) or (6), as applicable. This connection shall be made at any point on the separately derived system from the source to the first system disconnecting means. If the source is located outside the building or structure supplied, a grounding electrode connection shall be made in compliance with 250.30(C).

(2) Grounding Electrode. Except as permitted by 250.34 for portable and vehicle-mounted generators, the grounding electrode shall comply with 250.30(A)(4).

(3) Bonding Path and Conductor. A supply-side bonding jumper shall be installed from the source of a separately derived system to the first disconnecting means in compliance with 250.30(A)(2).

(C) Outdoor Source. If the source of the separately derived system is located outside the building or structure supplied, a grounding electrode connection shall be made at the source location to one or more grounding electrodes in compliance with 250.50. In addition, the installation shall comply with 250.30(A) for grounded systems or with 250.30(B) for ungrounded systems.

250.32 Buildings or Structures Supplied by a Feeder(s) or Branch Circuit(s).

(A) Grounding Electrode. Building(s) or structure(s) supplied by feeder(s) or branch circuit(s) shall have a grounding electrode or grounding electrode system installed in accordance with Part III of Article 250. The grounding electrode conductor(s) shall be connected in accordance with 250.32(B) or (C). Where there is no existing grounding electrode, the grounding electrode(s) required in 250.50 shall be installed.

Exception: A grounding electrode shall not be required where only a single branch circuit, including a multiwire branch circuit, supplies the building or structure and the branch circuit includes an equipment grounding conductor for grounding the normally non–current-carrying metal parts of equipment.

(B) Grounded Systems.

(1) Supplied by a Feeder or Branch Circuit. An equipment grounding conductor, as described in 250.118, shall be run with the supply conductors and be connected to the building or structure disconnecting means and to the grounding electrode(s). The equipment grounding conductor shall be used for grounding or bonding of equipment, structures, or frames required to be grounded or bonded. The equipment grounding conductor shall be sized in accordance with 250.122. Any installed grounded conductor shall not be connected to the equipment grounding conductor or to the grounding electrode(s).

Exception No. 1: For installations made in compliance with previous editions of this Code that permitted such connection, the grounded conductor run with the supply to the building or structure shall be permitted to serve as the ground-fault return path if all of the following requirements continue to be met:

(1) An equipment grounding conductor is not run with the supply to the building or structure.
(2) There are no continuous metallic paths bonded to the grounding system in each building or structure involved.
(3) Ground-fault protection of equipment has not been installed on the supply side of the feeder(s).

If the grounded conductor is used for grounding in accordance with the provision of this exception, the size of the grounded conductor shall not be smaller than the larger of either of the following:

(1) That required by 220.61
(2) That required by 250.122

Exception No. 2: If system bonding jumpers are installed in accordance with 250.30(A)(1), Exception No. 2, the feeder grounded circuit conductor at the building or structure served shall be connected to the equipment grounding conductors, grounding electrode conductor, and the enclosure for the first disconnecting means.

(2) Supplied by Separately Derived System.

(a) *With Overcurrent Protection.* If overcurrent protection is provided where the conductors originate, the installation shall comply with 250.32(B)(1).

(b) *Without Overcurrent Protection.* If overcurrent protection is not provided where the conductors originate, the installation shall comply with 250.30(A). If installed, the supply-side bonding jumper shall be connected to the building or structure disconnecting means and to the grounding electrode(s).

(D) Disconnecting Means Located in Separate Building or Structure on the Same Premises. Where one or more disconnecting means supply one or more additional buildings or structures under single management, and where these disconnecting means are located remote from those buildings or structures in accordance with the provisions of 225.32, Exception No. 1 and No. 2, 700.12(B)(6), 701.12(B)(5), or 702.12, all of the following conditions shall be met:

(1) The connection of the grounded conductor to the grounding electrode, to normally non–current-carrying metal parts of equipment, or to the equipment grounding conductor at a separate building or structure shall not be made.
(2) An equipment grounding conductor for grounding and bonding any normally non–current-carrying metal parts of equipment, interior metal piping systems, and building or structural metal frames is run with the circuit conductors to a separate building or structure and connected to

existing grounding electrode(s) required in Part III of this article, or, where there are no existing electrodes, the grounding electrode(s) required in Part III of this article shall be installed where a separate building or structure is supplied by more than one branch circuit.

(3) The connection between the equipment grounding conductor and the grounding electrode at a separate building or structure shall be made in a junction box, panelboard, or similar enclosure located immediately inside or outside the separate building or structure.

(E) Grounding Electrode Conductor. The size of the grounding electrode conductor to the grounding electrode(s) shall not be smaller than given in 250.66, based on the largest ungrounded supply conductor. The installation shall comply with Part III of this article.

250.35 Permanently Installed Generators. A conductor that provides an effective ground-fault current path shall be installed with the supply conductors from a permanently installed generator(s) to the first disconnecting mean(s) in accordance with (A) or (B).

(A) Separately Derived System. If the generator is installed as a separately derived system, the requirements in 250.30 shall apply.

(B) Nonseparately Derived System. If the generator is installed as a nonseparately derived system, and overcurrent protection is not integral with the generator assembly, a supply-side bonding jumper shall be installed between the generator equipment grounding terminal and the equipment grounding terminal, bar, or bus of the disconnecting mean(s). It shall be sized in accordance with 250.102(C) based on the size of the conductors supplied by the generator.

PART III. GROUNDING ELECTRODE SYSTEM AND GROUNDING ELECTRODE CONDUCTOR

250.50 Grounding Electrode System. All grounding electrodes as described in 250.52(A)(1) through (A)(7) that are present at each building or structure served shall be bonded together to form the grounding electrode system. Where none of these grounding electrodes exist, one or more of the grounding electrodes specified in 250.52(A)(4) through (A)(8) shall be installed and used.

Exception: Concrete-encased electrodes of existing buildings or structures shall not be required to be part of the grounding electrode system where the steel reinforcing bars or rods are not accessible for use without disturbing the concrete.

250.52 Grounding Electrodes.

(A) Electrodes Permitted for Grounding.

(1) Metal Underground Water Pipe. A metal underground water pipe in direct contact with the earth for 3.0 m (10 ft) or more (including any metal well casing bonded to the pipe) and electrically continuous (or made electrically continuous by bonding around insulating joints or insulating pipe) to the points of connection of the grounding electrode conductor and the bonding conductor(s) or jumper(s), if installed.

N **(2) Metal In-ground Support Structure(s).** One or more metal in-ground support structure(s) in direct contact with the earth vertically for 3.0 m (10 ft) or more, with or without concrete encasement. If multiple metal in-ground support structures are present at a building or a structure, it shall be permissible to bond only one into the grounding electrode system.

(3) Concrete-Encased Electrode. A concrete-encased electrode shall consist of at least 6.0 m (20 ft) of either (1) or (2):

(1) One or more bare or zinc galvanized or other electrically conductive coated steel reinforcing bars or rods of not less than 13 mm (½ in.) in diameter, installed in one continuous 6.0 m (20 ft) length, or if in multiple pieces connected together by the usual steel tie wires, exothermic welding, welding, or other effective means to create a 6.0 m (20 ft) or greater length; or
(2) Bare copper conductor not smaller than 4 AWG

Metallic components shall be encased by at least 50 mm (2 in.) of concrete and shall be located horizontally within that portion of a concrete foundation or footing that is in direct contact with the earth or within vertical foundations or structural components or members that are in direct contact with the earth. If multiple concrete-encased electrodes are present at a building or structure, it shall be permissible to bond only one into the grounding electrode system.

Informational Note: Concrete installed with insulation, vapor barriers, films or similar items separating the concrete from the earth is not considered to be in "direct contact" with the earth.

(4) Ground Ring. A ground ring encircling the building or structure, in direct contact with the earth, consisting of at least 6.0 m (20 ft) of bare copper conductor not smaller than 2 AWG.

(5) Rod and Pipe Electrodes. Rod and pipe electrodes shall not be less than 2.44 m (8 ft) in length and shall consist of the following materials.

(a) Grounding electrodes of pipe or conduit shall not be smaller than metric designator 21 (trade size ¾) and, where of steel, shall have the outer surface galvanized or otherwise metal-coated for corrosion protection.

(b) Rod-type grounding electrodes of stainless steel and copper or zinc coated steel shall be at least 15.87 mm (⅝ in.) in diameter, unless listed.

(6) Other Listed Electrodes. Other listed grounding electrodes shall be permitted.

(7) Plate Electrodes. Each plate electrode shall expose not less than 0.186 m^2 (2 ft^2) of surface to exterior soil. Electrodes of bare or electrically conductive coated iron or steel plates shall be at least 6.4 mm (¼ in.) in thickness. Solid, uncoated electrodes of nonferrous metal shall be at least 1.5 mm (0.06 in.) in thickness.

(8) Other Local Metal Underground Systems or Structures. Other local metal underground systems or structures such as piping systems, underground tanks, and underground metal well casings that are not bonded to a metal water pipe.

(B) Not Permitted for Use as Grounding Electrodes. The following systems and materials shall not be used as grounding electrodes:

(1) Metal underground gas piping systems
(2) Aluminum
(3) The structures and structural reinforcing steel described in 680.26(B)(1) and (B)(2)

250.53 Grounding Electrode System Installation.

(A) Rod, Pipe, and Plate Electrodes. Rod, pipe, and plate electrodes shall meet the requirements of 250.53(A)(1) through (A)(3).

(1) Below Permanent Moisture Level. If practicable, rod, pipe, and plate electrodes shall be embedded below permanent moisture level. Rod, pipe, and plate electrodes shall be free from nonconductive coatings such as paint or enamel.

(2) Supplemental Electrode Required. A single rod, pipe, or plate electrode shall be supplemented by an additional electrode of a type specified in 250.52(A)(2) through (A)(8). The supplemental electrode shall be permitted to be bonded to one of the following:

(1) Rod, pipe, or plate electrode
(2) Grounding electrode conductor
(3) Grounded service-entrance conductor
(4) Nonflexible grounded service raceway
(5) Any grounded service enclosure

Exception: If a single rod, pipe, or plate grounding electrode has a resistance to earth of 25 ohms or less, the supplemental electrode shall not be required.

(3) Supplemental Electrode. If multiple rod, pipe, or plate electrodes are installed to meet the requirements of this section, they shall not be less than 1.8 m (6 ft) apart.

Informational Note: The paralleling efficiency of rods is increased by spacing them twice the length of the longest rod.

(B) Electrode Spacing. Where more than one of the electrodes of the type specified in 250.52(A)(5) or (A)(7) are used, each electrode of one grounding system (including that used for strike termination devices) shall not be less than 1.83 m (6 ft) from any other electrode of another grounding system. Two or more grounding electrodes that are bonded together shall be considered a single grounding electrode system.

(C) Bonding Jumper. The bonding jumper(s) used to connect the grounding electrodes together to form the grounding electrode system shall be installed in accordance with 250.64(A), (B), and (E), shall be sized in accordance with 250.66, and shall be connected in the manner specified in 250.70.

(D) Metal Underground Water Pipe. If used as a grounding electrode, metal underground water pipe shall meet the requirements of 250.53(D)(1) and (D)(2).

(1) Continuity. Continuity of the grounding path or the bonding connection to interior piping shall not rely on water meters or filtering devices and similar equipment.

(2) Supplemental Electrode Required. A metal underground water pipe shall be supplemented by an additional electrode of a type specified in 250.52(A)(2) through (A)(8). If the supplemental electrode is of the rod, pipe, or plate type, it shall comply with 250.53(A). The supplemental electrode shall be bonded to one of the following:

(1) Grounding electrode conductor
(2) Grounded service-entrance conductor
(3) Nonflexible grounded service raceway
(4) Any grounded service enclosure
(5) As provided by 250.32(B)

Exception: The supplemental electrode shall be permitted to be bonded to the interior metal water piping at any convenient point as specified in 250.68(C)(1), Exception.

(E) Supplemental Electrode Bonding Connection Size. Where the supplemental electrode is a rod, pipe, or plate electrode, that portion of the bonding jumper that is the sole connection to the supplemental grounding electrode shall not be required to be larger than 6 AWG copper wire or 4 AWG aluminum wire.

(F) Ground Ring. The ground ring shall be installed not less than 750 mm (30 in.) below the surface of the earth.

(G) Rod and Pipe Electrodes. The electrode shall be installed such that at least 2.44 m (8 ft) of length is in contact with the soil. It shall be driven to a depth of not less than 2.44 m (8 ft) except that, where rock bottom is encountered, the electrode shall be driven at an oblique angle not to exceed 45 degrees from the vertical or, where rock bottom is encountered at an angle up to 45 degrees, the electrode shall be permitted to be buried in a trench that is at least 750 mm (30 in.) deep. The upper end of the electrode shall be flush with or below ground level unless the aboveground end and the grounding electrode conductor attachment are protected against physical damage as specified in 250.10.

(H) Plate Electrode. Plate electrodes shall be installed not less than 750 mm (30 in.) below the surface of the earth.

250.54 Auxiliary Grounding Electrodes. One or more grounding electrodes shall be permitted to be connected to the equipment grounding conductors specified in 250.118 and shall not be required to comply with the electrode bonding requirements of 250.50 or 250.53(C) or the resistance requirements of 250.53(A)(2) Exception, but the earth shall not be used as an effective ground-fault current path as specified in 250.4(A)(5) and 250.4(B)(4).

250.58 Common Grounding Electrode. Where an ac system is connected to a grounding electrode in or at a building or structure, the same electrode shall be used to ground conductor enclosures and equipment in or on that building or structure. Where separate services, feeders, or branch circuits supply a building and are required to be connected to a grounding electrode(s), the same grounding electrode(s) shall be used.

Two or more grounding electrodes that are bonded together shall be considered as a single grounding electrode system in this sense.

250.60 Use of Strike Termination Devices. Conductors and driven pipes, rods, or plate electrodes used for grounding strike termination devices shall not be used in lieu of the grounding electrodes required by 250.50 for grounding wiring systems and equipment. This provision shall not prohibit the required bonding together of grounding electrodes of different systems.

> Informational Note No. 1: See 250.106 for the bonding requirement of the lightning protection system components to the building or structure grounding electrode system.

> Informational Note No. 2: Bonding together of all separate grounding electrodes will limit voltage differences between them and between their associated wiring systems.

250.62 Grounding Electrode Conductor Material. The grounding electrode conductor shall be of copper, aluminum, copper-clad aluminum, or the items as permitted in 250.68(C). The material selected shall be resistant to any corrosive condition existing at the installation or shall be protected against corrosion. Conductors of the wire type shall be solid or stranded, insulated, covered, or bare.

250.64 Grounding Electrode Conductor Installation. Grounding electrode conductors at the service, at each building or structure where supplied by a feeder(s) or branch circuit(s), or at a separately derived system shall be installed as specified in 250.64(A) through (F).

(A) Aluminum or Copper-Clad Aluminum Conductors. Bare aluminum or copper-clad aluminum grounding electrode conductors shall not be used where in direct contact with masonry or the earth or where subject to corrosive conditions. Where used outside, aluminum or copper-clad aluminum grounding electrode conductors shall not be terminated within 450 mm (18 in.) of the earth.

(B) Securing and Protection Against Physical Damage. Where exposed, a grounding electrode conductor or its enclosure shall be securely fastened to the surface on which it is carried. Grounding electrode conductors shall be permitted to be installed on or through framing members.

(1) Not Exposed to Physical Damage. A 6 AWG or larger copper or aluminum grounding electrode conductor not exposed to physical damage shall be permitted to be run along the surface of the building construction without metal covering or protection.

(2) Exposed to Physical Damage. A 6 AWG or larger copper or aluminum grounding electrode conductor exposed to physical damage shall be protected in rigid metal conduit (RMC), intermediate metal conduit (IMC), rigid polyvinyl chloride conduit (PVC), reinforced thermosetting resin conduit Type XW (RTRC-XW), electrical metallic tubing (EMT), or cable armor.

(3) Smaller Than 6 AWG. Grounding electrode conductors smaller than 6 AWG shall be protected in RMC, IMC, PVC, RTRC-XW, EMT, or cable armor.

(4) In Contact with the Earth. Grounding electrode conductors and grounding electrode bonding jumpers in contact with the earth shall not be required to comply with 300.5, but shall be buried or otherwise protected if subject to physical damage.

(C) Continuous. Except as provided in 250.30(A)(5) and (A)(6), 250.30(B)(1), and 250.68(C), grounding electrode conductor(s) shall be installed in one continuous length without a splice or joint. If necessary, splices or connections shall be made as permitted in (1) through (4):

(1) Splicing of the wire-type grounding electrode conductor shall be permitted only by irreversible compression-type connectors listed as grounding and bonding equipment or by the exothermic welding process.
(2) Sections of busbars shall be permitted to be connected together to form a grounding electrode conductor.
(3) Bolted, riveted, or welded connections of structural metal frames of buildings or structures.
(4) Threaded, welded, brazed, soldered or bolted-flange connections of metal water piping.

(D) Building or Structure with Multiple Disconnecting Means in Separate Enclosures. If a building or structure is supplied by a service or feeder with two or more disconnecting means in separate enclosures, the grounding electrode connections shall be made in accordance with 250.64(D)(1), 250.64(D)(2), or 250.64(D)(3).

(1) Common Grounding Electrode Conductor and Taps. A common grounding electrode conductor and grounding electrode conductor taps shall be installed. The common grounding electrode conductor shall be sized in accordance with 250.66, based on the sum of the circular mil area of the largest ungrounded conductor(s) of each set of conductors that supplies the disconnecting means. If the service-entrance conductors connect directly to the overhead service conductors, service drop, underground service conductors, or service lateral, the common grounding electrode conductor shall be sized in accordance with Table 250.66, note 1.

A grounding electrode conductor tap shall extend to the inside of each disconnecting means enclosure. The grounding electrode conductor taps shall be sized in accordance with 250.66 for the largest service-entrance or feeder conductor serving the individual enclosure. The tap conductors shall be connected to the common grounding electrode conductor by one of the following methods in such a manner that the common grounding electrode conductor remains without a splice or joint:

(1) Exothermic welding.
(2) Connectors listed as grounding and bonding equipment.
(3) Connections to an aluminum or copper busbar not less than 6 mm thick × 50 mm wide (¼ in. thick × 2 in. wide) and of sufficient length to accommodate the number of terminations necessary for the installation. The busbar shall be securely fastened and shall be installed in an accessible location. Connections shall be made by a listed connector or by the exothermic welding process. If aluminum busbars are used, the installation shall comply with 250.64(A).

(2) Individual Grounding Electrode Conductors. A grounding electrode conductor shall be connected between the grounding electrode system and one or more of the following, as applicable:

(1) Grounded conductor in each service equipment disconnecting means enclosure
(2) Equipment grounding conductor installed with the feeder
(3) Supply-side bonding jumper

Each grounding electrode conductor shall be sized in accordance with 250.66 based on the service-entrance or feeder conductor(s) supplying the individual disconnecting means.

(3) Common Location. A grounding electrode conductor shall be connected in a wireway or other accessible enclosure on the supply side of the disconnecting means to one or more of the following, as applicable:

(1) Grounded service conductor(s)
(2) Equipment grounding conductor installed with the feeder
(3) Supply-side bonding jumper

The connection shall be made with exothermic welding or a connector listed as grounding and bonding equipment. The grounding electrode conductor shall be sized in accordance with 250.66 based on the service-entrance or feeder conductor(s) at the common location where the connection is made.

(E) Raceways and Enclosures for Grounding Electrode Conductors.

(1) General. Ferrous metal raceways and enclosures for grounding electrode conductors shall be electrically continuous from the point of attachment to cabinets or equipment to the grounding electrode and shall be securely fastened to the ground clamp or fitting. Ferrous metal raceways and enclosures shall be bonded at each end of the raceway or enclosure to the grounding electrode or grounding electrode conductor to create an electrically parallel path. Nonferrous metal raceways and enclosures shall not be required to be electrically continuous.

(2) Methods. Bonding shall be in compliance with 250.92(B) and ensured by one of the methods in 250.92(B)(2) through (B)(4).

(3) Size. The bonding jumper for a grounding electrode conductor raceway or cable armor shall be the same size as, or larger than, the enclosed grounding electrode conductor.

(4) Wiring Methods. If a raceway is used as protection for a grounding electrode conductor, the installation shall comply with the requirements of the appropriate raceway article.

(F) Installation to Electrode(s). Grounding electrode conductor(s) and bonding jumpers interconnecting grounding electrodes shall be installed in accordance with (1), (2), or (3). The grounding electrode conductor shall be sized for the largest grounding electrode conductor required among all the electrodes connected to it.

(1) The grounding electrode conductor shall be permitted to be run to any convenient grounding electrode available in the grounding electrode system where the other electrode(s), if any, is connected by bonding jumpers that are installed in accordance with 250.53(C).
(2) Grounding electrode conductor(s) shall be permitted to be run to one or more grounding electrode(s) individually.
(3) Bonding jumper(s) from grounding electrode(s) shall be permitted to be connected to an aluminum or copper busbar not less than 6 mm thick × 50 mm wide (¼ in. thick × 2 in wide.) and of sufficient length to accommodate the number of terminations necessary for the installation. The busbar shall be securely fastened and shall be installed in an accessible location. Connections shall be made by a listed connector or by the exothermic welding process. The grounding electrode conductor shall be permitted to be run to the busbar. Where aluminum busbars are used, the installation shall comply with 250.64(A).

250.66 Size of Alternating-Current Grounding Electrode Conductor. The size of the grounding electrode conductor at the service, at each building or structure where supplied by a feeder(s) or branch circuit(s), or at a separately derived system of a grounded or ungrounded ac system shall not be less than given in Table 250.66, except as permitted in 250.66(A) through (C).

Table 250.66 Grounding Electrode Conductor for Alternating-Current Systems

Size of Largest Ungrounded Service-Entrance Conductor or Equivalent Area for Parallel Conductors[a] (AWG/kcmil)		Size of Grounding Electrode Conductor (AWG/kcmil)	
Copper	Aluminum or Copper-Clad Aluminum	Copper	Aluminum or Copper-Clad Aluminum[b]
2 or smaller	1/0 or smaller	8	6
1 or 1/0	2/0 or 3/0	6	4
2/0 or 3/0	4/0 or 250	4	2
Over 3/0 through 350	Over 250 through 500	2	1/0
Over 350 through 600	Over 500 through 900	1/0	3/0
Over 600 through 1100	Over 900 through 1750	2/0	4/0
Over 1100	Over 1750	3/0	250

Notes:

1. If multiple sets of service-entrance conductors connect directly to a service drop, set of overhead service conductors, set of underground service conductors, or service lateral, the equivalent size of the largest service-entrance conductor shall be determined by the largest sum of the areas of the corresponding conductors of each set.

2. Where there are no service-entrance conductors, the grounding electrode conductor size shall be determined by the equivalent size of the largest service-entrance conductor required for the load to be served.

[a]This table also applies to the derived conductors of separately derived ac systems.

[b]See installation restrictions in 250.64(A).

(A) Connections to a Rod, Pipe, or Plate Electrode(s). If the grounding electrode conductor or bonding jumper connected to a single or multiple rod, pipe, or plate electrode(s), or any combination thereof, as described in 250.52(A)(5) or (A)(7), does not extend on to other types of electrodes that require a larger size conductor, the grounding electrode conductor shall not be required to be larger than 6 AWG copper wire or 4 AWG aluminum wire.

(B) Connections to Concrete-Encased Electrodes. If the grounding electrode conductor or bonding jumper connected to a single or multiple concrete-encased electrode(s), as described in 250.52(A)(3), does not extend on to other types of electrodes that require a larger size of conductor, the grounding electrode conductor shall not be required to be larger than 4 AWG copper wire.

(C) Connections to Ground Rings. If the grounding electrode conductor or bonding jumper connected to a ground ring, as described in 250.52(A)(4), does not extend on to other types of electrodes that require a larger size of conductor, the grounding electrode conductor shall not be required to be larger than the conductor used for the ground ring.

250.68 Grounding Electrode Conductor and Bonding Jumper Connection to Grounding Electrodes. The connection of a grounding electrode conductor at the service, at each building or structure where supplied by a feeder(s) or branch circuit(s), or at a separately derived system and associated bonding jumper(s) shall be made as specified 250.68(A) through (C).

(A) Accessibility. All mechanical elements used to terminate a grounding electrode conductor or bonding jumper to a grounding electrode shall be accessible.

Exception No. 1: An encased or buried connection to a concrete-encased, driven, or buried grounding electrode shall not be required to be accessible.

Exception No. 2: Exothermic or irreversible compression connections used at terminations, together with the mechanical means used to attach such terminations to fireproofed structural metal whether or not the mechanical means is reversible, shall not be required to be accessible.

(B) Effective Grounding Path. The connection of a grounding electrode conductor or bonding jumper to a grounding electrode shall be made in a manner that will ensure an effective grounding path. Where necessary to ensure the grounding path for a metal piping system used as a grounding electrode, bonding shall be provided around insulated joints and around any equipment likely to be disconnected for repairs or replacement. Bonding jumpers shall be of sufficient length to permit removal of such equipment while retaining the integrity of the grounding path.

(C) Grounding Electrode Conductor Connections. Grounding electrode conductors and bonding jumpers shall be permitted to be connected at the following locations and used to extend the connection to an electrode(s):

(1) Interior metal water piping that is electrically continuous with a metal underground water pipe electrode and is located not more than 1.52 m (5 ft) from the point of entrance to the building shall be permitted to extend the connection to an electrode(s). Interior metal water piping located more than 1.52 m (5 ft) from the point of entrance to the building shall not be used as a conductor to interconnect electrodes of the grounding electrode system.

Exception: In industrial, commercial, and institutional buildings or structures, if conditions of maintenance and supervision ensure that only qualified persons service the installation, interior metal water piping located more than 1.52 m (5 ft) from the point of entrance to the building shall be permitted as a bonding conductor to interconnect electrodes that are part of the grounding electrode system, or as a grounding electrode conductor, if the entire length, other than short sections passing perpendicularly through walls, floors, or ceilings, of the interior metal water pipe that is being used for the conductor is exposed.

(2) The metal structural frame of a building shall be permitted to be used as a conductor to interconnect electrodes that are part of the grounding electrode system, or as a grounding electrode conductor. Hold-down bolts securing the structural steel column that are connected to a concrete-encased electrode that complies with 250.52(A)(3) and is located in the support footing or foundation shall be permitted to connect the metal structural frame of a building or structure to the concrete encased grounding electrode. The hold-down bolts shall be connected to the concrete-encased electrode by welding, exothermic welding, the usual steel tie wires, or other approved means.

(3) A rebar-type concrete-encased electrode installed in accordance with 250.52(A)(3) with an additional rebar section extended from its location within the concrete to an accessible location that is not subject to corrosion shall be permitted for connection of grounding electrode conductors and bonding jumpers. The rebar extension shall not be exposed to contact with the earth without corrosion protection.

250.70 Methods of Grounding and Bonding Conductor Connection to Electrodes. The grounding or bonding conductor shall be connected to the grounding electrode by exothermic welding, listed lugs, listed pressure connectors, listed clamps, or other listed means. Connections depending on solder shall not be used. Ground clamps shall be listed for the materials of the grounding electrode and the grounding electrode conductor and, where used on pipe, rod, or other buried electrodes, shall also be listed for direct soil burial or concrete encasement. Not more than one conductor shall be connected to the grounding electrode by a single clamp or fitting unless the clamp or fitting is listed for multiple conductors. One of the following methods shall be used:

(1) A pipe fitting, pipe plug, or other approved device screwed into a pipe or pipe fitting
(2) A listed bolted clamp of cast bronze or brass, or plain or malleable iron
(3) For indoor communications purposes only, a listed sheet metal strap-type ground clamp having a rigid metal base that seats on the electrode and having a strap of such material and dimensions that it is not likely to stretch during or after installation
(4) An equally substantial approved means

PART IV. ENCLOSURE, RACEWAY, AND SERVICE CABLE CONNECTIONS

250.80 Service Raceways and Enclosures. Metal enclosures and raceways for service conductors and equipment shall be connected to the grounded system conductor if the electrical system is grounded or to the grounding electrode conductor for electrical systems that are not grounded.

Exception: Metal components that are installed in a run of underground nonmetallic raceway(s) and are isolated from possible contact by a minimum cover of 450 mm (18 in.) to all parts of the metal components shall not be required to be connected to the grounded system conductor, supply side bonding jumper, or grounding electrode conductor.

250.86 Other Conductor Enclosures and Raceways. Except as permitted by 250.112(I), metal enclosures and raceways for other than service conductors shall be connected to the equipment grounding conductor.

Exception No. 1: Metal enclosures and raceways for conductors added to existing installations of open wire, knob-and-tube wiring, and nonmetallic-sheathed cable shall not be required to be connected to the equipment grounding conductor where these enclosures or wiring methods comply with (1) through (4) as follows:

(1) Do not provide an equipment ground
(2) Are in runs of less than 7.5 m (25 ft)
(3) Are free from probable contact with ground, grounded metal, metal lath, or other conductive material
(4) Are guarded against contact by persons

Exception No. 2: Short sections of metal enclosures or raceways used to provide support or protection of cable assemblies from physical damage shall not be required to be connected to the equipment grounding conductor.

Exception No. 3: Metal components shall not be required to be connected to the equipment grounding conductor or supply-side bonding jumper where either of the following conditions exist:

(1) The metal components are installed in a run of nonmetallic raceway(s) and isolated from possible contact by a minimum cover of 450 mm (18 in.) to any part of the metal components.
(2) The metal components are part of an installation of nonmetallic raceway(s) and are isolated from possible contact to any part of the metal components by being encased in not less than 50 mm (2 in.) of concrete.

PART V. BONDING

250.90 General. Bonding shall be provided where necessary to ensure electrical continuity and the capacity to conduct safely any fault current likely to be imposed.

250.92 Services.

(A) Bonding of Equipment for Services. The normally non–current-carrying metal parts of equipment indicated in 250.92(A)(1) and (A)(2) shall be bonded together.

(1) All raceways, cable trays, cablebus framework, auxiliary gutters, or service cable armor or sheath that enclose, contain, or support service conductors, except as permitted in 250.80
(2) All enclosures containing service conductors, including meter fittings, boxes, or the like, interposed in the service raceway or armor

(B) Method of Bonding at the Service. Bonding jumpers meeting the requirements of this article shall be used around impaired connections, such as reducing washers or oversized, concentric, or eccentric knockouts. Standard locknuts or bushings shall not be the only means for the bonding required by this section but shall be permitted to be installed to make a mechanical connection of the raceway(s).

Electrical continuity at service equipment, service raceways, and service conductor enclosures shall be ensured by one of the following methods:

(1) Bonding equipment to the grounded service conductor in a manner provided in 250.8
(2) Connections utilizing threaded couplings or threaded hubs on enclosures if made up wrenchtight
(3) Threadless couplings and connectors if made up tight for metal raceways and metal-clad cables
(4) Other listed devices, such as bonding-type locknuts, bushings, or bushings with bonding jumpers

250.94 Bonding for Communication Systems. Communications system bonding terminations shall be connected in accordance with (A) or (B).

(A) The Intersystem Bonding Termination Device. An intersystem bonding termination (IBT) for connecting intersystem bonding conductors shall be provided external to enclosures at the service equipment or metering equipment enclosure and at the disconnecting means for any additional buildings or structures. If an IBT is used, it shall comply with the following:

(1) Be accessible for connection and inspection.
(2) Consist of a set of terminals with the capacity for connection of not less than three intersystem bonding conductors.
(3) Not interfere with opening the enclosure for a service, building or structure disconnecting means, or metering equipment.
(4) At the service equipment, be securely mounted and electrically connected to an enclosure for the service equipment, to the meter enclosure, or to an exposed nonflexible metallic service raceway, or be mounted at one of these enclosures and be connected to the enclosure or to the grounding electrode conductor with a minimum 6 AWG copper conductor
(5) At the disconnecting means for a building or structure, be securely mounted and electrically connected to the metallic enclosure for the building or structure disconnecting means, or be mounted at the disconnecting means and be connected to the metallic enclosure or to the grounding electrode conductor with a minimum 6 AWG copper conductor.
(6) The terminals shall be listed as grounding and bonding equipment.

Exception: In existing buildings or structures where any of the intersystem bonding and grounding electrode conductors required by 770.100(B)(2), 800.100(B)(2), 810.21(F)(2), 820.100(B)(2), and 830.100(B)(2) exist, installation of the intersystem bonding termination is not required. An accessible means external to enclosures for connecting intersystem bonding and grounding electrode conductors shall be permitted at the service equipment and at the disconnecting means for any additional buildings or structures by at least one of the following means:

(1) Exposed nonflexible metallic raceways
(2) An exposed grounding electrode conductor
(3) Approved means for the external connection of a copper or other corrosion-resistant bonding or grounding electrode conductor to the grounded raceway or equipment

N **(B) Other Means.** Connections to an aluminum or copper busbar not less than 6 mm thick × 50 mm wide (¼ in. thick × 2 in. wide) and of sufficient length to accommodate at least three terminations for communication systems in addition to other connections. The busbar shall be securely fastened and shall be installed in an accessible location. Connections shall be made by a listed connector. If aluminum busbars are used, the installation shall also comply with 250.64(A).

Exception to (A) and (B): Means for connecting intersystem bonding conductors are not required where communications systems are not likely to be used.

Informational Note: The use of an IBT can reduce electrical noise on communication systems.

250.96 Bonding Other Enclosures.

(A) General. Metal raceways, cable trays, cable armor, cable sheath, enclosures, frames, fittings, and other metal non–current-carrying parts that are to serve as equipment grounding conductors, with or without the use of supplementary equipment grounding conductors, shall be bonded where necessary to ensure electrical continuity and the capacity to conduct safely any fault current likely to be imposed on them. Any nonconductive paint, enamel, or similar coating shall be removed at threads, contact points, and contact surfaces or shall be connected by means of fittings designed so as to make such removal unnecessary.

(B) Isolated Grounding Circuits. Where installed for the reduction of electrical noise (electromagnetic interference) on the grounding circuit, an equipment enclosure supplied by a branch circuit shall be permitted to be isolated from a raceway containing circuits supplying only that equipment by one or more listed nonmetallic raceway fittings located at the point of attachment of the raceway to the equipment enclosure. The metal raceway shall comply with provisions of this article and shall be supplemented by an internal insulated equipment grounding conductor installed in accordance with 250.146(D) to ground the equipment enclosure.

250.97 Bonding for Over 250 Volts. For circuits of over 250 volts to ground, the electrical continuity of metal raceways and cables with metal sheaths that contain any conductor other than service conductors shall be ensured by one or more of the methods specified for services in 250.92(B), except for (B)(1).

Exception: Where oversized, concentric, or eccentric knockouts are not encountered, or where a box or enclosure with concentric or eccentric knockouts is listed to provide a reliable bonding connection, the following methods shall be permitted:

(1) Threadless couplings and connectors for cables with metal sheaths
(2) Two locknuts, on rigid metal conduit or intermediate metal conduit, one inside and one outside of boxes and cabinets
(3) Fittings with shoulders that seat firmly against the box or cabinet, such as electrical metallic tubing connectors, flexible metal conduit connectors, and cable connectors, with one locknut on the inside of boxes and cabinets
(4) Listed fittings

250.98 Bonding Loosely Jointed Metal Raceways. Expansion fittings and telescoping sections of metal raceways shall be made electrically continuous by equipment bonding jumpers or other means.

250.100 Bonding in Hazardous (Classified) Locations. Regardless of the voltage of the electrical system, the electrical continuity of non–current-carrying metal parts of equipment, raceways, and other enclosures in any hazardous (classified) location, as defined in 500.5, 505.5, and 506.5, shall be ensured by any of the bonding methods specified in 250.92(B)(2) through (B)(4). One or more of these bonding methods shall be used whether or not equipment grounding conductors of the wire type are installed.

250.102 Grounded Conductor, Bonding Conductors, and Jumpers.

(A) Material. Bonding jumpers shall be of copper, aluminum, copper-clad aluminum, or other corrosion-resistant material. A bonding jumper shall be a wire, bus, screw, or similar suitable conductor.

(B) Attachment. Bonding jumpers shall be attached in the manner specified by the applicable provisions of 250.8 for circuits and equipment and by 250.70 for grounding electrodes.

(C) Size — Supply-Side Bonding Jumper.

(1) Size for Supply Conductors in a Single Raceway or Cable. The supply-side bonding jumper shall not be smaller than specified in Table 250.102(C)(1).

(2) Size for Parallel Conductor Installations in Two or More Raceways or Cables. Where the ungrounded supply conductors are paralleled in two or more raceways or cables, and an individual supply-side bonding jumper is used for bonding these raceways or cables, the size of the supply-side bonding jumper for each raceway or cable shall be selected from Table 250.102(C)(1) based on the size of the ungrounded supply conductors in each raceway or cable. A single supply-side bonding jumper installed for bonding two or more raceways or cables shall be sized in accordance with 250.102(C)(1).

D) Size — Equipment Bonding Jumper on Load Side of an Overcurrent Device. The equipment bonding jumper on the load side of an overcurrent device(s) shall be sized in accordance with 250.122.

A single common continuous equipment bonding jumper shall be permitted to connect two or more raceways or cables if the bonding jumper is sized in accordance with 250.122 for the largest overcurrent device supplying circuits therein.

Table 250.102(C)(1) Grounded Conductor, Main Bonding Jumper, System Bonding Jumper, and Supply-Side Bonding Jumper for Alternating-Current Systems

Size of Largest Ungrounded Conductor or Equivalent Area for Parallel Conductors (AWG/kcmil)		Size of Grounded Conductor or Bonding Jumper* (AWG/kcmil)	
Copper	Aluminum or Copper-Clad Aluminum	Copper	Aluminum or Copper-Clad Aluminum
2 or smaller	1/0 or smaller	8	6
1 or 1/0	2/0 or 3/0	6	4
2/0 or 3/0	4/0 or 250	4	2
Over 3/0 through 350	Over 250 through 500	2	1/0
Over 350 through 600	Over 500 through 900	1/0	3/0
Over 600 through 1100	Over 900 through 1750	2/0	4/0
Over 1100	Over 1750	See Notes 1 and 2.	

Notes:

1. If the ungrounded supply conductors are larger than 1100 kcmil copper or 1750 kcmil aluminum, the grounded conductor or bonding jumper shall have an area not less than 12½ percent of the area of the largest ungrounded supply conductor or equivalent area for parallel supply conductors. The grounded conductor or bonding jumper shall not be required to be larger than the largest ungrounded conductor or set of ungrounded conductors.

2. If the ungrounded supply conductors are larger than 1100 kcmil copper or 1750 kcmil aluminum and if the ungrounded supply conductors and the bonding jumper are of different materials (copper, aluminum, or copper-clad aluminum), the minimum size of the grounded conductor or bonding jumper shall be based on the assumed use of ungrounded supply conductors of the same material as the grounded conductor or bonding jumper and will have an ampacity equivalent to that of the installed ungrounded supply conductors.

3. If multiple sets of service-entrance conductors are used as permitted in 230.40, Exception No. 2, or if multiple sets of ungrounded supply conductors are installed for a separately derived system, the equivalent size of the largest ungrounded supply conductor(s) shall be determined by the largest sum of the areas of the corresponding conductors of each set.

4. If there are no service-entrance conductors, the supply conductor size shall be determined by the equivalent size of the largest service-entrance conductor required for the load to be served.

*For the purposes of applying this table and its notes, the term *bonding jumper* refers to main bonding jumpers, system bonding jumpers, and supply-side bonding jumpers.

(E) Installation. Bonding jumpers or conductors and equipment bonding jumpers shall be permitted to be installed inside or outside of a raceway or an enclosure.

(1) Inside a Raceway or an Enclosure. If installed inside a raceway, equipment bonding jumpers and bonding jumpers or conductors shall comply with the requirements of 250.119 and 250.148.

(2) Outside a Raceway or an Enclosure. If installed on the outside, the length of the bonding jumper or conductor or equipment bonding jumper shall not exceed 1.8 m (6 ft) and shall be routed with the raceway or enclosure.

Exception: An equipment bonding jumper or supply-side bonding jumper longer than 1.8 m (6 ft) shall be permitted at outside pole locations for the purpose of bonding or grounding isolated sections of metal raceways or elbows installed in exposed risers of metal conduit or other metal raceway, and for bonding grounding electrodes, and shall not be required to be routed with a raceway or enclosure.

(3) Protection. Bonding jumpers or conductors and equipment bonding jumpers shall be installed in accordance with 250.64(A) and (B).

250.104 Bonding of Piping Systems and Exposed Structural Metal.

(A) Metal Water Piping. The metal water piping system shall be bonded as required in (A)(1), (A)(2), or (A)(3) of this section.

(1) General. Metal water piping system(s) installed in or attached to a building or structure shall be bonded to any of the following:

(1) Service equipment enclosure
(2) Grounded conductor at the service
(3) Grounding electrode conductor if of sufficient size
(4) One or more grounding electrodes used, if the grounding electrode conductor or bonding jumper to the grounding electrode is of sufficient size

The bonding jumper(s) shall be installed in accordance with 250.64(A), 250.64(B), and 250.64(E). The points of attachment of the bonding jumper(s) shall be accessible. The bonding jumper(s) shall be sized in accordance with Table 250.102(C)(1) except as permitted in 250.104(A)(2) and 250.104(A)(3).

(2) Buildings of Multiple Occupancy. In buildings of multiple occupancy where the metal water piping system(s) installed in or attached to a building or structure for the individual occupancies is metallically isolated from all other occupancies by use of nonmetallic water piping, the metal water piping system(s) for each occupancy shall be permitted to be bonded to the equipment grounding terminal of the switchgear, switchboard, or panelboard enclosure (other than service equipment) supplying that occupancy. The bonding jumper shall be sized in accordance with 250.102(D).

(3) Multiple Buildings or Structures Supplied by a Feeder(s) or Branch Circuit(s). The metal water piping system(s) installed in or attached to a building or structure shall be bonded to any of the following:

(1) Building or structure disconnecting means enclosure where located at the building or structure
(2) Equipment grounding conductor run with the supply conductors
(3) One or more grounding electrodes used

The bonding jumper(s) shall be sized in accordance with Table 250.102(C)(1), based on the size of the feeder or branch-circuit conductors that supply the building or structure. The bonding jumper shall not be required to be larger than the largest ungrounded feeder or branch-circuit conductor supplying the building or structure.

(B) Other Metal Piping. If installed in or attached to a building or structure, a metal piping system(s), including gas piping, that is likely to become energized shall be bonded to any of the following:

(1) Equipment grounding conductor for the circuit that is likely to energize the piping system
(2) Service equipment enclosure
(3) Grounded conductor at the service
(4) Grounding electrode conductor, if of sufficient size

(5) One or more grounding electrodes used, if the grounding electrode conductor or bonding jumper to the grounding electrode is of sufficient size

The bonding conductor(s) or jumper(s) shall be sized in accordance with Table 250.122, and equipment grounding conductors shall be sized in accordance with Table 250.122 using the rating of the circuit that is likely to energize the piping system(s). The points of attachment of the bonding jumper(s) shall be accessible.

Informational Note No. 1: Bonding all piping and metal air ducts within the premises will provide additional safety.

(C) Structural Metal. Exposed structural metal that is interconnected to form a metal building frame and is not intentionally grounded or bonded and is likely to become energized shall be bonded to any of the following:

(1) Service equipment enclosure
(2) Grounded conductor at the service
(3) Disconnecting means for buildings or structures supplied by a feeder or branch circuit
(4) Grounding electrode conductor, if of sufficient size
(5) One or more grounding electrodes used, if the grounding electrode conductor or bonding jumper to the grounding electrode is of sufficient size

The bonding conductor(s) or jumper(s) shall be sized in accordance with Table 250.102(C)(1) and installed in accordance with 250.64(A), 250.64(B), and 250.64(E). The points of attachment of the bonding jumper(s) shall be accessible unless installed in compliance with 250.68(A) Exception No. 2.

(D) Separately Derived Systems. Metal water piping systems and structural metal that is interconnected to form a building frame shall be bonded to separately derived systems in accordance with 250.104(D)(1) through 250.104(D)(3).

(1) Metal Water Piping System(s). The grounded conductor of each separately derived system shall be bonded to the nearest available point of the metal water piping system(s) in the area served by each separately derived system. This connection shall be made at the same point on the separately derived system where the grounding electrode conductor is connected. Each bonding jumper shall be sized in accordance with Table 250.102(C)(1) based on the largest ungrounded conductor of the separately derived system.

Exception No. 1: A separate bonding jumper to the metal water piping system shall not be required if the metal water piping system is used as the grounding electrode for the separately derived system and the water piping system is in the area served.

Exception No. 2: A separate water piping bonding jumper shall not be required if the metal frame of a building or structure is used as the grounding electrode for a separately derived system and is bonded to the metal water piping in the area served by the separately derived system.

(2) Structural Metal. If exposed structural metal that is interconnected to form the building frame exists in the area served by the separately derived system, it shall be bonded to the grounded conductor of each separately derived system. This connection shall be made at the same point on the separately derived system where the grounding electrode conductor is connected. Each bonding jumper shall be sized in accordance with Table 250.102(C)(1) based on the largest ungrounded conductor of the separately derived system.

Exception No. 1: A separate bonding jumper to the building structural metal shall not be required if the metal frame of a building or structure is used as the grounding electrode for the separately derived system.

Exception No. 2: A separate bonding jumper to the building structural metal shall not be required if the water piping of a building or structure is used as the grounding electrode for a separately derived system and is bonded to the building structural metal in the area served by the separately derived system.

(3) Common Grounding Electrode Conductor. If a common grounding electrode conductor is installed for multiple separately derived systems as permitted by 250.30(A)(6), and exposed structural metal that is interconnected to form the building frame or interior metal piping exists in the area served by the separately derived system, the metal piping and the structural metal member shall be bonded to the common grounding electrode conductor in the area served by the separately derived system.

Exception: A separate bonding jumper from each derived system to metal water piping and to structural metal members shall not be required if the metal water piping and the structural metal members in the area served by the separately derived system are bonded to the common grounding electrode conductor.

250.106 Lightning Protection Systems. The lightning protection system ground terminals shall be bonded to the building or structure grounding electrode system.

Informational Note No. 1: See 250.60 for use of strike termination devices. For further information, see NFPA 780-2014, *Standard for the Installation of Lightning Protection Systems*, which contains detailed information on grounding, bonding, and sideflash distance from lightning protection systems.

Informational Note No. 2: Metal raceways, enclosures, frames, and other non–current-carrying metal parts of electrical equipment installed on a building equipped with a lightning protection system may require bonding or spacing from the lightning protection conductors in accordance with NFPA 780-2014, *Standard for the Installation of Lightning Protection Systems*.

PART VI. EQUIPMENT GROUNDING AND EQUIPMENT GROUNDING CONDUCTORS

250.110 Equipment Fastened in Place (Fixed) or Connected by Permanent Wiring Methods. Exposed, normally non–current-carrying metal parts of fixed equipment supplied by or enclosing conductors or components that are likely to become energized shall be connected to an equipment grounding conductor under any of the following conditions:

(1) Where within 2.5 m (8 ft) vertically or 1.5 m (5 ft) horizontally of ground or grounded metal objects and subject to contact by persons
(2) Where located in a wet or damp location and not isolated
(3) Where in electrical contact with metal
(4) Where in a hazardous (classified) location as covered by Articles 500 through 517
(5) Where supplied by a wiring method that provides an equipment grounding conductor, except as permitted by 250.86, Exception No. 2, for short sections of metal enclosures
(6) Where equipment operates with any terminal at over 150 volts to ground

Exception No. 3: Listed equipment protected by a system of double insulation, or its equivalent, shall not be required to be connected to the equipment grounding conductor. Where such a system is employed, the equipment shall be distinctively marked.

250.112 Specific Equipment Fastened in Place (Fixed) or Connected by Permanent Wiring Methods. Except as permitted in 250.112(F) and (I), exposed, normally non–current-carrying metal parts of equipment described in 250.112(A) through (K), and normally non–current-carrying metal parts of equipment and enclosures described in 250.112(L) and (M), shall be connected to an equipment grounding conductor, regardless of voltage.

(A) Switchgear and Switchboard Frames and Structures. Switchgear or switchboard frames and structures supporting switching equipment, except frames of 2-wire dc switchgear or switchboards where effectively insulated from ground.

(C) Motor Frames. Motor frames, as provided by 430.242.

(D) Enclosures for Motor Controllers. Enclosures for motor controllers unless attached to ungrounded portable equipment.

(E) Elevators and Cranes. Electrical equipment for elevators and cranes.

(F) Garages, Theaters, and Motion Picture Studios. Electrical equipment in commercial garages, theaters, and motion picture studios, except pendant lampholders supplied by circuits not over 150 volts to ground.

(G) Electric Signs. Electric signs, outline lighting, and associated equipment as provided in 600.7.

(H) Motion Picture Projection Equipment. Motion picture projection equipment.

(I) Remote-Control, Signaling, and Fire Alarm Circuits. Equipment supplied by Class 1 circuits shall be grounded unless operating at less than 50 volts. Equipment supplied by Class 1 power-limited circuits, by Class 2 and Class 3 remote-control and signaling circuits, and by fire alarm circuits shall be grounded where system grounding is required by Part II or Part VIII of this article.

(J) Luminaires. Luminaires as provided in Part V of Article 410.

(K) Skid-Mounted Equipment. Permanently mounted electrical equipment and skids shall be connected to the equipment grounding conductor sized as required by 250.122.

(L) Motor-Operated Water Pumps. Motor-operated water pumps, including the submersible type.

(M) Metal Well Casings. Where a submersible pump is used in a metal well casing, the well casing shall be connected to the pump circuit equipment grounding conductor.

250.114 Equipment Connected by Cord and Plug. Under any of the conditions described in 250.114(1) through (4), exposed, normally non–current-carrying metal parts of cord-and-plug-connected equipment shall be connected to the equipment grounding conductor.

(2) Where operated at over 150 volts to ground

Exception No. 1: Motors, where guarded, shall not be required to be connected to an equipment grounding conductor.

Exception No. 2: Metal frames of electrically heated appliances, exempted by special permission, shall not be required to be connected to an equipment grounding conductor, in which case the frames shall be permanently and effectively insulated from ground.

(4) In other than residential occupancies:

  a. Refrigerators, freezers, and air conditioners
  b. Clothes-washing, clothes-drying, dish-washing machines; information technology equipment; sump pumps and electrical aquarium equipment
  c. Hand-held motor-operated tools, stationary and fixed motor-operated tools, and light industrial motor-operated tools
  d. Motor-operated appliances of the following types: hedge clippers, lawn mowers, snow blowers, and wet scrubbers
  e. Portable handlamps
  f. Cord-and-plug-connected appliances used in damp or wet locations or by persons standing on the ground or on metal floors or working inside of metal tanks or boilers
  g. Tools likely to be used in wet or conductive locations

Exception: Tools and portable handlamps likely to be used in wet or conductive locations shall not be required to be connected to an equipment grounding conductor where supplied through an isolating transformer with an ungrounded secondary of not over 50 volts.

250.116 Nonelectrical Equipment. The metal parts of the following nonelectrical equipment described in this section shall be connected to the equipment grounding conductor:

(1) Frames and tracks of electrically operated cranes and hoists
(2) Frames of nonelectrically driven elevator cars to which electrical conductors are attached
(3) Hand-operated metal shifting ropes or cables of electric elevators

250.118 Types of Equipment Grounding Conductors. The equipment grounding conductor run with or enclosing the circuit conductors shall be one or more or a combination of the following:

(1) A copper, aluminum, or copper-clad aluminum conductor. This conductor shall be solid or stranded; insulated,

covered, or bare; and in the form of a wire or a busbar of any shape.

(2) Rigid metal conduit.
(3) Intermediate metal conduit.
(4) Electrical metallic tubing.
(5) Listed flexible metal conduit meeting all the following conditions:

    a. The conduit is terminated in listed fittings.
    b. The circuit conductors contained in the conduit are protected by overcurrent devices rated at 20 amperes or less.
    c. The size of the conduit does not exceed metric designator 35 (trade size 1¼).
    d. The combined length of flexible metal conduit and flexible metallic tubing and liquidtight flexible metal conduit in the same ground-fault current path does not exceed 1.8 m (6 ft).
    e. If used to connect equipment where flexibility is necessary to minimize the transmission of vibration from equipment or to provide flexibility for equipment that requires movement after installation, an equipment grounding conductor shall be installed.

(6) Listed liquidtight flexible metal conduit meeting all the following conditions:

    a. The conduit is terminated in listed fittings.
    b. For metric designators 12 through 16 (trade sizes ⅜ through ½), the circuit conductors contained in the conduit are protected by overcurrent devices rated at 20 amperes or less.
    c. For metric designators 21 through 35 (trade sizes ¾ through 1¼), the circuit conductors contained in the conduit are protected by overcurrent devices rated not more than 60 amperes and there is no flexible metal conduit, flexible metallic tubing, or liquidtight flexible metal conduit in trade sizes metric designators 12 through 16 (trade sizes ⅜ through ½) in the ground-fault current path.
    d. The combined length of flexible metal conduit and flexible metallic tubing and liquidtight flexible metal

conduit in the same ground-fault current path does not exceed 1.8 m (6 ft).

e. If used to connect equipment where flexibility is necessary to minimize the transmission of vibration from equipment or to provide flexibility for equipment that requires movement after installation, an equipment grounding conductor shall be installed.

(7) Flexible metallic tubing where the tubing is terminated in listed fittings and meeting the following conditions:

a. The circuit conductors contained in the tubing are protected by overcurrent devices rated at 20 amperes or less.

b. The combined length of flexible metal conduit and flexible metallic tubing and liquidtight flexible metal conduit in the same ground-fault current path does not exceed 1.8 m (6 ft).

(8) Armor of Type AC cable as provided in 320.108.

(9) The copper sheath of mineral-insulated, metal-sheathed cable Type MI.

(10) Type MC cable that provides an effective ground-fault current path in accordance with one or more of the following:

a. It contains an insulated or uninsulated equipment grounding conductor in compliance with 250.118(1).

b. The combined metallic sheath and uninsulated equipment grounding/bonding conductor of interlocked metal tape–type MC cable that is listed and identified as an equipment grounding conductor

c. The metallic sheath or the combined metallic sheath and equipment grounding conductors of the smooth or corrugated tube-type MC cable that is listed and identified as an equipment grounding conductor

(11) Cable trays as permitted in 392.10 and 392.60.

(12) Cablebus framework as permitted in 370.60(1).

(13) Other listed electrically continuous metal raceways and listed auxiliary gutters.

(14) Surface metal raceways listed for grounding.

250.119 Identification of Equipment Grounding Conductors. Unless required elsewhere in this *Code,* equipment grounding conductors shall be permitted to be bare, covered, or insulated. Individually covered or insulated equipment grounding conductors shall have a continuous outer finish that is either green or green with one or more yellow stripes except as permitted in this section. Conductors with insulation or individual covering that is green, green with one or more yellow stripes, or otherwise identified as permitted by this section shall not be used for ungrounded or grounded circuit conductors.

Exception No. 1: Power-limited Class 2 or Class 3 cables, power-limited fire alarm cables, or communications cables containing only circuits operating at less than 50 volts where connected to equipment not required to be grounded in accordance with 250.112(I) shall be permitted to use a conductor with green insulation or green with one or more yellow stripes for other than equipment grounding purposes.

Exception No. 3: Conductors with green insulation shall be permitted to be used as ungrounded signal conductors where installed between the output terminations of traffic signal control and traffic signal indicating heads. Signaling circuits installed in accordance with this exception shall include an equipment grounding conductor in accordance with 250.118. Wire-type equipment grounding conductors shall be bare or have insulation or covering that is green with one or more yellow stripes.

(A) Conductors 4 AWG and Larger. Equipment grounding conductors 4 AWG and larger shall comply with 250.119(A)(1) and (A)(2).

(1) An insulated or covered conductor 4 AWG and larger shall be permitted, at the time of installation, to be permanently identified as an equipment grounding conductor at each end and at every point where the conductor is accessible.
(2) Identification shall encircle the conductor and shall be accomplished by one of the following:

a. Stripping the insulation or covering from the entire exposed length
b. Coloring the insulation or covering green at the termination
c. Marking the insulation or covering with green tape or green adhesive labels at the termination

(B) Multiconductor Cable. Where the conditions of maintenance and supervision ensure that only qualified persons service the installation, one or more insulated conductors in a multiconductor cable, at the time of installation, shall be permitted to be permanently identified as equipment grounding conductors at each end and at every point where the conductors are accessible by one of the following means:

(1) Stripping the insulation from the entire exposed length.
(2) Coloring the exposed insulation green.
(3) Marking the exposed insulation with green tape or green adhesive labels. Identification shall encircle the conductor.

(C) Flexible Cord. Equipment grounding conductors in flexible cords shall be insulated and shall have a continuous outer finish that is either green or green with one or more yellow stripes.

250.120 Equipment Grounding Conductor Installation. An equipment grounding conductor shall be installed in accordance with 250.120(A), (B), and (C).

(A) Raceway, Cable Trays, Cable Armor, Cablebus, or Cable Sheaths. Where it consists of a raceway, cable tray, cable armor, cablebus framework, or cable sheath or where it is a wire within a raceway or cable, it shall be installed in accordance with the applicable provisions in this *Code* using fittings for joints and terminations approved for use with the type raceway or cable used. All connections, joints, and fittings shall be made tight using suitable tools.

Informational Note: See the UL guide information on FHIT systems for equipment grounding conductors installed in a raceway that are part of an electrical circuit protective system or a fire-rated cable listed to maintain circuit integrity.

(B) Aluminum and Copper-Clad Aluminum Conductors. Equipment grounding conductors of bare or insulated aluminum or copper-clad aluminum shall be permitted. Bare conductors shall not come in direct contact with masonry or the earth or where subject to corrosive conditions. Aluminum or copper-clad aluminum conductors shall not be terminated within 450 mm (18 in.) of the earth.

(C) Equipment Grounding Conductors Smaller Than 6 AWG. Where not routed with circuit conductors as permitted in 250.130(C) and 250.134(B) Exception No. 2, equipment grounding conductors smaller than 6 AWG shall be protected from physical damage by an identified raceway or cable armor unless installed within hollow spaces of the framing members of buildings or structures and where not subject to physical damage.

250.121 Use of Equipment Grounding Conductors. An equipment grounding conductor shall not be used as a grounding electrode conductor.

Exception: A wire-type equipment grounding conductor installed in compliance with 250.6(A) and the applicable requirements for both the equipment grounding conductor and the grounding electrode conductor in Parts II, III, and VI of this article shall be permitted to serve as both an equipment grounding conductor and a grounding electrode conductor.

250.122 Size of Equipment Grounding Conductors.

(A) General. Copper, aluminum, or copper-clad aluminum equipment grounding conductors of the wire type shall not be smaller than shown in Table 250.122, but in no case shall they be required to be larger than the circuit conductors supplying the equipment. Where a cable tray, a raceway, or a cable armor or sheath is used as the equipment grounding conductor, as provided in 250.118 and 250.134(A), it shall comply with 250.4(A)(5) or (B)(4).

Equipment grounding conductors shall be permitted to be sectioned within a multiconductor cable, provided the combined circular mil area complies with Table 250.122.

(B) Increased in Size. Where ungrounded conductors are increased in size from the minimum size that has sufficient ampacity for the intended installation, wire-type equipment grounding conductors, where installed, shall be increased in size proportionately, according to the circular mil area of the ungrounded conductors.

(C) Multiple Circuits. Where a single equipment grounding conductor is run with multiple circuits in the same raceway, cable, or cable tray, it shall be sized for the largest overcurrent device protecting conductors in the raceway, cable, or cable tray. Equipment grounding conductors installed in cable trays shall meet the minimum requirements of 392.10(B)(1)(c).

(D) Motor Circuits. Equipment grounding conductors for motor circuits shall be sized in accordance with (D)(1) or (D)(2).

(1) General. The equipment grounding conductor size shall not be smaller than determined by 250.122(A) based on the rating of the branch-circuit short-circuit and ground-fault protective device.

(2) Instantaneous-Trip Circuit Breaker and Motor Short-Circuit Protector. Where the overcurrent device is an instantaneous-trip circuit breaker or a motor short-circuit protector, the equipment grounding conductor shall be sized not smaller than that given by 250.122(A) using the maximum permitted rating of a dual element time-delay fuse selected for branch-circuit short-circuit and ground-fault protection in accordance with 430.52(C)(1), Exception No. 1.

(E) Flexible Cord and Fixture Wire. The equipment grounding conductor in a flexible cord with the largest circuit conductor 10 AWG or smaller, and the equipment grounding conductor used with fixture wires of any size in accordance with 240.5, shall not be smaller than 18 AWG copper and shall not be smaller than the circuit conductors. The equipment grounding conductor in a flexible cord with a circuit conductor larger than 10 AWG shall be sized in accordance with Table 250.122.

(F) Conductors in Parallel. For circuits of parallel conductors as permitted in 310.10(H), the equipment grounding conductor shall be installed in accordance with (1) or (2).

N (1) Conductor Installations in Raceways, Auxiliary Gutters, or Cable Trays.

(a) *Single Raceway or Cable Tray*. If conductors are installed in parallel in the same raceway or cable tray, a single wire-type conductor shall be permitted as the equipment grounding conductor. The wire-type equipment grounding conductor shall be sized in accordance with 250.122, based on the overcurrent protective device for the feeder or branch circuit. Wire-type equipment grounding conductors installed in cable trays shall meet the minimum requirements of 392.10(B)(1)(c). Metal raceways or auxiliary gutters in accordance with 250.118 or cable trays complying with 392.60(B) shall be permitted as the equipment grounding conductor.

(b) *Multiple Raceways*. If conductors are installed in parallel in multiple raceways, wire-type equipment grounding conductors, where used, shall be installed in parallel in each raceway. The equipment grounding conductor installed in each raceway shall be sized in compliance with 250.122 based on the overcurrent protective device for the feeder or branch circuit. Metal raceways or auxiliary gutters in accordance with 250.118 or cable trays complying with 392.60(B) shall be permitted as the equipment grounding conductor.

N (2) Multiconductor Cables.

(a) If multiconductor cables are installed in parallel, the equipment grounding conductor(s) in each cable shall be connected in parallel.

(b) If multiconductor cables are installed in parallel in the same raceway, auxiliary gutter, or cable tray, a single equipment grounding conductor that is sized in accordance with 250.122 shall be permitted in combination with the equipment grounding conductors provided within the multiconductor cables and shall all be connected together.

(c) Equipment grounding conductors installed in cable trays shall meet the minimum requirements of 392.10(B)(1)(c). Cable trays complying with 392.60(B), metal raceways in accordance with 250.118, or auxiliary gutters shall be permitted as the equipment grounding conductor.

(d) Except as provided in 250.122(F)(2)(b) for raceway or cable tray installations, the equipment grounding conductor in each multiconductor cable shall be sized in accordance with 250.122 based on the overcurrent protective device for the feeder or branch circuit.

(G) Feeder Taps. Equipment grounding conductors run with feeder taps shall not be smaller than shown in Table 250.122 based on the rating of the overcurrent device ahead of the feeder but shall not be required to be larger than the tap conductors.

250.124 Equipment Grounding Conductor Continuity.

(B) Switches. No automatic cutout or switch shall be placed in the equipment grounding conductor of a premises wiring system unless the opening of the cutout or switch disconnects all sources of energy.

250.126 Identification of Wiring Device Terminals. The terminal for the connection of the equipment grounding conductor shall be identified by one of the following:

(1) A green, not readily removable terminal screw with a hexagonal head.
(2) A green, hexagonal, not readily removable terminal nut.
(3) A green pressure wire connector. If the terminal for the equipment grounding conductor is not visible, the conductor entrance hole shall be marked with the word *green* or *ground*, the letters *G* or *GR*, a grounding symbol, or otherwise identified by a distinctive green color. If the terminal for the equipment grounding conductor is readily removable, the area adjacent to the terminal shall be similarly marked.

    Informational Note: See Informational Note Figure 250.126.

Table 250.122 Minimum Size Equipment Grounding Conductors for Grounding Raceway and Equipment

Rating or Setting of Automatic Overcurrent Device in Circuit Ahead of Equipment, Conduit, etc., Not Exceeding (Amperes)	Size (AWG or kcmil)	
	Copper	Aluminum or Copper-Clad Aluminum*
15	14	12
20	12	10
60	10	8
100	8	6
200	6	4
300	4	2
400	3	1
500	2	1/0
600	1	2/0
800	1/0	3/0
1000	2/0	4/0
1200	3/0	250
1600	4/0	350
2000	250	400
2500	350	600
3000	400	600
4000	500	750
5000	700	1200
6000	800	1200

Note: Where necessary to comply with 250.4(A)(5) or (B)(4), the equipment grounding conductor shall be sized larger than given in this table.

*See installation restrictions in 250.120.

Informational Note Figure 250.126
One Example of a Symbol Used to Identify the Grounding Termination Point for an Equipment Grounding Conductor.

PART VII. METHODS OF EQUIPMENT GROUNDING

250.130 Equipment Grounding Conductor Connections. Equipment grounding conductor connections at the source of separately derived systems shall be made in accordance with 250.30(A)(1). Equipment grounding conductor connections at service equipment shall be made as indicated in 250.130(A) or (B). For replacement of non–grounding-type receptacles with grounding-type receptacles and for branch-circuit extensions only in existing installations that do not have an equipment grounding conductor in the branch circuit, connections shall be permitted as indicated in 250.130(C).

(A) For Grounded Systems. The connection shall be made by bonding the equipment grounding conductor to the grounded service conductor and the grounding electrode conductor.

(B) For Ungrounded Systems. The connection shall be made by bonding the equipment grounding conductor to the grounding electrode conductor.

(C) Nongrounding Receptacle Replacement or Branch Circuit Extensions. The equipment grounding conductor of a grounding-type receptacle or a branch-circuit extension shall be permitted to be connected to any of the following:

(1) Any accessible point on the grounding electrode system as described in 250.50
(2) Any accessible point on the grounding electrode conductor
(3) The equipment grounding terminal bar within the enclosure where the branch circuit for the receptacle or branch circuit originates
(4) An equipment grounding conductor that is part of another branch circuit that originates from the enclosure where the branch circuit for the receptacle or branch circuit originates
(5) For grounded systems, the grounded service conductor within the service equipment enclosure
(6) For ungrounded systems, the grounding terminal bar within the service equipment enclosure

250.132 Short Sections of Raceway. Isolated sections of metal raceway or cable armor, where required to be grounded, shall be connected to an equipment grounding conductor in accordance with 250.134.

250.134 Equipment Fastened in Place or Connected by Permanent Wiring Methods (Fixed) — Grounding. Unless grounded by connection to the grounded circuit conductor as permitted by 250.32, 250.140, and 250.142, non–current-carrying metal parts of equipment, raceways, and other enclosures, if grounded, shall be connected to an equipment grounding conductor by one of the methods specified in 250.134(A) or (B).

(A) Equipment Grounding Conductor Types. By connecting to any of the equipment grounding conductors permitted by 250.118.

(B) With Circuit Conductors. By connecting to an equipment grounding conductor contained within the same raceway, cable, or otherwise run with the circuit conductors.

Exception No. 1: As provided in 250.130(C), the equipment grounding conductor shall be permitted to be run separately from the circuit conductors.

Exception No. 2: For dc circuits, the equipment grounding conductor shall be permitted to be run separately from the circuit conductors.

250.136 Equipment Considered Grounded. Under the conditions specified in 250.136(A) and (B), the normally non–current-carrying metal parts of the equipment shall be considered grounded.

(A) Equipment Secured to Grounded Metal Supports. Electrical equipment secured to and in electrical contact with a metal rack or structure provided for its support and connected to an equipment grounding conductor by one of the means indicated in 250.134. The structural metal frame of a building shall not be used as the required equipment grounding conductor for ac equipment.

250.142 Use of Grounded Circuit Conductor for Grounding Equipment.

(A) Supply-Side Equipment. A grounded circuit conductor shall be permitted to ground non–current-carrying metal parts of equipment, raceways, and other enclosures at any of the following locations:

(1) On the supply side or within the enclosure of the ac service-disconnecting means
(2) On the supply side or within the enclosure of the main disconnecting means for separate buildings as provided in 250.32(B)
(3) On the supply side or within the enclosure of the main disconnecting means or overcurrent devices of a separately derived system where permitted by 250.30(A)(1)

(B) Load-Side Equipment. Except as permitted in 250.30(A)(1) and 250.32(B) Exception, a grounded circuit conductor shall not be used for grounding non–current-carrying metal parts of equipment on the load side of the service disconnecting means or on the load side of a separately derived system disconnecting means or the overcurrent devices for a separately derived system not having a main disconnecting means.

Exception No. 2: It shall be permissible to ground meter enclosures by connection to the grounded circuit conductor on the load side of the service disconnect where all of the following conditions apply:

(1) No service ground-fault protection is installed.
(2) All meter enclosures are located immediately adjacent to the service disconnecting means.
(3) The size of the grounded circuit conductor is not smaller than the size specified in Table 250.122 for equipment grounding conductors.

250.146 Connecting Receptacle Grounding Terminal to Box. An equipment bonding jumper shall be used to connect the grounding terminal of a grounding-type receptacle to a grounded box unless grounded as in 250.146(A) through (D). The equipment bonding jumper shall be sized in accordance with Table 250.122 based on the rating of the overcurrent device protecting the circuit conductors.

(A) Surface-Mounted Box. Where the box is mounted on the surface, direct metal-to-metal contact between the device yoke and the box or a contact yoke or device that complies with 250.146(B) shall be permitted to ground the receptacle to the box. At least one of the insulating washers shall be removed from receptacles that do not have a contact yoke or device that complies with 250.146(B) to ensure direct metal-to-metal contact. This provision shall not apply to cover-mounted receptacles unless the box and cover combination are listed as providing satisfactory ground continuity between the box and the receptacle. A listed exposed work cover shall be permitted to be the grounding and bonding means when (1) the device is attached to the cover with at least two fasteners that are permanent (such as a rivet) or have a thread locking or screw or nut locking means and (2) when the cover mounting holes are located on a flat non-raised portion of the cover.

(B) Contact Devices or Yokes. Contact devices or yokes designed and listed as self-grounding shall be permitted in conjunction with the supporting screws to establish equipment bonding between the device yoke and flush-type boxes.

(C) Floor Boxes. Floor boxes designed for and listed as providing satisfactory ground continuity between the box and the device shall be permitted.

(D) Isolated Ground Receptacles. Where installed for the reduction of electrical noise (electromagnetic interference) on the grounding circuit, a receptacle in which the grounding terminal is purposely insulated from the receptacle mounting means shall be permitted. The receptacle grounding terminal shall be connected to an insulated equipment grounding conductor run with the circuit conductors. This equipment grounding conductor shall be permitted to pass through one or more panelboards without a connection to the panelboard grounding terminal bar as permitted in 408.40, Exception, so as to terminate within the same building or structure directly at an equipment grounding conductor terminal of the applicable derived system or service. Where installed in accordance with the provisions of this section, this equipment grounding conductor shall also be permitted to pass through boxes, wireways, or other enclosures without being connected to such enclosures.

Informational Note: Use of an isolated equipment grounding conductor does not relieve the requirement for grounding the raceway system and outlet box.

250.148 Continuity and Attachment of Equipment Grounding Conductors to Boxes. If circuit conductors are spliced within a box or terminated on equipment within or supported by a box, all equipment grounding conductor(s) associated with any of those circuit conductors shall be connected within the box or to the box with devices suitable for the use in accordance with 250.8 and 250.148(A) through (E).

Exception: The equipment grounding conductor permitted in 250.146(D) shall not be required to be connected to the other equipment grounding conductors or to the box.

(A) Connections. Connections and splices shall be made in accordance with 110.14(B) except that insulation shall not be required.

(B) Grounding Continuity. The arrangement of grounding connections shall be such that the disconnection or the removal of a receptacle, luminaire, or other device fed from the box does not interfere with or interrupt the grounding continuity.

(C) Metal Boxes. A connection shall be made between the one or more equipment grounding conductors and a metal box by means of a grounding screw that shall be used for no other purpose, equipment listed for grounding, or a listed grounding device.

(D) Nonmetallic Boxes. One or more equipment grounding conductors brought into a nonmetallic outlet box shall be arranged such that a connection can be made to any fitting or device in that box requiring grounding.

(E) Solder. Connections depending solely on solder shall not be used.

PART VIII. DIRECT-CURRENT SYSTEMS

250.160 General. Direct-current systems shall comply with Part VIII and other sections of Article 250 not specifically intended for ac systems.

250.162 Direct-Current Circuits and Systems to Be Grounded. Direct-current circuits and systems shall be grounded as provided for in 250.162(A) and (B).

(A) Two-Wire, Direct-Current Systems. A 2-wire, dc system supplying premises wiring and operating at greater than 60 volts but not greater than 300 volts shall be grounded.

Exception No. 1: A system equipped with a ground detector and supplying only industrial equipment in limited areas shall not be required to be grounded where installed adjacent to or integral with the source of supply.

Exception No. 2: A rectifier-derived dc system supplied from an ac system complying with 250.20 shall not be required to be grounded.

Exception No. 3: Direct-current fire alarm circuits having a maximum current of 0.030 ampere as specified in Article 760, Part III, shall not be required to be grounded.

(B) Three-Wire, Direct-Current Systems. The neutral conductor of all 3-wire, dc systems supplying premises wiring shall be grounded.

250.164 Point of Connection for Direct-Current Systems.

(A) Off-Premises Source. Direct-current systems to be grounded and supplied from an off-premises source shall have the grounding connection made at one or more supply stations. A grounding connection shall not be made at individual services or at any point on the premises wiring.

(B) On-Premises Source. Where the dc system source is located on the premises, a grounding connection shall be made at one of the following:

(1) The source
(2) The first system disconnection means or overcurrent device

(3) By other means that accomplish equivalent system protection and that utilize equipment listed and identified for the use

250.166 Size of the Direct-Current Grounding Electrode Conductor. The size of the grounding electrode conductor for a dc system shall be as specified in 250.166(A) and (B), except as permitted by 250.166(C) through (E). The grounding electrode conductor for a dc system shall meet the sizing requirements in this section but shall not be required to be larger than 3/0 copper or 250 kcmil aluminum.

(A) Not Smaller Than the Neutral Conductor. Where the dc system consists of a 3-wire balancer set or a balancer winding with overcurrent protection as provided in 445.12(D), the grounding electrode conductor shall not be smaller than the neutral conductor and not smaller than 8 AWG copper or 6 AWG aluminum.

(B) Not Smaller Than the Largest Conductor. Where the dc system is other than as in 250.166(A), the grounding electrode conductor shall not be smaller than the largest conductor supplied by the system, and not smaller than 8 AWG copper or 6 AWG aluminum.

(C) Connected to Rod, Pipe, or Plate Electrodes. Where connected to rod, pipe, or plate electrodes as in 250.52(A)(5) or (A)(7), that portion of the grounding electrode conductor that is the sole connection to the grounding electrode shall not be required to be larger than 6 AWG copper wire or 4 AWG aluminum wire.

(D) Connected to a Concrete-Encased Electrode. Where connected to a concrete-encased electrode as in 250.52(A)(3), that portion of the grounding electrode conductor that is the sole connection to the grounding electrode shall not be required to be larger than 4 AWG copper wire.

(E) Connected to a Ground Ring. Where connected to a ground ring as in 250.52(A)(4), that portion of the grounding electrode conductor that is the sole connection to the grounding electrode shall not be required to be larger than the conductor used for the ground ring.

250.167 Direct-Current Ground-Fault Detection.

(A) Ungrounded Systems. Ground-fault detection systems shall be required for ungrounded systems.

(B) Grounded Systems. Ground-fault detection shall be permitted for grounded systems.

(C) Marking. Direct-current systems shall be legibly marked to indicate the grounding type at the dc source or the first disconnecting means of the system. The marking shall be of sufficient durability to withstand the environment involved.

250.168 Direct-Current System Bonding Jumper. For direct-current systems that are to be grounded, an unspliced bonding jumper shall be used to connect the equipment grounding conductor(s) to the grounded conductor at the source or the first system disconnecting means where the system is grounded. The size of the bonding jumper shall not be smaller than the system grounding electrode conductor specified in 250.166 and shall comply with the provisions of 250.28(A), (B), and (C).

PART X. GROUNDING OF SYSTEMS AND CIRCUITS OF OVER 1000 VOLTS

250.180 General. Where systems over 1000 volts are grounded, they shall comply with all applicable provisions of the preceding sections of this article and with 250.182 through 250.194, which supplement and modify the preceding sections.

CHAPTER 11

BOXES AND ENCLOSURES

INTRODUCTION

This chapter contains requirements from Article 314 — Outlet, Device, Pull and Junction Boxes; Conduit Bodies; Fittings; and Handhole Enclosures. This chapter also includes requirements from Article 312 — Cabinets, Cutout Boxes, and Meter Socket Enclosures. Article 314 requirements cover the installation of all boxes, conduit bodies, and handhole enclosures used as outlet, device, pull, or junction points in a wiring system. This article also includes installation requirements for fittings used to join raceways and to connect raceways (and cables) to boxes and conduit bodies. Article 312 covers the installation of cabinets, cutout boxes, and meter sockets.

Article 314's requirements range from selecting the size and type of box to installing the cover or canopy. Provisions for sizing boxes enclosing 18 through 6 AWG conductors are covered by 314.16. Enclosures (outlet, device, and junction boxes) covered by 314.16 are calculated based on the sizes and numbers of conductors, plus other devices and hardware that occupy space in the box. Requirements on sizing boxes containing larger conductors (4 AWG and larger) are located in 314.28. Pull and junction boxes covered by 314.28 are calculated using the size and number of raceways entering an enclosure. Where cables are used, the determination is based on transposing conductor size into the minimum size raceway required. The objective is to provide sufficient space to bend and route conductors within an enclosure without damaging conductor insulation. The requirements in Article 312 are also aimed at providing sufficient conductor space in enclosures where conductors are connected to terminals. Cabinets are the enclosures that are generally used to enclose panelboards covered in Article 408; thus it is necessary to apply the requirements of both articles to ensure a compliant installation. As an example of this concept, Section 408.55 is included in this chapter because it provides wire-bending space requirements for a cabinet enclosing a panelboard.

ARTICLE 314

Outlet, Device, Pull, and Junction Boxes; Conduit Bodies; Fittings; and Handhole Enclosures

PART I. SCOPE AND GENERAL

314.2 Round Boxes. Round boxes shall not be used where conduits or connectors requiring the use of locknuts or bushings are to be connected to the side of the box.

314.3 Nonmetallic Boxes. Nonmetallic boxes shall be permitted only with open wiring on insulators, concealed knob-and-tube wiring, cabled wiring methods with entirely nonmetallic sheaths, flexible cords, and nonmetallic raceways.

Exception No. 1: Where internal bonding means are provided between all entries, nonmetallic boxes shall be permitted to be used with metal raceways or metal-armored cables.

Exception No. 2: Where integral bonding means with a provision for attaching an equipment bonding jumper inside the box are provided between all threaded entries in nonmetallic boxes listed for the purpose, nonmetallic boxes shall be permitted to be used with metal raceways or metal-armored cables.

314.4 Metal Boxes. Metal boxes shall be grounded and bonded in accordance with Parts I, IV, V, VI, VII, and X of Article 250 as applicable, except as permitted in 250.112(I).

PART II. INSTALLATION

314.15 Damp or Wet Locations. In damp or wet locations, boxes, conduit bodies, outlet box hoods, and fittings shall be placed or equipped so as to prevent moisture from entering or accumulating within the box, conduit body, or fitting. Boxes, conduit bodies, outlet box hoods, and fittings installed in wet locations shall be listed for use in wet locations. Approved drainage openings not smaller than 3 mm (⅛ in.) and not larger than 6 mm (¼ in.) in diameter shall be permitted to be installed in the field in boxes or conduit bodies listed for use in damp or wet locations. For installation of listed drain fittings, larger openings are permitted to be installed in the field in accordance with manufacturer's instructions.

314.16 Number of Conductors in Outlet, Device, and Junction Boxes, and Conduit Bodies. Boxes and conduit bodies shall be of an approved size to provide free space for all enclosed conductors. In no case shall the volume of the box, as calculated in 314.16(A), be less than the fill calculation as calculated in 314.16(B). The minimum volume for conduit bodies shall be as calculated in 314.16(C).

The provisions of this section shall not apply to terminal housings supplied with motors or generators.

Boxes and conduit bodies enclosing conductors 4 AWG or larger shall also comply with the provisions of 314.28.

(A) Box Volume Calculations. The volume of a wiring enclosure (box) shall be the total volume of the assembled sections and, where used, the space provided by plaster rings, domed covers, extension rings, and so forth, that are marked with their volume or are made from boxes the dimensions of which are listed in Table 314.16(A). Where a box is provided with one or more securely installed barriers, the volume shall be apportioned to each of the resulting spaces. Each barrier, if not marked with its volume, shall be considered to take up 8.2 cm^3 (½ in^3) if metal, and 16.4 cm^3 (1.0 in^3) if nonmetallic.

(1) Standard Boxes. The volumes of standard boxes that are not marked with their volume shall be as given in Table 314.16(A).

(2) Other Boxes. Boxes 1650 cm^3 (100 $in.^3$) or less, other than those described in Table 314.16(A), and nonmetallic boxes shall be durably and legibly marked by the manufacturer with their volume(s). Boxes described in Table 314.16(A) that have a volume larger than is designated in the table shall be permitted to have their volume marked as required by this section.

(B) Box Fill Calculations. The volumes in paragraphs 314.16(B)(1) through (B)(5), as applicable, shall be added together. No allowance shall be required for small fittings such as locknuts and bushings. Each space within a box installed with a barrier shall be calculated separately.

(1) Conductor Fill. Each conductor that originates outside the box and terminates or is spliced within the box shall be counted once, and each conductor that passes through the box without splice or termination shall be counted once. Each loop or coil of unbroken conductor not less than twice the minimum length required for free conductors in 300.14 shall be counted twice. The conductor fill shall be calculated using Table 314.16(B). A conductor, no part of which leaves the box, shall not be counted.

Exception: An equipment grounding conductor or conductors or not over four fixture wires smaller than 14 AWG, or both, shall be permitted to be omitted from the calculations where they enter a box from a domed luminaire or similar canopy and terminate within that box.

(2) Clamp Fill. Where one or more internal cable clamps, whether factory or field supplied, are present in the box, a single volume allowance in accordance with Table 314.16(B) shall be made based on the largest conductor present in the box. No allowance shall be required for a cable connector with its clamping mechanism outside the box.

A clamp assembly that incorporates a cable termination for the cable conductors shall be listed and marked for use with specific nonmetallic boxes. Conductors that originate within the clamp assembly shall be included in conductor fill calculations covered in 314.16(B)(1) as though they entered from outside the box. The clamp assembly shall not require a fill allowance, but the volume of the portion of the assembly that remains within the box after installation shall be excluded from the box volume as marked in 314.16(A)(2).

(3) Support Fittings Fill. Where one or more luminaire studs or hickeys are present in the box, a single volume allowance in accordance with Table 314.16(B) shall be made for each type of fitting based on the largest conductor present in the box.

Table 314.16(A) Metal Boxes

Box Trade Size			Minimum Volume		Maximum Number of Conductors* (arranged by AWG size)						
mm	in.		cm³	in.³	18	16	14	12	10	8	6
100 × 32	(4 × 1¼)	round/octagonal	205	12.5	8	7	6	5	5	5	2
100 × 38	(4 × 1½)	round/octagonal	254	15.5	10	8	7	6	6	5	3
100 × 54	(4 × 2⅛)	round/octagonal	353	21.5	14	12	10	9	8	7	4
100 × 32	(4 × 1¼)	square	295	18.0	12	10	9	8	7	6	3
100 × 38	(4 × 1½)	square	344	21.0	14	12	10	9	8	7	4
100 × 54	(4 × 2⅛)	square	497	30.3	20	17	15	13	12	10	6
120 × 32	(4¹¹⁄₁₆ × 1¼)	square	418	25.5	17	14	12	11	10	8	5
120 × 38	(4¹¹⁄₁₆ × 1½)	square	484	29.5	19	16	14	13	11	9	5
120 × 54	(4¹¹⁄₁₆ × 2⅛)	square	689	42.0	28	24	21	18	16	14	8
75 × 50 × 38	(3 × 2 × 1½)	device	123	7.5	5	4	3	3	3	2	1
75 × 50 × 50	(3 × 2 × 2)	device	164	10.0	6	5	5	4	4	3	2
75 × 50 × 57	(3 × 2 × 2¼)	device	172	10.5	7	6	5	4	4	3	2
75 × 50 × 65	(3 × 2 × 2½)	device	205	12.5	8	7	6	5	5	4	2
75 × 50 × 70	(3 × 2 × 2¾)	device	230	14.0	9	8	7	6	5	4	2
75 × 50 × 90	(3 × 2 × 3½)	device	295	18.0	12	10	9	8	7	6	3

100 × 54 × 38	(4 × 2⅛ × 1½)	device	169	10.3	6	5	5	4	4	3	2
100 × 54 × 48	(4 × 2⅛ × 1⅞)	device	213	13.0	8	7	6	5	5	4	2
100 × 54 × 54	(4 × 2⅛ × 2⅛)	device	238	14.5	9	8	7	6	5	4	2
95 × 50 × 65	(3¾ × 2 × 2½)	masonry box	230	14.0	9	8	7	6	5	4	2
95 × 50 × 90	(3¾ × 2 × 3½)	masonry box	344	21.0	14	12	10	9	8	7	4
min. 44.5 depth	FS — single cover (1¾)		221	13.5	9	7	6	6	5	4	2
min. 60.3 depth	FD — single cover (2⅜)		295	18.0	12	10	9	8	7	6	3
min. 44.5 depth	FS — multiple cover (1¾)		295	18.0	12	10	9	8	7	6	3
min. 60.3 depth	FD — multiple cover (2⅜)		395	24.0	16	13	12	10	9	8	4

*Where no volume allowances are required by 314.16(B)(2) through (B)(5).

Table 314.16(B) Volume Allowance Required per Conductor

Size of Conductor (AWG)	Free Space Within Box for Each Conductor	
	cm³	in.³
18	24.6	1.50
16	28.7	1.75
14	32.8	2.00
12	36.9	2.25
10	41.0	2.50
8	49.2	3.00
6	81.9	5.00

(4) Device or Equipment Fill. For each yoke or strap containing one or more devices or equipment, a double volume allowance in accordance with Table 314.16(B) shall be made for each yoke or strap based on the largest conductor connected to a device(s) or equipment supported by that yoke or strap. A device or utilization equipment wider than a single 50 mm (2 in.) device box as described in Table 314.16(A) shall have double volume allowances provided for each gang required for mounting.

(5) Equipment Grounding Conductor Fill. Where one or more equipment grounding conductors or equipment bonding jumpers enter a box, a single volume allowance in accordance with Table 314.16(B) shall be made based on the largest equipment grounding conductor or equipment bonding jumper present in the box. Where an additional set of equipment grounding conductors, as permitted by 250.146(D), is present in the box, an additional volume allowance shall be made based on the largest equipment grounding conductor in the additional set.

(C) Conduit Bodies.

(1) General. Conduit bodies enclosing 6 AWG conductors or smaller, other than short-radius conduit bodies as described in 314.16(C)(3), shall have a cross-sectional area not less than twice the cross-sectional area of the largest conduit or tubing to which they can be attached. The maximum number of conductors permitted shall be the maximum number permitted by Table 1 of Chapter 9 for the conduit or tubing to which it is attached.

(2) With Splices, Taps, or Devices. Only those conduit bodies that are durably and legibly marked by the manufacturer with their volume shall be permitted to contain splices, taps, or devices. The maximum number of conductors shall be calculated in accordance with 314.16(B). Conduit bodies shall be supported in a rigid and secure manner.

(3) Short Radius Conduit Bodies. Conduit bodies such as capped elbows and service-entrance elbows that enclose conductors 6 AWG or smaller, and are only intended to enable the installation of the raceway and the contained conductors, shall not contain splices, taps, or devices and shall be of an approved size to provide free space for all conductors enclosed in the conduit body.

314.17 Conductors Entering Boxes, Conduit Bodies, or Fittings. Conductors entering boxes, conduit bodies, or fittings shall be protected from abrasion and shall comply with 314.17(A) through (D).

(A) Openings to Be Closed. Openings through which conductors enter shall be closed in an approved manner.

(B) Metal Boxes and Conduit Bodies. Where metal boxes or conduit bodies are installed with messenger-supported wiring, open wiring on insulators, or concealed knob-and-tube wiring, conductors shall enter through insulating bushings or, in dry locations, through flexible tubing extending from the last insulating support to not less than 6 mm (¼ in.) inside the box and beyond any cable clamps. Where nonmetallic-sheathed cable or multiconductor Type UF cable is used, the sheath shall extend not less than 6 mm (¼ in.) inside the box and beyond any cable clamp. Except as provided in 300.15(C), the wiring shall be firmly secured to the box or conduit body. Where raceway or cable is installed with metal boxes or conduit bodies, the raceway or cable shall be secured to such boxes and conduit bodies.

(C) Nonmetallic Boxes and Conduit Bodies. Nonmetallic boxes and conduit bodies shall be suitable for the lowest temperature-rated conductor entering the box. Where nonmetallic boxes and conduit bodies are used with messenger-supported wiring, open wiring on insulators, or concealed knob-and-tube wiring, the conductors shall enter the box through individual holes. Where flexible tubing is used to enclose the conductors, the tubing shall extend from the last insulating support to not less than 6 mm (¼ in.) inside the box and beyond any cable clamp. Where nonmetallic-sheathed cable or multiconductor Type UF cable is used, the sheath shall extend not less than 6 mm (¼ in.) inside the box and beyond any cable clamp. In all instances, all permitted wiring methods shall be secured to the boxes.

Exception: Where nonmetallic-sheathed cable or multiconductor Type UF cable is used with single gang boxes not larger than a nominal size 57 mm × 100 mm (2¼ in. × 4 in.) mounted in walls or ceilings, and where the cable is fastened within 200 mm (8 in.) of the box measured along the sheath and where the sheath extends through a cable knockout not less than 6 mm (¼ in.), securing the cable to the box shall not be required. Multiple cable entries shall be permitted in a single cable knockout opening.

314.19 Boxes Enclosing Flush Devices. Boxes used to enclose flush devices shall be of such design that the devices will be completely enclosed on back and sides and substantial support for the devices will be provided. Screws for supporting the box shall not also be used to attach a device.

314.20 Flush-Mounted Installations. Installations within or behind a surface of concrete, tile, gypsum, plaster, or other noncombustible material, including boxes employing a flush-type cover or faceplate, shall be made so that the front edge of the box, plaster ring, extension ring, or listed extender will not be set back of the finished surface more than 6 mm (¼ in.).

Installations within a surface of wood or other combustible surface material, boxes, plaster rings, extension rings, or listed extenders shall extend to the finished surface or project therefrom.

314.21 Repairing Noncombustible Surfaces. Noncombustible surfaces that are broken or incomplete around boxes employing a flush-type cover or faceplate shall be repaired so there will be no gaps or open spaces greater than 3 mm (⅛ in.) at the edge of the box.

314.22 Surface Extensions. Surface extensions shall be made by mounting and mechanically securing an extension ring over the box. Equipment grounding shall be in accordance with Part VI of Article 250.

Exception: A surface extension shall be permitted to be made from the cover of a box where the cover is designed so it is unlikely to fall off or be removed if its securing means becomes loose. The wiring method shall be flexible for an approved length that permits removal of the cover and provides access to the box interior and shall be arranged so that any grounding continuity is independent of the connection between the box and cover.

314.23 Supports. Enclosures within the scope of this article shall be supported in accordance with one or more of the provisions in 314.23(A) through (H).

(A) Surface Mounting. An enclosure mounted on a building or other surface shall be rigidly and securely fastened in place. If the surface does not provide rigid and secure support, additional support in accordance with other provisions of this section shall be provided.

(B) Structural Mounting. An enclosure supported from a structural member or from grade shall be rigidly supported either directly or by using a metal, polymeric, or wood brace.

(1) Nails and Screws. Nails and screws, where used as a fastening means, shall secure boxes by using brackets on the outside of the enclosure, or by using mounting holes in the back or in a single side of the enclosure, or they shall pass through the interior within 6 mm (¼ in.) of the back or ends of the enclosure. Screws shall not be permitted to pass through the box unless exposed threads in the box are protected using approved means to avoid abrasion of conductor insulation. Mounting holes made in the field shall be approved.

(2) Braces. Metal braces shall be protected against corrosion and formed from metal that is not less than 0.51 mm (0.020 in.) thick uncoated. Wood braces shall have a cross section not less than nominal 25 mm × 50 mm (1 in. × 2 in.). Wood braces in wet locations shall be treated for the conditions. Polymeric braces shall be identified as being suitable for the use.

(C) Mounting in Finished Surfaces. An enclosure mounted in a finished surface shall be rigidly secured thereto by clamps, anchors, or fittings identified for the application.

(D) Suspended Ceilings. An enclosure mounted to structural or supporting elements of a suspended ceiling shall be not more than 1650 cm^3 (100 $in.^3$) in size and shall be securely fastened in place in accordance with either 314.23(D)(1) or (D)(2).

(1) Framing Members. An enclosure shall be fastened to the framing members by mechanical means such as bolts, screws, or rivets, or by the use of clips or other securing means identified for use with the type of ceiling framing member(s) and enclosure(s) employed. The framing members shall be supported in an approved manner and securely fastened to each other and to the building structure.

(2) Support Wires. The installation shall comply with the provisions of 300.11(A). The enclosure shall be secured, using identified methods, to ceiling support wire(s), including any additional support wire(s) installed for ceiling support. Support wire(s) used for enclosure support shall be fastened at each end so as to be taut within the ceiling cavity.

(E) Raceway-Supported Enclosure, Without Devices, Luminaires, or Lampholders. An enclosure that does not contain a device(s), other than splicing devices, or supports a luminaire(s), a lampholder, or other equipment and is supported by entering raceways shall not exceed 1650 cm^3 (100 $in.^3$) in size. It shall have threaded entries or identified hubs. It shall be supported by two or more conduits threaded wrenchtight into the enclosure or hubs. Each conduit shall be secured within 900 mm (3 ft) of the enclosure, or within 450 mm (18 in.) of the enclosure if all conduit entries are on the same side.

Exception: The following wiring methods shall be permitted to support a conduit body of any size, including a conduit body constructed with only one conduit entry, provided that the trade size of the conduit body is not larger than the largest trade size of the conduit or tubing:

(1) Intermediate metal conduit, Type IMC
(2) Rigid metal conduit, Type RMC
(3) Rigid polyvinyl chloride conduit, Type PVC
(4) Reinforced thermosetting resin conduit, Type RTRC
(5) Electrical metallic tubing, Type EMT

(F) Raceway-Supported Enclosures, with Devices, Luminaires, or Lampholders. An enclosure that contains a device(s), other than splicing devices, or supports a luminaire(s), a lampholder, or other equipment and is supported by entering raceways shall not exceed 1650 cm^3 (100 $in.^3$) in size. It shall have threaded entries or identified hubs. It shall be supported by two or more conduits threaded wrenchtight into the enclosure or hubs. Each conduit shall be secured within 450 mm (18 in.) of the enclosure.

Exception No. 1: Rigid metal or intermediate metal conduit shall be permitted to support a conduit body of any size, including a conduit body constructed with only one conduit entry, provided the trade size of the conduit body is not larger than the largest trade size of the conduit.

Exception No. 2: An unbroken length(s) of rigid or intermediate metal conduit shall be permitted to support a box used for luminaire or lampholder support, or to support a wiring enclosure that is an integral part of a luminaire and used in lieu of a box in accordance with 300.15(B), where all of the following conditions are met:

(1) The conduit is securely fastened at a point so that the length of conduit beyond the last point of conduit support does not exceed 900 mm (3 ft).

(2) *The unbroken conduit length before the last point of conduit support is 300 mm (12 in.) or greater, and that portion of the conduit is securely fastened at some point not less than 300 mm (12 in.) from its last point of support.*
(3) *Where accessible to unqualified persons, the luminaire or lampholder, measured to its lowest point, is at least 2.5 m (8 ft) above grade or standing area and at least 900 mm (3 ft) measured horizontally to the 2.5 m (8 ft) elevation from windows, doors, porches, fire escapes, or similar locations.*
(4) *A luminaire supported by a single conduit does not exceed 300 mm (12 in.) in any direction from the point of conduit entry.*
(5) *The weight supported by any single conduit does not exceed 9 kg (20 lb).*
(6) *At the luminaire or lampholder end, the conduit(s) is threaded wrenchtight into the box, conduit body, integral wiring enclosure, or identified hubs. Where a box or conduit body is used for support, the luminaire shall be secured directly to the box or conduit body, or through a threaded conduit nipple not over 75 mm (3 in.) long.*

(G) Enclosures in Concrete or Masonry. An enclosure supported by embedment shall be identified as suitably protected from corrosion and securely embedded in concrete or masonry.

(H) Pendant Boxes. An enclosure supported by a pendant shall comply with 314.23(H)(1) or (H)(2).

(1) Flexible Cord. A box shall be supported from a multiconductor cord or cable in an approved manner that protects the conductors against strain, such as a strain-relief connector threaded into a box with a hub.

(2) Conduit. A box supporting lampholders or luminaires, or wiring enclosures within luminaires used in lieu of boxes in accordance with 300.15(B), shall be supported by rigid or intermediate metal conduit stems. For stems longer than

450 mm (18 in.), the stems shall be connected to the wiring system with flexible fittings suitable for the location. At the luminaire end, the conduit(s) shall be threaded wrenchtight into the box, wiring enclosure, or identified hubs.

Where supported by only a single conduit, the threaded joints shall be prevented from loosening by the use of setscrews or other effective means, or the luminaire, at any point, shall be at least 2.5 m (8 ft) above grade or standing area and at least 900 mm (3 ft) measured horizontally to the 2.5 m (8 ft) elevation from windows, doors, porches, fire escapes, or similar locations. A luminaire supported by a single conduit shall not exceed 300 mm (12 in.) in any horizontal direction from the point of conduit entry.

314.24 Depth of Boxes. Outlet and device boxes shall have an approved depth to allow equipment installed within them to be mounted properly and without likelihood of damage to conductors within the box.

(A) Outlet Boxes Without Enclosed Devices or Utilization Equipment. Outlet boxes that do not enclose devices or utilization equipment shall have a minimum internal depth of 12.7 mm (½ in.).

(B) Outlet and Device Boxes with Enclosed Devices or Utilization Equipment. Outlet and device boxes that enclose devices or utilization equipment shall have a minimum internal depth that accommodates the rearward projection of the equipment and the size of the conductors that supply the equipment. The internal depth shall include, where used, that of any extension boxes, plaster rings, or raised covers. The internal depth shall comply with all applicable provisions of 314.24(B)(1) through (B)(5).

(1) Large Equipment. Boxes that enclose devices or utilization equipment that projects more than 48 mm (1⅞ in.) rearward from the mounting plane of the box shall have a depth that is not less than the depth of the equipment plus 6 mm (¼ in.).

(2) Conductors Larger Than 4 AWG. Boxes that enclose devices or utilization equipment supplied by conductors larger than 4 AWG shall be identified for their specific function.

Exception to (2): Devices or utilization equipment supplied by conductors larger than 4 AWG shall be permitted to be mounted on or in junction and pull boxes larger than 1650 cm^3 (100 in.^{3}) if the spacing at the terminals meets the requirements of 312.6.

(3) Conductors 8, 6, or 4 AWG. Boxes that enclose devices or utilization equipment supplied by 8, 6, or 4 AWG conductors shall have an internal depth that is not less than 52.4 mm (2$\frac{1}{16}$ in.).

(4) Conductors 12 or 10 AWG. Boxes that enclose devices or utilization equipment supplied by 12 or 10 AWG conductors shall have an internal depth that is not less than 30.2 mm (1$\frac{3}{16}$ in.). Where the equipment projects rearward from the mounting plane of the box by more than 25 mm (1 in.), the box shall have a depth not less than that of the equipment plus 6 mm ($\frac{1}{4}$ in.).

(5) Conductors 14 AWG and Smaller. Boxes that enclose devices or utilization equipment supplied by 14 AWG or smaller conductors shall have a depth that is not less than 23.8 mm ($\frac{15}{16}$ in.).

Exception to (1) through (5): Devices or utilization equipment that is listed to be installed with specified boxes shall be permitted.

314.25 Covers and Canopies. In completed installations, each box shall have a cover, faceplate, lampholder, or luminaire canopy, except where the installation complies with 410.24(B). Screws used for the purpose of attaching covers, or other equipment, to the box shall be either machine screws matching the thread gauge or size that is integral to the box or shall be in accordance with the manufacturer's instructions.

(A) Nonmetallic or Metal Covers and Plates. Nonmetallic or metal covers and plates shall be permitted. Where metal covers or plates are used, they shall comply with the grounding requirements of 250.110.

(B) Exposed Combustible Wall or Ceiling Finish. Where a luminaire canopy or pan is used, any combustible wall or ceiling finish exposed between the edge of the canopy or pan and the outlet box shall be covered with noncombustible material if required by 410.23.

(C) Flexible Cord Pendants. Covers of outlet boxes and conduit bodies having holes through which flexible cord pendants pass shall be provided with identified bushings or shall have smooth, well-rounded surfaces on which the cords may bear. So-called hard rubber or composition bushings shall not be used.

314.27 Outlet Boxes.

(A) Boxes at Luminaire or Lampholder Outlets. Outlet boxes or fittings designed for the support of luminaires and lampholders, and installed as required by 314.23, shall be permitted to support a luminaire or lampholder.

(1) Vertical Surface Outlets. Boxes used at luminaire or lampholder outlets in or on a vertical surface shall be identified and marked on the interior of the box to indicate the maximum weight of the luminaire that is permitted to be supported by the box if other than 23 kg (50 lb).

Exception: A vertically mounted luminaire or lampholder weighing not more than 3 kg (6 lb) shall be permitted to be supported on other boxes or plaster rings that are secured to other boxes, provided that the luminaire or its supporting yoke, or the lampholder, is secured to the box with no fewer than two No. 6 or larger screws.

(2) Ceiling Outlets. At every outlet used exclusively for lighting, the box shall be designed or installed so that a luminaire or lampholder may be attached. Boxes shall be required to support a luminaire weighing a minimum of 23 kg (50 lb). A luminaire that weighs more than 23 kg (50 lb) shall be supported independently of the outlet box, unless the outlet box is listed for not less than the weight to be supported. The interior of the box shall be marked by the manufacturer to indicate the maximum weight the box shall be permitted to support.

(B) Floor Boxes. Boxes listed specifically for this application shall be used for receptacles located in the floor.

Exception: Where the authority having jurisdiction judges them free from likely exposure to physical damage, moisture, and dirt, boxes located in elevated floors of show windows and similar locations shall be permitted to be other than those listed for floor applications. Receptacles and covers shall be listed as an assembly for this type of location.

(C) Boxes at Ceiling-Suspended (Paddle) Fan Outlets. Outlet boxes or outlet box systems used as the sole support of a ceiling-suspended (paddle) fan shall be listed, shall be marked by their manufacturer as suitable for this purpose, and shall not support ceiling-suspended (paddle) fans that weigh more than 32 kg (70 lb). For outlet boxes or outlet box systems designed to support ceiling-suspended (paddle) fans that weigh more than 16 kg (35 lb), the required marking shall include the maximum weight to be supported.

(D) Utilization Equipment. Boxes used for the support of utilization equipment other than ceiling-suspended (paddle) fans shall meet the requirements of 314.27(A) for the support of a luminaire that is the same size and weight.

Exception: Utilization equipment weighing not more than 3 kg (6 lb) shall be permitted to be supported on other boxes or plaster rings that are secured to other boxes, provided the equipment or its supporting yoke is secured to the box with no fewer than two No. 6 or larger screws.

N **(E) Separable Attachment Fittings.** Outlet boxes required in 314.27 shall be permitted to support listed locking support and mounting receptacles used in combination with compatible attachment fittings. The combination shall be identified for the support of equipment within the weight and mounting orientation limits of the listing. Where the supporting receptacle is installed within a box, it shall be included in the fill calculation covered in 314.16(B)(4).

314.28 Pull and Junction Boxes and Conduit Bodies. Boxes and conduit bodies used as pull or junction boxes shall comply with 314.28(A) through (E).

Exception: Terminal housings supplied with motors shall comply with the provisions of 430.12.

(A) Minimum Size. For raceways containing conductors of 4 AWG or larger that are required to be insulated, and for cables containing conductors of 4 AWG or larger, the minimum dimensions of pull or junction boxes installed in a raceway or cable run shall comply with 314.28(A)(1) through (A)(3). Where an enclosure dimension is to be calculated based on the diameter of entering raceways, the diameter shall be the metric designator (trade size) expressed in the units of measurement employed.

(1) Straight Pulls. In straight pulls, the length of the box or conduit body shall not be less than eight times the metric designator (trade size) of the largest raceway.

(2) Angle or U Pulls, or Splices. Where splices or where angle or U pulls are made, the distance between each raceway entry inside the box or conduit body and the opposite wall of the box or conduit body shall not be less than six times the metric designator (trade size) of the largest raceway in a row. This distance shall be increased for additional entries by the amount of the sum of the diameters of all other raceway entries in the same row on the same wall of the box. Each row shall be calculated individually, and the single row that provides the maximum distance shall be used.

Exception: Where a raceway or cable entry is in the wall of a box or conduit body opposite a removable cover, the distance from that wall to the cover shall be permitted to comply with the distance required for one wire per terminal in Table 312.6(A).

The distance between raceway entries enclosing the same conductor shall not be less than six times the metric designator (trade size) of the larger raceway.

When transposing cable size into raceway size in 314.28(A)(1) and (A)(2), the minimum metric designator (trade size) raceway required for the number and size of conductors in the cable shall be used.

(3) Smaller Dimensions. Listed boxes or listed conduit bodies of dimensions less than those required in 314.28(A)(1) and (A)(2) shall be permitted for installations of combinations of conductors that are less than the maximum conduit or tubing fill (of conduits or tubing being used) permitted by Table 1 of Chapter 9.

Listed conduit bodies of dimensions less than those required in 314.28(A)(2), and having a radius of the curve to the centerline not less than that indicated in Table 2 of Chapter 9 for one-shot and full-shoe benders, shall be permitted for installations of combinations of conductors permitted by Table 1 of Chapter 9. These conduit bodies shall be marked to show they have been specifically evaluated in accordance with this provision.

Where the permitted combinations of conductors for which the box or conduit body has been listed are less than the maximum conduit or tubing fill permitted by Table 1 of Chapter 9, the box or conduit body shall be permanently marked with the maximum number and maximum size of conductors permitted. For other conductor sizes and combinations, the total cross-sectional area of the fill shall not exceed the cross-sectional area of the conductors specified in the marking, based on the type of conductor identified as part of the product listing.

(B) Conductors in Pull or Junction Boxes. In pull boxes or junction boxes having any dimension over 1.8 m (6 ft), all conductors shall be cabled or racked up in an approved manner.

(C) Covers. All pull boxes, junction boxes, and conduit bodies shall be provided with covers compatible with the box or conduit body construction and suitable for the conditions of use. Where used, metal covers shall comply with the grounding requirements of 250.110.

(D) Permanent Barriers. Where permanent barriers are installed in a box, each section shall be considered as a separate box.

(E) Power Distribution Blocks. Power distribution blocks shall be permitted in pull and junction boxes over 1650 cm^3 (100 $in.^3$) for connections of conductors where installed in boxes and where the installation complies with 314.28(E)(1) through (5).

Exception: Equipment grounding terminal bars shall be permitted in smaller enclosures.

(1) Installation. Power distribution blocks installed in boxes shall be listed. Power distribution blocks installed on the line side of the service equipment shall be listed and marked "suitable for use on the line side of service equipment" or equivalent.

(2) Size. In addition to the overall size requirement in the first sentence of 314.28(A)(2), the power distribution block shall be installed in a box with dimensions not smaller than specified in the installation instructions of the power distribution block.

(3) Wire Bending Space. Wire bending space at the terminals of power distribution blocks shall comply with 312.6.

(4) Live Parts. Power distribution blocks shall not have uninsulated live parts exposed within a box, whether or not the box cover is installed.

(5) Through Conductors. Where the pull or junction boxes are used for conductors that do not terminate on the power distribution block(s), the through conductors shall be arranged so the power distribution block terminals are unobstructed following installation.

314.29 Boxes, Conduit Bodies, and Handhole Enclosures to Be Accessible. Boxes, conduit bodies, and handhole enclosures shall be installed so that the wiring contained in them can be rendered accessible without removing any part of the building or structure or, in underground circuits, without excavating sidewalks, paving, earth, or other substance that is to be used to establish the finished grade.

Exception: Listed boxes and handhole enclosures shall be permitted where covered by gravel, light aggregate, or noncohesive granulated soil if their location is effectively identified and accessible for excavation.

314.30 Handhole Enclosures. Handhole enclosures shall be designed and installed to withstand all loads likely to be imposed on them. They shall be identified for use in underground systems.

(A) Size. Handhole enclosures shall be sized in accordance with 314.28(A) for conductors operating at 1000 volts or below, and in accordance with 314.71 for conductors operating at over 1000 volts. For handhole enclosures without bottoms where the provisions of 314.28(A)(2), Exception, or 314.71(B)(1), Exception No. 1, apply, the measurement to the removable cover shall be taken from the end of the conduit or cable assembly.

(B) Wiring Entries. Underground raceways and cable assemblies entering a handhole enclosure shall extend into the enclosure, but they shall not be required to be mechanically connected to the enclosure.

(C) Enclosed Wiring. All enclosed conductors and any splices or terminations, if present, shall be listed as suitable for wet locations.

(D) Covers. Handhole enclosure covers shall have an identifying mark or logo that prominently identifies the function of the enclosure, such as "electric." Handhole enclosure covers shall require the use of tools to open, or they shall weigh over 45 kg (100 lb). Metal covers and other exposed conductive surfaces shall be bonded in accordance with 250.92 if the conductors in the handhole are service conductors, or in accordance with 250.96(A) if the conductors in the handhole are feeder or branch-circuit conductors.

ARTICLE 312

Cabinets, Cutout Boxes, and Meter Socket Enclosures

PART I. SCOPE AND INSTALLATION

312.2 Damp and Wet Locations. In damp or wet locations, surface-type enclosures within the scope of this article shall be placed or equipped so as to prevent moisture or water from entering and accumulating within the cabinet or cutout box, and shall be mounted so there is at least 6-mm (¼-in.) airspace between the enclosure and the wall or other supporting surface. Enclosures installed in wet locations shall be weatherproof. For enclosures in wet locations, raceways or cables entering above the level of uninsulated live parts shall use fittings listed for wet locations.

Exception: Nonmetallic enclosures shall be permitted to be installed without the airspace on a concrete, masonry, tile, or similar surface.

312.3 Position in Wall. In walls of concrete, tile, or other noncombustible material, cabinets shall be installed so that the front edge of the cabinet is not set back of the finished surface more than 6 mm (¼ in.). In walls constructed of wood or other combustible material, cabinets shall be flush with the finished surface or project therefrom.

312.4 Repairing Noncombustible Surfaces. Noncombustible surfaces that are broken or incomplete shall be repaired so there will be no gaps or open spaces greater than 3 mm (⅛ in.) at the edge of the cabinet or cutout box employing a flush-type cover.

312.5 Cabinets, Cutout Boxes, and Meter Socket Enclosures. Conductors entering enclosures within the scope of this article shall be protected from abrasion and shall comply with 312.5(A) through (C).

(A) Openings to Be Closed. Openings through which conductors enter shall be closed in an approved manner.

(B) Metal Cabinets, Cutout Boxes, and Meter Socket Enclosures. Where metal enclosures within the scope of this article are installed with messenger-supported wiring, open wiring on insulators, or concealed knob-and-tube wiring, conductors shall enter through insulating bushings or, in dry locations, through flexible tubing extending from the last insulating support and firmly secured to the enclosure.

(C) Cables. Where cable is used, each cable shall be secured to the cabinet, cutout box, or meter socket enclosure.

Exception: Cables with entirely nonmetallic sheaths shall be permitted to enter the top of a surface-mounted enclosure through one or more nonflexible raceways not less than 450 mm (18 in.) and not more than 3.0 m (10 ft) in length, provided all of the following conditions are met:

(1) Each cable is fastened within 300 mm (12 in.), measured along the sheath, of the outer end of the raceway.
(2) The raceway extends directly above the enclosure and does not penetrate a structural ceiling.
(3) A fitting is provided on each end of the raceway to protect the cable(s) from abrasion and the fittings remain accessible after installation.
(4) The raceway is sealed or plugged at the outer end using approved means so as to prevent access to the enclosure through the raceway.
(5) The cable sheath is continuous through the raceway and extends into the enclosure beyond the fitting not less than 6 mm (¼ in.).
(6) The raceway is fastened at its outer end and at other points in accordance with the applicable article.
(7) Where installed as conduit or tubing, the cable fill does not exceed the amount that would be permitted for complete conduit or tubing systems by Table 1 of Chapter 9 of this Code and all applicable notes thereto. Note 2 to the tables in Chapter 9 does not apply to this condition.

312.6 Deflection of Conductors. Conductors at terminals or conductors entering or leaving cabinets or cutout boxes and the like shall comply with 312.6(A) through (C).

(A) Width of Wiring Gutters. Conductors shall not be deflected within a cabinet or cutout box unless a gutter having a width in accordance with Table 312.6(A) is provided.

Conductors in parallel in accordance with 310.10(H) shall be judged on the basis of the number of conductors in parallel.

(B) Wire-Bending Space at Terminals. Wire-bending space at each terminal shall be provided in accordance with 312.6(B)(1) or (B)(2).

(1) Conductors Not Entering or Leaving Opposite Wall. Table 312.6(A) shall apply where the conductor does not enter or leave the enclosure through the wall opposite its terminal.

(2) Conductors Entering or Leaving Opposite Wall. Table 312.6(B) shall apply where the conductor does enter or leave the enclosure through the wall opposite its terminal.

Exception No. 1: Where the distance between the wall and its terminal is in accordance with Table 312.6(A), a conductor shall be permitted to enter or leave an enclosure through the wall opposite its terminal, provided the conductor enters or leaves the enclosure where the gutter joins an adjacent gutter that has a width that conforms to Table 312.6(B) for the conductor.

408.55 Wire-Bending Space Within an Enclosure Containing a Panelboard.

(A) Top and Bottom Wire-Bending Space. The enclosure for a panelboard shall have the top and bottom wire-bending space sized in accordance with Table 312.6(B) for the largest conductor entering or leaving the enclosure.

(B) Side Wire-Bending Space. Side wire-bending space shall be in accordance with Table 312.6(A) for the largest conductor to be terminated in that space.

(C) Back Wire-Bending Space. Where a raceway or cable entry is in the wall of the enclosure opposite a removable cover, the distance from that wall to the cover shall be permitted to comply with the distance required for one wire per terminal in Table 312.6(A). The distance between the center of the rear entry and the nearest termination for the entering conductors shall not be less than the distance given in Table 312.6(B).

Table 312.6(A) Minimum Wire-Bending Space at Terminals and Minimum Width of Wiring Gutters

Wire Size (AWG or kcmil)		Wires per Terminal									
		1		2		3		4		5	
All Other Conductors	Compact Stranded AA-8000 Aluminum Alloy Conductors (see Note 2)	mm	in.	mm	in.	mm	in.	mm	in.	mm	in.
14–10	12–8	Not specified		—	—	—	—	—	—	—	—
8–6	6–4	38.1	1½	—	—	—	—	—	—	—	—
4–3	2–1	50.8	2	—	—	—	—	—	—	—	—
2	1/0	63.5	2½	—	—	—	—	—	—	—	—
1	2/0	76.2	3	—	—	—	—	—	—	—	—
1/0–2/0	3/0–4/0	88.9	3½	127	5	178	7	—	—	—	—
3/0–4/0	250–300	102	4	152	6	203	8	—	—	—	—
250	350	114	4½	152	6	203	8	254	10	—	—
300–350	400–500	127	5	203	8	254	10	305	12	—	—

400–500	600–750	152	6	203	8	254	10	305	12	356	14
600–700	800–1000	203	8	254	10	305	12	356	14	406	16
750–900	—	203	8	305	12	356	14	406	16	457	18
1000–1250	—	254	10	—	—	—	—	—	—	—	—
1500–2000	—	305	12	—	—	—	—	—	—	—	—

Notes:

1. Bending space at terminals shall be measured in a straight line from the end of the lug or wire connector (in the direction that the wire leaves the terminal) to the wall, barrier, or obstruction.

2. This column shall be permitted to be used to determine the minimum wire-bending space for compact stranded aluminum conductors in sizes up to 1000 kcmil and manufactured using AA-8000 series electrical grade aluminum alloy conductor material in accordance with 310.106(B). The minimum width of the wire gutter space shall be determined using the all other conductors value in this table.

Table 312.6(B) Minimum Wire-Bending Space at Terminals

Wire Size (AWG or kcmil)		Wires per Terminal							
		1		2		3		4 or More	
All Other Conductors	Compact Stranded AA-8000 Aluminum Alloy Conductors (See Note 3.)	mm	in.	mm	in.	mm	in.	mm	in.
14–10	12–8	Not specified		—	—	—		—	—
8	6	38.1	1½	—	—	—		—	—
6	4	50.8	2	—	—	—		—	—
4	2	76.2	3	—	—	—		—	—
3	1	76.2	3	—	—	—		—	—
2	1/0	88.9	3½	—	—	—		—	—
1	2/0	114	4½	—	—	—		—	—
1/0	3/0	140	5½	140	5½	178	7	—	—
2/0	4/0	152	6	152	6	190	7½	—	—
3/0	250	165[a]	6½[a]	165[a]	6½[a]	203	8	—	—
4/0	300	178[b]	7[b]	190[c]	7½[c]	216[a]	8½[a]	—	—
250	350	216[d]	8½[d]	229[d]	8½[d]	254[b]	9[b]	254	10
300	400	254[e]	10[e]	254[d]	10[d]	279[b]	11[b]	305	12
350	500	305[e]	12[e]	305[e]	12[e]	330[e]	13[e]	356[d]	14[d]

400	600	330[e]	13[e]	330[e]	13[e]	356[e]	14[e]	381[e]	15[e]
500	700–750	356[e]	14[e]	356[e]	14[e]	381[e]	15[e]	406[e]	16[e]
600	800–900	381[e]	15[e]	406[e]	16[e]	457[e]	18[e]	483[e]	19[e]
700	1000	406[e]	16[e]	457[e]	18[e]	508[e]	20[e]	559[e]	22[e]
750	—	432[e]	17[e]	483[e]	19[e]	559[e]	22[e]	610[e]	24[e]
800	—	457	18	508	20	559	22	610	24
900	—	483	19	559	22	610	24	610	24
1000	—	508	20	—	—	—		—	
1250	—	559	22	—	—	—		—	
1500	—	610	24	—	—	—		—	
1750	—	610	24	—	—	—		—	
2000	—	610	24	—	—	—		—	

Notes:

1. Bending space at terminals shall be measured in a straight line from the end of the lug or wire connector in a direction perpendicular to the enclosure wall.

2. For removable and lay-in wire terminals intended for only one wire, bending space shall be permitted to be reduced by the following number of millimeters (inches):

[a]12.7 mm (1/2 in.) [b]25.4 mm (1 in.) [c]38.1 mm (1 1/2 in.) [d]50.8 mm (2 in.) [e]76.2 mm (3 in.)

3. This column shall be permitted to determine the required wire-bending space for compact stranded aluminum conductors in sizes up to 1000 kcmil and manufactured using AA-8000 series electrical grade aluminum alloy conductor material in accordance with 310.106(B).

312.7 Space in Enclosures. Cabinets and cutout boxes shall have approved space to accommodate all conductors installed in them without crowding.

312.8 Switch and Overcurrent Device Enclosures. The wiring space within enclosures for switches and overcurrent devices shall be permitted for other wiring and equipment subject to limitations for specific equipment as provided in (A) and (B).

(A) Splices, Taps, and Feed-Through Conductors. The wiring space of enclosures for switches or overcurrent devices shall be permitted for conductors feeding through, spliced, or tapping off to other enclosures, switches, or overcurrent devices where all of the following conditions are met:

(1) The total of all conductors installed at any cross section of the wiring space does not exceed 40 percent of the cross-sectional area of that space.
(2) The total area of all conductors, splices, and taps installed at any cross section of the wiring space does not exceed 75 percent of the cross-sectional area of that space.
(3) A warning label complying with 110.21(B) is applied to the enclosure that identifies the closest disconnecting means for any feed-through conductors.

N **(B) Power Monitoring Equipment.** The wiring space of enclosures for switches or overcurrent devices shall be permitted to contain power monitoring equipment where all of the following conditions are met:

(1) The power monitoring equipment is identified as a field installable accessory as part of the listed equipment, or is a listed kit evaluated for field installation in switch or overcurrent device enclosures.
(2) The total area of all conductors, splices, taps, and equipment at any cross section of the wiring space does not exceed 75 percent of the cross-sectional area of that space.

CHAPTER 12

CABLES

INTRODUCTION

This chapter contains requirements from Article 320 — Armored Cable: Type AC; Article 330—Metal-Clad Cable: Type MC; Article 332 — Mineral-Insulated, Metal Sheathed Cable: Type MI; and Article 336 — Power and Control Tray Cable: Type TC. These four cable types are frequently used within the commercial and industrial electrical construction segment. Part I of the respective cable articles contains the definition for the respective cable types plus the requirement for the wiring method to be *listed* (see definition of *listed* in Article 100), and Part II provides the installation requirements. Some of the topics covered in Part II include uses permitted, uses not permitted, bending radius, securing and supporting, fittings, and ampacity. Each of these articles also contains construction requirements, but such requirements are outside the purview of this *Pocket Guide*.

ARTICLE 320

Armored Cable: Type AC

PART I. GENERAL

320.2 Definition.

Armored Cable, Type AC. A fabricated assembly of insulated conductors in a flexible interlocked metallic armor. See 320.100.

N **320.6 Listing Requirements.** Type AC cable and associated fittings shall be listed.

PART II. INSTALLATION

320.10 Uses Permitted. Type AC cable shall be permitted as follows:

(1) For feeders and branch circuits in both exposed and concealed installations
(2) In cable trays
(3) In dry locations
(4) Embedded in plaster finish on brick or other masonry, except in damp or wet locations
(5) To be run or fished in the air voids of masonry block or tile walls where such walls are not exposed or subject to excessive moisture or dampness

320.12 Uses Not Permitted. Type AC cable shall not be used as follows:

(1) Where subject to physical damage
(2) In damp or wet locations
(3) In air voids of masonry block or tile walls where such walls are exposed or subject to excessive moisture or dampness
(4) Where exposed to corrosive conditions
(5) Embedded in plaster finish on brick or other masonry in damp or wet locations

320.15 Exposed Work. Exposed runs of cable, except as provided in 300.11(A), shall closely follow the surface of the building finish or of running boards. Exposed runs shall also be permitted to be installed on the underside of joists where supported at each joist and located so as not to be subject to physical damage.

320.17 Through or Parallel to Framing Members. Type AC cable shall be protected in accordance with 300.4(A), (C), and (D) where installed through or parallel to framing members.

320.23 In Accessible Attics. Type AC cables in accessible attics or roof spaces shall be installed as specified in 320.23(A) and (B).

(A) Cables Run Across the Top of Floor Joists. Where run across the top of floor joists, or within 2.1 m (7 ft) of the floor or floor joists across the face of rafters or studding, the cable shall be protected by guard strips that are at least as high as the cable. Where this space is not accessible by permanent stairs or ladders, protection shall only be required within 1.8 m (6 ft) of the nearest edge of the scuttle hole or attic entrance.

(B) Cable Installed Parallel to Framing Members. Where the cable is installed parallel to the sides of rafters, studs, or ceiling or floor joists, neither guard strips nor running boards shall be required, and the installation shall also comply with 300.4(D).

320.24 Bending Radius. Bends in Type AC cable shall be made such that the cable is not damaged. The radius of the curve of the inner edge of any bend shall not be less than five times the diameter of the Type AC cable.

320.30 Securing and Supporting.

(A) General. Type AC cable shall be supported and secured by staples; cable ties listed and identified for securement and support; straps, hangers, or similar fittings; or other approved means designed and installed so as not to damage the cable.

(B) Securing. Unless otherwise permitted, Type AC cable shall be secured within 300 mm (12 in.) of every outlet box, junction box, cabinet, or fitting and at intervals not exceeding 1.4 m (4½ ft).

(C) Supporting. Unless otherwise permitted, Type AC cable shall be supported at intervals not exceeding 1.4 m (4½ ft).

Horizontal runs of Type AC cable installed in wooden or metal framing members or similar supporting means shall be considered supported where such support does not exceed 1.4 m (4½ ft) intervals.

320.40 Boxes and Fittings. At all points where the armor of AC cable terminates, a fitting shall be provided to protect wires from abrasion, unless the design of the outlet boxes or fittings is such as to afford equivalent protection, and, in addition, an insulating bushing or its equivalent protection shall be provided between the conductors and the armor. The connector or clamp by which the Type AC cable is fastened to boxes or cabinets shall be of such design that the insulating bushing or its equivalent will be visible for inspection. Where change is made from Type AC cable to other cable or raceway wiring methods, a box, fitting, or conduit body shall be installed at junction points as required in 300.15.

320.80 Ampacity. The ampacity shall be determined in accordance with 310.15.

(A) Thermal Insulation. Armored cable installed in thermal insulation shall have conductors rated at 90°C (194°F). The ampacity of cable installed in these applications shall not exceed that of a 60°C (140°F) rated conductor. The 90°C (194°F) rating shall be permitted to be used for ampacity adjustment and correction calculations; however, the ampacity shall not exceed that of a 60°C (140°F) rated conductor.

(B) Cable Tray. The ampacity of Type AC cable installed in cable tray shall be determined in accordance with 392.80(A).

ARTICLE 330
Metal-Clad Cable: Type MC

PART I. GENERAL

330.2 Definition.

Metal Clad Cable, Type MC. A factory assembly of one or more insulated circuit conductors with or without optical fiber members enclosed in an armor of interlocking metal tape, or a smooth or corrugated metallic sheath.

N 330.6 Listing Requirements. Type MC cable shall be listed. Fittings used for connecting Type MC cable to boxes, cabinets, or other equipment shall be listed and identified for such use.

PART II. INSTALLATION

330.10 Uses Permitted.

(A) General Uses. Type MC cable shall be permitted as follows:

(1) For services, feeders, and branch circuits.
(2) For power, lighting, control, and signal circuits.
(3) Indoors or outdoors.
(4) Exposed or concealed.
(5) To be direct buried where identified for such use.
(6) In cable tray where identified for such use.
(7) In any raceway.
(8) As aerial cable on a messenger.
(9) In hazardous (classified) locations where specifically permitted by other articles in this *Code*.
• (10) In dry locations and embedded in plaster finish on brick or other masonry except in damp or wet locations.
(11) In wet locations where a corrosion-resistant jacket is provided over the metallic covering and any of the following conditions are met:

   a. The metallic covering is impervious to moisture.
   b. A jacket resistant to moisture is provided under the metal covering.
   c. The insulated conductors under the metallic covering are listed for use in wet locations.

(12) Where single-conductor cables are used, all phase conductors and, where used, the grounded conductor shall be grouped together to minimize induced voltage on the sheath.

(B) Specific Uses. Type MC cable shall be permitted to be installed in compliance with Parts II and III of Article 725 and

770.133 as applicable and in accordance with 330.10(B)(1) through (B)(4).

(1) Cable Tray. Type MC cable installed in cable tray shall comply with 392.10, 392.12, 392.18, 392.20, 392.22, 392.30, 392.46, 392.56, 392.60(C), and 392.80.

(2) Direct Buried. Direct-buried cable shall comply with 300.5 or 300.50, as appropriate.

(3) Installed as Service-Entrance Cable. Type MC cable installed as service-entrance cable shall be permitted in accordance with 230.43.

(4) Installed Outside of Buildings or Structures or as Aerial Cable. Type MC cable installed outside of buildings or structures or as aerial cable shall comply with 225.10, 396.10, and 396.12.

330.12 Uses Not Permitted. Type MC cable shall not be used under either of the following conditions:

(1) Where subject to physical damage
(2) Where exposed to any of the destructive corrosive conditions in (a) or (b), unless the metallic sheath or armor is resistant to the conditions or is protected by material resistant to the conditions:
   a. Direct buried in the earth or embedded in concrete unless identified for direct burial
   b. Exposed to cinder fills, strong chlorides, caustic alkalis, or vapors of chlorine or of hydrochloric acids

N **330.15 Exposed Work.** Exposed runs of cable, except as provided in 300.11(A), shall closely follow the surface of the building finish or of running boards. Exposed runs shall also be permitted to be installed on the underside of joists where supported at each joist and located so as not to be subject to physical damage.

330.17 Through or Parallel to Framing Members. Type MC cable shall be protected in accordance with 300.4(A), (C), and (D) where installed through or parallel to framing members.

330.24 Bending Radius. Bends in Type MC cable shall be so made that the cable will not be damaged. The radius of the curve of the inner edge of any bend shall not be less than required in 330.24(A) through (C).

(A) Smooth Sheath.

(1) Ten times the external diameter of the metallic sheath for cable not more than 19 mm (¾ in.) in external diameter
(2) Twelve times the external diameter of the metallic sheath for cable more than 19 mm (¾ in.) but not more than 38 mm (1½ in.) in external diameter
(3) Fifteen times the external diameter of the metallic sheath for cable more than 38 mm (1½ in.) in external diameter

(B) Interlocked-Type Armor or Corrugated Sheath. Seven times the external diameter of the metallic sheath.

(C) Shielded Conductors. Twelve times the overall diameter of one of the individual conductors or seven times the overall diameter of the multiconductor cable, whichever is greater.

330.30 Securing and Supporting.

(A) General. Type MC cable shall be supported and secured by staples; cable ties listed and identified for securement and support; straps, hangers, or similar fittings; or other approved means designed and installed so as not to damage the cable.

(B) Securing. Unless otherwise provided, cables shall be secured at intervals not exceeding 1.8 m (6 ft). Cables containing four or fewer conductors sized no larger than 10 AWG shall be secured within 300 mm (12 in.) of every box, cabinet, fitting, or other cable termination. In vertical installations, listed cables with ungrounded conductors 250 kcmil and larger shall be permitted to be secured at intervals not exceeding 3 m (10 ft).

(C) Supporting. Unless otherwise provided, cables shall be supported at intervals not exceeding 1.8 m (6 ft).

Horizontal runs of Type MC cable installed in wooden or metal framing members or similar supporting means shall be considered supported and secured where such support does not exceed 1.8-m (6-ft) intervals.

(D) Unsupported Cables. Type MC cable shall be permitted to be unsupported and unsecured where the cable complies with any of the following:

(1) Is fished between access points through concealed spaces in finished buildings or structures and supporting is impractical.
(2) Is not more than 1.8 m (6 ft) in length from the last point of cable support to the point of connection to luminaires or other electrical equipment and the cable and point of connection are within an accessible ceiling.
(3) Is Type MC of the interlocked armor type in lengths not exceeding 900 mm (3 ft) from the last point where it is securely fastened and is used to connect equipment where flexibility is necessary to minimize the transmission of vibration from equipment or to provide flexibility for equipment that requires movement after installation.

For the purpose of this section, Type MC cable fittings shall be permitted as a means of cable support.

330.80 Ampacity. The ampacity of Type MC cable shall be determined in accordance with 310.15 or 310.60 for 14 AWG and larger conductors and in accordance with Table 402.5 for 18 AWG and 16 AWG conductors. The installation shall not exceed the temperature ratings of terminations and equipment.

ARTICLE 332

Mineral-Insulated, Metal-Sheathed Cable: Type MI

PART I. GENERAL

332.2 Definition.

Mineral-Insulated, Metal-Sheathed Cable, Type MI. A factory assembly of one or more conductors insulated with a highly compressed refractory mineral insulation and enclosed in a liquidtight and gastight continuous copper or alloy steel sheath.

N **332.6 Listing Requirements.** Type MI cable and associated fittings shall be listed.

PART II. INSTALLATION

332.10 Uses Permitted. Type MI cable shall be permitted as follows:

(1) For services, feeders, and branch circuits
(2) For power, lighting, control, and signal circuits
(3) In dry, wet, or continuously moist locations
(4) Indoors or outdoors
(5) Where exposed or concealed
(6) Where embedded in plaster, concrete, fill, or other masonry, whether above or below grade
(7) In hazardous (classified) locations where specifically permitted by other articles in this *Code*
(8) Where exposed to oil and gasoline
(9) Where exposed to corrosive conditions not deteriorating to its sheath
(10) In underground runs where suitably protected against physical damage and corrosive conditions
(11) In or attached to cable tray

332.12 Uses Not Permitted. Type MI cable shall not be used under the following conditions or in the following locations:

(1) In underground runs unless protected from physical damage, where necessary
(2) Where exposed to conditions that are destructive and corrosive to the metallic sheath, unless additional protection is provided

332.17 Through or Parallel to Framing Members. Type MI cable shall be protected in accordance with 300.4 where installed through or parallel to framing members.

332.24 Bending Radius. Bends in Type MI cable shall be so made that the cable will not be damaged. The radius of the inner edge of any bend shall not be less than required as follows:

(1) Five times the external diameter of the metallic sheath for cable not more than 19 mm (¾ in.) in external diameter

(2) Ten times the external diameter of the metallic sheath for cable greater than 19 mm (¾ in.) but not more than 25 mm (1 in.) in external diameter

332.30 Securing and Supporting. Type MI cable shall be supported and secured by staples, straps, hangers, or similar fittings, designed and installed so as not to damage the cable, at intervals not exceeding 1.8 m (6 ft).

(A) Horizontal Runs Through Holes and Notches. In other than vertical runs, cables installed in accordance with 300.4 shall be considered supported and secured where such support does not exceed 1.8 m (6 ft) intervals.

(B) Unsupported Cable. Type MI cable shall be permitted to be unsupported where the cable is fished between access points through concealed spaces in finished buildings or structures and supporting is impracticable.

(C) Cable Trays. All MI cable installed in cable trays shall comply with 392.30(A).

332.31 Single Conductors. Where single-conductor cables are used, all phase conductors and, where used, the neutral conductor shall be grouped together to minimize induced voltage on the sheath.

332.40 Boxes and Fittings.

(A) Fittings. Fittings used for connecting Type MI cable to boxes, cabinets, or other equipment shall be identified for such use.

(B) Terminal Seals. Where Type MI cable terminates, an end seal fitting shall be installed immediately after stripping to prevent the entrance of moisture into the insulation. The conductors extending beyond the sheath shall be individually provided with an insulating material.

332.80 Ampacity. The ampacity of Type MI cable shall be determined in accordance with 310.15. The conductor temperature at the end seal fitting shall not exceed the temperature rating of the listed end seal fitting, and the installation shall not exceed the temperature ratings of terminations or equipment.

(A) Type MI Cable Installed in Cable Tray. The ampacities for Type MI cable installed in cable tray shall be determined in accordance with 392.80(A).

(B) Single Type MI Conductors Grouped Together. Where single Type MI conductors are grouped together in a triangular or square configuration, as required by 332.31, and installed on a messenger or exposed with a maintained free air space of not less than 2.15 times one conductor diameter (2.15 × O.D.) of the largest conductor contained within the configuration and adjacent conductor configurations or cables, the ampacity of the conductors shall not exceed the allowable ampacities of Table 310.15(B)(17).

ARTICLE 336

Power and Control Tray Cable: Type TC

PART I. GENERAL

336.2 Definition.

Power and Control Tray Cable, Type TC. A factory assembly of two or more insulated conductors, with or without associated bare or covered grounding conductors, under a nonmetallic jacket.

N **336.6 Listing Requirements.** Type TC cables and associated fittings shall be listed.

PART II. INSTALLATION

336.10 Uses Permitted. Type TC cable shall be permitted to be used as follows:

(1) For power, lighting, control, and signal circuits.
(2) In cable trays, including those with mechanically discontinuous segments up to 300 mm (1 ft).
(3) In raceways.
(4) In outdoor locations supported by a messenger wire.
(5) For Class 1 circuits as permitted in Parts II and III of Article 725.

(6) For non–power-limited fire alarm circuits if conductors comply with the requirements of 760.49.

(7) Between a cable tray and the utilization equipment or device(s), provided all of the following apply:

a. The cable is Type TC-ER.
b. The cable is installed in industrial establishments where the conditions of maintenance and supervision ensure that only qualified persons service the installation.
c. The cable is continuously supported and protected against physical damage using mechanical protection such as struts, angles, or channels.
d. The cable that complies with the crush and impact requirements of Type MC cable and is identified with the marking "TC–ER."
e. The cable is secured at intervals not exceeding 1.8 m (6 ft).
f. Equipment grounding for the utilization equipment is provided by an equipment grounding conductor within the cable. In cables containing conductors sized 6 AWG or smaller, the equipment grounding conductor must be provided within the cable or, at the time of installation, one or more insulated conductors must be permanently identified as an equipment grounding conductor in accordance with 250.119(B).

(8) Where installed in wet locations, Type TC cable shall also be resistant to moisture and corrosive agents.

(10) Direct buried, where identified for such use

336.12 Uses Not Permitted. Type TC tray cable shall not be installed or used as follows:

(1) Installed where it will be exposed to physical damage
(2) Installed outside a raceway or cable tray system, except as permitted in 336.10(4), 336.10(7), 336.10(9), and 336.10(10)
(3) Used where exposed to direct rays of the sun, unless identified as sunlight resistant

336.24 Bending Radius. Bends in Type TC cable shall be made so as not to damage the cable. For Type TC cable without metal shielding, the minimum bending radius shall be as follows:

(1) Four times the overall diameter for cables 25 mm (1 in.) or less in diameter
(2) Five times the overall diameter for cables larger than 25 mm (1 in.) but not more than 50 mm (2 in.) in diameter
(3) Six times the overall diameter for cables larger than 50 mm (2 in.) in diameter

Type TC cables with metallic shielding shall have a minimum bending radius of not less than 12 times the cable overall diameter.

336.80 Ampacity. The ampacity of Type TC tray cable shall be determined in accordance with 392.80(A) for 14 AWG and larger conductors, in accordance with 402.5 for 18 AWG through 16 AWG conductors where installed in cable tray, and in accordance with 310.15 where installed in a raceway or as messenger-supported wiring.

CHAPTER 13

RACEWAYS AND BUSWAYS

INTRODUCTION

This chapter contains requirements from Article 342 — Intermediate Metal Conduit: Type IMC; Article 344 — Rigid Metal Conduit: Type RMC; Article 348 Flexible Metal Conduit: Type FMC; Article 350 Liquidtight Flexible Metal Conduit: Type LFMC; Article 352 — Rigid Polyvinyl Chloride Conduit: Type PVC; Article 356 — Liquidtight Flexible Nonmetallic Conduit: Type LFNC; Article 358—Electrical Metallic Tubing: Type EMT; Article 362 — Electrical Nonmetallic Tubing: Type ENT; Article 368 — Busways; Article 376 — Metal Wireways; Article 378 — Nonmetallic Wireways; Article 386 — Surface Metal Raceways; and Article 388 — Surface Nonmetallic Raceways. Contained in each of these articles are requirements pertaining to the use and installation for that particular wiring method. These articles cover metal and nonmetallic conduits and tubing and busway. Busway is manufactured with the conductors in place and is not designed to have conductors added or removed; therefore it is not a *raceway* as defined in Article 100.

Part I of each article provides a definition of the wiring method and may also provide a listing requirement (see definition of *listed* in Article 100). Not all of the wiring methods covered in this chapter are required to be *listed*. One example of this is metal wireways covered in Article 378. This is one reason for including construction specifications in Part III of the article. Part II covers the installation requirements for the respective wiring method. Some of the topics that are covered in most of these articles include uses permitted, uses not permitted, bends (how made and number in one run), trimming (or reaming and threading), securing and supporting, and joints (or couplings and connectors). Additional topics covered in Articles 376, 378, 386, and 388 include size of conductors, number of conductors, and splices and taps. Because there are application and construction differences between busways and raceways, there are several requirements in Article 368 that are unique busway installations.

Although the specific requirements in each article may be different, the arrangement and numbering of similar requirements in each article follow the same format. For example, in each article 3__.10 covers uses permitted, 3__.12 covers uses not permitted, 3__.24 covers bends—how made, 3__.26 covers bends— number in one run, 3__.28 covers trimming (or reaming and threading), 3__.30 covers securing and supporting, and 3__.42 covers couplings and connectors. This same convention is used throughout Chapter 3 as practicable as a tool for efficient use of the *Code.*

ARTICLE 342

Intermediate Metal Conduit: Type IMC

PART I. GENERAL

342.2 Definition.

Intermediate Metal Conduit (IMC). A steel threadable raceway of circular cross section designed for the physical protection and routing of conductors and cables and for use as an equipment grounding conductor when installed with its integral or associated coupling and appropriate fittings.

342.6 Listing Requirements. IMC, factory elbows and couplings, and associated fittings shall be listed.

PART II. INSTALLATION

342.10 Uses Permitted.

(A) All Atmospheric Conditions and Occupancies. Use of IMC shall be permitted under all atmospheric conditions and occupancies.

(B) Corrosion Environments. IMC, elbows, couplings, and fittings shall be permitted to be installed in concrete, in direct contact with the earth, or in areas subject to severe corrosive influences where protected by corrosion protection approved for the condition.

(C) Cinder Fill. IMC shall be permitted to be installed in or under cinder fill where subject to permanent moisture where protected on all sides by a layer of noncinder concrete not less than 50 mm (2 in.) thick; where the conduit is not less than 450 mm (18 in.) under the fill; or where protected by corrosion protection approved for the condition.

(D) Wet Locations. All supports, bolts, straps, screws, and so forth, shall be of corrosion-resistant materials or protected against corrosion by corrosion-resistant materials.

342.14 Dissimilar Metals. Where practicable, dissimilar metals in contact anywhere in the system shall be avoided to eliminate the possibility of galvanic action.

Aluminum fittings and enclosures shall be permitted to be used with galvanized steel IMC where not subject to severe corrosive influences. Stainless steel IMC shall only be used with stainless steel fittings and approved accessories, outlet boxes, and enclosures.

342.22 Number of Conductors. The number of conductors shall not exceed that permitted by the percentage fill specified in Table 1, Chapter 9.

Cables shall be permitted to be installed where such use is not prohibited by the respective cable articles. The number of cables shall not exceed the allowable percentage fill specified in Table 1, Chapter 9.

342.24 Bends — How Made. Bends of IMC shall be so made that the conduit will not be damaged and the internal diameter of the conduit will not be effectively reduced. The radius of the curve of any field bend to the centerline of the conduit shall not be less than indicated in Table 2, Chapter 9.

342.26 Bends — Number in One Run. There shall not be more than the equivalent of four quarter bends (360 degrees total) between pull points, for example, conduit bodies and boxes.

342.28 Reaming and Threading. All cut ends shall be reamed or otherwise finished to remove rough edges. Where conduit is threaded in the field, a standard cutting die with a taper of 1 in 16 (¾ in. taper per foot) shall be used.

342.30 Securing and Supporting. IMC shall be installed as a complete system in accordance with 300.18 and shall be securely fastened in place and supported in accordance with 342.30(A) and (B).

(A) Securely Fastened. IMC shall be secured in accordance with one of the following:

(1) IMC shall be securely fastened within 900 mm (3 ft) of each outlet box, junction box, device box, cabinet, conduit body, or other conduit termination.
(2) Where structural members do not readily permit fastening within 900 mm (3 ft), fastening shall be permitted to be increased to a distance of 1.5 m (5 ft).

(3) Where approved, conduit shall not be required to be securely fastened within 900 mm (3 ft) of the service head for above-the-roof termination of a mast.

(B) Supports. IMC shall be supported in accordance with one of the following:

(1) Conduit shall be supported at intervals not exceeding 3 m (10 ft).
(2) The distance between supports for straight runs of conduit shall be permitted in accordance with Table 344.30(B)(2), provided the conduit is made up with threaded couplings and supports that prevent transmission of stresses to termination where conduit is deflected between supports.
(3) Exposed vertical risers from industrial machinery or fixed equipment shall be permitted to be supported at intervals not exceeding 6 m (20 ft) if the conduit is made up with threaded couplings, the conduit is supported and securely fastened at the top and bottom of the riser, and no other means of intermediate support is readily available.
(4) Horizontal runs of IMC supported by openings through framing members at intervals not exceeding 3 m (10 ft) and securely fastened within 900 mm (3 ft) of termination points shall be permitted.

342.42 Couplings and Connectors.

(A) Threadless. Threadless couplings and connectors used with conduit shall be made tight. Where buried in masonry or concrete, they shall be the concretetight type. Where installed in wet locations, they shall comply with 314.15. Threadless couplings and connectors shall not be used on threaded conduit ends unless listed for the purpose.

(B) Running Threads. Running threads shall not be used on conduit for connection at couplings.

342.46 Bushings. Where a conduit enters a box, fitting, or other enclosure, a bushing shall be provided to protect the wires from abrasion unless the box, fitting, or enclosure is designed to provide such protection.

342.60 Grounding. IMC shall be permitted as an equipment grounding conductor.

ARTICLE 344
Rigid Metal Conduit: Type RMC

PART I. GENERAL

344.2 Definition.

Rigid Metal Conduit (RMC). A threadable raceway of circular cross section designed for the physical protection and routing of conductors and cables and for use as an equipment grounding conductor when installed with its integral or associated coupling and appropriate fittings.

344.6 Listing Requirements. RMC, factory elbows and couplings, and associated fittings shall be listed.

PART II. INSTALLATION

344.10 Uses Permitted.

(A) Atmospheric Conditions and Occupancies.

(1) Galvanized Steel and Stainless Steel RMC. Galvanized steel and stainless steel RMC shall be permitted under all atmospheric conditions and occupancies.

(2) Red Brass RMC. Red brass RMC shall be permitted to be installed for direct burial and swimming pool applications.

(3) Aluminum RMC. Aluminum RMC shall be permitted to be installed where approved for the environment. Rigid aluminum conduit encased in concrete or in direct contact with the earth shall be provided with approved supplementary corrosion protection.

(4) Ferrous Raceways and Fittings. Ferrous raceways and fittings protected from corrosion solely by enamel shall be permitted only indoors and in occupancies not subject to severe corrosive influences.

(B) Corrosive Environments.

(1) Galvanized Steel, Stainless Steel, and Red Brass RMC, Elbows, Couplings, and Fittings. Galvanized steel, stainless steel, and red brass RMC elbows, couplings, and fittings shall be permitted to be installed in concrete, in direct contact with the earth, or in areas subject to severe corrosive influences where protected by corrosion protection approved for the condition.

(2) Supplementary Protection of Aluminum RMC. Aluminum RMC shall be provided with approved supplementary corrosion protection where encased in concrete or in direct contact with the earth.

(C) Cinder Fill. Galvanized steel, stainless steel, and red brass RMC shall be permitted to be installed in or under cinder fill where subject to permanent moisture where protected on all sides by a layer of noncinder concrete not less than 50 mm (2 in.) thick; where the conduit is not less than 450 mm (18 in.) under the fill; or where protected by corrosion protection approved for the condition.

(D) Wet Locations. All supports, bolts, straps, screws, and so forth, shall be of corrosion-resistant materials or protected against corrosion by corrosion-resistant materials.

344.14 Dissimilar Metals. Where practicable, dissimilar metals in contact anywhere in the system shall be avoided to eliminate the possibility of galvanic action. Aluminum fittings and enclosures shall be permitted to be used with galvanized steel RMC, and galvanized steel fittings and enclosures shall be permitted to be used with aluminum RMC where not subject to severe corrosive influences. Stainless steel RMC shall only be used with stainless steel fittings and approved accessories, outlet boxes, and enclosures.

344.22 Number of Conductors. The number of conductors shall not exceed that permitted by the percentage fill specified in Table 1, Chapter 9.

Cables shall be permitted to be installed where such use is not prohibited by the respective cable articles. The number of cables shall not exceed the allowable percentage fill specified in Table 1, Chapter 9.

344.24 Bends — How Made. Bends of RMC shall be so made that the conduit will not be damaged and so that the internal diameter of the conduit will not be effectively reduced. The radius of the curve of any field bend to the centerline of the conduit shall not be less than indicated in Table 2, Chapter 9.

344.26 Bends — Number in One Run. There shall not be more than the equivalent of four quarter bends (360 degrees total) between pull points, for example, conduit bodies and boxes.

344.28 Reaming and Threading. All cut ends shall be reamed or otherwise finished to remove rough edges. Where conduit is threaded in the field, a standard cutting die with a 1 in 16 taper (¾ in. taper per foot) shall be used.

344.30 Securing and Supporting. RMC shall be installed as a complete system in accordance with 300.18 and shall be securely fastened in place and supported in accordance with 344.30(A) and (B).

(A) Securely Fastened. RMC shall be secured in accordance with one of the following:

(1) RMC shall be securely fastened within 900 mm (3 ft) of each outlet box, junction box, device box, cabinet, conduit body, or other conduit termination.
(2) Fastening shall be permitted to be increased to a distance of 1.5 m (5 ft) where structural members do not readily permit fastening within 900 mm (3 ft).
(3) Where approved, conduit shall not be required to be securely fastened within 900 mm (3 ft) of the service head for above-the-roof termination of a mast.

(B) Supports. RMC shall be supported in accordance with one of the following:

(1) Conduit shall be supported at intervals not exceeding 3 m (10 ft).
(2) The distance between supports for straight runs of conduit shall be permitted in accordance with Table 344.30(B)(2), provided the conduit is made up with threaded couplings and supports that prevent transmission of stresses to termination where conduit is deflected between supports.
(3) Exposed vertical risers from industrial machinery or fixed equipment shall be permitted to be supported at

Table 344.30(B)(2) Supports for Rigid Metal Conduit

Conduit Size		Maximum Distance Between Rigid Metal Conduit Supports	
Metric Designator	Trade Size	m	ft
16–21	½–¾	3.0	10
27	1	3.7	12
35–41	1¼–1½	4.3	14
53–63	2–2½	4.9	16
78 and larger	3 and larger	6.1	20

intervals not exceeding 6 m (20 ft) if the conduit is made up with threaded couplings, the conduit is supported and securely fastened at the top and bottom of the riser, and no other means of intermediate support is readily available.

(4) Horizontal runs of RMC supported by openings through framing members at intervals not exceeding 3 m (10 ft) and securely fastened within 900 mm (3 ft) of termination points shall be permitted.

344.42 Couplings and Connectors.

(A) Threadless. Threadless couplings and connectors used with conduit shall be made tight. Where buried in masonry or concrete, they shall be the concrete tight type. Where installed in wet locations, they shall comply with 314.15. Threadless couplings and connectors shall not be used on threaded conduit ends unless listed for the purpose.

(B) Running Threads. Running threads shall not be used on conduit for connection at couplings.

344.46 Bushings. Where a conduit enters a box, fitting, or other enclosure, a bushing shall be provided to protect the wires from abrasion unless the box, fitting, or enclosure is designed to provide such protection.

344.60 Grounding. RMC shall be permitted as an equipment grounding conductor.

ARTICLE 348

Flexible Metal Conduit: Type FMC

PART I. GENERAL

348.2 Definition.

Flexible Metal Conduit (FMC). A raceway of circular cross section made of helically wound, formed, interlocked metal strip.

348.6 Listing Requirements. FMC and associated fittings shall be listed.

PART II. INSTALLATION

348.10 Uses Permitted. FMC shall be permitted to be used in exposed and concealed locations.

348.12 Uses Not Permitted. FMC shall not be used in the following:

(1) In wet locations
(2) In hoistways, other than as permitted in 620.21(A)(1)
(3) In storage battery rooms
(4) In any hazardous (classified) location except as permitted by other articles in this *Code*
(5) Where exposed to materials having a deteriorating effect on the installed conductors, such as oil or gasoline
(6) Underground or embedded in poured concrete or aggregate
(7) Where subject to physical damage

348.22 Number of Conductors. The number of conductors shall not exceed that permitted by the percentage fill specified in Table 1, Chapter 9, or as permitted in Table 348.22, or for metric designator 12 (trade size ⅜).

Cables shall be permitted to be installed where such use is not prohibited by the respective cable articles. The number of cables shall not exceed the allowable percentage fill specified in Table 1, Chapter 9.

348.24 Bends — How Made. Bends in conduit shall be made so that the conduit is not damaged and the internal diameter of the conduit is not effectively reduced. Bends shall be permitted to be made manually without auxiliary equipment. The radius of the curve to the centerline of any bend shall not be less than shown in Table 2, Chapter 9 using the column "Other Bends."

348.26 Bends — Number in One Run. There shall not be more than the equivalent of four quarter bends (360 degrees total) between pull points, for example, conduit bodies and boxes.

348.28 Trimming. All cut ends shall be trimmed or otherwise finished to remove rough edges, except where fittings that thread into the convolutions are used.

348.30 Securing and Supporting. FMC shall be securely fastened in place and supported in accordance with 348.30(A) and (B).

(A) Securely Fastened. FMC shall be securely fastened in place by an approved means within 300 mm (12 in.) of each box, cabinet, conduit body, or other conduit termination and shall be supported and secured at intervals not to exceed 1.4 m (4½ ft). Where used, cable ties shall be listed and be identified for securement and support.

Exception No. 1: Where FMC is fished between access points through concealed spaces in finished buildings or structures and supporting is impracticable.

Exception No. 2: Where flexibility is necessary after installation, lengths from the last point where the raceway is securely fastened shall not exceed the following:

(1) 900 mm (3 ft) for metric designators 16 through 35 (trade sizes ½ through 1¼)
(2) 1200 mm (4 ft) for metric designators 41 through 53 (trade sizes 1½ through 2)
(3) 1500 mm (5 ft) for metric designators 63 (trade size 2½) and larger

Exception No. 3: Lengths not exceeding 1.8 m (6 ft) from a luminaire terminal connection for tap connections to luminaires as permitted in 410.117(C).

Exception No. 4: Lengths not exceeding 1.8 m (6 ft) from the last point where the raceway is securely fastened for connections within an accessible ceiling to a luminaire(s) or other equipment. For the purposes of this exception, listed flexible metal conduit fittings shall be permitted as a means of securement and support.

(B) Supports. Horizontal runs of FMC supported by openings through framing members at intervals not greater than 1.4 m (4½ ft) and securely fastened within 300 mm (12 in.) of termination points shall be permitted.

348.42 Couplings and Connectors. Angle connectors shall not be concealed.

348.60 Grounding and Bonding. If used to connect equipment where flexibility is necessary to minimize the transmission of vibration from equipment or to provide flexibility for equipment that requires movement after installation, an equipment grounding conductor shall be installed.

Where flexibility is not required after installation, FMC shall be permitted to be used as an equipment grounding conductor when installed in accordance with 250.118(5).

Where required or installed, equipment grounding conductors shall be installed in accordance with 250.134(B).

Where required or installed, equipment bonding jumpers shall be installed in accordance with 250.102.

ARTICLE 350
Liquidtight Flexible Metal Conduit: Type LFMC

PART I. GENERAL

350.2 Definition.

Liquidtight Flexible Metal Conduit (LFMC). A raceway of circular cross section having an outer liquidtight, nonmetallic, sunlight-resistant jacket over an inner flexible metal core with associated couplings, connectors, and fittings for the installation of electric conductors.

350.6 Listing Requirements. LFMC and associated fittings shall be listed.

PART II. INSTALLATION

350.10 Uses Permitted. LFMC shall be permitted to be used in exposed or concealed locations as follows:

(1) Where conditions of installation, operation, or maintenance require flexibility or protection from liquids, vapors, or solids
(2) In hazardous (classified) locations where specifically permitted by Chapter 5
(3) For direct burial where listed and marked for the purpose

350.12 Uses Not Permitted. LFMC shall not be used as follows:

(1) Where subject to physical damage
(2) Where any combination of ambient and conductor temperature produces an operating temperature in excess of that for which the material is approved

350.22 Number of Conductors or Cables.

(A) Metric Designators 16 through 103 (Trade Sizes ½ through 4). The number of conductors shall not exceed that permitted by the percentage fill specified in Table 1, Chapter 9.

Cables shall be permitted to be installed where such use is not prohibited by the respective cable articles. The number of cables shall not exceed the allowable percentage fill specified in Table 1, Chapter 9.

(B) Metric Designator 12 (Trade Size ⅜). The number of conductors shall not exceed that permitted in Table 348.22, "Fittings Outside Conduit" columns.

350.26 Bends — Number in One Run. There shall not be more than the equivalent of four quarter bends (360 degrees total) between pull points, for example, conduit bodies and boxes.

N **350.28 Trimming.** All cut ends of conduit shall be trimmed inside and outside to remove rough edges.

350.30 Securing and Supporting. LFMC shall be securely fastened in place and supported in accordance with 350.30(A) and (B).

(A) Securely Fastened. LFMC shall be securely fastened in place by an approved means within 300 mm (12 in.) of each box, cabinet, conduit body, or other conduit termination and shall be supported and secured at intervals not to exceed 1.4 m (4½ ft). Where used, cable ties shall be listed and be identified for securement and support.

Exception No. 1: Where LFMC is fished between access points through concealed spaces in finished buildings or structures and supporting is impractical.

Exception No. 2: Where flexibility is necessary after installation, lengths from the last point where the raceway is securely fastened shall not exceed the following:

(1) 900 mm (3 ft) for metric designators 16 through 35 (trade sizes ½ through 1¼)
(2) 1200 mm (4 ft) for metric designators 41 through 53 (trade sizes 1½ through 2)
(3) 1500 mm (5 ft) for metric designators 63 (trade size 2½) and larger

Exception No. 3: Lengths not exceeding 1.8 m (6 ft) from a luminaire terminal connection for tap conductors to luminaires, as permitted in 410.117(C).

Exception No. 4: Lengths not exceeding 1.8 m (6 ft) from the last point where the raceway is securely fastened for connections within an accessible ceiling to luminaire(s) or other equipment. For the purposes of 350.30, listed LFMC fittings shall be permitted as a means of securement and support.

(B) Supports. Horizontal runs of LFMC supported by openings through framing members at intervals not greater than 1.4 m (4½ ft) and securely fastened within 300 mm (12 in.) of termination points shall be permitted.

350.42 Couplings and Connectors. Only fittings listed for use with LFMC shall be used. Angle connectors shall not be concealed. Straight LFMC fittings shall be permitted for direct burial where marked.

350.60 Grounding and Bonding. If used to connect equipment where flexibility is necessary to minimize the transmission of vibration from equipment or to provide flexibility for equipment that requires movement after installation, an equipment grounding conductor shall be installed.

Where flexibility is not required after installation, LFMC shall be permitted to be used as an equipment grounding conductor when installed in accordance with 250.118(6).

Where required or installed, equipment grounding conductors shall be installed in accordance with 250.134(B).

Where required or installed, equipment bonding jumpers shall be installed in accordance with 250.102.

ARTICLE 352

Rigid Polyvinyl Chloride Conduit: Type PVC

PART I. GENERAL

352.2 Definition.

Rigid Polyvinyl Chloride Conduit (PVC). A rigid nonmetallic raceway of circular cross section, with integral or associated couplings, connectors, and fittings for the installation of electrical conductors and cables.

352.6 Listing Requirements. PVC conduit, factory elbows, and associated fittings shall be listed.

PART II. INSTALLATION

352.10 Uses Permitted. The use of PVC conduit shall be permitted in accordance with 352.10(A) through (I).

(A) Concealed. PVC conduit shall be permitted in walls, floors, and ceilings.

(B) Corrosive Influences. PVC conduit shall be permitted in locations subject to severe corrosive influences as covered in 300.6 and where subject to chemicals for which the materials are specifically approved.

(C) Cinders. PVC conduit shall be permitted in cinder fill.

(D) Wet Locations. PVC conduit shall be permitted in portions of dairies, laundries, canneries, or other wet locations, and in locations where walls are frequently washed, the entire conduit system, including boxes and fittings used therewith, shall be installed and equipped so as to prevent water from entering the conduit. All supports, bolts, straps, screws, and so forth, shall be of corrosion-resistant materials or be protected against corrosion by approved corrosion-resistant materials.

(E) Dry and Damp Locations. PVC conduit shall be permitted for use in dry and damp locations not prohibited by 352.12.

(F) Exposed. PVC conduit shall be permitted for exposed work. PVC conduit used exposed in areas of physical damage shall be identified for the use.

(G) Underground Installations. For underground installations, PVC shall be permitted for direct burial and underground encased in concrete. See 300.5 and 300.50.

(H) Support of Conduit Bodies. PVC conduit shall be permitted to support nonmetallic conduit bodies not larger than the largest trade size of an entering raceway. These conduit bodies shall not support luminaires or other equipment and shall not contain devices other than splicing devices as permitted by 110.14(B) and 314.16(C)(2).

(I) Insulation Temperature Limitations. Conductors or cables rated at a temperature higher than the listed temperature rating of PVC conduit shall be permitted to be installed in PVC conduit, provided the conductors or cables are not operated at a temperature higher than the listed temperature rating of the PVC conduit.

352.12 Uses Not Permitted. PVC conduit shall not be used under the conditions specified in 352.12(A) through (E).

(A) Hazardous (Classified) Locations. In any hazardous (classified) location, except as permitted by other articles of this *Code*.

(B) Support of Luminaires. For the support of luminaires or other equipment not described in 352.10(H).

(C) Physical Damage. Where subject to physical damage unless identified for such use.

(D) Ambient Temperatures. Where subject to ambient temperatures in excess of 50°C (122°F) unless listed otherwise.

(E) Theaters and Similar Locations. In theaters and similar locations, except as provided in 518.4 and 520.5.

352.22 Number of Conductors. The number of conductors shall not exceed that permitted by the percentage fill specified in Table 1, Chapter 9.

Cables shall be permitted to be installed where such use is not prohibited by the respective cable articles. The number of cables shall not exceed the allowable percentage fill specified in Table 1, Chapter 9.

352.24 Bends — How Made. Bends shall be so made that the conduit will not be damaged and the internal diameter of the conduit will not be effectively reduced. Field bends shall be made only with identified bending equipment. The radius of the curve to the centerline of such bends shall not be less than shown in Table 2, Chapter 9.

352.26 Bends — Number in One Run. There shall not be more than the equivalent of four quarter bends (360 degrees total) between pull points, for example, conduit bodies and boxes.

352.28 Trimming. All cut ends shall be trimmed inside and outside to remove rough edges.

352.30 Securing and Supporting. PVC conduit shall be installed as a complete system as provided in 300.18 and shall be fastened so that movement from thermal expansion or contraction is permitted. PVC conduit shall be securely fastened and supported in accordance with 352.30(A) and (B).

(A) Securely Fastened. PVC conduit shall be securely fastened within 900 mm (3 ft) of each outlet box, junction box, device box, conduit body, or other conduit termination. Conduit listed for securing at other than 900 mm (3 ft) shall be permitted to be installed in accordance with the listing.

(B) Supports. PVC conduit shall be supported as required in Table 352.30. Conduit listed for support at spacings other than as shown in Table 352.30 shall be permitted to be installed in accordance with the listing. Horizontal runs of

Table 352.30 Support of Rigid Polyvinyl Chloride Conduit (PVC)

Conduit Size		Maximum Spacing Between Supports	
Metric Designator	Trade Size	mm or m	ft
16–27	½–1	900 mm	3
35–53	1¼–2	1.5 m	5
63–78	2½–3	1.8 m	6
91–129	3½–5	2.1 m	7
155	6	2.5 m	8

PVC conduit supported by openings through framing members at intervals not exceeding those in Table 352.30 and securely fastened within 900 mm (3 ft) of termination points shall be permitted.

352.44 Expansion Fittings. Expansion fittings for PVC conduit shall be provided to compensate for thermal expansion and contraction where the length change, in accordance with Table 352.44, is expected to be 6 mm (¼ in.) or greater in a straight run between securely mounted items such as boxes, cabinets, elbows, or other conduit terminations.

352.46 Bushings. Where a conduit enters a box, fitting, or other enclosure, a bushing or adapter shall be provided to protect the wire from abrasion unless the box, fitting, or enclosure design provides equivalent protection.

352.48 Joints. All joints between lengths of conduit, and between conduit and couplings, fittings, and boxes, shall be made by an approved method.

352.56 Splices and Taps. Splices and taps shall be made in accordance with 300.15.

352.60 Grounding. Where equipment grounding is required, a separate equipment grounding conductor shall be installed in the conduit.

Table 352.44 Expansion Characteristics of PVC Rigid Nonmetallic Conduit Coefficient of Thermal Expansion = 6.084×10^{-5} mm/mm/°C (3.38×10^{-5} in./in./°F)

Temperature Change (°C)	Length Change of PVC Conduit (mm/m)	Temperature Change (°F)	Length Change of PVC Conduit (in./100 ft)	Temperature Change (°F)	Length Change of PVC Conduit (in./100 ft)
5	0.30	5	0.20	105	4.26
10	0.61	10	0.41	110	4.46
15	0.91	15	0.61	115	4.66
20	1.22	20	0.81	120	4.87
25	1.52	25	1.01	125	5.07
30	1.83	30	1.22	130	5.27
35	2.13	35	1.42	135	5.48
40	2.43	40	1.62	140	5.68
45	2.74	45	1.83	145	5.88
50	3.04	50	2.03	150	6.08
55	3.35	55	2.23	155	6.29
60	3.65	60	2.43	160	6.49
65	3.95	65	2.64	165	6.69
70	4.26	70	2.84	170	6.90
75	4.56	75	3.04	175	7.10
80	4.87	80	3.24	180	7.30
85	5.17	85	3.45	185	7.50
90	5.48	90	3.65	190	7.71
95	5.78	95	3.85	195	7.91
100	6.08	100	4.06	200	8.11

ARTICLE 356

Liquidtight Flexible Nonmetallic Conduit: Type LFNC

PART I. GENERAL

356.2 Definition.

Liquidtight Flexible Nonmetallic Conduit (LFNC). A raceway of circular cross section of various types as follows:

(1) A smooth seamless inner core and cover bonded together and having one or more reinforcement layers between the core and covers, designated as Type LFNC-A
(2) A smooth inner surface with integral reinforcement within the raceway wall, designated as Type LFNC-B
(3) A corrugated internal and external surface without integral reinforcement within the raceway wall, designated as LFNC-C

356.6 Listing Requirements. LFNC and associated fittings shall be listed.

PART II. INSTALLATION

356.10 Uses Permitted. LFNC shall be permitted to be used in exposed or concealed locations for the following purposes:

(1) Where flexibility is required for installation, operation, or maintenance.
(2) Where protection of the contained conductors is required from vapors, liquids, or solids.
(3) For outdoor locations where listed and marked as suitable for the purpose.
(4) For direct burial where listed and marked for the purpose.
(5) Type LFNC shall be permitted to be installed in lengths longer than 1.8 m (6 ft) where secured in accordance with 356.30.
(6) Type LFNC-B as a listed manufactured prewired assembly, metric designator 16 through 27 (trade size ½ through 1) conduit.

(7) For encasement in concrete where listed for direct burial and installed in compliance with 356.42.

356.12 Uses Not Permitted. LFNC shall not be used as follows:

(1) Where subject to physical damage
(2) Where any combination of ambient and conductor temperatures is in excess of that for which it is listed
(3) In lengths longer than 1.8 m (6 ft), except as permitted by 356.10(5) or where a longer length is approved as essential for a required degree of flexibility
(4) In any hazardous (classified) location, except as permitted by other articles in this *Code*

356.22 Number of Conductors. The number of conductors shall not exceed that permitted by the percentage fill specified in Table 1, Chapter 9.

Cables shall be permitted to be installed where such use is not prohibited by the respective cable articles. The number of cables shall not exceed the allowable percentage fill specified in Table 1, Chapter 9.

356.26 Bends — Number in One Run. There shall not be more than the equivalent of four quarter bends (360 degrees total) between pull points, for example, conduit bodies and boxes.

356.28 Trimming. All cut ends of conduit shall be trimmed inside and outside to remove rough edges.

356.30 Securing and Supporting. Type LFNC shall be securely fastened and supported in accordance with one of the following:

(1) Where installed in lengths exceeding 1.8 m (6 ft), the conduit shall be securely fastened at intervals not exceeding 900 mm (3 ft) and within 300 mm (12 in.) on each side of every outlet box, junction box, cabinet, or fitting. Where used, cable ties shall be listed as suitable for the application and for securing and supporting.
(2) Securing or supporting of the conduit shall not be required where it is fished, installed in lengths not

exceeding 900 mm (3 ft) at terminals where flexibility is required, or installed in lengths not exceeding 1.8 m (6 ft) from a luminaire terminal connection for tap conductors to luminaires permitted in 410.117(C).

(3) Horizontal runs of LFNC supported by openings through framing members at intervals not exceeding 900 mm (3 ft) and securely fastened within 300 mm (12 in.) of termination points shall be permitted.

(4) Securing or supporting of LFNC shall not be required where installed in lengths not exceeding 1.8 m (6 ft) from the last point where the raceway is securely fastened for connections within an accessible ceiling to a luminaire(s) or other equipment. For the purpose of 356.30, listed liquidtight flexible nonmetallic conduit fittings shall be permitted as a means of support.

356.42 Couplings and Connectors. Only fittings listed for use with LFNC shall be used. Angle connectors shall not be used for concealed raceway installations. Straight LFNC fittings are permitted for direct burial or encasement in concrete.

356.60 Grounding. Where equipment grounding is required, a separate equipment grounding conductor shall be installed in the conduit.

Exception No. 1: As permitted in 250.134(B), Exception No. 2, for dc circuits and 250.134(B), Exception No. 1, for separately run equipment grounding conductors.

Exception No. 2: Where the grounded conductor is used to ground equipment as permitted in 250.142.

ARTICLE 358
Electrical Metallic Tubing: Type EMT

PART I. GENERAL

358.2 Definition.

Electrical Metallic Tubing (EMT). An unthreaded thinwall raceway of circular cross section designed for the physical protection and routing of conductors and cables and for use as an equipment grounding conductor when installed utilizing appropriate fittings.

358.6 Listing Requirements. EMT, factory elbows, and associated fittings shall be listed.

PART II. INSTALLATION

358.10 Uses Permitted.

(A) Exposed and Concealed. The use of EMT shall be permitted for both exposed and concealed work for the following:

(1) In concrete, in direct contact with the earth or in areas subject to severe corrosive influences where installed in accordance with 358.10(B)
(2) In dry, damp, and wet locations
(3) In any hazardous (classified) location as permitted by other articles in this *Code*

(B) Corrosive Environments.

N (1) Galvanized Steel and Stainless Steel EMT, Elbows, and Fittings. Galvanized steel and stainless steel EMT, elbows, and fittings shall be permitted to be installed in concrete, in direct contact with the earth, or in areas subject to severe corrosive influences where protected by corrosion protection and approved as suitable for the condition.

N (2) Supplementary Protection of Aluminum EMT. Aluminum EMT shall be provided with approved supplementary corrosion protection where encased in concrete or in direct contact with the earth.

N (C) Cinder Fill. Galvanized steel and stainless steel EMT shall be permitted to be installed in cinder concrete or cinder fill where subject to permanent moisture when protected on all sides by a layer of noncinder concrete at least 50 mm (2 in.) thick or when the tubing is installed at least 450 mm (18 in.) under the fill.

(D) Wet Locations. All supports, bolts, straps, screws, and so forth shall be of corrosion-resistant materials or protected against corrosion by corrosion-resistant materials.

358.12 Uses Not Permitted. EMT shall not be used under the following conditions:

(1) Where subject to severe physical damage
(2) For the support of luminaires or other equipment except conduit bodies no larger than the largest trade size of the tubing

• N **358.14 Dissimilar Metals.** Where practicable, dissimilar metals in contact anywhere in the system shall be avoided to eliminate the possibility of galvanic action. Aluminum fittings and enclosures shall be permitted to be used with galvanized steel EMT, and galvanized steel fittings and enclosures shall be permitted to be used with aluminum EMT where not subject to severe corrosive influences. Stainless steel EMT shall only be used with stainless steel fittings and approved accessories, outlet boxes, and enclosures.

358.22 Number of Conductors. The number of conductors shall not exceed that permitted by the percentage fill specified in Table 1, Chapter 9.

Cables shall be permitted to be installed where such use is not prohibited by the respective cable articles. The number of cables shall not exceed the allowable percentage fill specified in Table 1, Chapter 9.

358.24 Bends — How Made. Bends shall be made so that the tubing is not damaged and the internal diameter of the tubing is not effectively reduced. The radius of the curve of any field bend to the centerline of the tubing shall not be less than shown in Table 2, Chapter 9 for one-shot and full shoe benders.

358.26 Bends — Number in One Run. There shall not be more than the equivalent of four quarter bends (360 degrees total) between pull points, for example, conduit bodies and boxes.

358.28 Reaming and Threading.

(A) Reaming. All cut ends of EMT shall be reamed or otherwise finished to remove rough edges.

358.30 Securing and Supporting. EMT shall be installed as a complete system in accordance with 300.18 and shall be securely fastened in place and supported in accordance with 358.30(A) and (B).

(A) Securely Fastened. EMT shall be securely fastened in place at intervals not to exceed 3 m (10 ft). In addition, each EMT run between termination points shall be securely fastened within 900 mm (3 ft) of each outlet box, junction box, device box, cabinet, conduit body, or other tubing termination.

Exception No. 1: Fastening of unbroken lengths shall be permitted to be increased to a distance of 1.5 m (5 ft) where structural members do not readily permit fastening within 900 mm (3 ft).

Exception No. 2: For concealed work in finished buildings or prefinished wall panels where such securing is impracticable, unbroken lengths (without coupling) of EMT shall be permitted to be fished.

(B) Supports. Horizontal runs of EMT supported by openings through framing members at intervals not greater than 3 m (10 ft) and securely fastened within 900 mm (3 ft) of termination points shall be permitted.

358.42 Couplings and Connectors. Couplings and connectors used with EMT shall be made up tight. Where buried in masonry or concrete, they shall be concretetight type. Where installed in wet locations, they shall comply with 314.15.

358.60 Grounding. EMT shall be permitted as an equipment grounding conductor.

ARTICLE 362
Electrical Nonmetallic Tubing: Type ENT

PART I. GENERAL

362.2 Definition.

Electrical Nonmetallic Tubing (ENT). A nonmetallic, pliable, corrugated raceway of circular cross section with integral or associated couplings, connectors, and fittings for the installation of electrical conductors. ENT is composed of a material that is resistant to moisture and chemical atmospheres and is flame retardant.

A pliable raceway is a raceway that can be bent by hand with a reasonable force but without other assistance.

362.6 Listing Requirements. ENT and associated fittings shall be listed.

PART II. INSTALLATION

362.10 Uses Permitted. For the purpose of this article, the first floor of a building shall be that floor that has 50 percent or more of the exterior wall surface area level with or above finished grade. One additional level that is the first level and not designed for human habitation and used only for vehicle parking, storage, or similar use shall be permitted. The use of ENT and fittings shall be permitted in the following:

(1) In any building not exceeding three floors above grade as follows:

    a. For exposed work, where not prohibited by 362.12
    b. Concealed within walls, floors, and ceilings

(2) In any building exceeding three floors above grade, ENT shall be concealed within walls, floors, and ceilings where the walls, floors, and ceilings provide a thermal barrier of material that has at least a 15-minute finish rating as identified in listings of fire-rated assemblies. The 15-minute-finish-rated thermal barrier shall be permitted to be used for combustible or noncombustible walls, floors, and ceilings.

Exception to (2): Where a fire sprinkler system(s) is installed in accordance with NFPA 13-2013, Standard for the Installation of Sprinkler Systems, on all floors, ENT shall be permitted to be used within walls, floors, and ceilings, exposed or concealed, in buildings exceeding three floors abovegrade.

(3) In locations subject to severe corrosive influences as covered in 300.6 and where subject to chemicals for which the materials are specifically approved.

(4) In concealed, dry, and damp locations not prohibited by 362.12.

(5) Above suspended ceilings where the suspended ceilings provide a thermal barrier of material that has at least a 15-minute finish rating as identified in listings of fire-rated assemblies, except as permitted in 362.10(1)a.

Exception to (5): ENT shall be permitted to be used above suspended ceilings in buildings exceeding three floors above grade where the building is protected throughout by a fire sprinkler system installed in accordance with NFPA 13-2013, Standard for the Installation of Sprinkler Systems.

(6) Encased in poured concrete, or embedded in a concrete slab on grade where ENT is placed on sand or approved screenings, provided fittings identified for this purpose are used for connections.
(7) For wet locations indoors as permitted in this section or in a concrete slab on or belowgrade, with fittings listed for the purpose.
(8) Metric designator 16 through 27 (trade size ½ through 1) as listed manufactured prewired assembly.
(9) Conductors or cables rated at a temperature higher than the listed temperature rating of ENT shall be permitted to be installed in ENT, if the conductors or cables are not operated at a temperature higher than the listed temperature rating of the ENT.

362.12 Uses Not Permitted. ENT shall not be used in the following:

(1) In any hazardous (classified) location, except as permitted by other articles in this *Code*
(2) For the support of luminaires and other equipment
(3) Where subject to ambient temperatures in excess of 50°C (122°F) unless listed otherwise
(4) For direct earth burial
(5) In exposed locations, except as permitted by 362.10(1), 362.10(5), and 362.10(7)
(6) In theaters and similar locations, except as provided in 518.4 and 520.5
(7) Where exposed to the direct rays of the sun, unless identified as sunlight resistant
(8) Where subject to physical damage

362.22 Number of Conductors. The number of conductors shall not exceed that permitted by the percentage fill in Table 1, Chapter 9.

Cables shall be permitted to be installed where such use is not prohibited by the respective cable articles. The number of cables shall not exceed the allowable percentage fill specified in Table 1, Chapter 9.

362.24 Bends — How Made. Bends shall be so made that the tubing will not be damaged and the internal diameter of the tubing will not be effectively reduced. Bends shall be permitted to be made manually without auxiliary equipment, and the radius of the curve to the centerline of such bends shall not be less than shown in Table 2, Chapter 9 using the column "Other Bends."

362.26 Bends — Number in One Run. There shall not be more than the equivalent of four quarter bends (360 degrees total) between pull points, for example, conduit bodies and boxes.

362.28 Trimming. All cut ends shall be trimmed inside and outside to remove rough edges.

362.30 Securing and Supporting. ENT shall be installed as a complete system in accordance with 300.18 and shall be securely fastened in place by an approved means and supported in accordance with 362.30(A) and (B).

(A) Securely Fastened. ENT shall be securely fastened at intervals not exceeding 900 mm (3 ft). In addition, ENT shall be securely fastened in place within 900 mm (3 ft) of each outlet box, device box, junction box, cabinet, or fitting where it terminates. Where used, cable ties shall be listed as suitable for the application and for securing and supporting.

Exception No. 1: Lengths not exceeding a distance of 1.8 m (6 ft) from a luminaire terminal connection for tap connections to lighting luminaires shall be permitted without being secured.

Exception No. 2: Lengths not exceeding 1.8 m (6 ft) from the last point where the raceway is securely fastened for connections within an accessible ceiling to luminaire(s) or other equipment.

Exception No. 3: For concealed work in finished buildings or prefinished wall panels where such securing is impracticable, unbroken lengths (without coupling) of ENT shall be permitted to be fished.

(B) Supports. Horizontal runs of ENT supported by openings in framing members at intervals not exceeding 900 mm (3 ft) and securely fastened within 900 mm (3 ft) of termination points shall be permitted.

362.46 Bushings. Where a tubing enters a box, fitting, or other enclosure, a bushing or adapter shall be provided to protect the wire from abrasion unless the box, fitting, or enclosure design provides equivalent protection.

362.48 Joints. All joints between lengths of tubing and between tubing and couplings, fittings, and boxes shall be by an approved method.

362.60 Grounding. Where equipment grounding is required, a separate equipment grounding conductor shall be installed in the raceway in compliance with Article 250, Part VI.

ARTICLE 368
Busways

PART I. GENERAL REQUIREMENTS

368.2 Definition.

Busway. A raceway consisting of a metal enclosure containing factory-mounted, bare or insulated conductors, which are usually copper or aluminum bars, rods, or tubes.

PART II. INSTALLATION

368.10 Uses Permitted. Busways shall be permitted to be installed where they are located in accordance with 368.10(A) through (C).

(A) Exposed. Busways shall be permitted to be located in the open where visible, except as permitted in 368.10(C).

(B) Behind Access Panels. Busways shall be permitted to be installed behind access panels, provided the busways are totally enclosed, of nonventilating-type construction, and installed so that the joints between sections and at fittings are accessible for maintenance purposes. Where installed behind access panels, means of access shall be provided, and either of the following conditions shall be met:

(1) The space behind the access panels shall not be used for air-handling purposes.
(2) Where the space behind the access panels is used for environmental air, other than ducts and plenums, there shall be no provisions for plug-in connections, and the conductors shall be insulated.

(C) Through Walls and Floors. Busways shall be permitted to be installed through walls or floors in accordance with (C)(1) and (C)(2).

(1) Walls. Unbroken lengths of busway shall be permitted to be extended through dry walls.

(2) Floors. Floor penetrations shall comply with (a) and (b):

(a) Busways shall be permitted to be extended vertically through dry floors if totally enclosed (unventilated) where passing through and for a minimum distance of 1.8 m (6 ft) above the floor to provide adequate protection from physical damage.

(b) In other than industrial establishments, where a vertical riser penetrates two or more dry floors, a minimum 100-mm (4-in.) high curb shall be installed around all floor openings for riser busways to prevent liquids from entering the opening. The curb shall be installed within 300 mm (12 in.) of the floor opening. Electrical equipment shall be located so that it will not be damaged by liquids that are retained by the curb.

368.12 Uses Not Permitted.

(A) Physical Damage. Busways shall not be installed where subject to severe physical damage or corrosive vapors.

(B) Hoistways. Busways shall not be installed in hoistways.

(C) Hazardous Locations. Busways shall not be installed in any hazardous (classified) location, unless specifically approved for such use.

(D) Wet Locations. Busways shall not be installed outdoors or in wet or damp locations unless identified for such use.

(E) Working Platform. Lighting busway and trolley busway shall not be installed less than 2.5 m (8 ft) above the floor or working platform unless provided with an identified cover.

368.17 Overcurrent Protection. Overcurrent protection shall be provided in accordance with 368.17(A) through (D).

(A) Rating of Overcurrent Protection — Feeders. A busway shall be protected against overcurrent in accordance with the allowable current rating of the busway.

Exception No. 1: The applicable provisions of 240.4 shall be permitted.

Exception No. 2: Where used as transformer secondary ties, the provisions of 450.6(A)(3) shall be permitted.

(B) Reduction in Ampacity Size of Busway. Overcurrent protection shall be required where busways are reduced in ampacity.

Exception: For industrial establishments only, omission of overcurrent protection shall be permitted at points where busways are reduced in ampacity, provided that the length of the busway having the smaller ampacity does not exceed 15 m (50 ft) and has an ampacity at least equal to one-third the rating or setting of the overcurrent device next back on the line, and provided that such busway is free from contact with combustible material.

(C) Feeder or Branch Circuits. Where a busway is used as a feeder, devices or plug-in connections for tapping off feeder or branch circuits from the busway shall contain the overcurrent devices required for the protection of the feeder or branch circuits. The plug-in device shall consist of an externally operable circuit breaker or an externally operable fusible switch. Where such devices are mounted out of reach and contain disconnecting means, suitable means such as ropes, chains, or sticks shall be provided for operating the disconnecting means from the floor.

Exception No. 1: As permitted in 240.21.

Exception No. 2: For fixed or semifixed luminaires, where the branch-circuit overcurrent device is part of the luminaire cord plug on cord-connected luminaires.

Exception No. 3: Where luminaires without cords are plugged directly into the busway and the overcurrent device is mounted on the luminaire.

Exception No. 4: Where the branch-circuit overcurrent plug-in device is directly supplying a readily accessible disconnect, a method of floor operation shall not be required.

(D) Rating of Overcurrent Protection — Branch Circuits. A busway used as a branch circuit shall be protected against overcurrent in accordance with 210.20.

368.30 Support. Busways shall be securely supported at intervals not exceeding 1.5 m (5 ft) unless otherwise designed and marked.

368.56 Branches from Busways. Branches from busways shall be permitted to be made in accordance with 368.56(A), (B), and (C).

(A) General. Branches from busways shall be permitted to use any of the following wiring methods:

(1) Type AC armored cable
(2) Type MC metal-clad cable
(3) Type MI mineral-insulated, metal-sheathed cable
(4) Type IMC intermediate metal conduit
(5) Type RMC rigid metal conduit
(6) Type FMC flexible metal conduit
(7) Type LFMC liquidtight flexible metal conduit
(8) Type PVC rigid polyvinyl chloride conduit
(9) Type RTRC reinforced thermosetting resin conduit
(10) Type LFNC liquidtight flexible nonmetallic conduit
(11) Type EMT electrical metallic tubing
(12) Type ENT electrical nonmetallic tubing
(13) Busways
(14) Strut-type channel raceway
(15) Surface metal raceway
(16) Surface nonmetallic raceway

Where a separate equipment grounding conductor is used, connection of the equipment grounding conductor to the busway shall comply with 250.8 and 250.12.

(B) Cord and Cable Assemblies. Suitable cord and cable assemblies approved for extra-hard usage or hard usage and listed bus drop cable shall be permitted as branches from busways for the connection of portable equipment or the connection of stationary equipment to facilitate their interchange in accordance with 400.10 and 400.12 and the following conditions:

(1) The cord or cable shall be attached to the building by an approved means.
(2) The length of the cord or cable from a busway plug-in device to a suitable tension take-up support device shall not exceed 1.8 m (6 ft).
(3) The cord and cable shall be installed as a vertical riser from the tension take-up support device to the equipment served.
(4) Strain relief cable grips shall be provided for the cord or cable at the busway plug-in device and equipment terminations.

Exception to (B)(2): In industrial establishments only, where the conditions of maintenance and supervision ensure that only qualified persons service the installation, lengths exceeding 1.8 m (6 ft) shall be permitted between the busway plug-in device and the tension take-up support device where the cord or cable is supported at intervals not exceeding 2.5 m (8 ft).

(C) Branches from Trolley-Type Busways. Suitable cord and cable assemblies approved for extra-hard usage or hard usage and listed bus drop cable shall be permitted as branches from trolley-type busways for the connection of movable equipment in accordance with 400.10 and 400.12.

368.58 Dead Ends. A dead end of a busway shall be closed.

368.60 Grounding. Busway shall be connected to an equipment grounding conductor(s), to an equipment bonding jumper, or to the grounded conductor where permitted or required by 250.92(B)(1) or 250.142.

ARTICLE 376
Metal Wireways

PART I. GENERAL

376.2 Definition.

Metal Wireways. Sheet metal troughs with hinged or removable covers for housing and protecting electrical wires and cable and in which conductors are laid in place after the raceway has been installed as a complete system.

PART II. INSTALLATION

376.10 Uses Permitted. The use of metal wireways shall be permitted as follows:

(1) For exposed work.
(2) In any hazardous (classified) location, as permitted by other articles in this *Code*.
(3) In wet locations where wireways are listed for the purpose.
(4) In concealed spaces as an extension that passes transversely through walls, if the length passing through the wall is unbroken. Access to the conductors shall be maintained on both sides of the wall.

376.12 Uses Not Permitted. Metal wireways shall not be used in the following:

(1) Where subject to severe physical damage
(2) Where subject to severe corrosive environments

N **376.20 Conductors Connected in Parallel.** Where single conductor cables comprising each phase, neutral, or grounded conductor of an alternating-current circuit are connected in parallel as permitted in 310.10(H), the conductors shall be installed in groups consisting of not more than one conductor per phase, neutral, or grounded conductor to prevent current imbalance in the paralleled conductors due to inductive reactance.

376.21 Size of Conductors. No conductor larger than that for which the wireway is designed shall be installed in any wireway.

376.22 Number of Conductors and Ampacity. The number of conductors or cables and their ampacity shall comply with 376.22(A) and (B).

(A) Cross-Sectional Areas of Wireway. The sum of the cross-sectional areas of all contained conductors and cables at any cross section of a wireway shall not exceed 20 percent of the interior cross-sectional area of the wireway.

(B) Adjustment Factors. The adjustment factors in 310.15(B)(3)(a) shall be applied only where the number of current-carrying conductors, including neutral conductors classified as current-carrying under the provisions of 310.15(B)(5), exceeds 30 at any cross section of the wireway. Conductors for signaling circuits or controller conductors between a motor and its starter and used only for starting duty shall not be considered as current-carrying conductors.

376.23 Insulated Conductors. Insulated conductors installed in a metal wireway shall comply with 376.23(A) and (B).

(A) Deflected Insulated Conductors. Where insulated conductors are deflected within a metal wireway, either at the ends or where conduits, fittings, or other raceways or cables enter or leave the metal wireway, or where the direction of the metal wireway is deflected greater than 30 degrees, dimensions corresponding to one wire per terminal in Table 312.6(A) shall apply.

(B) Metal Wireways Used as Pull Boxes. Where insulated conductors 4 AWG or larger are pulled through a wireway, the distance between raceway and cable entries enclosing the same conductor shall not be less than that required by 314.28(A)(1) for straight pulls and 314.28(A)(2) for angle pulls. When transposing cable size into raceway size, the minimum metric designator (trade size) raceway required for the number and size of conductors in the cable shall be used.

376.30 Securing and Supporting. Metal wireways shall be supported in accordance with 376.30(A) and (B).

(A) Horizontal Support. Wireways shall be supported where run horizontally at each end and at intervals not to exceed 1.5 m (5 ft) or for individual lengths longer than 1.5 m (5 ft) at each end or joint, unless listed for other support intervals. The distance between supports shall not exceed 3 m (10 ft).

(B) Vertical Support. Vertical runs of wireways shall be securely supported at intervals not exceeding 4.5 m (15 ft) and shall not have more than one joint between supports. Adjoining wireway sections shall be securely fastened together to provide a rigid joint.

376.56 Splices, Taps, and Power Distribution Blocks.

(A) Splices and Taps. Splices and taps shall be permitted within a wireway, provided they are accessible. The conductors, including splices and taps, shall not fill the wireway to more than 75 percent of its area at that point.

(B) Power Distribution Blocks.

(1) Installation. Power distribution blocks installed in metal wireways shall be listed. Power distribution blocks installed on the line side of the service equipment shall be marked "suitable for use on the line side of service equipment" or equivalent.

(2) Size of Enclosure. In addition to the wiring space requirement in 376.56(A), the power distribution block shall be installed in a wireway with dimensions not smaller than specified in the installation instructions of the power distribution block.

(3) Wire Bending Space. Wire bending space at the terminals of power distribution blocks shall comply with 312.6(B).

(4) Live Parts. Power distribution blocks shall not have uninsulated live parts exposed within a wireway, whether or not the wireway cover is installed.

(5) Conductors. Conductors shall be arranged so the power distribution block terminals are unobstructed following installation.

376.58 Dead Ends. Dead ends of metal wireways shall be closed.

376.70 Extensions from Metal Wireways. Extensions from wireways shall be made with cord pendants installed in accordance with 400.14 or with any wiring method in Chapter 3 that includes a means for equipment grounding. Where a separate equipment grounding conductor is employed, connection of the equipment grounding conductors in the wiring method to the wireway shall comply with 250.8 and 250.12.

ARTICLE 378
Nonmetallic Wireways

PART I. GENERAL

378.2 Definition.

Nonmetallic Wireways. Flame-retardant, nonmetallic troughs with removable covers for housing and protecting electrical wires and cables in which conductors are laid in place after the raceway has been installed as a complete system.

378.6 Listing Requirements. Nonmetallic wireways and associated fittings shall be listed.

PART II. INSTALLATION

378.10 Uses Permitted. The use of nonmetallic wireways shall be permitted in the following:

(1) Only for exposed work, except as permitted in 378.10(4).
(2) Where subject to corrosive environments where identified for the use.
(3) In wet locations where listed for the purpose.
(4) As extensions to pass transversely through walls if the length passing through the wall is unbroken. Access to the conductors shall be maintained on both sides of the wall.

378.12 Uses Not Permitted. Nonmetallic wireways shall not be used in the following:

(1) Where subject to physical damage
(2) In any hazardous (classified) location, except as permitted by other articles in this *Code*
(3) Where exposed to sunlight unless listed and marked as suitable for the purpose
(4) Where subject to ambient temperatures other than those for which nonmetallic wireway is listed
(5) For conductors whose insulation temperature limitations would exceed those for which the nonmetallic wireway is listed

N **378.20 Conductors Connected in Parallel.** Where single conductor cables comprising each phase, neutral, or grounded conductor of an alternating-current circuit are connected in parallel as permitted in 310.10(H), the conductors shall be installed in groups consisting of not more than one conductor per phase, neutral, or grounded conductor to prevent current imbalance in the paralleled conductors due to inductive reactance.

378.21 Size of Conductors. No conductor larger than that for which the nonmetallic wireway is designed shall be installed in any nonmetallic wireway.

378.22 Number of Conductors. The sum of cross-sectional areas of all contained conductors or cables at any cross section of the nonmetallic wireway shall not exceed 20 percent of the interior cross-sectional area of the nonmetallic wireway. Conductors for signaling circuits or controller conductors between a motor and its starter and used only for starting duty shall not be considered as current-carrying conductors.

The adjustment factors specified in 310.15(B)(3)(a) shall be applicable to the current-carrying conductors up to and including the 20 percent fill specified in the first paragraph of this section.

378.23 Insulated Conductors. Insulated conductors installed in a nonmetallic wireway shall comply with 378.23(A) and (B).

(A) Deflected Insulated Conductors. Where insulated conductors are deflected within a nonmetallic wireway, either at the ends or where conduits, fittings, or other raceways or cables enter or leave the nonmetallic wireway, or where the direction of the nonmetallic wireway is deflected greater than 30 degrees, dimensions corresponding to one wire per terminal in Table 312.6(A) shall apply.

(B) Nonmetallic Wireways Used as Pull Boxes. Where insulated conductors 4 AWG or larger are pulled through a wireway, the distance between raceway and cable entries enclosing the same conductor shall not be less than that required in 314.28(A)(1) for straight pulls and in 314.28(A)(2) for angle pulls. When transposing cable size into raceway size, the minimum metric designator (trade size) raceway required for the number and size of conductors in the cable shall be used.

378.30 Securing and Supporting. Nonmetallic wireway shall be supported in accordance with 378.30(A) and (B).

(A) Horizontal Support. Nonmetallic wireways shall be supported where run horizontally at intervals not to exceed 900 mm (3 ft), and at each end or joint, unless listed for other support intervals. In no case shall the distance between supports exceed 3 m (10 ft).

(B) Vertical Support. Vertical runs of nonmetallic wireway shall be securely supported at intervals not exceeding 1.2 m (4 ft), unless listed for other support intervals, and shall not have more than one joint between supports. Adjoining nonmetallic wireway sections shall be securely fastened together to provide a rigid joint.

378.44 Expansion Fittings. Expansion fittings for nonmetallic wireway shall be provided to compensate for thermal expansion and contraction where the length change is expected to be 6 mm (0.25 in.) or greater in a straight run.

378.56 Splices and Taps. Splices and taps shall be permitted within a nonmetallic wireway, provided they are accessible. The conductors, including splices and taps, shall not fill the nonmetallic wireway to more than 75 percent of its area at that point.

378.58 Dead Ends. Dead ends of nonmetallic wireway shall be closed using listed fittings.

378.60 Grounding. Where equipment grounding is required, a separate equipment grounding conductor shall be installed in the nonmetallic wireway. A separate equipment grounding conductor shall not be required where the grounded conductor is used to ground equipment as permitted in 250.142.

ARTICLE 386
Surface Metal Raceways

PART I. GENERAL

386.2 Definition.

Surface Metal Raceway. A metal raceway that is intended to be mounted to the surface of a structure, with associated couplings, connectors, boxes, and fittings for the installation of electrical conductors.

386.6 Listing Requirements. Surface metal raceway and associated fittings shall be listed.

PART II. INSTALLATION

386.10 Uses Permitted. The use of surface metal raceways shall be permitted in the following:

(1) In dry locations.
(2) In Class I, Division 2 hazardous (classified) locations as permitted in 501.10(B)(3).
(3) Under raised floors, as permitted in 645.5(E)(2).
(4) Extension through walls and floors. Surface metal raceway shall be permitted to pass transversely through dry walls, dry partitions, and dry floors if the length passing through is unbroken. Access to the conductors shall be maintained on both sides of the wall, partition, or floor.

386.12 Uses Not Permitted. Surface metal raceways shall not be used in the following:

(1) Where subject to severe physical damage, unless otherwise approved
(2) Where the voltage is 300 volts or more between conductors, unless the metal has a thickness of not less than 1.02 mm (0.040 in.) nominal

(3) Where subject to corrosive vapors
(4) In hoistways
(5) Where concealed, except as permitted in 386.10

386.21 Size of Conductors. No conductor larger than that for which the raceway is designed shall be installed in surface metal raceway.

386.22 Number of Conductors or Cables. The number of conductors or cables installed in surface metal raceway shall not be greater than the number for which the raceway is designed. Cables shall be permitted to be installed where such use is not prohibited by the respective cable articles.

The adjustment factors of 310.15(B)(3)(a) shall not apply to conductors installed in surface metal raceways where all of the following conditions are met:

(1) The cross-sectional area of the raceway exceeds 2500 mm^2 (4 $in.^2$).
(2) The current-carrying conductors do not exceed 30 in number.
(3) The sum of the cross-sectional areas of all contained conductors does not exceed 20 percent of the interior cross-sectional area of the surface metal raceway.

386.30 Securing and Supporting. Surface metal raceways and associated fittings shall be supported in accordance with the manufacturer's installation instructions.

386.56 Splices and Taps. Splices and taps shall be permitted in surface metal raceways having a removable cover that is accessible after installation. The conductors, including splices and taps, shall not fill the raceway to more than 75 percent of its area at that point. Splices and taps in surface metal raceways without removable covers shall be made only in boxes. All splices and taps shall be made by approved methods.

Taps of Type FC cable installed in surface metal raceway shall be made in accordance with 322.56(B).

386.60 Grounding. Surface metal raceway enclosures providing a transition from other wiring methods shall have a means for connecting an equipment grounding conductor.

ARTICLE 388

Surface Nonmetallic Raceways

PART I. GENERAL

388.2 Definition.

Surface Nonmetallic Raceway. A nonmetallic raceway that is intended to be mounted to the surface of a structure, with associated couplings, connectors, boxes, and fittings for the installation of electrical conductors.

388.6 Listing Requirements. Surface nonmetallic raceway and associated fittings shall be listed.

PART II. INSTALLATION

388.10 Uses Permitted. Surface nonmetallic raceways shall be permitted as follows:

(1) The use of surface nonmetallic raceways shall be permitted in dry locations.
(2) Extension through walls and floors shall be permitted. Surface nonmetallic raceway shall be permitted to pass transversely through dry walls, dry partitions, and dry floors if the length passing through is unbroken. Access to the conductors shall be maintained on both sides of the wall, partition, or floor.

388.12 Uses Not Permitted. Surface nonmetallic raceways shall not be used in the following:

(1) Where concealed, except as permitted in 388.10(2)
(2) Where subject to severe physical damage
(3) Where the voltage is 300 volts or more between conductors, unless listed for higher voltage
(4) In hoistways
(5) In any hazardous (classified) location, except as permitted by other articles in this *Code*
(6) Where subject to ambient temperatures exceeding those for which the nonmetallic raceway is listed

(7) For conductors whose insulation temperature limitations would exceed those for which the nonmetallic raceway is listed

388.21 Size of Conductors. No conductor larger than that for which the raceway is designed shall be installed in surface nonmetallic raceway.

388.22 Number of Conductors or Cables. The number of conductors or cables installed in surface nonmetallic raceway shall not be greater than the number for which the raceway is designed. Cables shall be permitted to be installed where such use is not prohibited by the respective cable articles.

388.30 Securing and Supporting. Surface nonmetallic raceways and associated fittings shall be supported in accordance with the manufacturer's installation instructions.

388.56 Splices and Taps. Splices and taps shall be permitted in surface nonmetallic raceways having a cover capable of being opened in place that is accessible after installation. The conductors, including splices and taps, shall not fill the raceway to more than 75 percent of its area at that point. Splices and taps in surface nonmetallic raceways without covers capable of being opened in place shall be made only in boxes. All splices and taps shall be made by approved methods.

388.60 Grounding. Where equipment grounding is required, a separate equipment grounding conductor shall be installed in the raceway.

388.70 Combination Raceways. When combination surface nonmetallic raceways are used both for signaling and for lighting and power circuits, the different systems shall be run in separate compartments identified by stamping, imprinting, or color coding of the interior finish.

CHAPTER 14

WIRING METHODS AND CONDUCTORS

INTRODUCTION

This chapter contains requirements from Articles 300 and 310. Article 300 covers wiring methods for all electrical installations, unless modified by other articles, and provides general installation requirements for cables, raceways, and other wiring systems. Some of the topics include conductors; protection against physical damage; underground installations; raceways exposed to different temperatures; securing and supporting; boxes, conduit bodies, and fittings; supporting conductors in vertical raceways; and induced currents in metal enclosures or metal raceways.

Article 310 provides the general requirements for conductors, including their ampacity ratings. For many commercial and industrial installations, the allowable ampacities for insulated conductors specified in Table 310.15(B)(16) are used. For some installations, the allowable ampacity of each conductor must be reduced because of the number of current carrying conductors in a raceway or cable, or because of the ambient temperature where the wiring method is installed. For some applications the conductor ampacity must be reduced for both of these conditions of use, both of which impact the operating temperature of insulated conductors.

ARTICLE 300

General Requirements for Wiring Methods and Materials

PART I. GENERAL REQUIREMENTS

300.1 Scope.

(B) Integral Parts of Equipment. The provisions of this article are not intended to apply to the conductors that form an integral part of equipment, such as motors, controllers, motor control centers, or factory-assembled control equipment or listed utilization equipment.

300.2 Limitations.

(A) Voltage. Wiring methods specified in Chapter 3 shall be used for 1000 volts, nominal, or less where not specifically limited in some section of Chapter 3. They shall be permitted for over 1000 volts, nominal, where specifically permitted elsewhere in this *Code*.

300.3 Conductors.

(A) Single Conductors. Single conductors specified in Table 310.104(A) shall only be installed where part of a recognized wiring method of Chapter 3.

(B) Conductors of the Same Circuit. All conductors of the same circuit and, where used, the grounded conductor and all equipment grounding conductors and bonding conductors shall be contained within the same raceway, auxiliary gutter, cable tray, cablebus assembly, trench, cable, or cord, unless otherwise permitted in accordance with 300.3(B)(1) through (B)(4).

(1) Paralleled Installations. Conductors shall be permitted to be run in parallel in accordance with the provisions of 310.10(H). The requirement to run all circuit conductors within the same raceway, auxiliary gutter, cable tray, trench, cable, or cord shall apply separately to each portion of the paralleled installation, and the equipment grounding conductors shall comply with the provisions of 250.122. Parallel runs in cable tray shall comply with the provisions of 392.20(C).

Exception: Conductors installed in nonmetallic raceways run underground shall be permitted to be arranged as isolated phase, neutral, and grounded conductor installations. The raceways shall be installed in close proximity, and the isolated phase, neutral, and grounded conductors shall comply with the provisions of 300.20(B).

(2) Grounding and Bonding Conductors. Equipment grounding conductors shall be permitted to be installed outside a raceway or cable assembly where in accordance with the provisions of 250.130(C) for certain existing installations or in accordance with 250.134(B), Exception No. 2, for dc circuits. Equipment bonding conductors shall be permitted to be installed on the outside of raceways in accordance with 250.102(E).

(C) Conductors of Different Systems.

(1) 1000 Volts, Nominal, or Less. Conductors of ac and dc circuits, rated 1000 volts, nominal, or less, shall be permitted to occupy the same equipment wiring enclosure, cable, or raceway. All conductors shall have an insulation rating equal to at least the maximum circuit voltage applied to any conductor within the enclosure, cable, or raceway.

Secondary wiring to electric-discharge lamps of 1000 volts or less, if insulated for the secondary voltage involved, shall be permitted to occupy the same luminaire, sign, or outline lighting enclosure as the branch-circuit conductors.

(2) Over 1000 Volts, Nominal. Conductors of circuits rated over 1000 volts, nominal, shall not occupy the same equipment wiring enclosure, cable, or raceway with conductors of circuits rated 1000 volts, nominal, or less unless otherwise permitted in 300.3(C)(2)(a) through 300.3(C)(2)(d).

(a) Primary leads of electric-discharge lamp ballasts insulated for the primary voltage of the ballast, where contained within the individual wiring enclosure, shall be permitted to occupy the same luminaire, sign, or outline lighting enclosure as the branch-circuit conductors.

(b) Excitation, control, relay, and ammeter conductors used in connection with any individual motor or starter shall be permitted to occupy the same enclosure as the motor-circuit conductors.

(c) In motors, transformers, switchgear, switchboards, control assemblies, and similar equipment, conductors of different voltage ratings shall be permitted.

(d) In manholes, if the conductors of each system are permanently and effectively separated from the conductors of the other systems and securely fastened to racks, insulators, or other approved supports, conductors of different voltage ratings shall be permitted.

Conductors having nonshielded insulation and operating at different voltage levels shall not occupy the same enclosure, cable, or raceway.

300.4 Protection Against Physical Damage. Where subject to physical damage, conductors, raceways, and cables shall be protected.

(A) Cables and Raceways Through Wood Members.

(1) Bored Holes. In both exposed and concealed locations, where a cable- or raceway-type wiring method is installed through bored holes in joists, rafters, or wood members, holes shall be bored so that the edge of the hole is not less than 32 mm (1¼ in.) from the nearest edge of the wood member. Where this distance cannot be maintained, the cable or raceway shall be protected from penetration by screws or nails by a steel plate(s) or bushing(s), at least 1.6 mm (1⁄16 in.) thick, and of appropriate length and width installed to cover the area of the wiring.

Exception No. 1: Steel plates shall not be required to protect rigid metal conduit, intermediate metal conduit, rigid nonmetallic conduit, or electrical metallic tubing.

Exception No. 2: A listed and marked steel plate less than 1.6 mm (1⁄16 in.) thick that provides equal or better protection against nail or screw penetration shall be permitted.

(2) Notches in Wood. Where there is no objection because of weakening the building structure, in both exposed and concealed locations, cables or raceways shall be permitted to be laid in notches in wood studs, joists, rafters, or other wood members where the cable or raceway at those points is protected against nails or screws by a steel plate at least 1.6 mm (1/16 in.) thick, and of appropriate length and width, installed to cover the area of the wiring. The steel plate shall be installed before the building finish is applied.

Exception No. 1: Steel plates shall not be required to protect rigid metal conduit, intermediate metal conduit, rigid nonmetallic conduit, or electrical metallic tubing.

Exception No. 2: A listed and marked steel plate less than 1.6 mm (1/16 in.) thick that provides equal or better protection against nail or screw penetration shall be permitted.

(B) Nonmetallic-Sheathed Cables and Electrical Nonmetallic Tubing Through Metal Framing Members.

(1) Nonmetallic-Sheathed Cable. In both exposed and concealed locations where nonmetallic-sheathed cables pass through either factory- or field-punched, cut, or drilled slots or holes in metal members, the cable shall be protected by listed bushings or listed grommets covering all metal edges that are securely fastened in the opening prior to installation of the cable.

(2) Nonmetallic-Sheathed Cable and Electrical Nonmetallic Tubing. Where nails or screws are likely to penetrate nonmetallic-sheathed cable or electrical nonmetallic tubing, a steel sleeve, steel plate, or steel clip not less than 1.6 mm (1/16 in.) in thickness shall be used to protect the cable or tubing.

Exception: A listed and marked steel plate less than 1.6 mm (1/16 in.) thick that provides equal or better protection against nail or screw penetration shall be permitted.

(C) Cables Through Spaces Behind Panels Designed to Allow Access. Cables or raceway-type wiring methods, installed behind panels designed to allow access, shall be supported according to their applicable articles.

(D) Cables and Raceways Parallel to Framing Members and Furring Strips. In both exposed and concealed locations, where a cable- or raceway-type wiring method is installed parallel to framing members, such as joists, rafters, or studs, or is installed parallel to furring strips, the cable or raceway shall be installed and supported so that the nearest outside surface of the cable or raceway is not less than 32 mm (1¼ in.) from the nearest edge of the framing member or furring strips where nails or screws are likely to penetrate. Where this distance cannot be maintained, the cable or raceway shall be protected from penetration by nails or screws by a steel plate, sleeve, or equivalent at least 1.6 mm (1⁄16 in.) thick.

Exception No. 1: Steel plates, sleeves, or the equivalent shall not be required to protect rigid metal conduit, intermediate metal conduit, rigid nonmetallic conduit, or electrical metallic tubing.

Exception No. 2: For concealed work in finished buildings, or finished panels for prefabricated buildings where such supporting is impracticable, it shall be permissible to fish the cables between access points.

Exception No. 3: A listed and marked steel plate less than 1.6 mm (1⁄16 in.) thick that provides equal or better protection against nail or screw penetration shall be permitted.

(E) Cables, Raceways, or Boxes Installed in or Under Roof Decking. A cable, raceway, or box, installed in exposed or concealed locations under metal-corrugated sheet roof decking, shall be installed and supported so there is not less than 38 mm (1½ in.) measured from the lowest surface of the roof decking to the top of the cable, raceway, or box. A cable, raceway, or box shall not be installed in concealed locations in metal-corrugated, sheet decking–type roof.

Exception: Rigid metal conduit and intermediate metal conduit shall not be required to comply with 300.4(E).

(F) Cables and Raceways Installed in Shallow Grooves. Cable- or raceway-type wiring methods installed in a groove, to be covered by wallboard, siding, paneling, carpeting, or similar finish, shall be protected by 1.6 mm (1⁄16 in.) thick steel plate, sleeve, or equivalent or by not less than 32-mm (1¼-in.) free space for the full length of the groove in which the cable or raceway is installed.

Exception No. 1: Steel plates, sleeves, or the equivalent shall not be required to protect rigid metal conduit, intermediate metal conduit, rigid nonmetallic conduit, or electrical metallic tubing.

Exception No. 2: A listed and marked steel plate less than 1.6 mm (1/16 in.) thick that provides equal or better protection against nail or screw penetration shall be permitted.

(G) Insulated Fittings. Where raceways contain 4 AWG or larger insulated circuit conductors, and these conductors enter a cabinet, a box, an enclosure, or a raceway, the conductors shall be protected by an identified fitting providing a smoothly rounded insulating surface, unless the conductors are separated from the fitting or raceway by identified insulating material that is securely fastened in place.

Exception: Where threaded hubs or bosses that are an integral part of a cabinet, box, enclosure, or raceway provide a smoothly rounded or flared entry for conductors.

Conduit bushings constructed wholly of insulating material shall not be used to secure a fitting or raceway. The insulating fitting or insulating material shall have a temperature rating not less than the insulation temperature rating of the installed conductors.

(H) Structural Joints. A listed expansion/deflection fitting or other approved means shall be used where a raceway crosses a structural joint intended for expansion, contraction or deflection, used in buildings, bridges, parking garages, or other structures.

300.5 Underground Installations.

(A) Minimum Cover Requirements. Direct-buried cable, conduit, or other raceways shall be installed to meet the minimum cover requirements of Table 300.5.

(B) Wet Locations. The interior of enclosures or raceways installed underground shall be considered to be a wet location. Insulated conductors and cables installed in these enclosures or raceways in underground installations shall comply with 310.10(C).

Table 300.5 Minimum Cover Requirements, 0 to 1000 Volts, Nominal, Burial in Millimeters (Inches)

Location of Wiring Method or Circuit	Type of Wiring Method or Circuit: Column 1 Direct Burial Cables or Conductors		Column 2 Rigid Metal Conduit or Intermediate Metal Conduit		Column 3 Nonmetallic Raceways Listed for Direct Burial Without Concrete Encasement or Other Approved Raceways		Column 4 Residential Branch Circuits Rated 120 Volts or Less with GFCI Protection and Maximum Overcurrent Protection of 20 Amperes		Column 5 Circuits for Control of Irrigation and Landscape Lighting Limited to Not More Than 30 Volts and Installed with Type UF or in Other Identified Cable or Raceway	
	mm	in.	mm	in.	mm	in.	mm	in.	mm	in.
All locations not specified below	600	24	150	6	450	18	300	12	150[a, b]	6[a, b]
In trench below 50 mm (2 in.) thick concrete or equivalent	450	18	150	6	300	12	150	6	150	6
Under a building	0 (in raceway or Type MC or Type MI cable identified for direct burial)	0	0	0	0	0	0 (in raceway or Type MC or Type MI cable identified for direct burial)	0	0 (in raceway or Type MC or Type MI cable identified for direct burial)	0
Under minimum of 102 mm (4 in.) thick concrete exterior slab with no vehicular traffic and the slab extending not less than 152 mm (6 in.) beyond the underground installation	450	18	100	4	100	4	150 (direct burial) 100 (in raceway)	6 4	150 (direct burial) 100 (in raceway)	6 4

Under streets, highways, roads, alleys, driveways, and parking lots	600	24	600	24	600	24	600	24	600	24
One- and two-family dwelling driveways and outdoor parking areas, and used only for dwelling-related purposes	450	18	450	18	450	18	300	12	450	18
In or under airport runways, including adjacent areas where trespassing prohibited	450	18	450	18	450	18	450	18	450	18

[a]A lesser depth shall be permitted where specified in the installation instructions of a listed low-voltage lighting system.
[b]A depth of 150 mm (6 in.) shall be permitted for pool, spa, and fountain lighting, installed in a nonmetallic raceway, limited to not more than 30 volts where part of a listed low-voltage lighting system.

Notes:
1. Cover is defined as the shortest distance in mm (in.) measured between a point on the top surface of any direct-buried conductor, cable, conduit, or other raceway and the top surface of finished grade, concrete, or similar cover.
2. Raceways approved for burial only where concrete encased shall require concrete envelope not less than 50 mm (2 in.) thick.
3. Lesser depths shall be permitted where cables and conductors rise for terminations or splices or where access is otherwise required.
4. Where one of the wiring method types listed in Columns 1 through 3 is used for one of the circuit types in Columns 4 and 5, the shallowest depth of burial shall be permitted.
5. Where solid rock prevents compliance with the cover depths specified in this table, the wiring shall be installed in a metal raceway, or a nonmetallic raceway permitted for direct burial. The raceways shall be covered by a minimum of 50 mm (2 in.) of concrete extending down to rock.

(C) Underground Cables and Conductors Under Buildings. Underground cable and conductors installed under a building shall be in a raceway.

Exception No. 1: Type MI cable shall be permitted under a building without installation in a raceway where embedded in concrete, fill, or other masonry in accordance with 332.10(6) or in underground runs where suitably protected against physical damage and corrosive conditions in accordance with 332.10(10).

Exception No. 2: Type MC cable listed for direct burial or concrete encasement shall be permitted under a building without installation in a raceway in accordance with 330.10(A)(5) and in wet locations in accordance with 330.10(A)(11).

(D) Protection from Damage. Direct-buried conductors and cables shall be protected from damage in accordance with 300.5(D)(1) through (D)(4).

(1) Emerging from Grade. Direct-buried conductors and cables emerging from grade and specified in columns 1 and 4 of Table 300.5 shall be protected by enclosures or raceways extending from the minimum cover distance below grade required by 300.5(A) to a point at least 2.5 m (8 ft) above finished grade. In no case shall the protection be required to exceed 450 mm (18 in.) below finished grade.

(2) Conductors Entering Buildings. Conductors entering a building shall be protected to the point of entrance.

(3) Service Conductors. Underground service conductors that are not encased in concrete and that are buried 450 mm (18 in.) or more below grade shall have their location identified by a warning ribbon that is placed in the trench at least 300 mm (12 in.) above the underground installation.

(4) Enclosure or Raceway Damage. Where the enclosure or raceway is subject to physical damage, the conductors shall be installed in electrical metallic tubing, rigid metal conduit, intermediate metal conduit, RTRC-XW, Schedule 80 PVC conduit, or equivalent.

(E) Splices and Taps. Direct-buried conductors or cables shall be permitted to be spliced or tapped without the use of splice boxes. The splices or taps shall be made in accordance with 110.14(B).

(F) Backfill. Backfill that contains large rocks, paving materials, cinders, large or sharply angular substances, or corrosive material shall not be placed in an excavation where materials may damage raceways, cables, conductors, or other substructures or prevent adequate compaction of fill or contribute to corrosion of raceways, cables, or other substructures.

Where necessary to prevent physical damage to the raceway, cable, or conductor, protection shall be provided in the form of granular or selected material, suitable running boards, suitable sleeves, or other approved means.

(G) Raceway Seals. Conduits or raceways through which moisture may contact live parts shall be sealed or plugged at either or both ends. Spare or unused raceways shall also be sealed. Sealants shall be identified for use with the cable insulation, conductor insulation, bare conductor, shield, or other components.

(H) Bushing. A bushing, or terminal fitting, with an integral bushed opening shall be used at the end of a conduit or other raceway that terminates underground where the conductors or cables emerge as a direct burial wiring method. A seal incorporating the physical protection characteristics of a bushing shall be permitted to be used in lieu of a bushing.

(I) Conductors of the Same Circuit. All conductors of the same circuit and, where used, the grounded conductor and all equipment grounding conductors shall be installed in the same raceway or cable or shall be installed in close proximity in the same trench.

Exception No. 1: Conductors shall be permitted to be installed in parallel in raceways, multiconductor cables, or direct-buried single conductor cables. Each raceway or multiconductor cable shall contain all conductors of the same circuit, including equipment grounding conductors. Each direct-buried single conductor cable shall be located in close proximity in the trench to the other single conductor cables in the same parallel set of conductors in the circuit, including equipment grounding conductors.

Exception No. 2: Isolated phase, polarity, grounded conductor, and equipment grounding and bonding conductor installations shall be permitted in nonmetallic raceways or cables with a nonmetallic covering or nonmagnetic sheath in close proximity where conductors are paralleled as permitted in 310.10(H), and where the conditions of 300.20(B) are met.

(J) Earth Movement. Where direct-buried conductors, raceways, or cables are subject to movement by settlement or frost, direct-buried conductors, raceways, or cables shall be arranged so as to prevent damage to the enclosed conductors or to equipment connected to the raceways.

(K) Directional Boring. Cables or raceways installed using directional boring equipment shall be approved for the purpose.

300.6 Protection Against Corrosion and Deterioration. Raceways, cable trays, cablebus, auxiliary gutters, cable armor, boxes, cable sheathing, cabinets, elbows, couplings, fittings, supports, and support hardware shall be of materials suitable for the environment in which they are to be installed.

(A) Ferrous Metal Equipment. Ferrous metal raceways, cable trays, cablebus, auxiliary gutters, cable armor, boxes, cable sheathing, cabinets, metal elbows, couplings, nipples, fittings, supports, and support hardware shall be suitably protected against corrosion inside and outside (except threads at joints) by a coating of approved corrosion-resistant material. Where corrosion protection is necessary and the conduit is threaded in the field, the threads shall be coated with an approved electrically conductive, corrosion-resistant compound.

Exception: Stainless steel shall not be required to have protective coatings.

(1) Protected from Corrosion Solely by Enamel. Where protected from corrosion solely by enamel, ferrous metal raceways, cable trays, cablebus, auxiliary gutters, cable armor, boxes, cable sheathing, cabinets, metal elbows, couplings, nipples, fittings, supports, and support hardware shall not be used outdoors or in wet locations as described in 300.6(D).

(2) Organic Coatings on Boxes or Cabinets. Where boxes or cabinets have an approved system of organic coatings and are marked "Raintight," "Rainproof," or "Outdoor Type," they shall be permitted outdoors.

(3) In Concrete or in Direct Contact with the Earth. Ferrous metal raceways, cable armor, boxes, cable sheathing, cabinets, elbows, couplings, nipples, fittings, supports, and support hardware shall be permitted to be installed in concrete or in direct contact with the earth, or in areas subject to severe corrosive influences where made of material approved for the condition, or where provided with corrosion protection approved for the condition.

(B) Aluminum Metal Equipment. Aluminum raceways, cable trays, cablebus, auxiliary gutters, cable armor, boxes, cable sheathing, cabinets, elbows, couplings, nipples, fittings, supports, and support hardware embedded or encased in concrete or in direct contact with the earth shall be provided with supplementary corrosion protection.

(C) Nonmetallic Equipment. Nonmetallic raceways, cable trays, cablebus, auxiliary gutters, boxes, cables with a nonmetallic outer jacket and internal metal armor or jacket, cable sheathing, cabinets, elbows, couplings, nipples, fittings, supports, and support hardware shall be made of material approved for the condition and shall comply with (C)(1) and (C)(2) as applicable to the specific installation.

(1) Exposed to Sunlight Where exposed to sunlight, the materials shall be listed as sunlight resistant or shall be identified as sunlight resistant.

(2) Chemical Exposure. Where subject to exposure to chemical solvents, vapors, splashing, or immersion, materials or coatings shall either be inherently resistant to chemicals based on their listing or be identified for the specific chemical reagent.

(D) Indoor Wet Locations. In portions of dairy processing facilities, laundries, canneries, and other indoor wet locations, and in locations where walls are frequently washed or where there are surfaces of absorbent materials, such as damp paper or wood, the entire wiring system, where installed exposed, including all boxes, fittings, raceways, and cable used therewith, shall be mounted so that there is at least a 6-mm (¼-in.) airspace between it and the wall or supporting surface.

Exception: Nonmetallic raceways, boxes, and fittings shall be permitted to be installed without the airspace on a concrete, masonry, tile, or similar surface.

300.7 Raceways Exposed to Different Temperatures.

(A) Sealing. Where portions of a raceway or sleeve are known to be subjected to different temperatures, and where condensation is known to be a problem, as in cold storage areas of buildings or where passing from the interior to the exterior of a building, the raceway or sleeve shall be filled with an approved material to prevent the circulation of warm air to a colder section of the raceway or sleeve. An explosionproof seal shall not be required for this purpose.

(B) Expansion, Expansion-Deflection, and Deflection Fittings. Raceways shall be provided with expansion, expansion-deflection, or deflection fittings where necessary to compensate for thermal expansion, deflection, and contraction.

300.8 Installation of Conductors with Other Systems. Raceways or cable trays containing electrical conductors shall not contain any pipe, tube, or equal for steam, water, air, gas, drainage, or any service other than electrical.

300.9 Raceways in Wet Locations Abovegrade. Where raceways are installed in wet locations abovegrade, the interior of these raceways shall be considered to be a wet location. Insulated conductors and cables installed in raceways in wet locations abovegrade shall comply with 310.10(C).

300.10 Electrical Continuity of Metal Raceways and Enclosures. Metal raceways, cable armor, and other metal enclosures for conductors shall be metallically joined together into a continuous electrical conductor and shall be connected to all boxes, fittings, and cabinets so as to provide effective electrical continuity. Unless specifically permitted elsewhere in this *Code*, raceways and cable assemblies shall be mechanically secured to boxes, fittings, cabinets, and other enclosures.

Exception No. 1: Short sections of raceways used to provide support or protection of cable assemblies from physical damage shall not be required to be made electrically continuous.

Exception No. 2: Equipment enclosures to be isolated, as permitted by 250.96(B), shall not be required to be metallically joined to the metal raceway.

300.11 Securing and Supporting.

(A) Secured in Place. Raceways, cable assemblies, boxes, cabinets, and fittings shall be securely fastened in place.

(B) Wiring Systems Installed Above Suspended Ceilings. Support wires that do not provide secure support shall not be permitted as the sole support. Support wires and associated fittings that provide secure support and that are installed in addition to the ceiling grid support wires shall be permitted as the sole support. Where independent support wires are used, they shall be secured at both ends. Cables and raceways shall not be supported by ceiling grids.

(1) Fire-Rated Assemblies. Wiring located within the cavity of a fire-rated floor–ceiling or roof–ceiling assembly shall not be secured to, or supported by, the ceiling assembly, including the ceiling support wires. An independent means of secure support shall be provided and shall be permitted to be attached to the assembly. Where independent support wires are used, they shall be distinguishable by color, tagging, or other effective means from those that are part of the fire-rated design.

Exception: The ceiling support system shall be permitted to support wiring and equipment that have been tested as part of the fire-rated assembly.

(2) Non–Fire-Rated Assemblies. Wiring located within the cavity of a non–fire-rated floor–ceiling or roof–ceiling assembly shall not be secured to, or supported by, the ceiling assembly, including the ceiling support wires. An independent means of secure support shall be provided and shall be permitted to be attached to the assembly. Where independent support wires are used, they shall be distinguishable by color, tagging, or other effective means.

Exception: The ceiling support system shall be permitted to support branch-circuit wiring and associated equipment where installed in accordance with the ceiling system manufacturer's instructions.

(C) Raceways Used as Means of Support. Raceways shall be used only as a means of support for other raceways, cables, or nonelectrical equipment under any of the following conditions:

(1) Where the raceway or means of support is identified as a means of support
(2) Where the raceway contains power supply conductors for electrically controlled equipment and is used to support Class 2 circuit conductors or cables that are solely for the purpose of connection to the equipment control circuits
(3) Where the raceway is used to support boxes or conduit bodies in accordance with 314.23 or to support luminaires in accordance with 410.36(E)

(D) Cables Not Used as Means of Support. Cable wiring methods shall not be used as a means of support for other cables, raceways, or nonelectrical equipment.

300.12 Mechanical Continuity — Raceways and Cables. Raceways, cable armors, and cable sheaths shall be continuous between cabinets, boxes, fittings, or other enclosures or outlets.

Exception No. 1: Short sections of raceways used to provide support or protection of cable assemblies from physical damage shall not be required to be mechanically continuous.

Exception No. 2: Raceways and cables installed into the bottom of open bottom equipment, such as switchboards, motor control centers, and floor or pad-mounted transformers, shall not be required to be mechanically secured to the equipment.

300.13 Mechanical and Electrical Continuity — Conductors.

(A) General. Conductors in raceways shall be continuous between outlets, boxes, devices, and so forth. There shall be no splice or tap within a raceway unless permitted by 300.15; 368.56(A); 376.56; 378.56; 384.56; 386.56; 388.56; or 390.7.

(B) Device Removal. In multiwire branch circuits, the continuity of a grounded conductor shall not depend on device connections such as lampholders, receptacles, and so forth, where the removal of such devices would interrupt the continuity.

300.14 Length of Free Conductors at Outlets, Junctions, and Switch Points. At least 150 mm (6 in.) of free conductor, measured from the point in the box where it emerges from its raceway or cable sheath, shall be left at each outlet, junction, and switch point for splices or the connection of luminaires or devices. Where the opening to an outlet, junction, or switch point is less than 200 mm (8 in.) in any dimension, each conductor shall be long enough to extend at least 75 mm (3 in.) outside the opening.

Exception: Conductors that are not spliced or terminated at the outlet, junction, or switch point shall not be required to comply with 300.14.

300.15 Boxes, Conduit Bodies, or Fittings — Where Required. A box shall be installed at each outlet and switch point for concealed knob-and-tube wiring.

Fittings and connectors shall be used only with the specific wiring methods for which they are designed and listed.

Where the wiring method is conduit, tubing, Type AC cable, Type MC cable, Type MI cable, nonmetallic-sheathed cable, or other cables, a box or conduit body shall be installed at each conductor splice point, outlet point, switch point, junction point, termination point, or pull point, unless otherwise permitted in 300.15(A) through (L).

(A) Wiring Methods with Interior Access. A box or conduit body shall not be required for each splice, junction, switch, pull, termination, or outlet points in wiring methods with removable covers, such as wireways, multioutlet assemblies, auxiliary gutters, and surface raceways. The covers shall be accessible after installation.

(B) Equipment. An integral junction box or wiring compartment as part of approved equipment shall be permitted in lieu of a box.

(C) Protection. A box or conduit body shall not be required where cables enter or exit from conduit or tubing that is used to provide cable support or protection against physical damage. A fitting shall be provided on the end(s) of the conduit or tubing to protect the cable from abrasion.

(D) Type MI Cable. A box or conduit body shall not be required where accessible fittings are used for straight-through splices in mineral-insulated metal-sheathed cable.

(E) Integral Enclosure. A wiring device with integral enclosure identified for the use, having brackets that securely fasten the device to walls or ceilings of conventional on-site frame construction, for use with nonmetallic-sheathed cable, shall be permitted in lieu of a box or conduit body.

(F) Fitting. A fitting identified for the use shall be permitted in lieu of a box or conduit body where conductors are not spliced or terminated within the fitting. The fitting shall be accessible after installation.

(G) Direct-Buried Conductors. As permitted in 300.5(E), a box or conduit body shall not be required for splices and taps in direct-buried conductors and cables.

(H) Insulated Devices. As permitted in 334.40(B), a box or conduit body shall not be required for insulated devices supplied by nonmetallic-sheathed cable.

(I) Enclosures. A box or conduit body shall not be required where a splice, switch, terminal, or pull point is in a cabinet or cutout box, in an enclosure for a switch or overcurrent device as permitted in 312.8, in a motor controller as permitted in 430.10(A), or in a motor control center.

(J) Luminaires. A box or conduit body shall not be required where a luminaire is used as a raceway as permitted in 410.64.

(K) Embedded. A box or conduit body shall not be required for splices where conductors are embedded as permitted in 424.40, 424.41(D), 426.22(B), 426.24(A), and 427.19(A).

(L) Manholes and Handhole Enclosures. A box or conduit body shall not be required for conductors in manholes or handhole enclosures, except where connecting to electrical equipment. The installation shall comply with the provisions of Part V of Article 110 for manholes, and 314.30 for handhole enclosures.

300.16 Raceway or Cable to Open or Concealed Wiring.

(A) Box, Conduit Body, or Fitting. A box, conduit body, or terminal fitting having a separately bushed hole for each conductor shall be used wherever a change is made from conduit, electrical metallic tubing, electrical nonmetallic tubing, nonmetallic-sheathed cable, Type AC cable, Type MC cable, or mineral-insulated, metal-sheathed cable and surface raceway wiring to open wiring or to concealed knob-and-tube wiring. A fitting used for this purpose shall contain no taps or splices and shall not be used at luminaire outlets. A conduit body used for this purpose shall contain no taps or splices, unless it complies with 314.16(C)(2).

(B) Bushing. A bushing shall be permitted in lieu of a box or terminal where the conductors emerge from a raceway and enter or terminate at equipment, such as open switchboards, unenclosed control equipment, or similar equipment. The bushing shall be of the insulating type for other than lead-sheathed conductors.

300.18 Raceway Installations.

(A) Complete Runs. Raceways, other than busways or exposed raceways having hinged or removable covers, shall be installed complete between outlet, junction, or splicing points prior to the installation of conductors. Where required to facilitate the installation of utilization equipment, the raceway shall be permitted to be initially installed without a terminating connection at the equipment. Prewired raceway assemblies shall be permitted only where specifically permitted in this *Code* for the applicable wiring method.

Exception: Short sections of raceways used to contain conductors or cable assemblies for protection from physical damage shall not be required to be installed complete between outlet, junction, or splicing points.

300.19 Supporting Conductors in Vertical Raceways.

(A) Spacing Intervals — Maximum. Conductors in vertical raceways shall be supported if the vertical rise exceeds the values in Table 300.19(A). At least one support method shall be provided for each conductor at the top of the vertical raceway or as close to the top as practical. Intermediate supports shall be provided as necessary to limit supported conductor lengths to not greater than those values specified in Table 300.19(A).

Exception: Steel wire armor cable shall be supported at the top of the riser with a cable support that clamps the steel wire armor. A safety device shall be permitted at the lower end of the riser to hold the cable in the event there is slippage of the cable in the wire-armored cable support. Additional wedge-type supports shall be permitted to relieve the strain on the equipment terminals caused by expansion of the cable under load.

(B) Fire-Rated Cables and Conductors. Support methods and spacing intervals for fire-rated cables and conductors shall comply with any restrictions provided in the listing of the electrical circuit protective system used and in no case shall exceed the values in Table 300.19(A).

Table 300.19(A) Spacings for Conductor Supports

Conductor Size	Support of Conductors in Vertical Raceways	Conductors			
		Aluminum or Copper-Clad Aluminum		Copper	
		m	ft	m	ft
18 AWG through 8 AWG	Not greater than	30	100	30	100
6 AWG through 1/0 AWG	Not greater than	60	200	30	100
2/0 AWG through 4/0 AWG	Not greater than	55	180	25	80
Over 4/0 AWG through 350 kcmil	Not greater than	41	135	18	60
Over 350 kcmil through 500 kcmil	Not greater than	36	120	15	50
Over 500 kcmil through 750 kcmil	Not greater than	28	95	12	40
Over 750 kcmil	Not greater than	26	85	11	35

(C) Support Methods. One of the following methods of support shall be used:

(1) By clamping devices constructed of or employing insulating wedges inserted in the ends of the raceways. Where clamping of insulation does not adequately support the cable, the conductor also shall be clamped.
(2) By inserting boxes at the required intervals in which insulating supports are installed and secured in an approved manner to withstand the weight of the conductors attached thereto, the boxes being provided with covers.
(3) In junction boxes, by deflecting the cables not less than 90 degrees and carrying them horizontally to a distance not less than twice the diameter of the cable, the cables being carried on two or more insulating supports and additionally secured thereto by tie wires if desired. Where this method is used, cables shall be supported at intervals not greater than 20 percent of those mentioned in the preceding tabulation.
(4) By other approved means.

300.20 Induced Currents in Ferrous Metal Enclosures or Ferrous Metal Raceways.

(A) Conductors Grouped Together. Where conductors carrying alternating current are installed in ferrous metal enclosures or ferrous metal raceways, they shall be arranged so as to avoid heating the surrounding ferrous metal by induction. To accomplish this, all phase conductors and, where used, the grounded conductor and all equipment grounding conductors shall be grouped together.

Exception No. 1: Equipment grounding conductors for certain existing installations shall be permitted to be installed separate from their associated circuit conductors where run in accordance with the provisions of 250.130(C).

Exception No. 2: A single conductor shall be permitted to be installed in a ferromagnetic enclosure and used for skin-effect heating in accordance with the provisions of 426.42 and 427.47.

(B) Individual Conductors. Where a single conductor carrying alternating current passes through metal with magnetic properties, the inductive effect shall be minimized by (1) cutting slots in the metal between the individual holes through which the individual conductors pass or (2) passing all the conductors in the circuit through an insulating wall sufficiently large for all of the conductors of the circuit.

Exception: In the case of circuits supplying vacuum or electric-discharge lighting systems or signs or X-ray apparatus, the currents carried by the conductors are so small that the inductive heating effect can be ignored where these conductors are placed in metal enclosures or pass through metal.

300.21 Spread of Fire or Products of Combustion. Electrical installations in hollow spaces, vertical shafts, and ventilation or air-handling ducts shall be made so that the possible spread of fire or products of combustion will not be substantially increased. Openings around electrical penetrations into or through fire-resistant-rated walls, partitions, floors, or ceilings shall be firestopped using approved methods to maintain the fire resistance rating.

300.22 Wiring in Ducts Not Used for Air Handling, Fabricated Ducts for Environmental Air, and Other Spaces for Environmental Air (Plenums). The provisions of this section shall apply to the installation and uses of electrical wiring and equipment in ducts used for dust, loose stock, or vapor removal; ducts specifically fabricated for environmental air; and other spaces used for environmental air (plenums).

Informational Note: See Article 424, Part VI, for duct heaters.

(A) Ducts for Dust, Loose Stock, or Vapor Removal. No wiring systems of any type shall be installed in ducts used to transport dust, loose stock, or flammable vapors. No wiring system of any type shall be installed in any duct, or shaft containing only such ducts, used for vapor removal or for ventilation of commercial-type cooking equipment.

(B) Ducts Specifically Fabricated for Environmental Air. Equipment, devices, and the wiring methods specified in this section shall be permitted within such ducts only if necessary for the direct action upon, or sensing of, the contained air. Where equipment or devices are installed and illumination is necessary to facilitate maintenance and repair, enclosed gasketed-type luminaires shall be permitted.

Only wiring methods consisting of Type MI cable without an overall nonmetallic covering, Type MC cable employing a smooth or corrugated impervious metal sheath without an overall nonmetallic covering, electrical metallic tubing, flexible metallic tubing, intermediate metal conduit, or rigid metal conduit without an overall nonmetallic covering shall be installed in ducts specifically fabricated to transport environmental air. Flexible metal conduit shall be permitted, in lengths not to exceed 1.2 m (4 ft), to connect physically adjustable equipment and devices permitted to be in these fabricated ducts. The connectors used with flexible metal conduit shall effectively close any openings in the connection.

Exception: Wiring methods and cabling systems, listed for use in other spaces used for environmental air (plenums), shall be permitted to be installed in ducts specifically fabricated for environmental air-handling purposes under the following conditions:

(1) The wiring methods or cabling systems shall be permitted only if necessary to connect to equipment or devices associated with the direct action upon or sensing of the contained air, and
(2) The total length of such wiring methods or cabling systems shall not exceed 1.2 m (4 ft).

(C) Other Spaces Used for Environmental Air (Plenums). This section shall apply to spaces not specifically fabricated for environmental air-handling purposes but used for air-handling purposes as a plenum. This section shall not apply to habitable rooms or areas of buildings, the prime purpose of which is not air handling.

Exception: This section shall not apply to the joist or stud spaces of dwelling units where the wiring passes through such spaces perpendicular to the long dimension of such spaces.

(1) Wiring Methods. The wiring methods for such other space shall be limited to totally enclosed, nonventilated, insulated busway having no provisions for plug-in connections, Type MI cable without an overall nonmetallic covering, Type MC cable without an overall nonmetallic covering, Type AC cable, or other factory-assembled multiconductor control or power cable that is specifically listed for use within an air-handling space, or listed prefabricated cable assemblies of metallic manufactured wiring systems without nonmetallic sheath. Other types of cables, conductors, and raceways shall be permitted to be installed in electrical metallic tubing, flexible metallic tubing, intermediate metal conduit, rigid metal conduit without an overall nonmetallic covering, flexible metal conduit, or, where accessible, surface metal raceway or metal wireway with metal covers.

Nonmetallic cable ties and other nonmetallic cable accessories used to secure and support cables shall be listed as having low smoke and heat release properties.

(2) Cable Tray Systems. The provisions in (a) or (b) shall apply to the use of metallic cable tray systems in other spaces used for environmental air (plenums), where accessible, as follows:

(a) *Metal Cable Tray Systems.* Metal cable tray systems shall be permitted to support the wiring methods in 300.22(C)(1).

(b) *Solid Side and Bottom Metal Cable Tray Systems.* Solid side and bottom metal cable tray systems with solid metal covers shall be permitted to enclose wiring methods and cables, not already covered in 300.22(C)(1), in accordance with 392.10(A) and (B).

(3) Equipment. Electrical equipment with a metal enclosure, or electrical equipment with a nonmetallic enclosure listed for use within an air-handling space and having low smoke and heat release properties, and associated wiring material suitable for the ambient temperature shall be permitted to be installed in such other space unless prohibited elsewhere in this *Code*.

Exception: Integral fan systems shall be permitted where specifically identified for use within an air-handling space.

(D) Information Technology Equipment. Electrical wiring in air-handling areas beneath raised floors for information technology equipment shall be permitted in accordance with Article 645.

300.23 Panels Designed to Allow Access. Cables, raceways, and equipment installed behind panels designed to allow access, including suspended ceiling panels, shall be arranged and secured so as to allow the removal of panels and access to the equipment.

PART II. REQUIREMENTS FOR OVER 1000 VOLTS, NOMINAL

300.37 Aboveground Wiring Methods. Aboveground conductors shall be installed in rigid metal conduit, in intermediate metal conduit, in electrical metallic tubing, in RTRC and PVC conduit, in cable trays, in auxiliary gutters, as busways, as cablebus, in other identified raceways, or as exposed runs of metal-clad cable suitable for the use and purpose. In locations accessible to qualified persons only, exposed runs of Type MV cables, bare conductors, and bare busbars shall also be permitted. Busbars shall be permitted to be either copper or aluminum.

Exception: Airfield lighting cable used in series circuits that are powered by regulators and installed in restricted airport lighting vaults shall be permitted as exposed cable installations.

300.50 Underground Installations.

(A) General. Underground conductors shall be identified for the voltage and conditions under which they are installed. Direct-burial cables shall comply with the provisions of 310.10(F). Underground cables shall be installed in accordance with 300.50(A)(1), (A)(2), or (A)(3), and the installation shall meet the depth requirements of Table 300.50.

Table 300.50 Minimum Cover[a] Requirements

	General Conditions (not otherwise specified)						Special Conditions (use if applicable)					
	Column 1		Column 2		Column 3		Column 4		Column 5		Column 6	
	Direct-Buried Cables[b]		RTRC, PVC, and HDPE Conduit[c]		Rigid Metal Conduit and Intermediate Metal Conduit		Raceways Under Buildings or Exterior Concrete Slabs, 100 mm (4 in.) Minimum Thickness[d]		Cables in Airport Runways or Adjacent Areas Where Trespass Is Prohibited		Areas Subject to Vehicular Traffic, Such as Thoroughfares and Commercial Parking Areas	
Circuit Voltage	mm	in.	mm	in.	mm	in.	mm	in.	mm	in.	mm	in.
Over 1000 V through 22 kV	750	30	450	18	150	6	100	4	450	18	600	24
Over 22 kV through 40 kV	900	36	600	24	150	6	100	4	450	18	600	24
Over 40 kV	1000	42	750	30	150	6	100	4	450	18	600	24

General Notes:
1. Lesser depths shall be permitted where cables and conductors rise for terminations or splices or where access is otherwise required.
2. Where solid rock prevents compliance with the cover depths specified in this table, the wiring shall be installed in a metal or nonmetallic raceway permitted for direct burial. The raceways shall be covered by a minimum of 50 mm (2 in.) of concrete extending down to rock.
3. In industrial establishments, where conditions of maintenance and supervision ensure that qualified persons will service the installation, the minimum cover requirements, for other than rigid metal conduit and intermediate metal conduit, shall be permitted to be reduced 150 mm (6 in.) for each 50 mm (2 in.) of concrete or equivalent placed entirely within the trench over the underground installation.

Specific Footnotes:
[a]Cover is defined as the shortest distance in millimeters (inches) measured between a point on the top surface of any direct-buried conductor, cable, conduit, or other raceway and the top surface of finished grade, concrete, or similar cover.
[b]Underground direct-buried cables that are not encased or protected by concrete and are buried 750 mm (30 in.) or more below grade shall have their location identified by a warning ribbon that is placed in the trench at least 300 mm (12 in.) above the cables.
[c]Listed by a qualified testing agency as suitable for direct burial without encasement. All other nonmetallic systems shall require 50 mm (2 in.) of concrete or equivalent above conduit in addition to the table depth.
[d]The slab shall extend a minimum of 150 mm (6 in.) beyond the underground installation, and a warning ribbon or other effective means suitable for the conditions shall be placed above the underground installation.

(1) Shielded Cables and Nonshielded Cables in Metal-Sheathed Cable Assemblies. Underground cables, including nonshielded, Type MC and moisture-impervious metal sheath cables, shall have those sheaths grounded through an effective grounding path meeting the requirements of 250.4(A)(5) or (B)(4). They shall be direct buried or installed in raceways identified for the use.

(2) Industrial Establishments. In industrial establishments, where conditions of maintenance and supervision ensure that only qualified persons service the installed cable, nonshielded single-conductor cables with insulation types up to 2000 volts that are listed for direct burial shall be permitted to be directly buried.

(3) Other Nonshielded Cables. Other nonshielded cables not covered in 300.50(A)(1) or (A)(2) shall be installed in rigid metal conduit, intermediate metal conduit, or rigid nonmetallic conduit encased in not less than 75 mm (3 in.) of concrete.

(B) Wet Locations. The interior of enclosures or raceways installed underground shall be considered to be a wet location. Insulated conductors and cables installed in these enclosures or raceways in underground installations shall be listed for use in wet locations and shall comply with 310.10(C). Any connections or splices in an underground installation shall be approved for wet locations.

(C) Protection from Damage. Conductors emerging from the ground shall be enclosed in listed raceways. Raceways installed on poles shall be of rigid metal conduit, intermediate metal conduit, RTRC-XW, Schedule 80 PVC conduit, or equivalent, extending from the minimum cover depth specified in Table 300.50 to a point 2.5 m (8 ft) above finished grade. Conductors entering a building shall be protected by an approved enclosure or raceway from the minimum cover depth to the point of entrance. Where direct-buried conductors, raceways, or cables are subject to movement by settlement or frost, they shall be installed to prevent damage to the enclosed conductors or to the equipment connected to the raceways. Metallic enclosures shall be grounded.

(D) Splices. Direct burial cables shall be permitted to be spliced or tapped without the use of splice boxes, provided they are installed using materials suitable for the application. The taps and splices shall be watertight and protected from mechanical damage. Where cables are shielded, the shielding shall be continuous across the splice or tap.

Exception: At splices of an engineered cabling system, metallic shields of direct-buried single-conductor cables with maintained spacing between phases shall be permitted to be interrupted and overlapped. Where shields are interrupted and overlapped, each shield section shall be grounded at one point.

(E) Backfill. Backfill containing large rocks, paving materials, cinders, large or sharply angular substances, or corrosive materials shall not be placed in an excavation where materials can damage or contribute to the corrosion of raceways, cables, or other substructures or where it may prevent adequate compaction of fill.

Protection in the form of granular or selected material or suitable sleeves shall be provided to prevent physical damage to the raceway or cable.

(F) Raceway Seal. Where a raceway enters from an underground system, the end within the building shall be sealed with an identified compound so as to prevent the entrance of moisture or gases, or it shall be so arranged to prevent moisture from contacting live parts.

ARTICLE 310
Conductors for General Wiring

PART I. GENERAL

310.1 Scope. This article covers general requirements for conductors and their type designations, insulations, markings, mechanical strengths, ampacity ratings, and uses. These requirements do not apply to conductors that form an integral part of equipment, such as motors, motor controllers, and similar equipment, or to conductors specifically provided for elsewhere in this *Code*.

PART II. INSTALLATION

310.10 Uses Permitted. The conductors described in 310.104 shall be permitted for use in any of the wiring methods covered in Chapter 3 and as specified in their respective tables or as permitted elsewhere in this *Code*.

(A) Dry Locations. Insulated conductors and cables used in dry locations shall be any of the types identified in this *Code*.

(B) Dry and Damp Locations. Insulated conductors and cables used in dry and damp locations shall be Types FEP, FEPB, MTW, PFA, RHH, RHW, RHW-2, SA, THHN, THW, THW-2, THHW, THWN, THWN-2, TW, XHH, XHHW, XHHW-2, Z, or ZW.

(C) Wet Locations. Insulated conductors and cables used in wet locations shall comply with one of the following:

(1) Be moisture-impervious metal-sheathed
(2) Be types MTW, RHW, RHW-2, TW, THW, THW-2, THHW, THWN, THWN-2, XHHW, XHHW-2, or ZW
(3) Be of a type listed for use in wet locations

(D) Locations Exposed to Direct Sunlight. Insulated conductors or cables used where exposed to direct rays of the sun shall comply with (D)(1) or (D)(2):

(1) Conductors and cables shall be listed, or listed and marked, as being sunlight resistant
(2) Conductors and cables shall be covered with insulating material, such as tape or sleeving, that is listed, or listed and marked, as being sunlight resistant

(E) Shielding. Nonshielded, ozone-resistant insulated conductors with a maximum phase-to-phase voltage of 5000 volts shall be permitted in Type MC cables in industrial establishments where the conditions of maintenance and supervision ensure that only qualified persons service the installation. For other establishments, solid dielectric insulated conductors operated above 2000 volts in permanent installations shall have ozone-resistant insulation and shall be shielded. All metallic insulation shields shall be connected to a grounding electrode conductor, a grounding busbar, an equipment grounding conductor, or a grounding electrode.

Exception No. 1: Nonshielded insulated conductors listed by a qualified testing laboratory shall be permitted for use up to 2400 volts under the following conditions:

(a) Conductors shall have insulation resistant to electric discharge and surface tracking, or the insulated conductor(s) shall be covered with a material resistant to ozone, electric discharge, and surface tracking.

(b) Where used in wet locations, the insulated conductor(s) shall have an overall nonmetallic jacket or a continuous metallic sheath.

(c) Insulation and jacket thicknesses shall be in accordance with Table 310.104(D).

Exception No. 2: Nonshielded insulated conductors listed by a qualified testing laboratory shall be permitted for use up to 5000 volts to replace existing nonshielded conductors, on existing equipment in industrial establishments only, under the following conditions:

(a) Where the condition of maintenance and supervision ensures that only qualified personnel install and service the installation.

(b) Conductors shall have insulation resistant to electric discharge and surface tracking, or the insulated conductor(s) shall be covered with a material resistant to ozone, electric discharge, and surface tracking.

(c) Where used in wet locations, the insulated conductor(s) shall have an overall nonmetallic jacket or a continuous metallic sheath.

(d) Insulation and jacket thicknesses shall be in accordance with Table 310.104(D).

Exception No. 3: Where permitted in 310.10(F), Exception No. 2.

(F) Direct-Burial Conductors. Conductors used for direct-burial applications shall be of a type identified for such use.

(G) Corrosive Conditions. Conductors exposed to oils, greases, vapors, gases, fumes, liquids, or other substances having a deleterious effect on the conductor or insulation shall be of a type suitable for the application.

(H) Conductors in Parallel.

(1) General. Aluminum, copper-clad aluminum, or copper conductors, for each phase, polarity, neutral, or grounded circuit shall be permitted to be connected in parallel (electrically joined at both ends) only in sizes 1/0 AWG and larger where installed in accordance with 310.10(H)(2) through (H)(6).

(2) Conductor and Installation Characteristics. The paralleled conductors in each phase, polarity, neutral, grounded circuit conductor, equipment grounding conductor, or equipment bonding jumper shall comply with all of the following:

(1) Be the same length.
(2) Consist of the same conductor material.
(3) Be the same size in circular mil area.
(4) Have the same insulation type.
(5) Be terminated in the same manner.

(3) Separate Cables or Raceways. Where run in separate cables or raceways, the cables or raceways with conductors shall have the same number of conductors and shall have the same electrical characteristics. Conductors of one phase, polarity, neutral, grounded circuit conductor, or equipment grounding conductor shall not be required to have the same physical characteristics as those of another phase, polarity, neutral, grounded circuit conductor, or equipment grounding conductor.

(4) Ampacity Adjustment. Conductors installed in parallel shall comply with the provisions of 310.15(B)(3)(a).

(5) Equipment Grounding Conductors. Where parallel equipment grounding conductors are used, they shall be sized in accordance with 250.122. Sectioned equipment grounding conductors smaller than 1/0 AWG shall be permitted in multiconductor cables, if the combined circular mil area of the sectioned equipment grounding conductors in each cable complies with 250.122.

(6) Bonding Jumpers. Where parallel equipment bonding jumpers or supply-side bonding jumpers are installed in raceways, they shall be sized and installed in accordance with 250.102.

310.15 Ampacities for Conductors Rated 0–2000 Volts.

(A) General.

(1) Tables or Engineering Supervision. Ampacities for conductors shall be permitted to be determined by tables as provided in 310.15(B) or under engineering supervision, as provided in 310.15(C).

(2) Selection of Ampacity. Where more than one ampacity applies for a given circuit length, the lowest value shall be used.

Exception: Where different ampacities apply to portions of a circuit, the higher ampacity shall be permitted to be used if the total portion(s) of the circuit with lower ampacity does not exceed the lesser of 3.0 m (10 ft) or 10 percent of the total circuit.

(3) Temperature Limitation of Conductors. No conductor shall be used in such a manner that its operating temperature exceeds that designated for the type of insulated conductor involved. In no case shall conductors be associated together in such a way, with respect to type of circuit, the wiring method employed, or the number of conductors, that the limiting temperature of any conductor is exceeded.

(B) Tables. Ampacities for conductors rated 0 to 2000 volts shall be as specified in the Allowable Ampacity Table 310.15(B)(16) through Table 310.15(B)(19), and Ampacity Table 310.15(B)(20) and Table 310.15(B)(21) as modified by 310.15(B)(1) through (B)(7).

The temperature correction and adjustment factors shall be permitted to be applied to the ampacity for the temperature rating of the conductor, if the corrected and adjusted ampacity does not exceed the ampacity for the temperature rating of the termination in accordance with the provisions of 110.14(C).

Informational Note: Table 310.15(B)(16) through Table 310.15(B)(19) are application tables for use in determining conductor sizes on loads calculated in accordance with Article 220. Allowable

Table 310.15(B)(2)(a) Ambient Temperature Correction Factors Based on 30°C (86°F)

For ambient temperatures other than 30°C (86°F), multiply the allowable ampacities specified in the ampacity tables by the appropriate correction factor shown below.

Ambient Temperature (°C)	Temperature Rating of Conductor			Ambient Temperature (°F)
	60°C	75°C	90°C	
10 or less	1.29	1.20	1.15	50 or less
11–15	1.22	1.15	1.12	51–59
16–20	1.15	1.11	1.08	60–68
21–25	1.08	1.05	1.04	69–77
26–30	1.00	1.00	1.00	78–86
31–35	0.91	0.94	0.96	87–95
36–40	0.82	0.88	0.91	96–104
41–45	0.71	0.82	0.87	105–113
46–50	0.58	0.75	0.82	114–122
51–55	0.41	0.67	0.76	123–131
56–60	—	0.58	0.71	132–140
61–65	—	0.47	0.65	141–149
66–70	—	0.33	0.58	150–158
71–75	—	—	0.50	159–167
76–80	—	—	0.41	168–176
81–85	—	—	0.29	177–185

ampacities result from consideration of one or more of the following:

(1) Temperature compatibility with connected equipment, especially the connection points.
(2) Coordination with circuit and system overcurrent protection.
(3) Compliance with the requirements of product listings or certifications. See 110.3(B).
(4) Preservation of the safety benefits of established industry practices and standardized procedures.

(2) Ambient Temperature Correction Factors. Ampacities for ambient temperatures other than those shown in the ampacity tables shall be corrected in accordance with Table 310.15(B)(2)(a).

Table 310.15(B)(3)(a) Adjustment Factors for More Than Three Current-Carrying Conductors

Number of Conductors[1]	Percent of Values in Table 310.15(B)(16) Through Table 310.15(B)(19) as Adjusted for Ambient Temperature if Necessary
4–6	80
7–9	70
10–20	50
21–30	45
31–40	40
41 and above	35

[1]Number of conductors is the total number of conductors in the raceway or cable, including spare conductors. The count shall be adjusted in accordance with 310.15(B)(5) and (6). The count shall not include conductors that are connected to electrical components that cannot be simultaneously energized.

(3) Adjustment Factors.

(a) *More than Three Current-Carrying Conductors.* Where the number of current-carrying conductors in a raceway or cable exceeds three, or where single conductors or multiconductor cables are installed without maintaining spacing for a continuous length longer than 600 mm (24 in.) and are not installed in raceways, the allowable ampacity of each conductor shall be reduced as shown in Table 310.15(B)(3)(a). Each current-carrying conductor of a paralleled set of conductors shall be counted as a current-carrying conductor.

Where conductors of different systems, as provided in 300.3, are installed in a common raceway or cable, the adjustment factors shown in Table 310.15(B)(3)(a) shall apply only to the number of power and lighting conductors (Articles 210, 215, 220, and 230).

(1) Where conductors are installed in cable trays, the provisions of 392.80 shall apply.
(2) Adjustment factors shall not apply to conductors in raceways having a length not exceeding 600 mm (24 in.).

(3) Adjustment factors shall not apply to underground conductors entering or leaving an outdoor trench if those conductors have physical protection in the form of rigid metal conduit, intermediate metal conduit, rigid polyvinyl chloride conduit (PVC), or reinforced thermosetting resin conduit (RTRC) having a length not exceeding 3.05 m (10 ft), and if the number of conductors does not exceed four.

(4) Adjustment factors shall not apply to Type AC cable or to Type MC cable under the following conditions:

   a. The cables do not have an overall outer jacket.
   b. Each cable has not more than three current-carrying conductors.
   c. The conductors are 12 AWG copper.
   d. Not more than 20 current-carrying conductors are installed without maintaining spacing, are stacked, or are supported on "bridle rings."

Exception to (4): If cables meeting the requirements in 310.15(B)(3)(4)a through c with more than 20 current-carrying conductors are installed longer than 600 mm (24 in.) without maintaining spacing, are stacked, or are supported on bridle rings, a 60 percent adjustment factor shall be applied.

(b) *Raceway Spacing.* Spacing between raceways shall be maintained.

(c) *Raceways and Cables Exposed to Sunlight on Rooftops.* Where raceways or cables are exposed to direct sunlight on or above rooftops, raceways or cables shall be installed a minimum distance above the roof to the bottom of the raceway or cable of 23 mm (⅞ in.). Where the distance above the roof to the bottom of the raceway is less than 23 mm (⅞ in.), a temperature adder of 33°C (60°F) shall be added to the outdoor temperature to determine the applicable ambient temperature for application of the correction factors in Table 310.15(B)(2)(a) or Table 310.15(B)(2)(b).

Exception: Type XHHW-2 insulated conductors shall not be subject to this ampacity adjustment.

(4) Bare or Covered Conductors. Where bare or covered conductors are installed with insulated conductors, the temperature rating of the bare or covered conductor shall be equal to the lowest temperature rating of the insulated conductors for the purpose of determining ampacity.

(5) Neutral Conductor.

(a) A neutral conductor that carries only the unbalanced current from other conductors of the same circuit shall not be required to be counted when applying the provisions of 310.15(B)(3)(a).

(b) In a 3-wire circuit consisting of two phase conductors and the neutral conductor of a 4-wire, 3-phase, wye-connected system, a common conductor carries approximately the same current as the line-to-neutral load currents of the other conductors and shall be counted when applying the provisions of 310.15(B)(3)(a).

(c) On a 4-wire, 3-phase wye circuit where the major portion of the load consists of nonlinear loads, harmonic currents are present in the neutral conductor; the neutral conductor shall therefore be considered a current-carrying conductor.

(6) Grounding or Bonding Conductor. A grounding or bonding conductor shall not be counted when applying the provisions of 310.15(B)(3)(a).

Table 310.15(B)(16) (formerly Table 310.16) Allowable Ampacities of Insulated Conductors Rated Up to and Including 2000 Volts, 60°C Through 90°C (140°F Through 194°F), Not More Than Three Current-Carrying Conductors in Raceway, Cable, or Earth (Directly Buried), Based on Ambient Temperature of 30°C (86°F)*

Size AWG or kcmil	Temperature Rating of Conductor [See Table 310.104(A).]						Size AWG or kcmil
	60°C (140°F)	75°C (167°F)	90°C (194°F)	60°C (140°F)	75°C (167°F)	90°C (194°F)	
	Types TW, UF	Types RHW, THHW, THW, THWN, XHHW, USE, ZW	Types TBS, SA, SIS, FEP, FEPB, MI, RHH, RHW-2, THHN, THHW, THW-2, THWN-2, USE-2, XHH, XHHW, XHHW-2, ZW-2	Types TW, UF	Types RHW, THHW, THW, THWN, XHHW, USE	Types TBS, SA, SIS, THHN, THHW, THW-2, THWN-2, RHH, RHW-2, USE-2, XHH, XHHW, XHHW-2, ZW-2	
	COPPER			ALUMINUM OR COPPER-CLAD ALUMINUM			
18**	—	—	14	—	—	—	—
16**	—	—	18	—	—	—	—
14**	15	20	25	—	—	—	—
12**	20	25	30	15	20	25	12**
10**	30	35	40	25	30	35	10**
8	40	50	55	35	40	45	8
6	55	65	75	40	50	55	6
4	70	85	95	55	65	75	4
3	85	100	115	65	75	85	3
2	95	115	130	75	90	100	2
1	110	130	145	85	100	115	1

1/0	125	150	170	100	120	135	1/0
2/0	145	175	195	115	135	150	2/0
3/0	165	200	225	130	155	175	3/0
4/0	195	230	260	150	180	205	4/0
250	215	255	290	170	205	230	250
300	240	285	320	195	230	260	300
350	260	310	350	210	250	280	350
400	280	335	380	225	270	305	400
500	320	380	430	260	310	350	500
600	350	420	475	285	340	385	600
700	385	460	520	315	375	425	700
750	400	475	535	320	385	435	750
800	410	490	555	330	395	445	800
900	435	520	585	355	425	480	900
1000	455	545	615	375	445	500	1000
1250	495	590	665	405	485	545	1250
1500	525	625	705	435	520	585	1500
1750	545	650	735	455	545	615	1750
2000	555	665	750	470	560	630	2000

*Refer to 310.15(B)(2) for the ampacity correction factors where the ambient temperature is other than 30°C (86°F). Refer to 310.15(B)(3)(a) for more than three current-carrying conductors.

**Refer to 240.4(D) for conductor overcurrent protection limitations.

PART III. CONSTRUCTION SPECIFICATIONS

310.106 Conductors.

(A) Minimum Size of Conductors. The minimum size of conductors shall be as shown in Table 310.106(A), except as permitted elsewhere in this *Code*.

(B) Conductor Material. Conductors in this article shall be of aluminum, copper-clad aluminum, or copper unless otherwise specified.

(C) Stranded Conductors. Where installed in raceways, conductors 8 AWG and larger, not specifically permitted or required elsewhere in this *Code* to be solid, shall be stranded.

(D) Insulated. Conductors, not specifically permitted elsewhere in this *Code* to be covered or bare, shall be insulated.

Table 310.106(A) Minimum Size of Conductors

Conductor Voltage Rating (Volts)	Minimum Conductor Size (AWG)	
	Copper	Aluminum or Copper-Clad Aluminum
0–2000	14	12
2001–5000	8	8
5001–8000	6	6
8001–15,000	2	2
15,001–28,000	1	1
28,001–35,000	1/0	1/0

CHAPTER 15

BRANCH CIRCUITS

INTRODUCTION

This chapter contains requirements from Article 210 — Branch Circuits. The requirements within Article 210 have been divided into three chapters of this *Pocket Guide.* This chapter covers Part I (general requirements) and Part II (branch-circuit ratings). Some of the general rules pertain to multiwire branch circuits, ground-fault circuit-interrupter protection for personnel, and required branch circuits. Part II covers the branch circuit ratings for conductors, equipment, and outlets.

Chapter 16 contains the receptacle outlet requirements from Article 210, Chapter 18 contains the requirements for lighting outlets, Chapter 19 provides requirements for branch circuits supplying electric motors, and the Chapter 20 requirements cover branch circuits supplying air-conditioning and refrigeration equipment that employs hermetic refrigerant motor-compressors and equipment.

ARTICLE 210

Branch Circuits

PART I. GENERAL PROVISIONS

210.3 Other Articles for Specific-Purpose Branch Circuits. Table 210.3 lists references for specific equipment and applications not located in Chapters 5, 6, and 7 that amend or supplement the requirements of this article.

210.4 Multiwire Branch Circuits.

(A) General. Branch circuits recognized by this article shall be permitted as multiwire circuits. A multiwire circuit shall be permitted to be considered as multiple circuits. All conductors of a multiwire branch circuit shall originate from the same panelboard or similar distribution equipment.

(B) Disconnecting Means. Each multiwire branch circuit shall be provided with a means that will simultaneously disconnect all ungrounded conductors at the point where the branch circuit originates.

Table 210.3 Specific-Purpose Branch Circuits

Equipment	Article	Section
Air-conditioning and refrigerating equipment		440.6, 440.31, 440.32
Busways		368.17
Central heating equipment other than fixed electric space-heating equipment		422.12
Fixed electric heating equipment for pipelines and vessels		427.4
Fixed electric space-heating equipment		424.3
Fixed outdoor electrical deicing and snow-melting equipment		426.4
Infrared lamp industrial heating equipment		422.48, 424.3
Motors, motor circuits, and controllers	430	
Switchboards and panelboards		408.52

Informational Note: See 240.15(B) for information on the use of single-pole circuit breakers as the disconnecting means.

(C) Line-to-Neutral Loads. Multiwire branch circuits shall supply only line-to-neutral loads.

Exception No. 1: A multiwire branch circuit that supplies only one utilization equipment.

Exception No. 2: Where all ungrounded conductors of the multiwire branch circuit are opened simultaneously by the branch-circuit overcurrent device.

(D) Grouping. The ungrounded and grounded circuit conductors of each multiwire branch circuit shall be grouped in accordance with 200.4(B).

• **210.5 Identification for Branch Circuits.**

(C) Identification of Ungrounded Conductors. Ungrounded conductors shall be identified in accordance with 210.5(C)(1) or (2), as applicable.

(1) Branch Circuits Supplied from More Than One Nominal Voltage System. Where the premises wiring system has branch circuits supplied from more than one nominal voltage system, each ungrounded conductor of a branch circuit shall be identified by phase or line and system at all termination, connection, and splice points in compliance with 210.5(C)(1)(a) and (b).

(a) *Means of Identification.* The means of identification shall be permitted to be by separate color coding, marking tape, tagging, or other approved means.

(b) *Posting of Identification Means.* The method utilized for conductors originating within each branch-circuit panelboard or similar branch-circuit distribution equipment shall be documented in a manner that is readily available or shall be permanently posted at each branch-circuit panelboard or similar branch-circuit distribution equipment. The label shall be of sufficient durability to withstand the environment involved and shall not be handwritten.

Exception: In existing installations where a voltage system(s) already exists and a different voltage system is being added, it shall be permissible to mark only the new system voltage.

Existing unidentified systems shall not be required to be identified at each termination, connection, and splice point in compliance with 210.5(C)(1)(a) and (b). Labeling shall be required at each voltage system distribution equipment to identify that only one voltage system has been marked for a new system(s). The new system label(s) shall include the words "other unidentified systems exist on the premises."

(2) Branch Circuits Supplied from Direct-Current Systems. Where a branch circuit is supplied from a dc system operating at more than 60 volts, each ungrounded conductor of 4 AWG or larger shall be identified by polarity at all termination, connection, and splice points by marking tape, tagging, or other approved means; each ungrounded conductor of 6 AWG or smaller shall be identified by polarity at all termination, connection, and splice points in compliance with 210.5(C)(2)(a) and (b). The identification methods utilized for conductors originating within each branch-circuit panelboard or similar branch-circuit distribution equipment shall be documented in a manner that is readily available or shall be permanently posted at each branch-circuit panelboard or similar branch-circuit distribution equipment.

(a) *Positive Polarity, Sizes 6 AWG or Smaller.* Where the positive polarity of a dc system does not serve as the connection point for the grounded conductor, each positive ungrounded conductor shall be identified by one of the following means:

(1) A continuous red outer finish
(2) A continuous red stripe durably marked along the conductor's entire length on insulation of a color other than green, white, gray, or black
(3) Imprinted plus signs (+) or the word POSITIVE or POS durably marked on insulation of a color other than green, white, gray, or black and repeated at intervals not exceeding 610 mm (24 in.) in accordance with 310.120(B)
(4) An approved permanent marking means such as sleeving or shrink-tubing that is suitable for the conductor size, at all termination, connection, and splice points, with imprinted plus signs (+) or the word POSITIVE or POS durably marked on insulation of a color other than green, white, gray, or black

(b) *Negative Polarity, Sizes 6 AWG or Smaller.* Where the negative polarity of a dc system does not serve as the connection point for the grounded conductor, each negative ungrounded conductor shall be identified by one of the following means:

(1) A continuous black outer finish
(2) A continuous black stripe durably marked along the conductor's entire length on insulation of a color other than green, white, gray, or red
(3) Imprinted minus signs (–) or the word NEGATIVE or NEG durably marked on insulation of a color other than green, white, gray, or red and repeated at intervals not exceeding 610 mm (24 in.) in accordance with 310.120(B)
(4) An approved permanent marking means such as sleeving or shrink-tubing that is suitable for the conductor size, at all termination, connection, and splice points, with imprinted minus signs (–) or the word NEGATIVE or NEG durably marked on insulation of a color other than green, white, gray, or red

210.6 Branch-Circuit Voltage Limitations. The nominal voltage of branch circuits shall not exceed the values permitted by 210.6(A) through (E).

(A) Occupancy Limitation. In dwelling units and guest rooms or guest suites of hotels, motels, and similar occupancies, the voltage shall not exceed 120 volts, nominal, between conductors that supply the terminals of the following:

(1) Luminaires
(2) Cord-and-plug-connected loads 1440 volt-amperes, nominal, or less or less than ¼ hp

(B) 120 Volts Between Conductors. Circuits not exceeding 120 volts, nominal, between conductors shall be permitted to supply the following:

(1) The terminals of lampholders applied within their voltage ratings
(2) Auxiliary equipment of electric-discharge lamps
(3) Cord-and-plug-connected or permanently connected utilization equipment

(C) 277 Volts to Ground. Circuits exceeding 120 volts, nominal, between conductors and not exceeding 277 volts, nominal, to ground shall be permitted to supply the following:

(1) Listed electric-discharge or listed light-emitting diode-type luminaires
(2) Listed incandescent luminaires, where supplied at 120 volts or less from the output of a stepdown autotransformer that is an integral component of the luminaire and the outer shell terminal is electrically connected to a grounded conductor of the branch circuit
(3) Luminaires equipped with mogul-base screw shell lampholders
(4) Lampholders, other than the screw shell type, applied within their voltage ratings
(5) Auxiliary equipment of electric-discharge lamps
(6) Cord-and-plug-connected or permanently connected utilization equipment

(D) 600 Volts Between Conductors. Circuits exceeding 277 volts, nominal, to ground and not exceeding 600 volts, nominal, between conductors shall be permitted to supply the following:

(1) The auxiliary equipment of electric-discharge lamps mounted in permanently installed luminaires where the luminaires are mounted in accordance with one of the following:

    a. Not less than a height of 6.7 m (22 ft) on poles or similar structures for the illumination of outdoor areas such as highways, roads, bridges, athletic fields, or parking lots
    b. Not less than a height of 5.5 m (18 ft) on other structures such as tunnels

(2) Cord-and-plug-connected or permanently connected utilization equipment other than luminaires
(3) Luminaires powered from direct-current systems where either of the following apply:

    a. The luminaire contains a listed, dc-rated ballast that provides isolation between the dc power source and the lamp circuit and protection from electric shock when changing lamps.

b. The luminaire contains a listed, dc-rated ballast and has no provision for changing lamps.

Exception No. 1 to (B), (C), and (D): For lampholders of infrared industrial heating appliances as provided in 425.14.

Exception No. 2 to (B), (C), and (D): For railway properties as described in 110.19.

210.7 Multiple Branch Circuits. Where two or more branch circuits supply devices or equipment on the same yoke or mounting strap, a means to simultaneously disconnect the ungrounded supply conductors shall be provided at the point at which the branch circuits originate.

210.8 Ground-Fault Circuit-Interrupter Protection for Personnel. Ground-fault circuit-interrupter protection for personnel shall be provided as required in 210.8(A) through (E). The ground-fault circuit interrupter shall be installed in a readily accessible location.

For the purposes of this section, when determining distance from receptacles the distance shall be measured as the shortest path the cord of an appliance connected to the receptacle would follow without piercing a floor, wall, ceiling, or fixed barrier, or passing through a door, doorway, or window.

(B) Other Than Dwelling Units. All single-phase receptacles rated 150 volts to ground or less, 50 amperes or less and three-phase receptacles rated 150 volts to ground or less, 100 amperes or less installed in the following locations shall have ground-fault circuit-interrupter protection for personnel.

(1) Bathrooms
(2) Kitchens
(3) Rooftops

Exception: Receptacles on rooftops shall not be required to be readily accessible other than from the rooftop.

(4) Outdoors

Exception No. 1 to (3) and (4): Receptacles that are not readily accessible and are supplied by a branch circuit dedicated to electric snow-melting, deicing, or pipeline and vessel heating equipment shall be permitted to be installed in accordance with 426.28 or 427.22, as applicable.

Exception No. 2 to (4): In industrial establishments only, where the conditions of maintenance and supervision ensure that only qualified personnel are involved, an assured equipment grounding conductor program as specified in 590.6(B)(3) shall be permitted for only those receptacle outlets used to supply equipment that would create a greater hazard if power is interrupted or having a design that is not compatible with GFCI protection.

(5) Sinks — where receptacles are installed within 1.8 m (6 ft) from the top inside edge of the bowl of the sink

Exception No. 1 to (5): In industrial laboratories, receptacles used to supply equipment where removal of power would introduce a greater hazard shall be permitted to be installed without GFCI protection.

Exception No. 2 to (5): For receptacles located in patient bed locations of general care (Category 2) or critical care (Category 1) spaces of health care facilities other than those covered under 210.8(B)(1), GFCI protection shall not be required.

(6) Indoor wet locations
(7) Locker rooms with associated showering facilities
(8) Garages, service bays, and similar areas other than vehicle exhibition halls and showrooms
(9) Crawl spaces — at or below grade level
(10) Unfinished portions or areas of the basement not intended as habitable rooms

• N **(E) Crawl Space Lighting Outlets.** GFCI protection shall be provided for lighting outlets not exceeding 120 volts installed in crawl spaces.

210.9 Circuits Derived from Autotransformers. Branch circuits shall not be derived from autotransformers unless the circuit supplied has a grounded conductor that is electrically connected to a grounded conductor of the system supplying the autotransformer.

Exception No. 1: An autotransformer shall be permitted without the connection to a grounded conductor where transforming from a nominal 208 volts to a nominal 240-volt supply or similarly from 240 volts to 208 volts.

Exception No. 2: In industrial occupancies, where conditions of maintenance and supervision ensure that only qualified persons service the installation, autotransformers shall be permitted to supply nominal 600-volt loads from nominal 480-volt systems, and 480-volt loads from nominal 600-volt systems, without the connection to a similar grounded conductor.

210.11 Branch Circuits Required. Branch circuits for lighting and for appliances, including motor-operated appliances, shall be provided to supply the loads calculated in accordance with 220.10. In addition, branch circuits shall be provided for specific loads not covered by 220.10 where required elsewhere in this *Code* and for dwelling unit loads as specified in 210.11(C).

(A) Number of Branch Circuits. The minimum number of branch circuits shall be determined from the total calculated load and the size or rating of the circuits used. In all installations, the number of circuits shall be sufficient to supply the load served. In no case shall the load on any circuit exceed the maximum specified by 220.18.

(B) Load Evenly Proportioned Among Branch Circuits. Where the load is calculated on the basis of volt-amperes per square meter or per square foot, the wiring system up to and including the branch-circuit panelboard(s) shall be provided to serve not less than the calculated load. This load shall be evenly proportioned among multioutlet branch circuits within the panelboard(s). Branch-circuit overcurrent devices and circuits shall be required to be installed only to serve the connected load.

210.12 Arc-Fault Circuit-Interrupter Protection. Arc-fault circuit-interrupter protection shall be provided as required in 210.12(A), (B), and (C). The arc-fault circuit interrupter shall be installed in a readily accessible location.

(A) Dwelling Units.

(1) A listed combination-type arc-fault circuit interrupter, installed to provide protection of the entire branch circuit

(2) A listed branch/feeder-type AFCI installed at the origin of the branch-circuit in combination with a listed outlet branch-circuit type arc-fault circuit interrupter installed at the first outlet box on the branch circuit. The first outlet box in the branch circuit shall be marked to indicate that it is the first outlet of the circuit.

(3) A listed supplemental arc protection circuit breaker installed at the origin of the branch circuit in combination with a listed outlet branch-circuit type arc-fault circuit interrupter installed at the first outlet box on the branch circuit where all of the following conditions are met:

a. The branch-circuit wiring shall be continuous from the branch-circuit overcurrent device to the outlet branch-circuit arc-fault circuit interrupter.
b. The maximum length of the branch-circuit wiring from the branch-circuit overcurrent device to the first outlet shall not exceed 15.2 m (50 ft) for a 14 AWG conductor or 21.3 m (70 ft) for a 12 AWG conductor.
c. The first outlet box in the branch circuit shall be marked to indicate that it is the first outlet of the circuit.

(4) A listed outlet branch-circuit type arc-fault circuit interrupter installed at the first outlet on the branch circuit in combination with a listed branch-circuit overcurrent protective device where all of the following conditions are met:

a. The branch-circuit wiring shall be continuous from the branch-circuit overcurrent device to the outlet branch-circuit arc-fault circuit interrupter.
b. The maximum length of the branch-circuit wiring from the branch-circuit overcurrent device to the first outlet shall not exceed 15.2 m (50 ft) for a 14 AWG conductor or 21.3 m (70 ft) for a 12 AWG conductor.

c. The first outlet box in the branch circuit shall be marked to indicate that it is the first outlet of the circuit.
d. The combination of the branch-circuit overcurrent device and outlet branch-circuit AFCI shall be identified as meeting the requirements for a system combination–type AFCI and shall be listed as such.

(5) If RMC, IMC, EMT, Type MC, or steel-armored Type AC cables meeting the requirements of 250.118, metal wireways, metal auxiliary gutters, and metal outlet and junction boxes are installed for the portion of the branch circuit between the branch-circuit overcurrent device and the first outlet, it shall be permitted to install a listed outlet branch-circuit type AFCI at the first outlet to provide protection for the remaining portion of the branch circuit.
(6) Where a listed metal or nonmetallic conduit or tubing or Type MC cable is encased in not less than 50 mm (2 in.) of concrete for the portion of the branch circuit between the branch-circuit overcurrent device and the first outlet, it shall be permitted to install a listed outlet branch-circuit type AFCI at the first outlet to provide protection for the remaining portion of the branch circuit.

Exception: Where an individual branch circuit to a fire alarm system installed in accordance with 760.41(B) or 760.121(B) is installed in RMC, IMC, EMT, or steel-sheathed cable, Type AC or Type MC, meeting the requirements of 250.118, with metal outlet and junction boxes, AFCI protection shall be permitted to be omitted.

(B) Dormitory Units. All 120-volt, single-phase, 15- and 20-ampere branch circuits supplying outlets and devices installed in dormitory unit bedrooms, living rooms, hallways, closets, bathrooms, and similar rooms shall be protected by any of the means described in 210.12(A)(1) through (6).

N (C) Guest Rooms and Guest Suites. All 120-volt, single-phase, 15- and 20-ampere branch circuits supplying outlets and devices installed in guest rooms and guest suites of hotels and motels shall be protected by any of the means described in 210.12(A)(1) through (6).

(D) Branch Circuit Extensions or Modifications — Dwelling Units and Dormitory Units. In any of the areas specified in 210.12(A) or (B), where branch-circuit wiring is modified, replaced, or extended, the branch circuit shall be protected by one of the following:

(1) A listed combination-type AFCI located at the origin of the branch circuit

(2) A listed outlet branch-circuit-type AFCI located at the first receptacle outlet of the existing branch circuit

Exception: AFCI protection shall not be required where the extension of the existing conductors is not more than 1.8 m (6 ft) and does not include any additional outlets or devices.

210.13 Ground-Fault Protection of Equipment. Each branch-circuit disconnect rated 1000 A or more and installed on solidly grounded wye electrical systems of more than 150 volts to ground, but not exceeding 600 volts phase-to-phase, shall be provided with ground-fault protection of equipment in accordance with the provisions of 230.95.

Exception No. 1: The provisions of this section shall not apply to a disconnecting means for a continuous industrial process where a nonorderly shutdown will introduce additional or increased hazards.

Exception No. 2: The provisions of this section shall not apply if ground-fault protection of equipment is provided on the supply side of the branch circuit and on the load side of any transformer supplying the branch circuit.

• **210.17 Guest Rooms and Guest Suites.** Guest rooms and guest suites that are provided with permanent provisions for cooking shall have branch circuits installed to meet the rules for dwelling units.

PART II. BRANCH-CIRCUIT RATINGS

N **210.18 Rating.** Branch circuits recognized by this article shall be rated in accordance with the maximum permitted ampere rating or setting of the overcurrent device. The rating for other than individual branch circuits shall be 15, 20, 30, 40, and 50 amperes. Where conductors of higher ampacity are used for any reason, the ampere rating or setting of the specified overcurrent device shall determine the circuit rating.

Exception: Multioutlet branch circuits greater than 50 amperes shall be permitted to supply nonlighting outlet loads on industrial premises where conditions of maintenance and supervision ensure that only qualified persons service the equipment.

210.19 Conductors — Minimum Ampacity and Size.

(A) Branch Circuits Not More Than 600 Volts.

(1) General. Branch-circuit conductors shall have an ampacity not less than the maximum load to be served. Conductors shall be sized to carry not less than the larger of 210.19(A)(1)(a) or (b).

(a) Where a branch circuit supplies continuous loads or any combination of continuous and noncontinuous loads, the minimum branch-circuit conductor size shall have an allowable ampacity not less than the noncontinuous load plus 125 percent of the continuous load.

(b) The minimum branch-circuit conductor size shall have an allowable ampacity not less than the maximum load to be served after the application of any adjustment or correction factors.

Exception: If the assembly, including the overcurrent devices protecting the branch circuit(s), is listed for operation at 100 percent of its rating, the allowable ampacity of the branch-circuit conductors shall be permitted to be not less than the sum of the continuous load plus the noncontinuous load.

(2) Branch Circuits with More than One Receptacle. Conductors of branch circuits supplying more than one receptacle for cord-and-plug-connected portable loads shall have an ampacity of not less than the rating of the branch circuit.

(4) Other Loads. Branch-circuit conductors that supply loads other than those specified in 210.3 and other than cooking appliances as covered in 210.19(A)(3) shall have an ampacity sufficient for the loads served and shall not be smaller than 14 AWG.

Exception No. 1: Tap conductors shall have an ampacity sufficient for the load served. In addition, they shall have an ampacity of not less than 15 for circuits rated less than 40 amperes and not less than 20 for circuits rated at 40 or 50 amperes and only where these tap conductors supply any of the following loads:

(a) Individual lampholders or luminaires with taps extending not longer than 450 mm (18 in.) beyond any portion of the lampholder or luminaire

(b) A luminaire having tap conductors as provided in 410.117

(c) Individual outlets, other than receptacle outlets, with taps not over 450 mm (18 in.) long

(d) Infrared lamp industrial heating appliances

(e) Nonheating leads of deicing and snow-melting cables and mats

Exception No. 2: Fixture wires and flexible cords shall be permitted to be smaller than 14 AWG as permitted by 240.5.

(B) Branch Circuits Over 600 Volts. The ampacity of conductors shall be in accordance with 310.15 and 310.60, as applicable. Branch-circuit conductors over 600 volts shall be sized in accordance with 210.19(B)(1) or (B)(2).

(1) General. The ampacity of branch-circuit conductors shall not be less than 125 percent of the designed potential load of utilization equipment that will be operated simultaneously.

(2) Supervised Installations. For supervised installations, branch-circuit conductor sizing shall be permitted to be determined by qualified persons under engineering supervision. Supervised installations are defined as those portions of a facility where both of the following conditions are met:

(1) Conditions of design and installation are provided under engineering supervision.
(2) Qualified persons with documented training and experience in over 600-volt systems provide maintenance, monitoring, and servicing of the system.

210.20 Overcurrent Protection. Branch-circuit conductors and equipment shall be protected by overcurrent protective devices that have a rating or setting that complies with 210.20(A) through (D).

(A) Continuous and Noncontinuous Loads. Where a branch circuit supplies continuous loads or any combination of continuous and noncontinuous loads, the rating of the overcurrent device shall not be less than the noncontinuous load plus 125 percent of the continuous load.

Exception: Where the assembly, including the overcurrent devices protecting the branch circuit(s), is listed for operation at 100 percent of its rating, the ampere rating of the overcurrent device shall be permitted to be not less than the sum of the continuous load plus the noncontinuous load.

210.21 Outlet Devices. Outlet devices shall have an ampere rating that is not less than the load to be served and shall comply with 210.21(A) and (B).

(A) Lampholders. Where connected to a branch circuit having a rating in excess of 20 amperes, lampholders shall be of the heavy-duty type. A heavy-duty lampholder shall have a rating of not less than 660 watts if of the admedium type, or not less than 750 watts if of any other type.

(B) Receptacles.

(1) Single Receptacle on an Individual Branch Circuit. A single receptacle installed on an individual branch circuit shall have an ampere rating not less than that of the branch circuit.

Exception No. 1: A receptacle installed in accordance with 430.81(B).

Exception No. 2: A receptacle installed exclusively for the use of a cord-and-plug-connected arc welder shall be permitted to have an ampere rating not less than the minimum branch-circuit conductor ampacity determined by 630.11(A) for arc welders.

(2) Total Cord-and-Plug-Connected Load. Where connected to a branch circuit supplying two or more receptacles or outlets, a receptacle shall not supply a total

Table 210.21(B)(2) Maximum Cord-and-Plug-Connected Load to Receptacle

Circuit Rating (Amperes)	Receptacle Rating (Amperes)	Maximum Load (Amperes)
15 or 20	15	12
20	20	16
30	30	24

cord-and-plug-connected load in excess of the maximum specified in Table 210.21(B)(2).

(3) Receptacle Ratings. Where connected to a branch circuit supplying two or more receptacles or outlets, receptacle ratings shall conform to the values listed in Table 210.21(B)(3), or, where rated higher than 50 amperes, the receptacle rating shall not be less than the branch-circuit rating.

Exception No. 1: Receptacles installed exclusively for the use of one or more cord-and-plug-connected arc welders shall be permitted to have ampere ratings not less than the minimum branch-circuit conductor ampacity determined by 630.11(A) or (B) for arc welders.

Exception No. 2: The ampere rating of a receptacle installed for electric discharge lighting shall be permitted to be based on 410.62(C).

210.22 Permissible Loads, Individual Branch Circuits. An individual branch circuit shall be permitted to supply any load for which it is rated, but in no case shall the load exceed the branch-circuit ampere rating.

210.23 Permissible Loads, Multiple-Outlet Branch Circuits. In no case shall the load exceed the branch-circuit ampere rating. A branch circuit supplying two or more outlets or receptacles shall supply only the loads specified according to its size as specified in 210.23(A) through (D) and as summarized in 210.24 and Table 210.24.

(A) 15- and 20-Ampere Branch Circuits. A 15- or 20-ampere branch circuit shall be permitted to supply lighting

Table 210.21(B)(3) Receptacle Ratings for Various Size Circuits

Circuit Rating (Amperes)	Receptacle Rating (Amperes)
15	Not over 15
20	15 or 20
30	30
40	40 or 50
50	50

units or other utilization equipment, or a combination of both, and shall comply with 210.23(A)(1) and (A)(2).

(1) Cord-and-Plug-Connected Equipment Not Fastened in Place. The rating of any one cord-and-plug-connected utilization equipment not fastened in place shall not exceed 80 percent of the branch-circuit ampere rating.

(2) Utilization Equipment Fastened in Place. The total rating of utilization equipment fastened in place, other than luminaires, shall not exceed 50 percent of the branch-circuit ampere rating where lighting units, cord-and-plug-connected utilization equipment not fastened in place, or both, are also supplied.

(C) 40- and 50-Ampere Branch Circuits. A 40- or 50-ampere branch circuit shall be permitted to supply cooking appliances that are fastened in place in any occupancy. In other than dwelling units, such circuits shall be permitted to supply fixed lighting units with heavy-duty lampholders, infrared heating units, or other utilization equipment.

(D) Branch Circuits Larger Than 50 Amperes. Branch circuits larger than 50 amperes shall supply only nonlighting outlet loads.

210.24 Branch-Circuit Requirements — Summary. The requirements for circuits that have two or more outlets or receptacles, other than the receptacle circuits of 210.11(C)(1), (C)(2), and (C)(3), are summarized in Table 210.24. This table provides only a summary of minimum requirements. See 210.19, 210.20, and 210.21 for the specific requirements applying to branch circuits.

Table 210.24 Summary of Branch-Circuit Requirements

Circuit Rating	15 A	20 A	30 A	40 A	50 A
Conductors (min. size):					
Circuit wires[1]	14	12	10	8	6
Taps	14	14	14	12	12
Fixture wires and cords — see 240.5					
Overcurrent Protection	**15 A**	**20 A**	**30 A**	**40 A**	**50 A**
Outlet devices:					
Lampholders permitted	Any type	Any type	Heavy duty	Heavy duty	Heavy duty
Receptacle rating[2]	15 max. A	15 or 20 A	30 A	40 or 50 A	50 A
MaximumLoad	**15 A**	**20 A**	**30 A**	**40 A**	**50 A**
Permissible load	See 210.23(A)	See 210.23(A)	See 210.23(B)	See 210.23(C)	See 210.23(C)

[1]These gauges are for copper conductors.

[2]For receptacle rating of cord-connected electric-discharge luminaires, see 410.62(C).

210.25 Branch Circuits in Buildings with More Than One Occupancy.

(B) Common Area Branch Circuits. Branch circuits installed for the purpose of lighting, central alarm, signal, communications, or other purposes for public or common areas of a two-family dwelling, a multifamily dwelling, or a multi-occupancy building shall not be supplied from equipment that supplies an individual dwelling unit or tenant space.

PART III. REQUIRED OUTLETS

(See Chapter 16 for required receptacles, and Chapter 18 for required lighting outlets.)

CHAPTER 16

RECEPTACLE PROVISIONS

INTRODUCTION

This chapter contains requirements from Article 210 — Branch Circuits, and Article 406 — Receptacles, Cord Connectors, and Attachment Plugs (Caps). The requirements from Article 210 covered in this chapter include 210.60, 210.62, 210.63, 210.64, 210.71, and the first part of 210.50. These sections cover cord pendants, cord connections, rooms in hotels, motels and dormitories, show windows, and required receptacle outlet(s) for servicing of mechanical and electrical equipment. With the increase demand for receptacles in meeting rooms, rules for installation of receptacle outlets in those spaces have also been established. In contrast to the numerous receptacle placement rules for dwelling units, commercial and industrial occupancies have relatively few such requirements.

Article 406 covers the rating, type, and installation of receptacles, cord connectors, and attachment plugs (cord caps). Some of the topics covered include receptacle rating and type, general installation requirements, receptacle mounting, receptacles in damp or wet locations, and connecting the receptacle's grounding terminal to the box. The types of requirements contained in Article 406 provide for safe use of receptacles that are installed in any type of occupancy.

ARTICLE 210
Branch Circuits

PART III. REQUIRED OUTLETS

210.50 General. Receptacle outlets shall be installed as specified in 210.52 through 210.64.

(A) Cord Pendants. A cord connector that is supplied by a permanently connected cord pendant shall be considered a receptacle outlet.

(B) Cord Connections. A receptacle outlet shall be installed wherever flexible cords with attachment plugs are used. Where flexible cords are permitted to be permanently connected, receptacles shall be permitted to be omitted for such cords.

(C) Appliance Receptacle Outlets. Appliance receptacle outlets installed in a dwelling unit for specific appliances, such as laundry equipment, shall be installed within 1.8 m (6 ft) of the intended location of the appliance.

210.60 Guest Rooms, Guest Suites, Dormitories, and Similar Occupancies.

(A) General. Guest rooms or guest suites in hotels, motels, sleeping rooms in dormitories, and similar occupancies shall have receptacle outlets installed in accordance with 210.52(A) and (D). Guest rooms or guest suites provided with permanent provisions for cooking shall have receptacle outlets installed in accordance with all of the applicable rules in 210.52.

(B) Receptacle Placement. In applying the provisions of 210.52(A), the total number of receptacle outlets shall not be less than the minimum number that would comply with the provisions of that section. These receptacle outlets shall be permitted to be located conveniently for permanent furniture layout. At least two receptacle outlets shall be readily accessible. Where receptacles are installed behind the bed, the receptacle shall be located to prevent the bed from contacting any attachment plug that may be installed or the receptacle shall be provided with a suitable guard.

210.62 Show Windows. At least one 125-volt, single-phase, 15- or 20-ampere-rated receptacle outlet shall be installed within 450 mm (18 in.) of the top of a show window for each 3.7 linear m (12 linear ft) or major fraction thereof of show window area measured horizontally at its maximum width.

210.63 Heating, Air-Conditioning, and Refrigeration Equipment Outlet. A 125-volt, single-phase, 15- or 20-ampere-rated receptacle outlet shall be installed at an accessible location for the servicing of heating, air-conditioning, and refrigeration equipment. The receptacle shall be located on the same level and within 7.5 m (25 ft) of the heating, air-conditioning, and refrigeration equipment. The receptacle outlet shall not be connected to the load side of the equipment disconnecting means.

Informational Note: See 210.8 for ground-fault circuit-interrupter requirements.

210.64 Electrical Service Areas. At least one 125-volt, single-phase, 15- or 20-ampere-rated receptacle outlet shall be installed in an accessible location within 7.5 m (25 ft) of the indoor electrical service equipment. The required receptacle outlet shall be located within the same room or area as the service equipment.

Exception No. 2: Where the service voltage is greater than 120 volts to ground, a receptacle outlet shall not be required for services dedicated to equipment covered in Articles 675 and 682.

N 210.71 Meeting Rooms.

(A) General. Each meeting room of not more than 93 m² (1000 ft²) in other than dwelling units shall have outlets for nonlocking-type, 125-volt, 15- or 20-ampere receptacles. The outlets shall be installed in accordance with 210.71(B). Where a room or space is provided with movable partition(s), each room size shall be determined with the partition in the position that results in the smallest size meeting room.

(B) Receptacle Outlets Required. The total number of receptacle outlets, including floor outlets and receptacle outlets in fixed furniture, shall not be less than as determined in (1) and (2). These receptacle outlets shall be permitted to be located as determined by the designer or building owner.

(1) Receptacle Outlets in Fixed Walls. Receptacle outlets shall be installed in accordance with 210.52(A)(1) through (A)(4).

(2) Floor Receptacle Outlets. A meeting room that is at least 3.7 m (12 ft) wide and that has a floor area of at least 20 m^2 (215 ft^2) shall have at least one receptacle outlet located in the floor at a distance not less than 1.8 m (6 ft) from any fixed wall for each 20 m^2 (215 ft^2) or major portion of floor space.

ARTICLE 406

Receptacles, Cord Connectors, and Attachment Plugs (Caps)

406.2 Definitions.

Child Care Facility. A building or structure, or portion thereof, for educational, supervisory, or personal care services for more than four children 7 years old or less.

N **Outlet Box Hood.** A housing shield intended to fit over a faceplate for flush-mounted wiring devices, or an integral component of an outlet box or of a faceplate for flush-mounted wiring devices. The hood does not serve to complete the electrical enclosure; it reduces the risk of water coming in contact with electrical components within the hood, such as attachment plugs, current taps, surge protective devices, direct plug-in transformer units, or wiring devices.

406.3 Receptacle Rating and Type.

(C) Receptacles for Aluminum Conductors. Receptacles rated 20 amperes or less and designed for the direct connection of aluminum conductors shall be marked CO/ALR.

(D) Isolated Ground Receptacles. Receptacles incorporating an isolated grounding conductor connection intended for the reduction of electrical noise (electromagnetic interference) as permitted in 250.146(D) shall be identified by an orange triangle located on the face of the receptacle.

(1) Isolated Equipment Grounding Conductor Required. Receptacles so identified shall be used only with equipment grounding conductors that are isolated in accordance with 250.146(D).

(2) Installation in Nonmetallic Boxes. Isolated ground receptacles installed in nonmetallic boxes shall be covered with a nonmetallic faceplate.

Exception: Where an isolated ground receptacle is installed in a nonmetallic box, a metal faceplate shall be permitted if the box contains a feature or accessory that permits the effective grounding of the faceplate.

(E) Controlled Receptacle Marking. All nonlocking-type, 125-volt, 15- and 20-ampere receptacles that are controlled by an automatic control device, or that incorporate control features that remove power from the receptacle for the purpose of energy management or building automation, shall be permanently marked with the symbol shown in Figure 406.3(E) and the word "controlled."

For receptacles controlled by an automatic control device, the marking shall be located on the receptacle face and visible after installation.

In both cases where a multiple receptacle device is used, the required marking of the word "controlled" and symbol shall denote which contact device(s) are controlled.

Figure 406.3(E) Controlled Receptacle Marking Symbol.

406.4 General Installation Requirements. Receptacle outlets shall be located in branch circuits in accordance with Part III of Article 210. General installation requirements shall be in accordance with 406.4(A) through (F).

(A) Grounding Type. Except as provided in 406.4(D), receptacles installed on 15- and 20-ampere branch circuits shall be of the grounding type. Grounding-type receptacles shall be installed only on circuits of the voltage class and current for which they are rated, except as provided in Table 210.21(B)(2) and Table 210.21(B)(3).

(B) To Be Grounded. Receptacles and cord connectors that have equipment grounding conductor contacts shall have those contacts connected to an equipment grounding conductor.

Exception No. 2: Replacement receptacles as permitted by 406.4(D).

(C) Methods of Grounding. The equipment grounding conductor contacts of receptacles and cord connectors shall be grounded by connection to the equipment grounding conductor of the circuit supplying the receptacle or cord connector.

The branch-circuit wiring method shall include or provide an equipment grounding conductor to which the equipment grounding conductor contacts of the receptacle or cord connector are connected.

(D) Replacements. Replacement of receptacles shall comply with 406.4(D)(1) through (D)(6), as applicable. Arc-fault circuit-interrupter type and ground-fault circuit-interrupter type receptacles shall be installed in a readily accessible location.

(1) Grounding-Type Receptacles. Where a grounding means exists in the receptacle enclosure or an equipment grounding conductor is installed in accordance with 250.130(C), grounding-type receptacles shall be used and shall be connected to the equipment grounding conductor in accordance with 406.4(C) or 250.130(C).

(2) Non–Grounding-Type Receptacles. Where attachment to an equipment grounding conductor does not exist in the receptacle enclosure, the installation shall comply with (D)(2)(a), (D)(2)(b), or (D)(2)(c).

(a) A non–grounding-type receptacle(s) shall be permitted to be replaced with another non–grounding-type receptacle(s).

(b) A non–grounding-type receptacle(s) shall be permitted to be replaced with a ground-fault circuit interrupter-type of receptacle(s). These receptacles or their cover plates shall be marked "No Equipment Ground." An equipment grounding conductor shall not be connected from the ground-fault circuit-interrupter-type receptacle to any outlet supplied from the ground-fault circuit-interrupter receptacle.

(c) A non–grounding-type receptacle(s) shall be permitted to be replaced with a grounding-type receptacle(s) where supplied through a ground-fault circuit interrupter. Where grounding-type receptacles are supplied through the ground-fault circuit interrupter, grounding-type receptacles or their cover plates shall be marked "GFCI Protected" and "No Equipment Ground," visible after installation. An equipment grounding conductor shall not be connected between the grounding-type receptacles.

(3) Ground-Fault Circuit Interrupters. Ground-fault circuit-interrupter protected receptacles shall be provided where replacements are made at receptacle outlets that are required to be so protected elsewhere in this *Code*.

Exception: Where replacement of the receptacle type is impracticable, such as where the outlet box size will not permit the installation of the GFCI receptacle, the receptacle shall be permitted to be replaced with a new receptacle of the existing type, where GFCI protection is provided and the receptacle is marked "GFCI Protected" and "No Equipment Ground," in accordance with 406.4(D)(2)(a), (b), or (c), as applicable.

(5) Tamper-Resistant Receptacles. Listed tamper-resistant receptacles shall be provided where replacements are made at receptacle outlets that are required to be tamper-resistant elsewhere in this *Code*, except where a non-grounding receptacle is replaced with another non-grounding receptacle.

(6) Weather-Resistant Receptacles. Weather-resistant receptacles shall be provided where replacements are made at receptacle outlets that are required to be so protected elsewhere in this *Code*.

(F) Noninterchangeable Types. Receptacles connected to circuits that have different voltages, frequencies, or types of current (ac or dc) on the same premises shall be of such design that the attachment plugs used on these circuits are not interchangeable.

406.5 Receptacle Mounting. Receptacles shall be mounted in identified boxes or assemblies. The boxes or assemblies shall be securely fastened in place unless otherwise permitted elsewhere in this *Code*. Screws used for the purpose of attaching receptacles to a box shall be of the type provided with a listed receptacle, or shall be machine screws having 32 threads per inch or part of listed assemblies or systems, in accordance with the manufacturer's instructions.

(A) Boxes That Are Set Back. Receptacles mounted in boxes that are set back from the finished surface as permitted in 314.20 shall be installed such that the mounting yoke or strap of the receptacle is held rigidly at the finished surface.

(B) Boxes That Are Flush. Receptacles mounted in boxes that are flush with the finished surface or project therefrom shall be installed such that the mounting yoke or strap of the receptacle is held rigidly against the box or box cover.

(C) Receptacles Mounted on Covers. Receptacles mounted to and supported by a cover shall be held rigidly against the cover by more than one screw or shall be a device assembly or box cover listed and identified for securing by a single screw.

(D) Position of Receptacle Faces. After installation, receptacle faces shall be flush with or project from faceplates of insulating material and shall project a minimum of 0.4 mm (0.015 in.) from metal faceplates.

Exception: Listed kits or assemblies encompassing receptacles and nonmetallic faceplates that cover the receptacle face, where the plate cannot be installed on any other receptacle, shall be permitted.

(E) Receptacles in Countertops. Receptacle assemblies for installation in countertop surfaces shall be listed for countertop applications. Where receptacle assemblies for countertop applications are required to provide ground-fault circuit-interrupter protection for personnel in accordance with 210.8, such assemblies shall be permitted to be listed as GFCI receptacle assemblies for countertop applications.

N **(F) Receptacles in Work Surfaces.** Receptacle assemblies and GFCI receptacle assemblies listed for work surface or countertop applications shall be permitted to be installed in work surfaces.

N **(G) Receptacle Orientation.** Receptacles shall not be installed in a face-up position in or on countertop surfaces or work surfaces unless listed for countertop or work surface applications.

(H) Receptacles in Seating Areas and Other Similar Surfaces. In seating areas or similar surfaces, receptacles shall not be installed in a face-up position unless the receptacle is any of the following:

(1) Part of an assembly listed as a furniture power distribution unit
(2) Part of an assembly listed either as household furnishings or as commercial furnishings
(3) Listed either as a receptacle assembly for countertop applications or as a GFCI receptacle assembly for countertop applications
(4) Installed in a listed floor box

(J) Voltage Between Adjacent Devices. A receptacle shall not be grouped or ganged in enclosures with other receptacles, snap switches, or similar devices, unless they are arranged so that the voltage between adjacent devices does not exceed 300 volts, or unless they are installed in enclosures equipped with identified, securely installed barriers between adjacent devices.

406.6 Receptacle Faceplates (Cover Plates). Receptacle faceplates shall be installed so as to completely cover the opening and seat against the mounting surface.

Receptacle faceplates mounted inside a box having a recess-mounted receptacle shall effectively close the opening and seat against the mounting surface.

406.7 Attachment Plugs, Cord Connectors, and Flanged Surface Devices. All attachment plugs, cord connectors, and flanged surface devices (inlets and outlets) shall be listed and marked with the manufacturer's name or identification and voltage and ampere ratings.

(B) Connection of Attachment Plugs. Attachment plugs shall be installed so that their prongs, blades, or pins are not energized unless inserted into an energized receptacle or cord connectors. No receptacle shall be installed so as to require the insertion of an energized attachment plug as its source of supply.

(D) Flanged Surface Inlet. A flanged surface inlet shall be installed such that the prongs, blades, or pins are not energized unless an energized cord connector is inserted into it.

406.9 Receptacles in Damp or Wet Locations.

(A) Damp Locations. A receptacle installed outdoors in a location protected from the weather or in other damp locations shall have an enclosure for the receptacle that is weatherproof when the receptacle is covered (attachment plug cap not inserted and receptacle covers closed).

An installation suitable for wet locations shall also be considered suitable for damp locations.

A receptacle shall be considered to be in a location protected from the weather where located under roofed open porches, canopies, marquees, and the like, and will not be subjected to a beating rain or water runoff. All 15- and 20-ampere, 125- and 250-volt nonlocking receptacles shall be a listed weather-resistant type.

(B) Wet Locations.

(1) Receptacles of 15 and 20 Amperes in a Wet Location. Receptacles of 15 and 20 amperes, 125 and 250 volts installed in a wet location shall have an enclosure that is weatherproof whether or not the attachment plug cap is inserted. An outlet box hood installed for this purpose shall be listed and shall be identified as "extra-duty." Other listed products, enclosures, or assemblies providing weatherproof protection that do not utilize an outlet box hood need not be marked "extra duty."

Exception: 15- and 20-ampere, 125- through 250-volt receptacles installed in a wet location and subject to routine high-pressure spray washing shall be permitted to have an enclosure that is weatherproof when the attachment plug is removed.

All 15- and 20-ampere, 125- and 250-volt nonlocking-type receptacles shall be listed and so identified as the weather-resistant type.

(2) Other Receptacles. All other receptacles installed in a wet location shall comply with (B)(2)(a) or (B)(2)(b).

(a) A receptacle installed in a wet location, where the product intended to be plugged into it is not attended while in use, shall have an enclosure that is weatherproof with the attachment plug cap inserted or removed.

(b) A receptacle installed in a wet location where the product intended to be plugged into it will be attended while in use (e.g., portable tools) shall have an enclosure that is weatherproof when the attachment plug is removed.

(C) Bathtub and Shower Space. Receptacles shall not be installed within or directly over a bathtub or shower stall.

(D) Protection for Floor Receptacles. Standpipes of floor receptacles shall allow floor-cleaning equipment to be operated without damage to receptacles.

(E) Flush Mounting with Faceplate. The enclosure for a receptacle installed in an outlet box flush-mounted in a finished surface shall be made weatherproof by means of a weatherproof faceplate assembly that provides a watertight connection between the plate and the finished surface.

406.12 Tamper-Resistant Receptacles. All 15- and 20-ampere, 125- and 250-volt nonlocking-type receptacles in the areas specified in 406.12(1) through (7) shall be listed tamper-resistant receptacles.

(1) Dwelling units in all areas specified in 210.52 and 550.13
(2) Guest rooms and guest suites of hotels and motels
(3) Child care facilities
(4) Preschools and elementary education facilities
(5) Business offices, corridors, waiting rooms and the like in clinics, medical and dental offices and outpatient facilities
(6) Subset of assembly occupancies described in 518.2 to include places of waiting transportation, gymnasiums, skating rinks, and auditoriums
(7) Dormitories

Exception to (1), (2), (3), (4), (5), (6), and (7): Receptacles in the following locations shall not be required to be tamper resistant:

(1) Receptacles located more than 1.7 m (5½ ft) above the floor
(2) Receptacles that are part of a luminaire or appliance
(3) A single receptacle or a duplex receptacle for two appliances located within the dedicated space for each appliance that, in normal use, is not easily moved from one place to another and that is cord-and-plug-connected in accordance with 400.10(A)(6), (A)(7), or (A)(8)
(4) Nongrounding receptacles used for replacements as permitted in 406.4(D)(2)(a)

CHAPTER 17

SWITCHES

INTRODUCTION

This chapter contains requirements from Article 404 — Switches. Unlike the disconnect switches covered in Chapter 7, this chapter covers general-use snap switches. These switches are generally installed in device boxes or on box covers and are most commonly used for local control of lighting or other outlets. General-use snap switches can also be used as disconnecting means for smaller motors, air-conditioning equipment, and appliances. The disconnecting means requirements in Articles 404, 422, 430, and 440 provide the limitations on this use of general-use snap switches. Requirements from Article 404 also cover general-use snap switches (single-pole, three-way, four-way, double-pole, etc.), dimmers, and similar control switches. The *Code* contains few rules on the actual placement of switches that control lighting load and those rules are contained in Article 210 (see Chapter 18). Many modern switches that react to movement or a room being occupied require power for operation, and the requirements to have a circuit available to power these devices is covered in Article 404.

ARTICLE 404
Switches

PART I. INSTALLATION

404.2 Switch Connections.

(A) Three-Way and Four-Way Switches. Three-way and four-way switches shall be wired so that all switching is done only in the ungrounded circuit conductor. Where in metal raceways or metal-armored cables, wiring between switches and outlets shall be in accordance with 300.20(A).

Exception: Switch loops shall not require a grounded conductor.

(B) Grounded Conductors. Switches or circuit breakers shall not disconnect the grounded conductor of a circuit.

Exception: A switch or circuit breaker shall be permitted to disconnect a grounded circuit conductor where all circuit conductors are disconnected simultaneously, or where the device is arranged so that the grounded conductor cannot be disconnected until all the ungrounded conductors of the circuit have been disconnected.

(C) Switches Controlling Lighting Loads. The grounded circuit conductor for the controlled lighting circuit shall be installed at the location where switches control lighting loads that are supplied by a grounded general-purpose branch circuit serving bathrooms, hallways, stairways, or rooms suitable for human habitation or occupancy as defined in the applicable building code. Where multiple switch locations control the same lighting load such that the entire floor area of the room or space is visible from the single or combined switch locations, the grounded circuit conductor shall only be required at one location. A grounded conductor shall not be required to be installed at lighting switch locations under any of the following conditions:

(1) Where conductors enter the box enclosing the switch through a raceway, provided that the raceway is large

enough for all contained conductors, including a grounded conductor

(2) Where the box enclosing the switch is accessible for the installation of an additional or replacement cable without removing finish materials
(3) Where snap switches with integral enclosures comply with 300.15(E)
(4) Where lighting in the area is controlled by automatic means
(5) Where a switch controls a receptacle load

The grounded conductor shall be extended to any switch location as necessary and shall be connected to switching devices that require line-to-neutral voltage to operate the electronics of the switch in the standby mode and shall meet the requirements of 404.22.

Exception: The connection requirement shall become effective on January 1, 2020. It shall not apply to replacement or retrofit switches installed in locations prior to local adoption of 404.2(C) and where the grounded conductor cannot be extended without removing finish materials. The number of electronic lighting control switches on a branch circuit shall not exceed five, and the number connected to any feeder on the load side of a system or main bonding jumper shall not exceed 25. For the purpose of this exception, a neutral busbar, in compliance with 200.2(B) and to which a main or system bonding jumper is connected shall not be limited as to the number of electronic lighting control switches connected.

404.3 Enclosure.

(B) Used as a Raceway. Enclosures shall not be used as junction boxes, auxiliary gutters, or raceways for conductors feeding through or tapping off to other switches or overcurrent devices, unless the enclosure complies with 312.8.

404.4 Damp or Wet Locations.

(A) Surface-Mounted Switch or Circuit Breaker. A surface-mounted switch or circuit breaker shall be enclosed in a weatherproof enclosure or cabinet that complies with 312.2.

(B) Flush-Mounted Switch or Circuit Breaker. A flush-mounted switch or circuit breaker shall be equipped with a weatherproof cover.

(C) Switches in Tub or Shower Spaces. Switches shall not be installed within tubs or shower spaces unless installed as part of a listed tub or shower assembly.

404.8 Accessibility and Grouping.

(A) Location. All switches and circuit breakers used as switches shall be located so that they may be operated from a readily accessible place. They shall be installed such that the center of the grip of the operating handle of the switch or circuit breaker, when in its highest position, is not more than 2.0 m (6 ft 7 in.) above the floor or working platform.

Exception No. 1: On busway installations, fused switches and circuit breakers shall be permitted to be located at the same level as the busway. Suitable means shall be provided to operate the handle of the device from the floor.

Exception No. 2: Switches and circuit breakers installed adjacent to motors, appliances, or other equipment that they supply shall be permitted to be located higher than 2.0 m (6 ft 7 in.) and to be accessible by portable means.

Exception No. 3: Hookstick operable isolating switches shall be permitted at greater heights.

(B) Voltage Between Adjacent Devices. A snap switch shall not be grouped or ganged in enclosures with other snap switches, receptacles, or similar devices, unless they are arranged so that the voltage between adjacent devices does not exceed 300 volts, or unless they are installed in enclosures equipped with identified, securely installed barriers between adjacent devices.

(C) Multipole Snap Switches. A multipole, general-use snap switch shall not be permitted to be fed from more than a single circuit unless it is listed and marked as a two-circuit or three-circuit switch.

404.9 Provisions for General-Use Snap Switches.

(A) Faceplates. Faceplates provided for snap switches mounted in boxes and other enclosures shall be installed so as to completely cover the opening and, where the switch is flush mounted, seat against the finished surface.

(B) Grounding. Snap switches, including dimmer and similar control switches, shall be connected to an equipment grounding conductor and shall provide a means to connect metal faceplates to the equipment grounding conductor, whether or not a metal faceplate is installed. Metal faceplates shall be grounded. Snap switches shall be considered to be part of an effective ground-fault current path if either of the following conditions is met:

(1) The switch is mounted with metal screws to a metal box or metal cover that is connected to an equipment grounding conductor or to a nonmetallic box with integral means for connecting to an equipment grounding conductor.
(2) An equipment grounding conductor or equipment bonding jumper is connected to an equipment grounding termination of the snap switch.

Exception No. 1 to (B): Where no means exists within the snap-switch enclosure for connecting to the equipment grounding conductor, or where the wiring method does not include or provide an equipment grounding conductor, a snap switch without a connection to an equipment grounding conductor shall be permitted for replacement purposes only. A snap switch wired under the provisions of this exception and located within 2.5 m (8 ft) vertically, or 1.5 m (5 ft) horizontally, of ground or exposed grounded metal objects shall be provided with a faceplate of nonconducting noncombustible material with nonmetallic attachment screws, unless the switch mounting strap or yoke is nonmetallic or the circuit is protected by a ground-fault circuit interrupter.

Exception No. 2 to (B): Listed kits or listed assemblies shall not be required to be connected to an equipment grounding conductor if all of the following conditions are met:

(1) The device is provided with a nonmetallic faceplate that cannot be installed on any other type of device,
(2) The device does not have mounting means to accept other configurations of faceplates,
(3) The device is equipped with a nonmetallic yoke, and

(4) All parts of the device that are accessible after installation of the faceplate are manufactured of nonmetallic materials.

Exception No. 3 to (B): A snap switch with integral nonmetallic enclosure complying with 300.15(E) shall be permitted without a connection to an equipment grounding conductor.

404.10 Mounting of Snap Switches.

(B) Box Mounted. Flush-type snap switches mounted in boxes that are set back of the finished surface as permitted in 314.20 shall be installed so that the extension plaster ears are seated against the surface. Flush-type snap switches mounted in boxes that are flush with the finished surface or project from it shall be installed so that the mounting yoke or strap of the switch is seated against the box. Screws used for the purpose of attaching a snap switch to a box shall be of the type provided with a listed snap switch, or shall be machine screws having 32 threads per inch or part of listed assemblies or systems, in accordance with the manufacturer's instructions.

404.12 Grounding of Enclosures. Metal enclosures for switches or circuit breakers shall be connected to an equipment grounding conductor as specified in Part IV of Article 250. Metal enclosures for switches or circuit breakers used as service equipment shall comply with the provisions of Part V of Article 250. Where nonmetallic enclosures are used with metal raceways or metal-armored cables, provision shall be made for connecting the equipment grounding conductor(s).

Except as covered in 404.9(B), Exception No. 1, nonmetallic boxes for switches shall be installed with a wiring method that provides or includes an equipment grounding conductor.

404.14 Rating and Use of Switches. Switches shall be used within their ratings and as indicated in 404.14(A) through (F).

(A) Alternating-Current General-Use Snap Switch. A form of general-use snap switch suitable only for use on ac circuits for controlling the following:

(1) Resistive and inductive loads not exceeding the ampere rating of the switch at the voltage applied
(2) Tungsten-filament lamp loads not exceeding the ampere rating of the switch at 120 volts
(3) Motor loads not exceeding 80 percent of the ampere rating of the switch at its rated voltage

(B) Alternating-Current or Direct-Current General-Use Snap Switch. A form of general-use snap switch suitable for use on either ac or dc circuits for controlling the following:

(1) Resistive loads not exceeding the ampere rating of the switch at the voltage applied.
(2) Inductive loads not exceeding 50 percent of the ampere rating of the switch at the applied voltage. Switches rated in horsepower are suitable for controlling motor loads within their rating at the voltage applied.
(3) Tungsten-filament lamp loads not exceeding the ampere rating of the switch at the applied voltage if T-rated.

(C) CO/ALR Snap Switches. Snap switches rated 20 amperes or less directly connected to aluminum conductors shall be listed and marked CO/ALR.

(D) Alternating-Current Specific-Use Snap Switches Rated for 347 Volts. Snap switches rated 347 volts ac shall be listed and shall be used only for controlling the loads permitted by (D)(1) and (D)(2).

(1) Noninductive Loads. Noninductive loads other than tungsten-filament lamps not exceeding the ampere and voltage ratings of the switch.

(2) Inductive Loads. Inductive loads not exceeding the ampere and voltage ratings of the switch. Where particular load characteristics or limitations are specified as a condition of the listing, those restrictions shall be observed regardless of the ampere rating of the load.

The ampere rating of the switch shall not be less than 15 amperes at a voltage rating of 347 volts ac. Flush-type snap switches rated 347 volts ac shall not be readily interchangeable in box mounting with switches identified in 404.14(A) and (B).

(E) Dimmer Switches. General-use dimmer switches shall be used only to control permanently installed incandescent luminaires unless listed for the control of other loads and installed accordingly.

(F) Cord- and Plug-Connected Loads. Where a snap switch or control device is used to control cord- and plug-connected equipment on a general-purpose branch circuit, each snap switch or control device controlling receptacle outlets or cord connectors that are supplied by permanently connected cord pendants shall be rated at not less than the rating of the maximum permitted ampere rating or setting of the overcurrent device protecting the receptacles or cord connectors, as provided in 210.21(B).

Exception: Where a snap switch or control device is used to control not more than one receptacle on a branch circuit, the switch or control device shall be permitted to be rated at not less than the rating of the receptacle.

404.22 Electronic Lighting Control Switches. Electronic lighting control switches shall be listed. Electronic lighting control switches shall not introduce current on the equipment grounding conductor during normal operation. The requirement to not introduce current on the equipment grounding conductor shall take effect on January 1, 2020.

Exception: Electronic lighting control switches that introduce current on the equipment grounding conductor shall be permitted for applications covered by 404.2(C), Exception. Electronic lighting control switches that introduce current on the equipment grounding conductor shall be listed and marked for use in replacement or retrofit applications only.

CHAPTER 18

LIGHTING REQUIREMENTS

INTRODUCTION

This chapter contains requirements from Article 210 — Branch Circuits; Article 410 — Luminaires, Lampholders, and Lamps; and Article 411 — Low-Voltage Lighting. While dwelling units have numerous switch-controlled lighting outlet requirements, commercial and industrial locations have their lighting outlets arranged based on room, furniture, or equipment layout, and design considerations for the necessary amount of illumination required for a specific use or operation.

Article 410 covers luminaires, portable luminaires, lampholders, pendants, incandescent filament lamps, arc lamps, electric-discharge lamps, decorative lighting products, lighting accessories for temporary seasonal and holiday use, portable flexible lighting products, and the wiring and equipment forming part of such products and lighting installations. Contained in this *Pocket Guide* are ten of the fifteen parts that are included in Article 410: Part I General; Part II Luminaire Locations; Part III Provisions at Luminaire Outlet Boxes, Canopies, and Pans; Part IV Luminaire Supports; Part V Grounding; Part VI Wiring of Luminaires; Part X Special Provisions for Flush and Recessed Luminaires; Part XII Special Provisions for Electric-Discharge Lighting Systems of 1000 Volts or less; Part XIII Special Provisions for Electric-Discharge Lighting Systems of More than 1000 Volts; and Part XIV Lighting Track.

Article 411 covers lighting systems operating at 30 volts or less and their associated components. While these systems have historically been used for landscape and other outdoor accent uses, they have become extremely popular for interior display and task lighting. Under counter low-voltage LED systems are used extensively in residential kitchens.

ARTICLE 210
Branch Circuits

PART III. REQUIRED OUTLETS

210.70 Lighting Outlets Required.

(C) All Occupancies. For attics and underfloor spaces, utility rooms, and basements, at least one lighting outlet containing a switch or controlled by a wall switch shall be installed where these spaces are used for storage or contain equipment requiring servicing. At least one point of control shall be at the usual point of entry to these spaces. The lighting outlet shall be provided at or near the equipment requiring servicing.

ARTICLE 410
Luminaires, Lampholders, and Lamps

PART I. GENERAL

410.6 Listing Required. All luminaires, lampholders, and retrofit kits shall be listed.

410.8 Inspection. Luminaires shall be installed such that the connections between the luminaire conductors and the circuit conductors can be inspected without requiring the disconnection of any part of the wiring unless the luminaires are connected by attachment plugs and receptacles.

PART II. LUMINAIRE LOCATIONS

410.10 Luminaires in Specific Locations.

(A) Wet and Damp Locations. Luminaires installed in wet or damp locations shall be installed such that water cannot enter or accumulate in wiring compartments, lampholders, or other electrical parts. All luminaires installed in wet locations shall be marked, "Suitable for Wet Locations." All luminaires installed in damp locations shall be marked "Suitable for Wet Locations" or "Suitable for Damp Locations."

(B) Corrosive Locations. Luminaires installed in corrosive locations shall be of a type suitable for such locations.

(C) In Ducts or Hoods. Luminaires shall be permitted to be installed in commercial cooking hoods where all of the following conditions are met:

(1) The luminaire shall be identified for use within commercial cooking hoods and installed such that the temperature limits of the materials used are not exceeded.
(2) The luminaire shall be constructed so that all exhaust vapors, grease, oil, or cooking vapors are excluded from the lamp and wiring compartment. Diffusers shall be resistant to thermal shock.
(3) Parts of the luminaire exposed within the hood shall be corrosion resistant or protected against corrosion, and the surface shall be smooth so as not to collect deposits and to facilitate cleaning.
(4) Wiring methods and materials supplying the luminaire(s) shall not be exposed within the cooking hood.

(D) Bathtub and Shower Areas. No parts of cord-connected luminaires, chain-, cable-, or cord-suspended luminaires, lighting track, pendants, or ceiling-suspended (paddle) fans shall be located within a zone measured 900 mm (3 ft) horizontally and 2.5 m (8 ft) vertically from the top of the bathtub rim or shower stall threshold. This zone is all encompassing and includes the space directly over the tub or shower stall. Luminaires located within the actual outside dimension of the bathtub or shower to a height of 2.5 m (8 ft) vertically from the top of the bathtub rim or shower threshold shall be marked for damp locations, or marked for wet locations where subject to shower spray.

(E) Luminaires in Indoor Sports, Mixed-Use, and All-Purpose Facilities. Luminaires subject to physical damage, using a mercury vapor or metal halide lamp, installed in playing and spectator seating areas of indoor sports, mixed-use, or all-purpose facilities shall be of the type that protects the lamp with a glass or plastic lens. Such luminaires shall be permitted to have an additional guard.

(F) Luminaires Installed in or Under Roof Decking. Luminaires installed in exposed or concealed locations under metal-corrugated sheet roof decking shall be installed and supported so there is not less than 38 mm (1½ in.) measured from the lowest surface of the roof decking to the top of the luminaire.

410.11 Luminaires Near Combustible Material. Luminaires shall be constructed, installed, or equipped with shades or guards so that combustible material is not subjected to temperatures in excess of 90°C (194°F).

410.12 Luminaires over Combustible Material. Lampholders installed over highly combustible material shall be of the unswitched type. Unless an individual switch is provided for each luminaire, lampholders shall be located at least 2.5 m (8 ft) above the floor or shall be located or guarded so that the lamps cannot be readily removed or damaged.

410.14 Luminaires in Show Windows. Chain-supported luminaires used in a show window shall be permitted to be externally wired. No other externally wired luminaires shall be used.

410.18 Space for Cove Lighting. Coves shall have adequate space and shall be located so that lamps and equipment can be properly installed and maintained.

PART III. PROVISIONS AT LUMINAIRE OUTLET BOXES, CANOPIES, AND PANS

410.20 Space for Conductors. Canopies and outlet boxes taken together shall provide sufficient space so that luminaire conductors and their connecting devices are capable of being installed in accordance with 314.16.

410.21 Temperature Limit of Conductors in Outlet Boxes. Luminaires shall be of such construction or installed so that the conductors in outlet boxes shall not be subjected to temperatures greater than that for which the conductors are rated.

Branch-circuit wiring, other than 2-wire or multiwire branch circuits supplying power to luminaires connected together, shall not be passed through an outlet box that is an integral part of a luminaire unless the luminaire is identified for through-wiring.

410.22 Outlet Boxes to Be Covered. In a completed installation, each outlet box shall be provided with a cover unless covered by means of a luminaire canopy, lampholder, receptacle, or similar device.

410.23 Covering of Combustible Material at Outlet Boxes. Any combustible wall or ceiling finish exposed between the edge of a luminaire canopy or pan and an outlet box having a surface area of 1160 mm^2 (180 $in.^2$) or more shall be covered with noncombustible material.

410.24 Connection of Electric-Discharge and LED Luminaires.

(A) Independent of the Outlet Box. Electric-discharge and LED luminaires supported independently of the outlet box shall be connected to the branch circuit through metal raceway, nonmetallic raceway, Type MC cable, Type AC cable, Type MI cable, nonmetallic sheathed cable, or by flexible cord as permitted in 410.62(B) or 410.62(C).

(B) Access to Boxes. Electric-discharge and LED luminaires surface mounted over concealed outlet, pull, or junction boxes and designed not to be supported solely by the outlet box shall be provided with suitable openings in the back of the luminaire to provide access to the wiring in the box.

PART IV. LUMINAIRE SUPPORTS

410.30 Supports.

(A) General. Luminaires and lampholders shall be securely supported. A luminaire that weighs more than 3 kg (6 lb) or exceeds 400 mm (16 in.) in any dimension shall not be supported by the screw shell of a lampholder.

(B) Metal or Nonmetallic Poles Supporting Luminaires. Metal or nonmetallic poles shall be permitted to be used to support luminaires and as a raceway to enclose supply conductors, provided the following conditions are met:

(1) A pole shall have a handhole not less than 50 mm × 100 mm (2 in. × 4 in.) with a cover suitable for use in wet locations to provide access to the supply terminations within the pole or pole base.

Exception No. 1: No handhole shall be required in a pole 2.5 m (8 ft) or less in height abovegrade where the supply wiring method continues without splice or pull point, and where the interior of the pole and any splices are accessible by removing the luminaire.

Exception No. 2: No handhole shall be required in a pole 6.0 m (20 ft) or less in height abovegrade that is provided with a hinged base.

(2) Where raceway risers or cable is not installed within the pole, a threaded fitting or nipple shall be brazed, welded, or attached to the pole opposite the handhole for the supply connection.
(3) A metal pole shall be provided with an equipment grounding terminal as follows:

    a. A pole with a handhole shall have the equipment grounding terminal accessible from the handhole.
    b. A pole with a hinged base shall have the equipment grounding terminal accessible within the base.

Exception to (3): No grounding terminal shall be required in a pole 2.5 m (8 ft) or less in height abovegrade where the supply wiring method continues without splice or pull, and where the interior of the pole and any splices are accessible by removing the luminaire.

(4) A metal pole with a hinged base shall have the hinged base and pole bonded together.
(5) Metal raceways or other equipment grounding conductors shall be bonded to the metal pole with an equipment grounding conductor recognized by 250.118 and sized in accordance with 250.122.
(6) Conductors in vertical poles used as raceway shall be supported as provided in 300.19.

410.36 Means of Support.

(A) Outlet Boxes. Outlet boxes or fittings installed as required by 314.23 and complying with the provisions of 314.27(A)(1) and 314.27(A)(2) shall be permitted to support luminaires.

(B) Suspended Ceilings. Framing members of suspended ceiling systems used to support luminaires shall be securely fastened to each other and shall be securely attached to the building structure at appropriate intervals. Luminaires shall be securely fastened to the ceiling framing member by mechanical means such as bolts, screws, or rivets. Listed clips identified for use with the type of ceiling framing member(s) and luminaire(s) shall also be permitted.

(C) Luminaire Studs. Luminaire studs that are not a part of outlet boxes, hickeys, tripods, and crowfeet shall be made of steel, malleable iron, or other material suitable for the application.

(D) Insulating Joints. Insulating joints that are not designed to be mounted with screws or bolts shall have an exterior metal casing, insulated from both screw connections.

(E) Raceway Fittings. Raceway fittings used to support a luminaire(s) shall be capable of supporting the weight of the complete fixture assembly and lamp(s).

(F) Busways. Luminaires shall be permitted to be connected to busways in accordance with 368.17(C).

(G) Trees. Outdoor luminaires and associated equipment shall be permitted to be supported by trees.

PART V. GROUNDING

410.40 General. Luminaires and lighting equipment shall be grounded as required in Article 250 and Part V of this article.

410.42 Luminaire(s) with Exposed Conductive Parts. Exposed metal parts shall be connected to an equipment grounding conductor or insulated from the equipment grounding conductor and other conducting surfaces or be inaccessible to unqualified personnel. Lamp tie wires, mounting screws, clips, and decorative bands on glass spaced at least 38 mm (1½ in.) from lamp terminals shall not be required to be grounded.

410.44 Methods of Grounding. Luminaires and equipment shall be mechanically connected to an equipment grounding conductor as specified in 250.118 and sized in accordance with 250.122.

Exception No. 1: Luminaires made of insulating material that is directly wired or attached to outlets supplied by a wiring method that does not provide a ready means for grounding attachment to an equipment grounding conductor shall be made of insulating material and shall have no exposed conductive parts.

Exception No. 2: Replacement luminaires shall be permitted to connect an equipment grounding conductor from the outlet in compliance with 250.130(C). The luminaire shall then comply with 410.42.

Exception No. 3: Where no equipment grounding conductor exists at the outlet, replacement luminaires that are GFCI protected shall not be required to be connected to an equipment grounding conductor.

410.46 Equipment Grounding Conductor Attachment. Luminaires with exposed metal parts shall be provided with a means for connecting an equipment grounding conductor for such luminaires.

PART VI. WIRING OF LUMINAIRES

410.48 Luminaire Wiring — General. Wiring on or within luminaires shall be neatly arranged and shall not be exposed to physical damage. Excess wiring shall be avoided. Conductors shall be arranged so that they are not subjected to temperatures above those for which they are rated.

410.50 Polarization of Luminaires. Luminaires shall be wired so that the screw shells of lampholders are connected to the same luminaire or circuit conductor or terminal. The grounded conductor, where connected to a screw shell lampholder, shall be connected to the screw shell.

410.52 Conductor Insulation. Luminaires shall be wired with conductors having insulation suitable for the environmental conditions, current, voltage, and temperature to which the conductors will be subjected.

410.56 Protection of Conductors and Insulation.

(A) Properly Secured. Conductors shall be secured in a manner that does not tend to cut or abrade the insulation.

(B) Protection Through Metal. Conductor insulation shall be protected from abrasion where it passes through metal.

(C) Luminaire Stems. Splices and taps shall not be located within luminaire arms or stems.

(D) Splices and Taps. No unnecessary splices or taps shall be made within or on a luminaire.

(E) Stranding. Stranded conductors shall be used for wiring on luminaire chains and on other movable or flexible parts.

(F) Tension. Conductors shall be arranged so that the weight of the luminaire or movable parts does not put tension on the conductors.

410.59 Cord-Connected Showcases. Individual showcases, other than fixed, shall be permitted to be connected by flexible cord to permanently installed receptacles, and groups of not more than six such showcases shall be permitted to be coupled together by flexible cord and separable locking-type connectors with one of the group connected by flexible cord to a permanently installed receptacle.

The installation shall comply with 410.59(A) through (E).

(A) Cord Requirements. Flexible cord shall be of the hard-service type, having conductors not smaller than the branch-circuit conductors, having ampacity at least equal to the branch-circuit overcurrent device, and having an equipment grounding conductor.

(B) Receptacles, Connectors, and Attachment Plugs. Receptacles, connectors, and attachment plugs shall be of a listed grounding type rated 15 or 20 amperes.

(C) Support. Flexible cords shall be secured to the undersides of showcases such that all of the following conditions are ensured:

(1) The wiring is not exposed to physical damage.
(2) The separation between cases is not in excess of 50 mm (2 in.), or more than 300 mm (12 in.) between the first case and the supply receptacle.
(3) The free lead at the end of a group of showcases has a female fitting not extending beyond the case.

(D) No Other Equipment. Equipment other than showcases shall not be electrically connected to showcases.

(E) Secondary Circuit(s). Where showcases are cord-connected, the secondary circuit(s) of each electric-discharge lighting ballast shall be limited to one showcase.

410.62 Cord-Connected Lampholders and Luminaires.

(B) Adjustable Luminaires. Luminaires that require adjusting or aiming after installation shall not be required to be equipped with an attachment plug or cord connector, provided the exposed cord is suitable for hard-usage or extra-hard-usage and is not longer than that required for maximum adjustment. The cord shall not be subject to strain or physical damage.

(C) Electric-Discharge and LED Luminaires. Electric-discharge and LED luminaires shall comply with (1), (2), and (3) as applicable.

(1) Cord-Connected Installation. A luminaire or a listed assembly in compliance with any of the conditions in (a) through (c) shall be permitted to be cord connected provided the luminaire is located directly below the outlet or busway, the cord is not subject to strain or physical damage, and the cord is visible over its entire length except at terminations.

(a) A luminaire shall be permitted to be connected with a cord terminating in a grounding-type attachment plug or busway plug.

(b) A luminaire assembly equipped with a strain relief and canopy shall be permitted to use a cord connection between the luminaire assembly and the canopy. The canopy shall be permitted to include a section of raceway not over 150 mm (6 in.) in length and intended to facilitate the connection to an outlet box mounted above a suspended ceiling.

• (c) Listed luminaires connected using listed assemblies that incorporate manufactured wiring system connectors in accordance with 604.100(C) shall be permitted to be cord connected.

(2) Provided with Mogul-Base, Screw Shell Lampholders. Electric-discharge luminaires provided with mogul-base, screw shell lampholders shall be permitted to be connected to branch circuits of 50 amperes or less by cords complying with 240.5. Receptacles and attachment plugs shall be permitted to be of a lower ampere rating than the branch circuit but not less than 125 percent of the luminaire full-load current.

(3) Equipped with Flanged Surface Inlet. Electric-discharge luminaires equipped with a flanged surface inlet shall be permitted to be supplied by cord pendants equipped with cord connectors. Inlets and connectors shall be permitted to be of a lower ampere rating than the branch circuit but not less than 125 percent of the luminaire load current.

410.64 Luminaires as Raceways. Luminaires shall not be used as a raceway for circuit conductors unless they comply with 410.64(A), (B), or (C).

(A) Listed. Luminaires listed and marked for use as a raceway shall be permitted to be used as a raceway.

(B) Through-Wiring. Luminaires identified for through-wiring, as permitted by 410.21, shall be permitted to be used as a raceway.

(C) Luminaires Connected Together. Luminaires designed for end-to-end connection to form a continuous assembly, or luminaires connected together by recognized wiring methods, shall be permitted to contain the conductors of a 2-wire branch circuit, or one multiwire branch circuit, supplying the connected luminaires and shall not be required to be listed as a raceway. One additional 2-wire branch circuit separately supplying one or more of the connected luminaires shall also be permitted.

410.68 Feeder and Branch-Circuit Conductors and Ballasts. Feeder and branch-circuit conductors within 75 mm (3 in.) of a ballast, LED driver, power supply, or transformer shall have an insulation temperature rating not lower than 90°C (194°F), unless supplying a luminaire marked as suitable for a different insulation temperature.

PART X. SPECIAL PROVISIONS FOR FLUSH AND RECESSED LUMINAIRES

410.110 General. Luminaires installed in recessed cavities in walls or ceilings, including suspended ceilings, shall comply with 410.115 through 410.122.

410.115 Temperature.

(A) Combustible Material. Luminaires shall be installed so that adjacent combustible material will not be subjected to temperatures in excess of 90°C (194°F).

(B) Fire-Resistant Construction. Where a luminaire is recessed in fire-resistant material in a building of fire-resistant construction, a temperature higher than 90°C (194°F) but not higher than 150°C (302°F) shall be considered acceptable if the luminaire is plainly marked for that service.

(C) Recessed Incandescent Luminaires. Incandescent luminaires shall have thermal protection and shall be identified as thermally protected.

Exception No. 1: Thermal protection shall not be required in a recessed luminaire identified for use and installed in poured concrete.

Exception No. 2: Thermal protection shall not be required in a recessed luminaire whose design, construction, and thermal performance characteristics are equivalent to a thermally protected luminaire and are identified as inherently protected.

410.116 Clearance and Installation.

(A) Clearance.

(1) Non-Type IC. A recessed luminaire that is not identified for contact with insulation shall have all recessed parts spaced not less than 13 mm (½ in.) from combustible materials. The points of support and the trim finishing off the openings in the ceiling, wall, or other finished surface shall be permitted to be in contact with combustible materials.

(2) Type IC. A recessed luminaire that is identified for contact with insulation, Type IC, shall be permitted to be in contact with combustible materials at recessed parts, points of support, and portions passing through or finishing off the opening in the building structure.

(B) Installation. Thermal insulation shall not be installed above a recessed luminaire or within 75 mm (3 in.) of the recessed luminaire's enclosure, wiring compartment, ballast, transformer, LED driver, or power supply unless the luminaire is identified as Type IC for insulation contact.

410.117 Wiring.

(A) General. Conductors that have insulation suitable for the temperature encountered shall be used.

(B) Circuit Conductors. Branch-circuit conductors that have an insulation suitable for the temperature encountered shall be permitted to terminate in the luminaire.

(C) Tap Conductors. Tap conductors of a type suitable for the temperature encountered shall be permitted to run from the luminaire terminal connection to an outlet box placed at least 300 mm (1 ft) from the luminaire. Such tap conductors shall be in suitable raceway or Type AC or MC cable of at least 450 mm (18 in.) but not more than 1.8 m (6 ft) in length.

PART XII. SPECIAL PROVISIONS FOR ELECTRIC-DISCHARGE LIGHTING SYSTEMS OF 1000 VOLTS OR LESS

410.130 General.

(A) Open-Circuit Voltage of 1000 Volts or Less. Equipment for use with electric-discharge lighting systems and designed for an open-circuit voltage of 1000 volts or less shall be of a type identified for such service.

(F) High-Intensity Discharge Luminaires.

(1) Recessed. Recessed high-intensity luminaires designed to be installed in wall or ceiling cavities shall have thermal protection and be identified as thermally protected.

(2) Inherently Protected. Thermal protection shall not be required in a recessed high-intensity luminaire whose design, construction, and thermal performance characteristics are equivalent to a thermally protected luminaire and are identified as inherently protected.

(3) Installed in Poured Concrete. Thermal protection shall not be required in a recessed high-intensity discharge luminaire identified for use and installed in poured concrete.

(4) Recessed Remote Ballasts. A recessed remote ballast for a high-intensity discharge luminaire shall have thermal protection that is integral with the ballast and shall be identified as thermally protected.

(5) Metal Halide Lamp Containment. Luminaires that use a metal halide lamp other than a thick-glass parabolic reflector lamp (PAR) shall be provided with a containment barrier that encloses the lamp, or shall be provided with a physical means that only allows the use of a lamp that is Type O.

(G) Disconnecting Means.

(1) General. In indoor locations other than dwellings and associated accessory structures, fluorescent luminaires that utilize double-ended lamps and contain ballast(s) that can be serviced in place shall have a disconnecting means either internal or external to each luminaire. For existing installed luminaires without disconnecting means, at the time a ballast is replaced, a disconnecting means shall be installed. The line side terminals of the disconnecting means shall be guarded.

Exception No. 1: A disconnecting means shall not be required for luminaires installed in hazardous (classified) location(s).

Exception No. 2: A disconnecting means shall not be required for luminaires that provide emergency illumination required in 700.16.

Exception No. 3: For cord-and-plug-connected luminaires, an accessible separable connector or an accessible plug and receptacle shall be permitted to serve as the disconnecting means.

Exception No. 4: Where more than one luminaire is installed and supplied by other than a multiwire branch circuit, a disconnecting means shall not be required for every luminaire when the design of the installation includes disconnecting means, such that the illuminated space cannot be left in total darkness.

(2) Multiwire Branch Circuits. When connected to multiwire branch circuits, the disconnecting means shall simultaneously break all the supply conductors to the ballast, including the grounded conductor.

(3) Location. The disconnecting means shall be located so as to be accessible to qualified persons before servicing or maintaining the ballast. Where the disconnecting means is external to the luminaire, it shall be a single device, and shall be attached to the luminaire or the luminaire shall be located within sight of the disconnecting means.

410.134 Direct-Current Equipment. Luminaires installed on dc circuits shall be equipped with auxiliary equipment and resistors designed for dc operation. The luminaires shall be marked for dc operation.

410.136 Luminaire Mounting.

(A) Exposed Components. Luminaires that have exposed ballasts, transformers, LED drivers, or power supplies shall be installed such that ballasts, transformers, LED drivers, or power supplies shall not be in contact with combustible material unless listed for such condition.

(B) Combustible Low-Density Cellulose Fiberboard. Where a surface-mounted luminaire containing a ballast, transformer, LED driver, or power supply is to be installed on combustible low-density cellulose fiberboard, it shall be marked for this condition or shall be spaced not less than 38 mm (1½ in.) from the surface of the fiberboard. Where such luminaires are partially or wholly recessed, the provisions of 410.110 through 410.122 shall apply.

PART XIII. SPECIAL PROVISIONS FOR ELECTRIC-DISCHARGE LIGHTING SYSTEMS OF MORE THAN 1000 VOLTS

410.140 General.

(A) Listing. Electric-discharge lighting systems with an open-circuit voltage exceeding 1000 volts shall be listed and installed in conformance with that listing.

410.141 Control.

(A) Disconnection. Luminaires or lamp installation shall be controlled either singly or in groups by an externally operable switch or circuit breaker that opens all ungrounded primary conductors.

(B) Within Sight or Locked Type. The switch or circuit breaker shall be located within sight from the luminaires or lamps, or it shall be permitted to be located elsewhere if it is lockable in accordance with 110.25.

410.144 Transformer Locations.

(A) Accessible. Transformers shall be accessible after installation.

(C) Adjacent to Combustible Materials. Transformers shall be located so that adjacent combustible materials are not subjected to temperatures in excess of 90°C (194°F).

PART XIV. LIGHTING TRACK

410.151 Installation.

(A) Lighting Track. Lighting track shall be permanently installed and permanently connected to a branch circuit. Only lighting track fittings shall be installed on lighting track. Lighting track fittings shall not be equipped with general-purpose receptacles.

(B) Connected Load. The connected load on lighting track shall not exceed the rating of the track. Lighting track shall be supplied by a branch circuit having a rating not more than that of the track. The load calculation in 220.43(B) shall not be required to limit the length of track on a single branch circuit, and it shall not be required to limit the number of luminaires on a single track.

(C) Locations Not Permitted. Lighting track shall not be installed in the following locations:

(1) Where likely to be subjected to physical damage
(2) In wet or damp locations
(3) Where subject to corrosive vapors
(4) In storage battery rooms
(5) In hazardous (classified) locations
(6) Where concealed
(7) Where extended through walls or partitions
(8) Less than 1.5 m (5 ft) above the finished floor except where protected from physical damage or track operating at less than 30 volts rms open-circuit voltage
(9) Where prohibited by 410.10(D)

(D) Support. Fittings identified for use on lighting track shall be designed specifically for the track on which they are to be installed. They shall be securely fastened to the track, shall maintain polarization and connections to the equipment grounding conductor, and shall be designed to be suspended directly from the track.

410.153 Heavy-Duty Lighting Track. Heavy-duty lighting track is lighting track identified for use exceeding 20 amperes. Each fitting attached to a heavy-duty lighting track shall have individual overcurrent protection.

410.154 Fastening. Lighting track shall be securely mounted so that each fastening is suitable for supporting the maximum weight of luminaires that can be installed. Unless identified for supports at greater intervals, a single section 1.2 m (4 ft) or shorter in length shall have two supports, and, where installed in a continuous row, each individual section of not more than 1.2 m (4 ft) in length shall have one additional support.

410.155 Construction Requirements.

(B) Grounding. Lighting track shall be grounded in accordance with Article 250, and the track sections shall be securely coupled to maintain continuity of the circuitry, polarization, and grounding throughout.

ARTICLE 411

Low-Voltage Lighting

411.1 Scope. This article covers lighting systems and their associated components operating at no more than 30 volts ac or 60 volts dc. Where wet contact is likely to occur, the limits are 15 volts ac or 30 volts dc.

411.3 Low-Voltage Lighting Systems. Low voltage lighting systems shall consist of an isolating power supply, low-voltage luminaires, and associated equipment that are all identified for the use. The output circuits of the power supply shall be rated for 25 amperes maximum under all load conditions.

411.4 Listing Required. Low-voltage lighting systems shall comply with 411.4(A) or 411.4(B).

(A) Listed System. The luminaires, power supply, and luminaire fittings (including the exposed bare conductors) of an exposed bare conductor lighting system shall be listed for the use as part of the same identified lighting system.

(B) Assembly of Listed Parts. A lighting system assembled from the following listed parts shall be permitted:

(1) Low-voltage luminaires
(2) Power supply
(3) Low-voltage luminaire fittings
(4) Suitably rated cord, cable, conductors in conduit, or other fixed Chapter 3 wiring method for the secondary circuit

411.5 Specific Location Requirements.

(A) Walls, Floors, and Ceilings. Conductors concealed or extended through a wall, floor, or ceiling shall be in accordance with (1) or (2):

(1) Installed using any of the wiring methods specified in Chapter 3
(2) Installed using wiring supplied by a listed Class 2 power source and installed in accordance with 725.130

(B) Pools, Spas, Fountains, and Similar Locations. Lighting systems shall be installed not less than 3 m (10 ft) horizontally from the nearest edge of the water, unless permitted by Article 680.

411.6 Secondary Circuits.

(A) Grounding. Secondary circuits shall not be grounded.

(B) Isolation. The secondary circuit shall be insulated from the branch circuit by an isolating transformer.

(C) Bare Conductors. Exposed bare conductors and current-carrying parts shall be permitted for indoor installations only. Bare conductors shall not be installed less than 2.1 m (7 ft) above the finished floor, unless specifically listed for a lower installation height.

(D) Insulated Conductors. Insulated secondary circuit conductors shall be of the type, and installed as, described in (1), (2), or (3):

(1) Class 2 cable supplied by a Class 2 power source and installed in accordance with Parts I and III of Article 725.

(2) Conductors, cord, or cable of the listed system and installed not less than 2.1 m (7 ft) above the finished floor unless the system is specifically listed for a lower installation height.
(3) Wiring methods described in Chapter 3.

411.7 Branch Circuit. Lighting systems covered by this article shall be supplied from a maximum 20-ampere branch circuit.

411.8 Hazardous (Classified) Locations. Where installed in hazardous (classified) locations, these systems shall conform with Articles 500 through 517 in addition to this article.

CHAPTER 19

MOTORS, MOTOR CIRCUITS, AND CONTROLLERS

INTRODUCTION

This chapter contains requirements from Article 430 — Motors, Motor Circuits, and Controllers. Article 430 contains requirements for motors, motor branch-circuit and feeder conductors and their protection, motor overload protection, motor control circuits, motor controllers, motor control centers, and grounding of motors operating at any voltage. Overcurrent protection of motor circuits is approached differently than the typical lighting and appliance branch circuit. The general approach is to locate any overcurrent protective device at the beginning of a circuit, and that device provides the requisite level of protection for all overcurrent conditions. Inrush currents that occur when a motor starts necessitate an intentional delay in the operation of the overcurrent protective device, which in many cases is accomplished by increasing the size of the device.

The device at the point the circuit is supplied provides short-circuit and line-to-ground fault level protection, while a thermal or other type of current sensing device located elsewhere in the circuit protects the circuit conductors and the motor from overload conditions. In smaller motors these levels of protection can often be provided by a single device in the branch circuit. However, as motor size increases, so does the inrush current, and the allowable two-part approach to overcurrent protection is necessary. All of these rules are contained in the various parts of Article 430, and many are based on sizing the device based on a certain percentage of the motor current or the current values assigned in the Article 430 tables.

Contained in this *Pocket Guide* are twelve of the fourteen parts that are included in Article 430: Part I General; Part II

Motor Circuit Conductors; Part III Motor and Branch-Circuit Overload Protection; Part IV Motor Branch-Circuit Short-Circuit and Ground-Fault Protection; Part V Motor Feeder Short-Circuit and Ground-Fault Protection; Part VI Motor Control Circuits; Part VII Motor Controllers; Part VIII Motor Control Centers; Part IX Disconnecting Means; Part X Adjustable-Speed Drive Systems; Part XIII Grounding — All Voltages; and Part XIV Tables.

ARTICLE 430

Motors, Motor Circuits, and Controllers

PART I. GENERAL

430.2 Definitions.

Valve Actuator Motor (VAM) Assemblies. A manufactured assembly, used to operate a valve, consisting of an actuator motor and other components such as controllers, torque switches, limit switches, and overload protection.

430.6 Ampacity and Motor Rating Determination. The size of conductors supplying equipment covered by Article 430 shall be selected from the allowable ampacity tables in accordance with 310.15(B) or shall be calculated in accordance with 310.15(C). Where flexible cord is used, the size of the conductor shall be selected in accordance with 400.5. The required ampacity and motor ratings shall be determined as specified in 430.6(A), (B), (C), and (D).

(A) General Motor Applications. For general motor applications, current ratings shall be determined based on (A)(1) and (A)(2).

(1) Table Values. Other than for motors built for low speeds (less than 1200 RPM) or high torques, and for multispeed motors, the values given in Table 430.247, Table 430.248, Table 430.249, and Table 430.250 shall be used to determine the ampacity of conductors or ampere ratings of switches, branch-circuit short-circuit and ground-fault protection, instead of the actual current rating marked on the motor nameplate. Where a motor is marked in amperes, but not horsepower, the horsepower rating shall be assumed to be that corresponding to the value given in Table 430.247, Table 430.248, Table 430.249, and Table 430.250, interpolated if necessary. Motors built for low speeds (less than 1200 RPM) or high torques may have higher full-load currents, and multispeed motors will have full-load current varying with speed, in which case the nameplate current ratings shall be used.

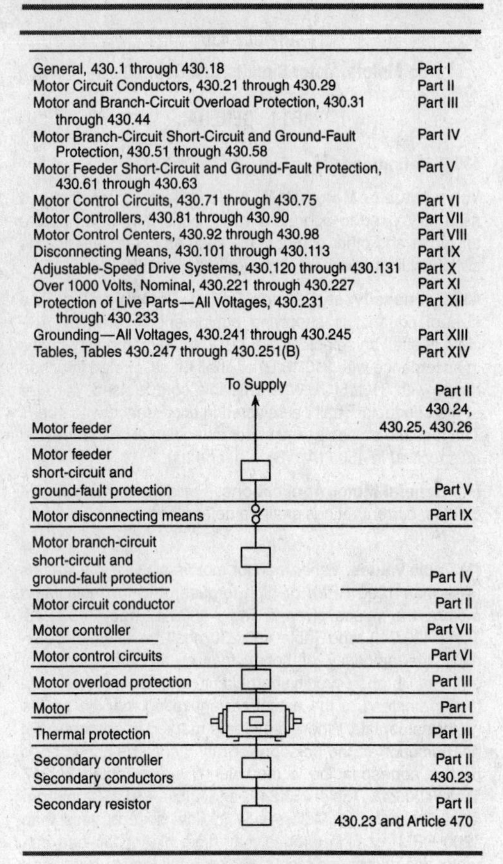

General, 430.1 through 430.18	Part I
Motor Circuit Conductors, 430.21 through 430.29	Part II
Motor and Branch-Circuit Overload Protection, 430.31 through 430.44	Part III
Motor Branch-Circuit Short-Circuit and Ground-Fault Protection, 430.51 through 430.58	Part IV
Motor Feeder Short-Circuit and Ground-Fault Protection, 430.61 through 430.63	Part V
Motor Control Circuits, 430.71 through 430.75	Part VI
Motor Controllers, 430.81 through 430.90	Part VII
Motor Control Centers, 430.92 through 430.98	Part VIII
Disconnecting Means, 430.101 through 430.113	Part IX
Adjustable-Speed Drive Systems, 430.120 through 430.131	Part X
Over 1000 Volts, Nominal, 430.221 through 430.227	Part XI
Protection of Live Parts—All Voltages, 430.231 through 430.233	Part XII
Grounding—All Voltages, 430.241 through 430.245	Part XIII
Tables, Tables 430.247 through 430.251(B)	Part XIV

Figure 430.1 Article 430 Contents.

Exception No. 1: Multispeed motors shall be in accordance with 430.22(B) and 430.52.

Exception No. 2: For equipment that employs a shaded-pole or permanent-split capacitor-type fan or blower motor that is marked with the motor type, the full load current for such motor marked on the nameplate of the equipment in which the fan or blower motor is employed shall be used instead of the horsepower rating to determine the ampacity or rating of the disconnecting means, the branch-circuit conductors, the controller, the branch-circuit short-circuit and ground-fault protection, and the separate overload protection. This marking on the equipment nameplate shall not be less than the current marked on the fan or blower motor nameplate.

Exception No. 3: For a listed motor-operated appliance that is marked with both motor horsepower and full-load current, the motor full-load current marked on the nameplate of the appliance shall be used instead of the horsepower rating on the appliance nameplate to determine the ampacity or rating of the disconnecting means, the branch-circuit conductors, the controller, the branch-circuit short-circuit and ground-fault protection, and any separate overload protection.

(2) Nameplate Values. Separate motor overload protection shall be based on the motor nameplate current rating.

(D) Valve Actuator Motor Assemblies. For valve actuator motor assemblies (VAMs), the rated current shall be the nameplate full-load current, and this current shall be used to determine the maximum rating or setting of the motor branch-circuit short-circuit and ground-fault protective device and the ampacity of the conductors.

430.9 Terminals.

(B) Conductors. Motor controllers and terminals of control circuit devices shall be connected with copper conductors unless identified for use with a different conductor.

(C) Torque Requirements. Control circuit devices with screw-type pressure terminals used with 14 AWG or smaller copper conductors shall be torqued to a minimum of 0.8 N·m (7 lb-in.) unless identified for a different torque value.

430.10 Wiring Space in Enclosures.

(A) General. Enclosures for motor controllers and disconnecting means shall not be used as junction boxes, auxiliary gutters, or raceways for conductors feeding through or tapping off to the other apparatus unless designs are employed that provide adequate space for this purpose.

430.11 Protection Against Liquids. Suitable guards or enclosures shall be provided to protect exposed current-carrying parts of motors and the insulation of motor leads where installed directly under equipment, or in other locations where dripping or spraying oil, water, or other liquid is capable of occurring, unless the motor is designed for the existing conditions.

430.14 Location of Motors.

(A) Ventilation and Maintenance. Motors shall be located so that adequate ventilation is provided and so that maintenance, such as lubrication of bearings and replacing of brushes, can be readily accomplished.

430.16 Exposure to Dust Accumulations. In locations where dust or flying material collects on or in motors in such quantities as to seriously interfere with the ventilation or cooling of motors and thereby cause dangerous temperatures, suitable types of enclosed motors that do not overheat under the prevailing conditions shall be used.

430.17 Highest Rated or Smallest Rated Motor. In determining compliance with 430.24, 430.53(B), and 430.53(C), the highest rated or smallest rated motor shall be based on the rated full-load current as selected from Table 430.247, Table 430.248, Table 430.249, and Table 430.250.

PART II. MOTOR CIRCUIT CONDUCTORS

430.22 Single Motor. Conductors that supply a single motor used in a continuous duty application shall have an ampacity of not less than 125 percent of the motor full-load current rating, as determined by 430.6(A)(1), or not less than specified in 430.22(A) through (G).

(A) Direct-Current Motor-Rectifier Supplied. For dc motors operating from a rectified power supply, the conductor ampacity on the input of the rectifier shall not be less than 125 percent of the rated input current to the rectifier. For dc motors operating from a rectified single-phase power supply, the conductors between the field wiring output terminals of the rectifier and the motor shall have an ampacity of not less than the following percentages of the motor full-load current rating:

(1) Where a rectifier bridge of the single-phase, half-wave type is used, 190 percent.
(2) Where a rectifier bridge of the single-phase, full-wave type is used, 150 percent.

(B) Multispeed Motor. For a multispeed motor, the selection of branch-circuit conductors on the line side of the controller shall be based on the highest of the full-load current ratings shown on the motor nameplate. The ampacity of the branch-circuit conductors between the controller and the motor shall not be less than 125 percent of the current rating of the winding(s) that the conductors energize.

(C) Wye-Start, Delta-Run Motor. For a wye-start, delta-run connected motor, the ampacity of the branch-circuit conductors on the line side of the controller shall not be less than 125 percent of the motor full-load current as determined by 430.6(A)(1). The ampacity of the conductors between the controller and the motor shall not be less than 72 percent of the motor full-load current rating as determined by 430.6(A)(1).

(D) Part-Winding Motor. For a part-winding connected motor, the ampacity of the branch-circuit conductors on the line side of the controller shall not be less than 125 percent of the motor full-load current as determined by 430.6(A)(1). The ampacity of the conductors between the controller and the motor shall not be less than 62.5 percent of the motor full-load current rating as determined by 430.6(A)(1).

(E) Other Than Continuous Duty. Conductors for a motor used in a short-time, intermittent, periodic, or varying duty application shall have an ampacity of not less than the percentage of the motor nameplate current rating shown in Table 430.22(E), unless the authority having jurisdiction grants special permission for conductors of lower ampacity.

Table 430.22(E) Duty-Cycle Service

Classification of Service	Nameplate Current Rating Percentages			
	5-Minute Rated Motor	15-Minute Rated Motor	30- & 60-Minute Rated Motor	Continuous Rated Motor
Short-time duty operating valves, raising or lowering rolls, etc.	110	120	150	—
Intermittent duty freight and passenger elevators, tool heads, pumps, drawbridges, turntables, etc. (for arc welders, see 630.11)	85	85	90	140
Periodic duty rolls, ore- and coal-handling machines, etc.	85	90	95	140
Varying duty	110	120	150	200

Note: Any motor application shall be considered as continuous duty unless the nature of the apparatus it drives is such that the motor will not operate continuously with load under any condition of use.

(G) Conductors for Small Motors. Conductors for small motors shall not be smaller than 14 AWG unless otherwise permitted in 430.22(G)(1) or (G)(2).

(1) 18 AWG Copper. 18 AWG individual copper conductors installed in a cabinet or enclosure, copper conductors that are part of a jacketed multiconductor cable assembly, or copper conductors in a flexible cord shall be permitted, under either of the following sets of conditions:

(1) The circuit supplies a motor with a full-load current rating, as determined by 430.6(A)(1), of greater than 3.5 amperes, and less than or equal to 5 amperes, and all the following conditions are met:

   a. The circuit is protected in accordance with 430.52.
   b. The circuit is provided with maximum Class 10 or Class 10A overload protection in accordance with 430.32.
   c. Overcurrent protection is provided in accordance with 240.4(D)(1)(2).

(2) The circuit supplies a motor with a full-load current rating, as determined by 430.6(A)(1), of 3.5 amperes or less, and all the following conditions are met:

   a. The circuit is protected in accordance with 430.52.
   b. The circuit is provided with maximum Class 20 overload protection in accordance with 430.32.
   c. Overcurrent protection is provided in accordance with 240.4(D)(1)(2).

(2) 16 AWG Copper. 16 AWG individual copper conductors installed in a cabinet or enclosure, copper conductors that are part of a jacketed multiconductor cable assembly, or copper conductors in a flexible cord shall be permitted under either of the following sets of conditions:

(1) The circuit supplies a motor with a full-load current rating, as determined by 430.6(A)(1), of greater than 5.5 amperes,

and less than or equal to 8 amperes, and all the following conditions are met:

a. The circuit is protected in accordance with 430.52.
b. The circuit is provided with maximum Class 10 or Class 10A overload protection in accordance with 430.32.
c. Overcurrent protection is provided in accordance with 240.4(D)(2)(2).

(2) The circuit supplies a motor with a full-load current rating, as determined by 430.6(A)(1), of 5.5 amperes or less, and all the following conditions are met:

a. The circuit is protected in accordance with 430.52.
b. The circuit is provided with maximum Class 20 overload protection in accordance with 430.32.
c. Overcurrent protection is provided in accordance with 240.4(D)(2)(2).

430.23 Wound-Rotor Secondary.

(A) Continuous Duty. For continuous duty, the conductors connecting the secondary of a wound-rotor ac motor to its controller shall have an ampacity not less than 125 percent of the full-load secondary current of the motor.

(B) Other Than Continuous Duty. For other than continuous duty, these conductors shall have an ampacity, in percent of full-load secondary current, not less than that specified in Table 430.22(E).

(C) Resistor Separate from Controller. Where the secondary resistor is separate from the controller, the ampacity of the conductors between controller and resistor shall not be less than that shown in Table 430.23(C).

430.24 Several Motors or a Motor(s) and Other Load(s). Conductors supplying several motors, or a motor(s) and other load(s), shall have an ampacity not less than the sum of each of the following:

(1) 125 percent of the full-load current rating of the highest rated motor, as determined by 430.6(A)

Table 430.23(C) Secondary Conductor

Resistor Duty Classification	Ampacity of Conductor in Percent of Full-Load Secondary Current
Light starting duty	35
Heavy starting duty	45
Extra-heavy starting duty	55
Light intermittent duty	65
Medium intermittent duty	75
Heavy intermittent duty	85
Continuous duty	110

(2) Sum of the full-load current ratings of all the other motors in the group, as determined by 430.6(A)
(3) 100 percent of the noncontinuous non-motor load
(4) 125 percent of the continuous non-motor load.

Exception No. 1: Where one or more of the motors of the group are used for short-time, intermittent, periodic, or varying duty, the ampere rating of such motors to be used in the summation shall be determined in accordance with 430.22(E). For the highest rated motor, the greater of either the ampere rating from 430.22(E) or the largest continuous duty motor full-load current multiplied by 1.25 shall be used in the summation.

Exception No. 2: The ampacity of conductors supplying motor-operated fixed electric space-heating equipment shall comply with 424.3(B).

Exception No. 3: Where the circuitry is interlocked so as to prevent simultaneous operation of selected motors or other loads, the conductor ampacity shall be permitted to be based on the summation of the currents of the motors and other loads to be operated simultaneously that results in the highest total current.

430.25 Multimotor and Combination-Load Equipment. The ampacity of the conductors supplying multimotor and combination-load equipment shall not be less than the minimum circuit ampacity marked on the equipment in accordance with 430.7(D). Where the equipment is not factory-wired and the individual nameplates are visible in accordance with 430.7(D)(2), the conductor ampacity shall be determined in accordance with 430.24.

430.26 Feeder Demand Factor. Where reduced heating of the conductors results from motors operating on duty-cycle, intermittently, or from all motors not operating at one time, the authority having jurisdiction may grant permission for feeder conductors to have an ampacity less than specified in 430.24, provided the conductors have sufficient ampacity for the maximum load determined in accordance with the sizes and number of motors supplied and the character of their loads and duties.

430.28 Feeder Taps. Feeder tap conductors shall have an ampacity not less than that required by Part II, shall terminate in a branch-circuit protective device, and, in addition, shall meet one of the following requirements:

(1) Be enclosed either by an enclosed controller or by a raceway, be not more than 3.0 m (10 ft) in length, and, for field installation, be protected by an overcurrent device on the line side of the tap conductor, the rating or setting of which shall not exceed 1000 percent of the tap conductor ampacity
(2) Have an ampacity of at least one-third that of the feeder conductors, be suitably protected from physical damage or enclosed in a raceway, and be not more than 7.5 m (25 ft) in length
(3) Have an ampacity not less than the feeder conductors

Exception: Feeder taps over 7.5 m (25 ft) long. In high-bay manufacturing buildings [over 11 m (35 ft) high at walls], where conditions of maintenance and supervision ensure that only qualified persons service the systems, conductors tapped to a feeder shall be permitted to be not over 7.5 m (25 ft) long horizontally and not over 30.0 m (100 ft) in total length where all of the following conditions are met:

(1) The ampacity of the tap conductors is not less than one-third that of the feeder conductors.

(2) The tap conductors terminate with a single circuit breaker or a single set of fuses complying with (1) Part IV, where the load-side conductors are a branch circuit, or (2) Part V, where the load-side conductors are a feeder.

(3) The tap conductors are suitably protected from physical damage and are installed in raceways.

(4) The tap conductors are continuous from end-to-end and contain no splices.

(5) The tap conductors shall be 6 AWG copper or 4 AWG aluminum or larger.

(6) The tap conductors shall not penetrate walls, floors, or ceilings.

(7) The tap shall not be made less than 9.0 m (30 ft) from the floor.

PART III. MOTOR AND BRANCH-CIRCUIT OVERLOAD PROTECTION

430.31 General. Part III specifies overload devices intended to protect motors, motor-control apparatus, and motor branch-circuit conductors against excessive heating due to motor overloads and failure to start.

These provisions shall not require overload protection where a power loss would cause a hazard, such as in the case of fire pumps.

The provisions of Part III shall not apply to motor circuits rated over 1000 volts, nominal.

430.32 Continuous-Duty Motors.

(A) More Than 1 Horsepower. Each motor used in a continuous duty application and rated more than 1 hp shall be protected against overload by one of the means in 430.32(A)(1) through (A)(4).

(1) Separate Overload Device. A separate overload device that is responsive to motor current. This device shall be selected to trip or shall be rated at no more than the following percent of the motor nameplate full-load current rating:

Motors with a marked service factor 1.15 or greater	125%
Motors with a marked temperature rise 40°C or less	125%
All other motors	115%

Modification of this value shall be permitted as provided in 430.32(C). For a multispeed motor, each winding connection shall be considered separately.

Where a separate motor overload device is connected so that it does not carry the total current designated on the motor nameplate, such as for wye-delta starting, the proper percentage of nameplate current applying to the selection or setting of the overload device shall be clearly designated on the equipment, or the manufacturer's selection table shall take this into account.

Informational Note: Where power factor correction capacitors are installed on the load side of the motor overload device, see 460.9.

(2) Thermal Protector. A thermal protector integral with the motor, approved for use with the motor it protects on the basis that it will prevent dangerous overheating of the motor due to overload and failure to start. The ultimate trip current of a thermally protected motor shall not exceed the following percentage of motor full-load current given in Table 430.248, Table 430.249, and Table 430.250:

Motor full-load current 9 amperes or less	170%
Motor full-load current from 9.1 to, and including, 20 amperes	156%
Motor full-load current greater than 20 amperes	140%

If the motor current-interrupting device is separate from the motor and its control circuit is operated by a protective device integral with the motor, it shall be arranged so that the opening of the control circuit will result in interruption of current to the motor.

(C) Selection of Overload Device. Where the sensing element or setting or sizing of the overload device selected in accordance with 430.32(A)(1) and 430.32(B)(1) is not sufficient to start the motor or to carry the load, higher size sensing elements or incremental settings or sizing shall be permitted to be used, provided the trip current of the overload device does not exceed the following percentage of motor nameplate full-load current rating:

Motors with marked service factor 1.15 or greater	140%
Motors with a marked temperature rise 40°C or less	140%
All other motors	130%

If not shunted during the starting period of the motor as provided in 430.35, the overload device shall have sufficient time delay to permit the motor to start and accelerate its load.

Informational Note: A Class 20 overload relay will provide a longer motor acceleration time than a Class 10 or Class 10A overload relay. A Class 30 overload relay will provide a longer motor acceleration time than a Class 20 overload relay. Use of a higher class overload relay may preclude the need for selection of a higher trip current.

(D) One Horsepower or Less, Nonautomatically Started.

(1) Permanently Installed. Overload protection shall be in accordance with 430.32(B).

(2) Not Permanently Installed.

(a) *Within Sight from Controller.* Overload protection shall be permitted to be furnished by the branch-circuit short-circuit and ground-fault protective device; such device, however, shall not be larger than that specified in Part IV of Article 430.

Exception: Any such motor shall be permitted on a nominal 120-volt branch circuit protected at not over 20 amperes.

(b) *Not Within Sight from Controller.* Overload protection shall be in accordance with 430.32(B).

(E) Wound-Rotor Secondaries. The secondary circuits of wound-rotor ac motors, including conductors, controllers, resistors, and so forth, shall be permitted to be protected against overload by the motor-overload device.

430.36 Fuses — In Which Conductor. Where fuses are used for motor overload protection, a fuse shall be inserted in each ungrounded conductor and also in the grounded conductor if the supply system is 3-wire, 3-phase ac with one conductor grounded.

430.37 Devices Other Than Fuses — In Which Conductor. Where devices other than fuses are used for motor overload protection, Table 430.37 shall govern the minimum allowable number and location of overload units such as trip coils or relays.

430.40 Overload Relays. Overload relays and other devices for motor overload protection that are not capable of opening short circuits or ground faults shall be protected by fuses or

Table 430.37 Overload Units

Kind of Motor	Supply System	Number and Location of Overload Units, Such as Trip Coils or Relays
1-phase ac or dc	2-wire, 1-phase ac or dc ungrounded	1 in either conductor
1-phase ac or dc	2-wire, 1-phase ac or dc, one conductor grounded	1 in ungrounded conductor
1-phase ac or dc	3-wire, 1-phase ac or dc, grounded neutral conductor	1 in either ungrounded conductor
1-phase ac	Any 3-phase	1 in ungrounded conductor
2-phase ac	3-wire, 2-phase ac, ungrounded	2, one in each phase
2-phase ac	3-wire, 2-phase ac, one conductor grounded	2 in ungrounded conductors
2-phase ac	4-wire, 2-phase ac, grounded or ungrounded	2, one for each phase in ungrounded conductors
2-phase ac	Grounded neutral or 5-wire, 2-phase ac, ungrounded	2, one for each phase in any ungrounded phase wire
3-phase ac	Any 3-phase	3, one in each phase*

**Exception: An overload unit in each phase shall not be required where overload protection is provided by other approved means.*

circuit breakers with ratings or settings in accordance with 430.52 or by a motor short-circuit protector in accordance with 430.52.

Exception: Where approved for group installation and marked to indicate the maximum size of fuse or inverse time circuit breaker by which they must be protected, the overload devices shall be protected in accordance with this marking.

430.42 Motors on General-Purpose Branch Circuits. Overload protection for motors used on general-purpose branch circuits as permitted in Article 210 shall be provided as specified in 430.42(A), (B), (C), or (D).

(A) Not over 1 Horsepower. One or more motors without individual overload protection shall be permitted to be connected to a general-purpose branch circuit only where the installation complies with the limiting conditions specified in 430.32(B) and 430.32(D) and 430.53(A)(1) and (A)(2).

(B) Over 1 Horsepower. Motors of ratings larger than specified in 430.53(A) shall be permitted to be connected to general-purpose branch circuits only where each motor is protected by overload protection selected to protect the motor as specified in 430.32. Both the controller and the motor overload device shall be approved for group installation with the short-circuit and ground-fault protective device selected in accordance with 430.53.

430.43 Automatic Restarting. A motor overload device that can restart a motor automatically after overload tripping shall not be installed unless approved for use with the motor it protects. A motor overload device that can restart a motor automatically after overload tripping shall not be installed if automatic restarting of the motor can result in injury to persons.

430.44 Orderly Shutdown. If immediate automatic shutdown of a motor by a motor overload protective device(s) would introduce additional or increased hazard(s) to a person(s) and continued motor operation is necessary for safe shutdown of equipment or process, a motor overload sensing device(s) complying with the provisions of Part III of this article shall be permitted to be connected to a supervised alarm instead of causing immediate interruption of the motor circuit, so that corrective action or an orderly shutdown can be initiated.

PART IV. MOTOR BRANCH-CIRCUIT SHORT-CIRCUIT AND GROUND-FAULT PROTECTION

430.51 General. Part IV specifies devices intended to protect the motor branch-circuit conductors, the motor control apparatus, and the motors against overcurrent due to short circuits or ground faults. These rules add to or amend the provisions of Article 240. The devices specified in Part IV do not include the types of devices required by 210.8, 230.95, and 590.6.

The provisions of Part IV shall not apply to motor circuits rated over 1000 volts, nominal.

430.52 Rating or Setting for Individual Motor Circuit.

(A) General. The motor branch-circuit short-circuit and ground-fault protective device shall comply with 430.52(B) and either 430.52(C) or (D), as applicable.

(B) All Motors. The motor branch-circuit short-circuit and ground-fault protective device shall be capable of carrying the starting current of the motor.

(C) Rating or Setting.

(1) In Accordance with Table 430.52. A protective device that has a rating or setting not exceeding the value calculated according to the values given in Table 430.52 shall be used.

Exception No. 1: Where the values for branch-circuit short-circuit and ground-fault protective devices determined by Table 430.52 do not correspond to the standard sizes or ratings of fuses, nonadjustable circuit breakers, thermal protective devices, or possible settings of adjustable circuit breakers, a higher size, rating, or possible setting that does not exceed the next higher standard ampere rating shall be permitted.

(2) Overload Relay Table. Where maximum branch-circuit short-circuit and ground-fault protective device ratings are

Table 430.52 Maximum Rating or Setting of Motor Branch-Circuit Short-Circuit and Ground-Fault Protective Devices

	Percentage of Full-Load Current			
Type of Motor	Nontime Delay Fuse[1]	Dual Element (Time-Delay) Fuse[1]	Instantaneous Trip Breaker	Inverse Time Breaker[2]
Single-phase motors	300	175	800	250
AC polyphase motors other than wound-rotor	300	175	800	250
Squirrel cage — other than Design B energy-efficient	300	175	800	250
Design B energy-efficient	300	175	1100	250
Synchronous[3]	300	175	800	250
Wound-rotor	150	150	800	150
DC (constant voltage)	150	150	250	150

Note: For certain exceptions to the values specified, see 430.54.

[1]The values in the Nontime Delay Fuse column apply to time-delay Class CC fuses.

[2]The values given in the last column also cover the ratings of nonadjustable inverse time types of circuit breakers that may be modified as in 430.52(C)(1), Exceptions No. 1 and No. 2.

[3]Synchronous motors of the low-torque, low-speed type (usually 450 rpm or lower), such as are used to drive reciprocating compressors, pumps, and so forth, that start unloaded, do not require a fuse rating or circuit-breaker setting in excess of 200 percent of full-load current.

shown in the manufacturer's overload relay table for use with a motor controller or are otherwise marked on the equipment, they shall not be exceeded even if higher values are allowed as shown above.

(3) Instantaneous Trip Circuit Breaker. An instantaneous trip circuit breaker shall be used only if adjustable and if part of a listed combination motor controller having coordinated motor overload and short-circuit and ground-fault protection in each conductor, and the setting is adjusted to no more than the value specified in Table 430.52.

Exception No. 1: Where the setting specified in Table 430.52 is not sufficient for the starting current of the motor, the setting of an instantaneous trip circuit breaker shall be permitted to be increased but shall in no case exceed 1300 percent of the motor full-load current for other than Design B energy-efficient motors and no more than 1700 percent of full-load motor current for Design B energy-efficient motors. Trip settings above 800 percent for other than Design B energy-efficient motors and above 1100 percent for Design B energy-efficient motors shall be permitted where the need has been demonstrated by engineering evaluation. In such cases, it shall not be necessary to first apply an instantaneous-trip circuit breaker at 800 percent or 1100 percent.

Exception No. 2: Where the motor full-load current is 8 amperes or less, the setting of the instantaneous-trip circuit breaker with a continuous current rating of 15 amperes or less in a listed combination motor controller that provides coordinated motor branch-circuit overload and short-circuit and ground-fault protection shall be permitted to be increased to the value marked on the controller.

(4) Multispeed Motor. For a multispeed motor, a single short-circuit and ground-fault protective device shall be permitted for two or more windings of the motor, provided the rating of the protective device does not exceed the above applicable percentage of the nameplate rating of the smallest winding protected.

Exception: For a multispeed motor, a single short-circuit and ground-fault protective device shall be permitted to be used and sized according to the full-load current of the highest current winding, where all of the following conditions are met:

(a) Each winding is equipped with individual overload protection sized according to its full-load current.

(b) The branch-circuit conductors supplying each winding are sized according to the full-load current of the highest full-load current winding.

(c) The controller for each winding has a horsepower rating not less than that required for the winding having the highest horsepower rating.

430.53 Several Motors or Loads on One Branch Circuit. Two or more motors or one or more motors and other loads shall be permitted to be connected to the same branch circuit under conditions specified in 430.53(D) and in 430.53(A),

(B), or (C). The branch-circuit protective device shall be fuses or inverse time circuit breakers.

(A) Not Over 1 Horsepower. Several motors, each not exceeding 1 hp in rating, shall be permitted on a nominal 120-volt branch circuit protected at not over 20 amperes or a branch circuit of 1000 volts, nominal, or less, protected at not over 15 amperes, if all of the following conditions are met:

(1) The full-load rating of each motor does not exceed 6 amperes.
(2) The rating of the branch-circuit short-circuit and ground-fault protective device marked on any of the controllers is not exceeded.
(3) Individual overload protection conforms to 430.32.

(B) If Smallest Rated Motor Protected. If the branch-circuit short-circuit and ground-fault protective device is selected not to exceed that allowed by 430.52 for the smallest rated motor, two or more motors or one or more motors and other load(s), with each motor having individual overload protection, shall be permitted to be connected to a branch circuit where it can be determined that the branch-circuit short-circuit and ground-fault protective device will not open under the most severe normal conditions of service that might be encountered.

(C) Other Group Installations. Two or more motors of any rating or one or more motors and other load(s), with each motor having individual overload protection, shall be permitted to be connected to one branch circuit where the motor controller(s) and overload device(s) are (1) installed as a listed factory assembly and the motor branch-circuit short-circuit and ground-fault protective device either is provided as part of the assembly or is specified by a marking on the assembly, or (2) the motor branch-circuit short-circuit and ground-fault protective device, the motor controller(s), and overload device(s) are field-installed as separate assemblies listed for such use and provided with manufacturers' instructions for use with each other, and (3) all of the following conditions are complied with:

(1) Each motor overload device is either (a) listed for group installation with a specified maximum rating of fuse, inverse time circuit breaker, or both, or (b) selected such that the ampere rating of the motor-branch short-circuit and ground-fault protective device does not exceed that permitted by 430.52 for that individual motor overload device and corresponding motor load.
(2) Each motor controller is either (a) listed for group installation with a specified maximum rating of fuse, circuit breaker, or both, or (b) selected such that the ampere rating of the motor-branch short-circuit and ground-fault protective device does not exceed that permitted by 430.52 for that individual controller and corresponding motor load.
(3) Each circuit breaker is listed and is of the inverse time type.
(4) The branch circuit shall be protected by fuses or inverse time circuit breakers having a rating not exceeding that specified in 430.52 for the highest rated motor connected to the branch circuit plus an amount equal to the sum of the full-load current ratings of all other motors and the ratings of other loads connected to the circuit. Where this calculation results in a rating less than the ampacity of the branch-circuit conductors, it shall be permitted to increase the maximum rating of the fuses or circuit breaker to a value not exceeding that permitted by 240.4(B).
(5) The branch-circuit fuses or inverse time circuit breakers are not larger than allowed by 430.40 for the overload relay protecting the smallest rated motor of the group.
(6) Overcurrent protection for loads other than motor loads shall be in accordance with Parts I through VII of Article 240.

(D) Single Motor Taps. For group installations described above, the conductors of any tap supplying a single motor shall not be required to have an individual branch-circuit short-circuit and ground-fault protective device, provided they comply with one of the following:

(1) No conductor to the motor shall have an ampacity less than that of the branch-circuit conductors.

(2) No conductor to the motor shall have an ampacity less than one-third that of the branch-circuit conductors, with a minimum in accordance with 430.22. The conductors from the point of the tap to the motor overload device shall be not more than 7.5 m (25 ft) long and be protected from physical damage by being enclosed in an approved raceway or by use of other approved means.

(3) Conductors from the point of the tap from the branch circuit to a listed manual motor controller additionally marked "Suitable for Tap Conductor Protection in Group Installations," or to a branch-circuit protective device, shall be permitted to have an ampacity not less than one-tenth the rating or setting of the branch-circuit short-circuit and ground-fault protective device. The conductors from the controller to the motor shall have an ampacity in accordance with 430.22. The conductors from the point of the tap to the controller(s) shall (1) be suitably protected from physical damage and enclosed either by an enclosed controller or by a raceway and be not more than 3 m (10 ft) long or (2) have an ampacity not less than that of the branch-circuit conductors.

(4) Conductors from the point of the tap from the branch circuit to a listed manual motor controller additionally marked "Suitable for Tap Conductor Protection in Group Installations," or to a branch-circuit protective device, shall be permitted to have an ampacity not less than one-third that of the branch-circuit conductors. The conductors from the controller to the motor shall have an ampacity in accordance with 430.22. The conductors from the point of the tap to the controller(s) shall (1) be suitably protected from physical damage and enclosed either by an enclosed controller or by a raceway and be not more than 7.5 m (25 ft) long or (2) have an ampacity not less than that of the branch-circuit conductors.

430.55 Combined Overcurrent Protection. Motor branch-circuit short-circuit and ground-fault protection and motor overload protection shall be permitted to be combined in a single protective device where the rating or setting of the device provides the overload protection specified in 430.32.

PART V. MOTOR FEEDER SHORT-CIRCUIT AND GROUND-FAULT PROTECTION

430.62 Rating or Setting — Motor Load.

(A) Specific Load. A feeder supplying a specific fixed motor load(s) and consisting of conductor sizes based on 430.24 shall be provided with a protective device having a rating or setting not greater than the largest rating or setting of the branch-circuit short-circuit and ground-fault protective device for any motor supplied by the feeder [based on the maximum permitted value for the specific type of a protective device in accordance with 430.52, or 440.22(A) for hermetic refrigerant motor-compressors], plus the sum of the full-load currents of the other motors of the group.

Where the same rating or setting of the branch-circuit short-circuit and ground-fault protective device is used on two or more of the branch circuits supplied by the feeder, one of the protective devices shall be considered the largest for the above calculations.

Exception No. 1: Where one or more instantaneous trip circuit breakers or motor short-circuit protectors are used for motor branch-circuit short-circuit and ground-fault protection as permitted in 430.52(C), the procedure provided above for determining the maximum rating of the feeder protective device shall apply with the following provision: For the purpose of the calculation, each instantaneous trip circuit breaker or motor short-circuit protector shall be assumed to have a rating not exceeding the maximum percentage of motor full-load current permitted by Table 430.52 for the type of feeder protective device employed.

Exception No. 2: Where the feeder overcurrent protective device also provides overcurrent protection for a motor control center, the provisions of 430.94 shall apply.

(B) Other Installations. Where feeder conductors have an ampacity greater than required by 430.24, the rating or setting of the feeder overcurrent protective device shall be permitted to be based on the ampacity of the feeder conductors.

430.63 Rating or Setting — Motor Load and Other Load(s). Where a feeder supplies a motor load and other load(s), the feeder protective device shall have a rating not less than that required for the sum of the other load(s) plus the following:

(1) For a single motor, the rating permitted by 430.52
(2) For a single hermetic refrigerant motor-compressor, the rating permitted by 440.22
(3) For two or more motors, the rating permitted by 430.62

Exception: Where the feeder overcurrent device provides the overcurrent protection for a motor control center, the provisions of 430.94 shall apply.

PART VI. MOTOR CONTROL CIRCUITS

430.72 Overcurrent Protection.

(A) General. A motor control circuit tapped from the load side of a motor branch-circuit short-circuit and ground-fault protective device(s) and functioning to control the motor(s) connected to that branch circuit shall be protected against overcurrent in accordance with 430.72. Such a tapped control circuit shall not be considered to be a branch circuit and shall be permitted to be protected by either a supplementary or branch-circuit overcurrent protective device(s). A motor control circuit other than such a tapped control circuit shall be protected against overcurrent in accordance with 725.43 or the notes to Table 11(A) and Table 11(B) in Chapter 9, as applicable.

(B) Conductor Protection. The overcurrent protection for conductors shall be provided as specified in 430.72(B)(1) or (B)(2).

Exception No. 1: Where the opening of the control circuit would create a hazard as, for example, the control circuit of a fire pump motor, and the like, conductors of control circuits shall require only short-circuit and ground-fault protection and shall be permitted to be protected by the motor branch-circuit short-circuit and ground-fault protective device(s).

Exception No. 2: Conductors supplied by the secondary side of a single-phase transformer having only a two-wire (single-voltage) secondary shall be permitted to be protected by overcurrent protection provided on the primary (supply) side of the transformer, provided this protection does not exceed the value determined by multiplying the appropriate maximum rating of the overcurrent device for the secondary conductor from Table 430.72(B) by the secondary-to-primary voltage ratio. Transformer secondary conductors (other than two-wire) shall not be considered to be protected by the primary overcurrent protection.

(1) Separate Overcurrent Protection. Where the motor branch-circuit short-circuit and ground-fault protective device does not provide protection in accordance with 430.72(B)(2), separate overcurrent protection shall be provided. The overcurrent protection shall not exceed the values specified in Column A of Table 430.72(B).

(2) Branch-Circuit Overcurrent Protective Device. Conductors shall be permitted to be protected by the motor branch-circuit short-circuit and ground-fault protective device and shall require only short-circuit and ground-fault protection. Where the conductors do not extend beyond the motor control equipment enclosure, the rating of the protective device(s) shall not exceed the value specified in Column B of Table 430.72(B). Where the conductors extend beyond the motor control equipment enclosure, the rating of the protective device(s) shall not exceed the value specified in Column C of Table 430.72(B).

(C) Control Circuit Transformer. Where a motor control circuit transformer is provided, the transformer shall be protected in accordance with 430.72(C)(1), (C)(2), (C)(3), (C)(4), or (C)(5).

Exception: Overcurrent protection shall be omitted where the opening of the control circuit would create a hazard as, for example, the control circuit of a fire pump motor and the like.

(1) Compliance with Article 725. Where the transformer supplies a Class 1 power-limited circuit, Class 2, or Class 3 remote-control circuit complying with the requirements of Article 725, protection shall comply with Article 725.

Table 430.72(B) Maximum Rating of Overcurrent Protective Device in Amperes

			Protection Provided by Motor Branch-Circuit Protective Device(s)			
	Column A Separate Protection Provided		Column B Conductors Within Enclosure		Column C Conductors Extend Beyond Enclosure	
Control Circuit Conductor Size (AWG)	Copper	Aluminum or Copper-Clad Aluminum	Copper	Aluminum or Copper-Clad Aluminum	Copper	Aluminum or Copper-Clad Aluminum
18	7	—	25	—	7	—
16	10	—	40	—	10	—
14	(Note 1)	—	100	—	45	—
12	(Note 1)	(Note 1)	120	100	60	45
10	(Note 1)	(Note 1)	160	140	90	75
Larger than 10	(Note 1)	(Note 1)	(Note 2)	(Note 2)	(Note 3)	(Note 3)

Notes:

1. Value specified in 310.15 as applicable.
2. 400 percent of value specified in Table 310.15(B)(17) for 60°C conductors.
3. 300 percent of value specified in Table 310.15(B)(16) for 60°C conductors.

(2) Compliance with Article 450. Protection shall be permitted to be provided in accordance with 450.3.

(3) Less Than 50 Volt-Amperes. Control circuit transformers rated less than 50 volt-amperes (VA) and that are an integral part of the motor controller and located within the motor controller enclosure shall be permitted to be protected by primary overcurrent devices, impedance limiting means, or other inherent protective means.

(4) Primary Less Than 2 Amperes. Where the control circuit transformer rated primary current is less than 2 amperes, an overcurrent device rated or set at not more than 500 percent of the rated primary current shall be permitted in the primary circuit.

(5) Other Means. Protection shall be permitted to be provided by other approved means.

430.73 Protection of Conductors from Physical Damage. Where damage to a motor control circuit would constitute a hazard, all conductors of such a remote motor control circuit that are outside the control device itself shall be installed in a raceway or be otherwise protected from physical damage.

430.74 Electrical Arrangement of Control Circuits. Where one conductor of the motor control circuit is grounded, the motor control circuit shall be arranged so that a ground fault in the control circuit remote from the motor controller will (1) not start the motor and (2) not bypass manually operated shutdown devices or automatic safety shutdown devices.

430.75 Disconnection.

(A) General. Motor control circuits shall be arranged so that they will be disconnected from all sources of supply when the disconnecting means is in the open position. The disconnecting means shall be permitted to consist of two or more separate devices, one of which disconnects the motor and the controller from the source(s) of power supply for the motor, and the other(s), the motor control circuit(s) from its power supply. Where separate devices are used, they shall be located immediately adjacent to each other.

Exception No. 1: Where more than 12 motor control circuit conductors are required to be disconnected, the disconnecting means shall be permitted to be located other than immediately adjacent to each other where all of the following conditions are complied with:

(a) Access to energized parts is limited to qualified persons in accordance with Part XII of this article.

(b) A warning sign is permanently located on the outside of each equipment enclosure door or cover permitting access to the live parts in the motor control circuit(s), warning that motor control circuit disconnecting means are remotely located and specifying the location and identification of each disconnect. Where energized parts are not in an equipment enclosure as permitted by 430.232 and 430.233, an additional warning sign(s) shall be located where visible to persons who may be working in the area of the energized parts.

Exception No. 2: The motor control circuit disconnecting means shall be permitted to be remote from the motor controller power supply disconnecting means where the opening of one or more motor control circuit disconnecting means is capable of resulting in potentially unsafe conditions for personnel or property and the conditions of items (a) and (b) of Exception No. 1 are complied with.

(B) Control Transformer in Controller Enclosure. Where a transformer or other device is used to obtain a reduced voltage for the motor control circuit and is located in the controller enclosure, such transformer or other device shall be connected to the load side of the disconnecting means for the motor control circuit.

PART VII. MOTOR CONTROLLERS

430.81 General. Part VII is intended to require suitable controllers for all motors.

(A) Stationary Motor of ⅛ Horsepower or Less. For a stationary motor rated at ⅛ hp or less that is normally left running and is constructed so that it cannot be damaged by overload or failure to start, such as clock motors and the like, the branch-circuit disconnecting means shall be permitted to serve as the controller.

(B) Portable Motor of ⅓ Horsepower or Less. For a portable motor rated at ⅓ hp or less, the controller shall be permitted to be an attachment plug and receptacle or cord connector.

430.83 Ratings. The controller shall have a rating as specified in 430.83(A), unless otherwise permitted in 430.83(B) or (C), or as specified in (D), under the conditions specified.

(A) General.

(1) Horsepower Ratings. Controllers, other than inverse time circuit breakers and molded case switches, shall have horsepower ratings at the application voltage not lower than the horsepower rating of the motor.

(2) Circuit Breaker. A branch-circuit inverse time circuit breaker rated in amperes shall be permitted as a controller for all motors. Where this circuit breaker is also used for overload protection, it shall conform to the appropriate provisions of this article governing overload protection.

(3) Molded Case Switch. A molded case switch rated in amperes shall be permitted as a controller for all motors.

(B) Small Motors. Devices as specified in 430.81(A) and (B) shall be permitted as a controller.

(C) Stationary Motors of 2 Horsepower or Less. For stationary motors rated at 2 hp or less and 300 volts or less, the controller shall be permitted to be either of the following:

(1) A general-use switch having an ampere rating not less than twice the full-load current rating of the motor
(2) On ac circuits, a general-use snap switch suitable only for use on ac (not general-use ac–dc snap switches) where the motor full-load current rating is not more than 80 percent of the ampere rating of the switch

(E) Voltage Rating. A controller with a straight voltage rating, for example, 240 volts or 480 volts, shall be permitted to be applied in a circuit in which the nominal voltage between any two conductors does not exceed the controller's voltage rating. A controller with a slash rating, for example, 120/240 volts or 480Y/277 volts, shall only be applied in a solidly grounded circuit in which the nominal voltage to ground from any conductor does not exceed the lower of the two values of the controller's voltage rating and the nominal voltage between any two conductors does not exceed the higher value of the controller's voltage rating.

430.84 Need Not Open All Conductors. The controller shall not be required to open all conductors to the motor.

Exception: Where the controller serves also as a disconnecting means, it shall open all ungrounded conductors to the motor as provided in 430.111.

430.87 Number of Motors Served by Each Controller. Each motor shall be provided with an individual controller.

Exception No. 1: For motors rated 1000 volts or less, a single controller rated at not less than the equivalent horsepower, as determined in accordance with 430.110(C)(1), of all the motors in the group shall be permitted to serve the group under any of the following conditions:

(a) Where a number of motors drive several parts of a single machine or piece of apparatus, such as metal and woodworking machines, cranes, hoists, and similar apparatus

(b) Where a group of motors is under the protection of one overcurrent device as permitted in 430.53(A)

(c) Where a group of motors is located in a single room within sight from the controller location

Exception No. 2: A branch-circuit disconnecting means serving as the controller as allowed in 430.81(A) shall be permitted to serve more than one motor.

PART VIII. MOTOR CONTROL CENTERS

430.94 Overcurrent Protection. Motor control centers shall be provided with overcurrent protection in accordance with Parts I, II, and VIII of Article 240. The ampere rating or setting of the overcurrent protective device shall not exceed the rating of the common power bus. This protection shall be provided by (1) an overcurrent protective device located ahead of the motor control center or (2) a main overcurrent protective device located within the motor control center.

430.95 Service Equipment. Where used as service equipment, each motor control center shall be provided with a single main disconnecting means to disconnect all ungrounded service conductors.

Exception: A second service disconnect shall be permitted to supply additional equipment.

Where a grounded conductor is provided, the motor control center shall be provided with a main bonding jumper, sized in accordance with 250.28(D), within one of the sections for connecting the grounded conductor, on its supply side, to the motor control center equipment ground bus.

Exception: High-impedance grounded neutral systems shall be permitted to be connected as provided in 250.36.

N **430.99 Available Fault Current.** The available short circuit current at the motor control center and the date the short circuit current calculation was performed shall be documented and made available to those authorized to inspect the installation.

PART IX. DISCONNECTING MEANS

430.102 Location.

(A) Controller. An individual disconnecting means shall be provided for each controller and shall disconnect the controller. The disconnecting means shall be located in sight from the controller location.

Exception No. 1: For motor circuits over 1000 volts, nominal, a controller disconnecting means lockable in accordance with 110.25 shall be permitted to be out of sight of the controller, provided that the controller is marked with a warning label giving the location of the disconnecting means.

Exception No. 2: A single disconnecting means shall be permitted for a group of coordinated controllers that drive several parts of a single machine or piece of apparatus. The disconnecting means shall be located in sight from the controllers, and both the disconnecting means and the controllers shall be located in sight from the machine or apparatus.

Exception No. 3: The disconnecting means shall not be required to be in sight from valve actuator motor (VAM) assemblies containing the controller where such a location introduces additional or increased hazards to persons or property and conditions (a) and (b) are met.

(a) The valve actuator motor assembly is marked with a warning label giving the location of the disconnecting means.

(b) The disconnecting means is lockable in accordance with 110.25.

(B) Motor. A disconnecting means shall be provided for a motor in accordance with (B)(1) or (B)(2).

(1) Separate Motor Disconnect. A disconnecting means for the motor shall be located in sight from the motor location and the driven machinery location.

(2) Controller Disconnect. The controller disconnecting means required in accordance with 430.102(A) shall be permitted to serve as the disconnecting means for the motor if it is in sight from the motor location and the driven machinery location.

Exception to (1) and (2): The disconnecting means for the motor shall not be required under either condition (a) or condition (b), which follow, provided that the controller disconnecting means required in 430.102(A) is lockable in accordance with 110.25.

(a) Where such a location of the disconnecting means for the motor is impracticable or introduces additional or increased hazards to persons or property

(b) In industrial installations, with written safety procedures, where conditions of maintenance and supervision ensure that only qualified persons service the equipment

430.103 Operation. The disconnecting means shall open all ungrounded supply conductors and shall be designed so that no pole can be operated independently. The disconnecting means shall be permitted in the same enclosure with the controller. The disconnecting means shall be designed so that it cannot be closed automatically.

Informational Note: See 430.113 for equipment receiving energy from more than one source.

430.104 To Be Indicating. The disconnecting means shall plainly indicate whether it is in the open (off) or closed (on) position.

430.107 Readily Accessible. At least one of the disconnecting means shall be readily accessible.

430.108 Every Disconnecting Means. Every disconnecting means in the motor circuit between the point of attachment to the feeder or branch circuit and the point of connection to the motor shall comply with the requirements of 430.109 and 430.110.

430.109 Type. The disconnecting means shall be a type specified in 430.109(A), unless otherwise permitted in 430.109(B) through (G), under the conditions specified.

(A) General.

(1) Motor Circuit Switch. A listed motor-circuit switch rated in horsepower.

(2) Molded Case Circuit Breaker. A listed molded case circuit breaker.

(3) Molded Case Switch. A listed molded case switch.

(4) Instantaneous Trip Circuit Breaker. An instantaneous trip circuit breaker that is part of a listed combination motor controller.

(5) Self-Protected Combination Controller. Listed self-protected combination controller.

(6) Manual Motor Controller. Listed manual motor controllers additionally marked "Suitable as Motor Disconnect" shall be permitted as a disconnecting means where installed between the final motor branch-circuit short-circuit protective device and the motor. Listed manual motor controllers additionally marked "Suitable as Motor Disconnect" shall be permitted as disconnecting means on the line side of the fuses permitted in 430.52(C)(5). In this case, the fuses permitted in 430.52(C)(5) shall be considered supplementary fuses, and suitable branch-circuit short-circuit and ground-fault protective devices shall be installed on the line side of the manual motor controller additionally marked "Suitable as Motor Disconnect."

(7) System Isolation Equipment. System isolation equipment shall be listed for disconnection purposes. System isolation equipment shall be installed on the load side of the overcurrent protection and its disconnecting means. The disconnecting means shall be one of the types permitted by 430.109(A)(1) through (A)(3).

(B) Stationary Motors of ⅛ Horsepower or Less. For stationary motors of ⅛ hp or less, the branch-circuit overcurrent device shall be permitted to serve as the disconnecting means.

(C) Stationary Motors of 2 Horsepower or Less. For stationary motors rated at 2 hp or less and 300 volts or less, the disconnecting means shall be permitted to be one of the devices specified in (1), (2), or (3):

(1) A general-use switch having an ampere rating not less than twice the full-load current rating of the motor
(2) On ac circuits, a general-use snap switch suitable only for use on ac (not general-use ac–dc snap switches) where the motor full-load current rating is not more than 80 percent of the ampere rating of the switch
(3) A listed manual motor controller having a horsepower rating not less than the rating of the motor and marked "Suitable as Motor Disconnect"

(D) Autotransformer-Type Controlled Motors. For motors of over 2 hp to and including 100 hp, the separate disconnecting means required for a motor with an autotransformer-type controller shall be permitted to be a general-use switch where all of the following provisions are met:

(1) The motor drives a generator that is provided with overload protection.
(2) The controller is capable of interrupting the locked-rotor current of the motors, is provided with a no voltage release, and is provided with running overload protection

not exceeding 125 percent of the motor full-load current rating.

(3) Separate fuses or an inverse time circuit breaker rated or set at not more than 150 percent of the motor full-load current is provided in the motor branch circuit.

(E) Isolating Switches. For stationary motors rated at more than 40 hp dc or 100 hp ac, the disconnecting means shall be permitted to be a general-use or isolating switch where plainly marked "Do not operate under load."

(F) Cord-and-Plug-Connected Motors. For a cord-and-plug-connected motor, a horsepower-rated attachment plug and receptacle, flanged surface inlet and cord connector, or attachment plug and cord connector having ratings no less than the motor ratings shall be permitted to serve as the disconnecting means. Horsepower-rated attachment plugs, flanged surface inlets, receptacles, or cord connectors shall not be required for cord-and-plug-connected appliances in accordance with 422.33, room air conditioners in accordance with 440.63, or portable motors rated ⅓ hp or less.

(G) Torque Motors. For torque motors, the disconnecting means shall be permitted to be a general-use switch.

430.110 Ampere Rating and Interrupting Capacity.

(A) General. The disconnecting means for motor circuits rated 1000 volts, nominal, or less shall have an ampere rating not less than 115 percent of the full-load current rating of the motor.

Exception: A listed unfused motor-circuit switch having a horsepower rating not less than the motor horsepower shall be permitted to have an ampere rating less than 115 percent of the full-load current rating of the motor.

(B) For Torque Motors. Disconnecting means for a torque motor shall have an ampere rating of at least 115 percent of the motor nameplate current.

(C) For Combination Loads. Where two or more motors are used together or where one or more motors are used in combination with other loads, such as resistance heaters, and where the combined load may be simultaneous on a single disconnecting means, the ampere and horsepower ratings of the combined load shall be determined as follows.

(1) Horsepower Rating. The rating of the disconnecting means shall be determined from the sum of all currents, including resistance loads, at the full-load condition and also at the locked-rotor condition. The combined full-load current and the combined locked-rotor current so obtained shall be considered as a single motor for the purpose of this requirement as follows.

The full-load current equivalent to the horsepower rating of each motor shall be selected from Table 430.247, Table 430.248, Table 430.249, or Table 430.250. These full-load currents shall be added to the rating in amperes of other loads to obtain an equivalent full-load current for the combined load.

The locked-rotor current equivalent to the horsepower rating of each motor shall be selected from Table 430.251(A) or Table 430.251(B). The locked-rotor currents shall be added to the rating in amperes of other loads to obtain an equivalent locked-rotor current for the combined load. Where two or more motors or other loads cannot be started simultaneously, the largest sum of locked-rotor currents of a motor or group of motors that can be started simultaneously and the full-load currents of other concurrent loads shall be permitted to be used to determine the equivalent locked-rotor current for the simultaneous combined loads. In cases where different current ratings are obtained when applying these tables, the largest value obtained shall be used.

Exception: Where part of the concurrent load is resistance load, and where the disconnecting means is a switch rated in horsepower and amperes, the switch used shall be permitted to have a horsepower rating that is not less than the combined load of the motor(s), if the ampere rating of the switch is not less than the locked-rotor current of the motor(s) plus the resistance load.

(2) Ampere Rating. The ampere rating of the disconnecting means shall not be less than 115 percent of the sum of all currents at the full-load condition determined in accordance with 430.110(C)(1).

Exception: A listed nonfused motor-circuit switch having a horsepower rating equal to or greater than the equivalent horsepower of the combined loads, determined in accordance with 430.110(C)(1), shall be permitted to have an ampere rating less than 115 percent of the sum of all currents at the full-load condition.

(3) Small Motors. For small motors not covered by Table 430.247, Table 430.248, Table 430.249, or Table 430.250, the locked-rotor current shall be assumed to be six times the full-load current.

430.111 Switch or Circuit Breaker as Both Controller and Disconnecting Means. A switch or circuit breaker shall be permitted to be used as both the controller and disconnecting means if it complies with 430.111(A) and is one of the types specified in 430.111(B).

(A) General. The switch or circuit breaker complies with the requirements for controllers specified in 430.83, opens all ungrounded conductors to the motor, and is protected by an overcurrent device in each ungrounded conductor (which shall be permitted to be the branch-circuit fuses). The overcurrent device protecting the controller shall be permitted to be part of the controller assembly or shall be permitted to be separate. An autotransformer-type controller shall be provided with a separate disconnecting means.

(B) Type. The device shall be one of the types specified in 430.111(B)(1), (B)(2), or (B)(3).

(1) Air-Break Switch. An air-break switch, operable directly by applying the hand to a lever or handle.

(2) Inverse Time Circuit Breaker. An inverse time circuit breaker operable directly by applying the hand to a lever or handle. The circuit breaker shall be permitted to be both power and manually operable.

(3) Oil Switch. An oil switch used on a circuit whose rating does not exceed 1000 volts or 100 amperes, or by special permission on a circuit exceeding this capacity where under expert supervision. The oil switch shall be permitted to be both power and manually operable.

430.112 Motors Served by Single Disconnecting Means. Each motor shall be provided with an individual disconnecting means.

Exception: A single disconnecting means shall be permitted to serve a group of motors under any one of the conditions of (a), (b), and (c). The single disconnecting means shall be rated in accordance with 430.110(C).

(a) Where a number of motors drive several parts of a single machine or piece of apparatus, such as metal- and woodworking machines, cranes, and hoists.

(b) Where a group of motors is under the protection of one set of branch-circuit protective devices as permitted by 430.53(A).

(c) Where a group of motors is in a single room within sight from the location of the disconnecting means.

430.113 Energy from More Than One Source. Motor and motor-operated equipment receiving electric energy from more than one source shall be provided with disconnecting means from each source of electric energy immediately adjacent to the equipment served. Each source shall be permitted to have a separate disconnecting means. Where multiple disconnecting means are provided, a permanent warning sign shall be provided on or adjacent to each disconnecting means.

Exception No. 1: Where a motor receives electric energy from more than one source, the disconnecting means for the main power supply to the motor shall not be required to be immediately adjacent to the motor, provided that the controller disconnecting means is lockable in accordance with 110.25.

Exception No. 2: A separate disconnecting means shall not be required for a Class 2 remote-control circuit conforming with Article 725, rated not more than 30 volts, and isolated and ungrounded.

PART X. ADJUSTABLE-SPEED DRIVE SYSTEMS

430.120 General. The installation provisions of Part I through Part IX are applicable unless modified or supplemented by Part X.

430.122 Conductors — Minimum Size and Ampacity.

(A) Branch/Feeder Circuit Conductors. Circuit conductors supplying power conversion equipment included as part of an adjustable-speed drive system shall have an ampacity not less than 125 percent of the rated input current to the power conversion equipment.

> Informational Note: Power conversion equipment can have multiple power ratings and corresponding input currents.

(B) Bypass Device. For an adjustable-speed drive system that utilizes a bypass device, the conductor ampacity shall not be less than required by 430.6. The ampacity of circuit conductors supplying power conversion equipment included as part of an adjustable-speed drive system that utilizes a bypass device shall be the larger of either of the following:

(1) 125 percent of the rated input current to the power conversion equipment
(2) 125 percent of the motor full-load current rating as determined by 430.6

430.124 Overload Protection. Overload protection of the motor shall be provided.

(A) Included in Power Conversion Equipment. Where the power conversion equipment is marked to indicate that motor overload protection is included, additional overload protection shall not be required.

(B) Bypass Circuits. For adjustable-speed drive systems that utilize a bypass device to allow motor operation at rated full-load speed, motor overload protection as described in Article 430, Part III, shall be provided in the bypass circuit.

(C) Multiple Motor Applications. For multiple motor application, individual motor overload protection shall be provided in accordance with Article 430, Part III.

430.126 Motor Overtemperature Protection.

(A) General. Adjustable-speed drive systems shall protect against motor overtemperature conditions where the motor is not rated to operate at the nameplate rated current over the speed range required by the application. This protection shall be provided in addition to the conductor protection required in 430.32. Protection shall be provided by one of the following means.

(1) Motor thermal protector in accordance with 430.32
(2) Adjustable-speed drive system with load and speed-sensitive overload protection and thermal memory retention upon shutdown or power loss

Exception to (2): Thermal memory retention upon shutdown or power loss is not required for continuous duty loads.

(3) Overtemperature protection relay utilizing thermal sensors embedded in the motor and meeting the requirements of 430.126(A)(2) or (B)(2)
(4) Thermal sensor embedded in the motor whose communications are received and acted upon by an adjustable-speed drive system

(B) Multiple Motor Applications. For multiple motor applications, individual motor overtemperature protection shall be provided as required in 430.126(A).

(C) Automatic Restarting and Orderly Shutdown. The provisions of 430.43 and 430.44 shall apply to the motor overtemperature protection means.

430.128 Disconnecting Means. The disconnecting means shall be permitted to be in the incoming line to the conversion equipment and shall have a rating not less than 115 percent of the rated input current of the conversion unit.

430.130 Branch-Circuit Short-Circuit and Ground-Fault Protection for Single Motor Circuits Containing Power Conversion Equipment.

(A) Circuits Containing Power Conversion Equipment. Circuits containing power conversion equipment shall be protected by a branch-circuit short-circuit and ground-fault protective device in accordance with the following:

(1) The rating and type of protection shall be determined by 430.52(C)(1), (C)(3), (C)(5), or (C)(6), using the full-load current rating of the motor load as determined by 430.6.
(2) Where maximum branch-circuit short-circuit and ground-fault protective ratings are stipulated for specific device types in the manufacturer's instructions for the power conversion equipment or are otherwise marked on the equipment, they shall not be exceeded even if higher values are permitted by 430.130(A)(1).
(3) A self-protected combination controller shall only be permitted where specifically identified in the manufacturer's instructions for the power conversion equipment or if otherwise marked on the equipment.
(4) Where an instantaneous trip circuit breaker or semiconductor fuses are permitted in accordance with the drive manufacturer's instructions for use as the branch-circuit short-circuit and ground-fault protective device for listed power conversion equipment, they shall be provided as an integral part of a single listed assembly incorporating both the protective device and power conversion equipment.

(B) Bypass Circuit/Device. Branch-circuit short-circuit and ground-fault protection shall also be provided for a bypass circuit/device(s). Where a single branch-circuit short-circuit and ground-fault protective device is provided for circuits containing both power conversion equipment and a bypass circuit, the branch-circuit protective device type and its rating or setting shall be in accordance with those determined for the power conversion equipment and for the bypass circuit/device(s) equipment.

430.131 Several Motors or Loads on One Branch Circuit Including Power Conversion Equipment. For installations meeting all the requirements of 430.53 that include one or more power converters, the branch-circuit short-circuit and ground-fault protective fuses or inverse time circuit breakers shall be of a type and rating or setting permitted for use with the power conversion equipment using the full-load current rating of the connected motor load in accordance with 430.53. For the purposes of 430.53 and 430.131, power conversion equipment shall be considered to be a motor controller.

PART XI. OVER 1000 VOLTS, NOMINAL

430.221 General. Part XI recognizes the additional hazard due to the use of higher voltages. It adds to or amends the other provisions of this article.

PART XIII. GROUNDING — ALL VOLTAGES

430.242 Stationary Motors. The frames of stationary motors shall be grounded under any of the following conditions:

(1) Where supplied by metal-enclosed wiring
(2) Where in a wet location and not isolated or guarded
(3) If in a hazardous (classified) location
(4) If the motor operates with any terminal at over 150 volts to ground

Where the frame of the motor is not grounded, it shall be permanently and effectively insulated from the ground.

430.243 Portable Motors. The frames of portable motors that operate over 150 volts to ground shall be guarded or grounded.

Exception No. 1: Listed motor-operated tools, listed motor-operated appliances, and listed motor-operated equipment shall not be required to be grounded where protected by a system of double insulation or its equivalent. Double-insulated equipment shall be distinctively marked.

Exception No. 2: Listed motor-operated tools, listed motor-operated appliances, and listed motor-operated equipment connected by a cord and attachment plug other than those required to be grounded in accordance with 250.114.

430.244 Controllers. Controller enclosures shall be connected to the equipment grounding conductor regardless of voltage. Controller enclosures shall have means for attachment of an equipment grounding conductor termination in accordance with 250.8.

Exception: Enclosures attached to ungrounded portable equipment shall not be required to be grounded.

430.245 Method of Grounding. Connection to the equipment grounding conductor shall be done in the manner specified in Part VI of Article 250.

(A) Grounding Through Terminal Housings. Where the wiring to motors is metal-enclosed cable or in metal raceways, junction boxes to house motor terminals shall be provided, and the armor of the cable or the metal raceways shall be connected to them in the manner specified in 250.96(A) and 250.97.

(B) Separation of Junction Box from Motor. The junction box required by 430.245(A) shall be permitted to be separated from the motor by not more than 1.8 m (6 ft), provided the leads to the motor are stranded conductors within Type AC cable, interlocked metal tape Type MC cable where listed and identified in accordance with 250.118(10)(a), or armored cord or are stranded leads enclosed in liquidtight flexible metal conduit, flexible metal conduit, intermediate metal conduit, rigid metal conduit, or electrical metallic tubing not smaller than metric designator 12 (trade size ⅜), the armor or raceway being connected both to the motor and to the box.

Liquidtight flexible nonmetallic conduit and rigid nonmetallic conduit shall be permitted to enclose the leads to the motor, provided the leads are stranded and the required equipment grounding conductor is connected to both the motor and to the box.

Where stranded leads are used, protected as specified above, each strand within the conductor shall be not larger than 10 AWG and shall comply with other requirements of this *Code* for conductors to be used in raceways.

(C) Grounding of Controller-Mounted Devices. Instrument transformer secondaries and exposed non–current-carrying metal or other conductive parts or cases of instrument transformers, meters, instruments, and relays shall be grounded as specified in 250.170 through 250.178.

PART XIV. TABLES

Table 430.247 Full-Load Current in Amperes, Direct-Current Motors

The following values of full-load currents* are for motors running at base speed.

	Armature Voltage Rating*					
Horsepower	90 Volts	120 Volts	180 Volts	240 Volts	500 Volts	550 Volts
¼	4.0	3.1	2.0	1.6	—	—
⅓	5.2	4.1	2.6	2.0	—	—
½	6.8	5.4	3.4	2.7	—	—
¾	9.6	7.6	4.8	3.8	—	—
1	12.2	9.5	6.1	4.7	—	—
1½	—	13.2	8.3	6.6	—	—
2	—	17	10.8	8.5	—	—
3	—	25	16	12.2	—	—
5	—	40	27	20	—	—
7½	—	58	—	29	13.6	12.2
10	—	76	—	38	18	16
15	—	—	—	55	27	24
20	—	—	—	72	34	31
25	—	—	—	89	43	38
30	—	—	—	106	51	46
40	—	—	—	140	67	61
50	—	—	—	173	83	75
60	—	—	—	206	99	90
75	—	—	—	255	123	111
100	—	—	—	341	164	148
125	—	—	—	425	205	185
150	—	—	—	506	246	222
200	—	—	—	675	330	294

*These are average dc quantities.

Table 430.248 Full-Load Currents in Amperes, Single-Phase Alternating-Current Motors

The following values of full-load currents are for motors running at usual speeds and motors with normal torque characteristics. The voltages listed are rated motor voltages. The currents listed shall be permitted for system voltage ranges of 110 to 120 and 220 to 240 volts.

Horsepower	115 Volts	200 Volts	208 Volts	230 Volts
1/6	4.4	2.5	2.4	2.2
1/4	5.8	3.3	3.2	2.9
1/3	7.2	4.1	4.0	3.6
1/2	9.8	5.6	5.4	4.9
3/4	13.8	7.9	7.6	6.9
1	16	9.2	8.8	8.0
1 1/2	20	11.5	11.0	10
2	24	13.8	13.2	12
3	34	19.6	18.7	17
5	56	32.2	30.8	28
7 1/2	80	46.0	44.0	40
10	100	57.5	55.0	50

Table 430.250 Full-Load Current, Three-Phase Alternating-Current Motors

The following values of full-load currents are typical for motors running at speeds usual for belted motors and motors with normal torque characteristics. The voltages listed are rated motor voltages. The currents listed shall be permitted for system voltage ranges of 110 to 120, 220 to 240, 440 to 480, and 550 to 600 volts.

	Induction-Type Squirrel Cage and Wound Rotor (Amperes)							Synchronous-Type Unity Power Factor* (Amperes)			
Horsepower	115 Volts	200 Volts	208 Volts	230 Volts	460 Volts	575 Volts	2300 Volts	230 Volts	460 Volts	575 Volts	2300 Volts
½	4.4	2.5	2.4	2.2	1.1	0.9	—	—	—	—	—
¾	6.4	3.7	3.5	3.2	1.6	1.3	—	—	—	—	—
1	8.4	4.8	4.6	4.2	2.1	1.7	—	—	—	—	—
1½	12.0	6.9	6.6	6.0	3.0	2.4	—	—	—	—	—
2	13.6	7.8	7.5	6.8	3.4	2.7	—	—	—	—	—
3	—	11.0	10.6	9.6	4.8	3.9	—	—	—	—	—
5	—	17.5	16.7	15.2	7.6	6.1	—	—	—	—	—
7½	—	25.3	24.2	22	11	9	—	—	—	—	—
10	—	32.2	30.8	28	14	11	—	—	—	—	—
15	—	48.3	46.2	42	21	17	—	—	—	—	—

(continued)

Table 430.250 Full-Load Current, Three-Phase Alternating-Current Motors ***Continued***

The following values of full-load currents are typical for motors running at speeds usual for belted motors and motors with normal torque characteristics. The voltages listed are rated motor voltages. The currents listed shall be permitted for system voltage ranges of 110 to 120, 220 to 240, 440 to 480, and 550 to 600 volts.

	Induction-Type Squirrel Cage and Wound Rotor (Amperes)							Synchronous-Type Unity Power Factor* (Amperes)			
Horsepower	115 Volts	200 Volts	208 Volts	230 Volts	460 Volts	575 Volts	2300 Volts	230 Volts	460 Volts	575 Volts	2300 Volts
20	—	62.1	59.4	54	27	22	—	—	—	—	—
25	—	78.2	74.8	68	34	27	—	53	26	21	—
30	—	92	88	80	40	32	—	63	32	26	—
40	—	120	114	104	52	41	—	83	41	33	—
50	—	150	143	130	65	52	—	104	52	42	—
60	—	177	169	154	77	62	16	123	61	49	12
75	—	221	211	192	96	77	20	155	78	62	15
100	—	285	273	248	124	99	26	202	101	81	20
125	—	359	343	312	156	125	31	253	126	101	25
150	—	414	396	360	180	144	37	302	151	121	30
200		552	528	480	240	192	49	400	201	161	40

250	—	—	—	—	302	242	60	—	—	—	—
300	—	—	—	—	361	289	72	—	—	—	—
350	—	—	—	—	414	336	83	—	—	—	—
400	—	—	—	—	477	382	95	—	—	—	—
450	—	—	—	—	515	412	103	—	—	—	—
500	—	—	—	—	590	472	118	—	—	—	—

*For 90 and 80 percent power factor, the figures shall be multiplied by 1.1 and 1.25, respectively.

Table 430.251(A) Conversion Table of Single-Phase Locked-Rotor Currents for Selection of Disconnecting Means and Controllers as Determined from Horsepower and Voltage Rating

For use only with 430.110, 440.12, 440.41, and 455.8(C).

Rated Horsepower	Maximum Locked-Rotor Current in Amperes, Single Phase		
	115 Volts	208 Volts	230 Volts
½	58.8	32.5	29.4
¾	82.8	45.8	41.4
1	96	53	48
1½	120	66	60
2	144	80	72
3	204	113	102
5	336	186	168
7½	480	265	240
10	1000	332	300

Table 430.251(B) Conversion Table of Polyphase Design B, C, and D Maximum Locked-Rotor Currents for Selection of Disconnecting Means and Controllers as Determined from Horsepower and Voltage Rating and Design Letter

For use only with 430.110, 440.12, 440.41, and 455.8(C).

Rated Horsepower	Maximum Motor Locked-Rotor Current in Amperes, Two- and Three-Phase, Design B, C, and D*					
	115 Volts	200 Volts	208 Volts	230 Volts	460 Volts	575 Volts
	B, C, D	B, C, D	B, C, D	B, C, D	B, C, D	B, C, D
½	40	23	22.1	20	10	8
¾	50	28.8	27.6	25	12.5	10
1	60	34.5	33	30	15	12
1½	80	46	44	40	20	16
2	100	57.5	55	50	25	20
3	—	73.6	71	64	32	25.6
5	—	105.8	102	92	46	36.8
7½	—	146	140	127	63.5	50.8
10	—	186.3	179	162	81	64.8
15	—	267	257	232	116	93

(continued)

Table 430.251(B) Conversion Table of Polyphase Design B, C, and D Maximum Locked-Rotor Currents for Selection of Disconnecting Means and Controllers as Determined from Horsepower and Voltage Rating and Design Letter ***Continued***

For use only with 430.110, 440.12, 440.41 and 455.8(C).

Rated Horsepower	Maximum Motor Locked-Rotor Current in Amperes, Two- and Three-Phase, Design B, C, and D*					
	115 Volts	200 Volts	208 Volts	230 Volts	460 Volts	575 Volts
	B, C, D	B, C, D	B, C, D	B, C, D	B, C, D	B, C, D
20	—	334	321	290	145	116
25	—	420	404	365	183	146
30	—	500	481	435	218	174
40	—	667	641	580	290	232
50	—	834	802	725	363	290
60	—	1001	962	870	435	348
75	—	1248	1200	1085	543	434
100	—	1668	1603	1450	725	580
125	—	2087	2007	1815	908	726

150	—	2496	2400	2170	1085	868
200	—	3335	3207	2900	1450	1160
250	—	—	—	—	1825	1460
300	—	—	—	—	2200	1760
350	—	—	—	—	2550	2040
400	—	—	—	—	2900	2320
450	—	—	—	—	3250	2600
500	—	—	—	—	3625	2900

*Design A motors are not limited to a maximum starting current or locked rotor current.

CHAPTER 20

AIR-CONDITIONING AND REFRIGERATING EQUIPMENT

INTRODUCTION

This chapter contains requirements from Article 440 — Air-Conditioning and Refrigerating Equipment. Article 440 covers electric motor–driven air-conditioning and refrigerating equipment, including their branch circuits and controllers. Provided in Article 440 are special considerations necessary for branch circuits supplying hermetic refrigerant motor-compressors and for any air-conditioning or refrigerating equipment that is supplied from a branch circuit that supplies a hermetic refrigerant motor-compressor. The requirements of Article 430 are used for feeders and services that supply multiple air-conditioning and refrigeration units that employ hermetic refrigerant motor compressors.

Similar to motor circuits covered in Article 430, the requirements of Article 440 recognize the operational characteristics of the motors associated with air-conditioning and refrigeration equipment, much of which contains multiple motors. Motors that are sealed and immersed in the refrigerant (i.e., hermetically sealed) are the unique feature of the equipment covered in Article 440. However, the approach to overcurrent protection takes on a two-parts: one device sized to accommodate the starting current that protects against the effects of short-circuits and line-to-ground faults, and a device that is responsive to motor or equipment operating current that protects against overloads.

Installers of the equipment covered by Article 440 need to be aware of the markings that are found on the equipment that in most cases will provide specific guidance on maximum size and type of overcurrent protective devices in the supply circuit, the minimum size for circuit conductors and the short-circuit current rating of internal motor control and protection components. These markings are essential to providing a Code-compliant and operationally safe installation.

Contained in this *Pocket Guide* are six of the seven parts that are included in Article 440. The six parts are: I General; II Disconnecting Means; III Branch-Circuit Short-Circuit and Ground-Fault Protection; IV Branch-Circuit Conductors; Controllers for Motor-Compressors; and VI Motor-Compressor and Branch-Circuit Overload Protection. The requirements covering supply circuits, disconnecting means, length of supply cords, grounding and protection of room air conditioners are located in Part VII of Article 440. Article 440, Part VII is not included in this *Pocket Guide*.

ARTICLE 440

Air-Conditioning and Refrigerating Equipment

PART I. GENERAL

440.2 Definitions.

Branch-Circuit Selection Current. The value in amperes to be used instead of the rated-load current in determining the ratings of motor branch-circuit conductors, disconnecting means, controllers, and branch-circuit short-circuit and ground-fault protective devices wherever the running overload protective device permits a sustained current greater than the specified percentage of the rated-load current. The value of branch-circuit selection current will always be equal to or greater than the marked rated-load current.

440.6 Ampacity and Rating. The size of conductors for equipment covered by this article shall be selected from Table 310.15(B)(16) through Table 310.15(B)(19) or calculated in accordance with 310.15 as applicable. The required ampacity of conductors and rating of equipment shall be determined according to 440.6(A) and 440.6(B).

(A) Hermetic Refrigerant Motor-Compressor. For a hermetic refrigerant motor-compressor, the rated-load current marked on the nameplate of the equipment in which the motor-compressor is employed shall be used in determining the rating or ampacity of the disconnecting means, the branch-circuit conductors, the controller, the branch-circuit short-circuit and ground-fault protection, and the separate motor overload protection. Where no rated-load current is shown on the equipment nameplate, the rated-load current shown on the compressor nameplate shall be used.

Exception No. 1: Where so marked, the branch-circuit selection current shall be used instead of the rated-load current to determine the rating or ampacity of the disconnecting means, the branch-circuit conductors, the controller, and the branch-circuit short-circuit and ground-fault protection.

Exception No. 2: For cord-and-plug-connected equipment, the nameplate marking shall be used in accordance with 440.22(B), Exception No. 2.

(B) Multimotor Equipment. For multimotor equipment employing a shaded-pole or permanent split-capacitor-type fan or blower motor, the full-load current for such motor marked on the nameplate of the equipment in which the fan or blower motor is employed shall be used instead of the horsepower rating to determine the ampacity or rating of the disconnecting means, the branch-circuit conductors, the controller, the branch-circuit short-circuit and ground-fault protection, and the separate overload protection. This marking on the equipment nameplate shall not be less than the current marked on the fan or blower motor nameplate.

440.7 Highest Rated (Largest) Motor. In determining compliance with this article and with 430.24, 430.53(B) and 430.53(C), and 430.62(A), the highest rated (largest) motor shall be considered to be the motor that has the highest rated-load current. Where two or more motors have the same highest rated-load current, only one of them shall be considered as the highest rated (largest) motor. For other than hermetic refrigerant motor-compressors, and fan or blower motors as covered in 440.6(B), the full-load current used to determine the highest rated motor shall be the equivalent value corresponding to the motor horsepower rating selected from Table 430.248, Table 430.249, or Table 430.250.

Exception: Where so marked, the branch-circuit selection current shall be used instead of the rated-load current in determining the highest rated (largest) motor-compressor.

440.8 Single Machine. An air-conditioning or refrigerating system shall be considered to be a single machine under the provisions of 430.87, Exception No. 1, and 430.112, Exception. The motors shall be permitted to be located remotely from each other.

N **440.9 Grounding and Bonding.** Where multimotor and combination-load equipment is installed outdoors on a roof, an equipment grounding conductor of the wire type shall be installed in outdoor portions of metallic raceway systems that use non-threaded fittings.

N **440.10 Short-Circuit Current Rating.**

(A) Installation. Motor controllers of multimotor and combination-load equipment shall not be installed where the available short-circuit current exceeds its short-circuit current rating as marked in accordance with 440.4(B).

(B) Documentation. When motor controllers or industrial control panels of multimotor and combination load equipment are required to be marked with a short circuit current rating, the available short circuit current and the date the short circuit current calculation was performed shall be documented and made available to those authorized to inspect the installation.

PART II. DISCONNECTING MEANS

440.12 Rating and Interrupting Capacity.

(A) Hermetic Refrigerant Motor-Compressor. A disconnecting means serving a hermetic refrigerant motor-compressor shall be selected on the basis of the nameplate rated-load current or branch-circuit selection current, whichever is greater, and locked-rotor current, respectively, of the motor-compressor as follows.

(1) Ampere Rating. The ampere rating shall be at least 115 percent of the nameplate rated-load current or branch-circuit selection current, whichever is greater.

Exception: A listed unfused motor circuit switch, without fuseholders, having a horsepower rating not less than the equivalent horsepower determined in accordance with 440.12(A)(2) shall be permitted to have an ampere rating less than 115 percent of the specified current.

(2) Equivalent Horsepower. To determine the equivalent horsepower in complying with the requirements of 430.109, the horsepower rating shall be selected from Table 430.248, Table 430.249, or Table 430.250 corresponding to the rated-load current or branch-circuit selection current, whichever is greater, and also the horsepower rating from Table 430.251(A) or Table 430.251(B) corresponding to the locked-rotor current. In case the nameplate rated-load current or branch-circuit selection current and locked-rotor current do not correspond to the currents shown in Table 430.248, Table 430.249, Table 430.250, Table 430.251(A), or Table 430.251(B), the horsepower rating corresponding to the next higher value shall be selected. In case different horsepower ratings are obtained when applying these tables, a horsepower rating at least equal to the larger of the values obtained shall be selected.

(B) Combination Loads. Where the combined load of two or more hermetic refrigerant motor-compressors or one or more hermetic refrigerant motor-compressor with other motors or loads may be simultaneous on a single disconnecting means, the rating for the disconnecting means shall be determined in accordance with 440.12(B)(1) and (B)(2).

(1) Horsepower Rating. The horsepower rating of the disconnecting means shall be determined from the sum of all currents, including resistance loads, at the rated-load condition and also at the locked-rotor condition. The combined rated-load current and the combined locked-rotor current so obtained shall be considered as a single motor for the purpose of this requirement as required by (1)(a) and (1)(b).

(a) The full-load current equivalent to the horsepower rating of each motor, other than a hermetic refrigerant motor-compressor, and fan or blower motors as covered in 440.6(B) shall be selected from Table 430.248, Table 430.249, or Table 430.250. These full-load currents shall be added to the motor-compressor rated-load current(s) or branch-circuit selection current(s), whichever is greater, and to the rating in amperes of other loads to obtain an equivalent full-load current for the combined load.

(b) The locked-rotor current equivalent to the horsepower rating of each motor, other than a hermetic refrigerant motor-compressor, shall be selected from Table 430.251(A) or Table 430.251(B), and, for fan and blower motors of the shaded-pole or permanent split-capacitor type marked with the locked-rotor current, the marked value shall be used. The locked-rotor currents shall be added to the motor-compressor locked-rotor current(s) and to the rating in amperes of other loads to obtain an equivalent locked-rotor current for the combined load. Where two or more motors or other loads such as resistance heaters, or both, cannot be started simultaneously, appropriate combinations of locked-rotor and rated-load current or branch-circuit selection current, whichever is greater, shall be an acceptable means of determining the equivalent locked-rotor current for the simultaneous combined load.

Exception: Where part of the concurrent load is a resistance load and the disconnecting means is a switch rated in horsepower and amperes, the switch used shall be permitted to have a horsepower rating not less than the combined load to the motor-compressor(s) and other motor(s) at the locked-rotor condition, if the ampere rating of the switch is not less than this locked-rotor load plus the resistance load.

(2) Full-Load Current Equivalent. The ampere rating of the disconnecting means shall be at least 115 percent of the sum of all currents at the rated-load condition determined in accordance with 440.12(B)(1).

Exception: A listed unfused motor circuit switch, without fuseholders, having a horsepower rating not less than the equivalent horsepower determined by 440.12(B)(1) shall be permitted to have an ampere rating less than 115 percent of the sum of all currents.

(C) Small Motor-Compressors. For small motor-compressors not having the locked-rotor current marked on the nameplate, or for small motors not covered by Table 430.247, Table 430.248, Table 430.249, or Table 430.250, the locked-rotor current shall be assumed to be six times the rated-load current.

(D) Disconnecting Means. Every disconnecting means in the refrigerant motor-compressor circuit between the point of attachment to the feeder and the point of connection to the refrigerant motor-compressor shall comply with the requirements of 440.12.

(E) Disconnecting Means Rated in Excess of 100 Horsepower. Where the rated-load or locked-rotor current as determined above would indicate a disconnecting means rated in excess of 100 hp, the provisions of 430.109(E) shall apply.

440.13 Cord-Connected Equipment. For cord-connected equipment such as room air conditioners, household refrigerators and freezers, drinking water coolers, and beverage dispensers, a separable connector or an attachment plug and receptacle shall be permitted to serve as the disconnecting means.

440.14 Location. Disconnecting means shall be located within sight from, and readily accessible from the air-conditioning or refrigerating equipment. The disconnecting means shall be permitted to be installed on or within the air-conditioning or refrigerating equipment.

The disconnecting means shall not be located on panels that are designed to allow access to the air-conditioning or refrigeration equipment or to obscure the equipment nameplate(s).

Exception No. 1: Where the disconnecting means provided in accordance with 430.102(A) is lockable in accordance with 110.25 and the refrigerating or air-conditioning equipment is essential to an industrial process in a facility with written safety procedures, and where the conditions of maintenance and supervision ensure that only qualified persons service the equipment, a disconnecting means within sight from the equipment shall not be required.

Exception No. 2: Where an attachment plug and receptacle serve as the disconnecting means in accordance with 440.13, their location shall be accessible but shall not be required to be readily accessible.

PART III. BRANCH-CIRCUIT SHORT-CIRCUIT AND GROUND-FAULT PROTECTION

440.22 Application and Selection.

(A) Rating or Setting for Individual Motor-Compressor. The motor-compressor branch-circuit short-circuit and ground-fault protective device shall be capable of carrying the starting current of the motor. A protective device having a rating or setting not exceeding 175 percent of the motor-compressor rated-load current or branch-circuit selection current, whichever is greater, shall be permitted, provided that, where the protection specified is not sufficient for the starting current of the motor, the rating or setting shall be permitted to be increased but shall not exceed 225 percent of the motor rated-load current or branch-circuit selection current, whichever is greater.

Exception: The rating of the branch-circuit short-circuit and ground-fault protective device shall not be required to be less than 15 amperes.

(B) Rating or Setting for Equipment. The equipment branch-circuit short-circuit and ground-fault protective device shall be capable of carrying the starting current of the equipment. Where the hermetic refrigerant motor-compressor is the only load on the circuit, the protection shall comply with 440.22(A). Where the equipment incorporates more than one hermetic refrigerant motor-compressor or a hermetic refrigerant motor-compressor and other motors or other loads, the equipment short-circuit and ground-fault protection shall comply with 430.53 and 440.22(B)(1) and (B)(2).

(1) Motor-Compressor Largest Load. Where a hermetic refrigerant motor-compressor is the largest load connected to the circuit, the rating or setting of the branch-circuit short-circuit and ground-fault protective device shall not exceed the value specified in 440.22(A) for the largest motor-compressor plus the sum of the rated-load current or branch-circuit selection current, whichever is greater, of the other motor-compressor(s) and the ratings of the other loads supplied.

(2) Motor-Compressor Not Largest Load. Where a hermetic refrigerant motor-compressor is not the largest load connected to the circuit, the rating or setting of the branch-circuit short-circuit and ground-fault protective device shall not exceed a value equal to the sum of the rated-load current or branch-circuit selection current, whichever is greater, rating(s) for the motor-compressor(s) plus the value specified in 430.53(C)(4) where other motor loads are supplied, or the value specified in 240.4 where only nonmotor loads are supplied in addition to the motor-compressor(s).

Exception No. 1: Equipment that starts and operates on a 15- or 20-ampere 120-volt, or 15-ampere 208- or 240-volt single-phase branch circuit, shall be permitted to be protected by the 15- or 20-ampere overcurrent device protecting the branch circuit, but if the maximum branch-circuit short-circuit and ground-fault protective device rating marked on the equipment is less than these values, the circuit protective device shall not exceed the value marked on the equipment nameplate.

Exception No. 2: The nameplate marking of cord-and-plug-connected equipment rated not greater than 250 volts, single-phase, such as household refrigerators and freezers, drinking water coolers, and beverage dispensers, shall be used in determining the branch-circuit requirements, and each unit shall be considered as a single motor unless the nameplate is marked otherwise.

(C) Protective Device Rating Not to Exceed the Manufacturer's Values. Where maximum protective device ratings shown on a manufacturer's overload relay table for use with a motor controller are less than the rating or setting selected in accordance with 440.22(A) and (B), the protective device rating shall not exceed the manufacturer's values marked on the equipment.

PART IV. BRANCH-CIRCUIT CONDUCTORS

440.32 Single Motor-Compressor. Branch-circuit conductors supplying a single motor-compressor shall have an ampacity not less than 125 percent of either the motor-compressor rated-load current or the branch-circuit selection current, whichever is greater.

For a wye-start, delta-run connected motor-compressor, the selection of branch-circuit conductors between the controller and the motor-compressor shall be permitted to be based on 72 percent of either the motor-compressor rated-load current or the branch-circuit selection current, whichever is greater.

440.33 Motor-Compressor(s) With or Without Additional Motor Loads. Conductors supplying one or more motor-compressor(s) with or without an additional motor load(s) shall have an ampacity not less than the sum of each of the following:

(1) The sum of the rated-load or branch-circuit selection current, whichever is greater, of all motor-compressor(s)
(2) The sum of the full-load current rating of all other motors
(3) 25 percent of the highest motor-compressor or motor full load current in the group

Exception No. 1: Where the circuitry is interlocked so as to prevent the starting and running of a second motor-compressor or group of motor-compressors, the conductor size shall be determined from the largest motor-compressor or group of motor-compressors that is to be operated at a given time.

Exception No. 2: The branch-circuit conductors for room air conditioners shall be in accordance with Part VII of Article 440.

440.34 Combination Load. Conductors supplying a motor-compressor load in addition to other load(s) as calculated from Article 220 and other applicable articles shall have an ampacity sufficient for the other load(s) plus the required ampacity for the motor-compressor load determined in accordance with 440.33 or, for a single motor-compressor, in accordance with 440.32.

440.35 Multimotor and Combination-Load Equipment. The ampacity of the conductors supplying multimotor and combination-load equipment shall not be less than the minimum circuit ampacity marked on the equipment in accordance with 440.4(B).

PART V. CONTROLLERS FOR MOTOR-COMPRESSORS

440.41 Rating.

(A) Motor-Compressor Controller. A motor-compressor controller shall have both a continuous-duty full-load current rating and a locked-rotor current rating not less than the nameplate rated-load current or branch-circuit selection current, whichever is greater, and locked-rotor current, respectively, of the compressor. In case the motor controller is rated in horsepower but is without one or both of the foregoing current ratings, equivalent currents shall be determined from the ratings as follows. Table 430.248, Table 430.249, and Table 430.250 shall be used to determine the equivalent full-load current rating. Table 430.251(A) and Table 430.251(B) shall be used to determine the equivalent locked-rotor current ratings.

(B) Controller Serving More Than One Load. A controller serving more than one motor-compressor or a motor-compressor and other loads shall have a continuous-duty full-load current rating and a locked-rotor current rating not less than the combined load as determined in accordance with 440.12(B).

PART VI. MOTOR-COMPRESSOR AND BRANCH-CIRCUIT OVERLOAD PROTECTION

440.52 Application and Selection.

(A) Protection of Motor-Compressor. Each motor-compressor shall be protected against overload and failure to start by one of the following means:

(1) A separate overload relay that is responsive to motor-compressor current. This device shall be selected to trip at not more than 140 percent of the motor-compressor rated-load current.
(2) A thermal protector integral with the motor-compressor, approved for use with the motor-compressor that it protects on the basis that it will prevent dangerous overheating of the motor-compressor due to overload and failure to start. If the current-interrupting device is separate from the motor-compressor and its control circuit is operated by a protective device integral with the motor-compressor, it shall be arranged so that the opening of the control circuit will result in interruption of current to the motor-compressor.

(3) A fuse or inverse time circuit breaker responsive to motor current, which shall also be permitted to serve as the branch-circuit short-circuit and ground-fault protective device. This device shall be rated at not more than 125 percent of the motor-compressor rated-load current. It shall have sufficient time delay to permit the motor-compressor to start and accelerate its load. The equipment or the motor-compressor shall be marked with this maximum branch-circuit fuse or inverse time circuit breaker rating.

(4) A protective system, furnished or specified and approved for use with the motor-compressor that it protects on the basis that it will prevent dangerous overheating of the motor-compressor due to overload and failure to start. If the current-interrupting device is separate from the motor-compressor and its control circuit is operated by a protective device that is not integral with the current-interrupting device, it shall be arranged so that the opening of the control circuit will result in interruption of current to the motor-compressor.

(B) Protection of Motor-Compressor Control Apparatus and Branch-Circuit Conductors. The motor-compressor controller(s), the disconnecting means, and the branch-circuit conductors shall be protected against overcurrent due to motor overload and failure to start by one of the following means, which shall be permitted to be the same device or system protecting the motor-compressor in accordance with 440.52(A):

Exception: Overload protection of motor-compressors and equipment on 15- and 20-ampere, single-phase, branch circuits shall be permitted to be in accordance with 440.54 and 440.55.

(1) An overload relay selected in accordance with 440.52(A)(1)

(2) A thermal protector applied in accordance with 440.52(A)(2), that will not permit a continuous current in excess of 156 percent of the marked rated-load current or branch-circuit selection current

(3) A fuse or inverse time circuit breaker selected in accordance with 440.52(A)(3)

(4) A protective system, in accordance with 440.52(A)(4), that will not permit a continuous current in excess of 156 percent of the marked rated-load current or branch-circuit selection current

440.53 Overload Relays. Overload relays and other devices for motor overload protection that are not capable of opening short circuits shall be protected by fuses or inverse time circuit breakers with ratings or settings in accordance with Part III unless identified for group installation or for part-winding motors and marked to indicate the maximum size of fuse or inverse time circuit breaker by which they shall be protected.

Exception: The fuse or inverse time circuit breaker size marking shall be permitted on the nameplate of the equipment in which the overload relay or other overload device is used.

440.54 Motor-Compressors and Equipment on 15- or 20-Ampere Branch Circuits — Not Cord- and Attachment-Plug-Connected. Overload protection for motor-compressors and equipment used on 15- or 20-ampere 120-volt, or 15-ampere 208- or 240-volt single-phase branch circuits as permitted in Article 210 shall be permitted as indicated in 440.54(A) and 440.54(B).

(A) Overload Protection. The motor-compressor shall be provided with overload protection selected as specified in 440.52(A). Both the controller and motor overload protective device shall be identified for installation with the short-circuit and ground-fault protective device for the branch circuit to which the equipment is connected.

(B) Time Delay The short-circuit and ground-fault protective device protecting the branch circuit shall have sufficient time delay to permit the motor-compressor and other motors to start and accelerate their loads.

440.55 Cord- and Attachment-Plug-Connected Motor-Compressors and Equipment on 15- or 20-Ampere Branch Circuits. Overload protection for motor-compressors and equipment that are cord- and attachment-plug-connected and used on 15- or 20-ampere 120-volt, or 15-ampere 208- or

240-volt, single-phase branch circuits as permitted in Article 210 shall be permitted as indicated in 440.55(A), (B), and (C).

(A) Overload Protection. The motor-compressor shall be provided with overload protection as specified in 440.52(A). Both the controller and the motor overload protective device shall be identified for installation with the short-circuit and ground-fault protective device for the branch circuit to which the equipment is connected.

(B) Attachment Plug and Receptacle or Cord Connector Rating. The rating of the attachment plug and receptacle or cord connector shall not exceed 20 amperes at 125 volts or 15 amperes at 250 volts.

(C) Time Delay. The short-circuit and ground-fault protective device protecting the branch circuit shall have sufficient time delay to permit the motor-compressor and other motors to start and accelerate their loads.

PART VII. PROVISIONS FOR ROOM AIR CONDITIONERS

440.62 Branch-Circuit Requirements.

(A) Room Air Conditioner as a Single Motor Unit. A room air conditioner shall be considered as a single motor unit in determining its branch-circuit requirements where all the following conditions are met:

(1) It is cord- and attachment-plug-connected.
(2) Its rating is not more than 40 amperes and 250 volts, single phase.
(3) Total rated-load current is shown on the room air-conditioner nameplate rather than individual motor currents.
(4) The rating of the branch-circuit short-circuit and ground-fault protective device does not exceed the ampacity of the branch-circuit conductors or the rating of the receptacle, whichever is less.

(B) Where No Other Loads Are Supplied. The total marked rating of a cord- and attachment-plug-connected room air conditioner shall not exceed 80 percent of the rating of a branch circuit where no other loads are supplied.

(C) Where Lighting Units or Other Appliances Are Also Supplied. The total marked rating of a cord- and attachment-plug-connected room air conditioner shall not exceed 50 percent of the rating of a branch circuit where lighting outlets, other appliances, or general-use receptacles are also supplied. Where the circuitry is interlocked to prevent simultaneous operation of the room air conditioner and energization of other outlets on the same branch circuit, a cord- and attachment-plug-connected room air conditioner shall not exceed 80 percent of the branch-circuit rating.

440.63 Disconnecting Means. An attachment plug and receptacle or cord connector shall be permitted to serve as the disconnecting means for a single-phase room air conditioner rated 250 volts or less if (1) the manual controls on the room air conditioner are readily accessible and located within 1.8 m (6 ft) of the floor, or (2) an approved manually operable disconnecting means is installed in a readily accessible location within sight from the room air conditioner.

CHAPTER 21

GENERATORS

INTRODUCTION

This chapter contains requirements from Article 445 — Generators. Article 445 covers the protection and installation of generators. Included in the requirements contained in this article are those covering: location, overcurrent protection, ampacities of conductors, bushings, and disconnecting means for generators. The conductors and system supplied by the generator are covered elsewhere in the *Code,* such as in Articles 240 and 250.

A generator installed as a separately derived system (with or without a transfer equipment interface) must be grounded in accordance with the requirements of 250.30. An alternating ac power source such as an on-site generator is not a separately derived system if the neutral is solidly interconnected to a service or other supply system grounded or neutral conductor. Portable and vehicle-mounted generators have unique system and equipment grounding requirements in Article 250. While not covered in this *Pocket Guide,* portable generators rated 15 kW or less are covered by GFCI requirements for the receptacles that are installed as part of the generator. These rules ensure safe conditions for personnel using the generator as a portable power source for cord- and plug-connected appliances and tools or as a portable source to supply an optional standby system (see Article 702 in Chapter 25) at a dwelling or other occupancy.

ARTICLE 445
Generators

445.10 Location. Generators shall be of a type suitable for the locations in which they are installed. They shall also meet the requirements for motors in 430.14.

445.11 Marking. Marking shall be provided by the manufacturer to indicate whether or not the generator neutral is bonded to its frame. Where the bonding is modified in the field, additional marking shall be required to indicate whether the neutral is bonded to the frame.

445.13 Ampacity of Conductors.

(A) General. The ampacity of the conductors from the generator output terminals to the first distribution device(s) containing overcurrent protection shall not be less than 115 percent of the nameplate current rating of the generator. It shall be permitted to size the neutral conductors in accordance with 220.61. Conductors that must carry ground-fault currents shall not be smaller than required by 250.30(A). Neutral conductors of dc generators that must carry ground-fault currents shall not be smaller than the minimum required size of the largest conductor.

Exception: Where the design and operation of the generator prevent overloading, the ampacity of the conductors shall not be less than 100 percent of the nameplate current rating of the generator.

N **(B) Overcurrent Protection Provided.** Where the generator set is equipped with a listed overcurrent protective device or a combination of a current transformer and overcurrent relay, conductors shall be permitted to be tapped from the load side of the protected terminals in accordance with 240.21(B).

Tapped conductors shall not be permitted for portable generators rated 15 kW or less where field wiring connection terminals are not accessible.

445.16 Bushings. Where field-installed wiring passes through an opening in an enclosure, a conduit box, or a barrier, a bushing shall be used to protect the conductors from the edges of an opening having sharp edges. The bushing shall have smooth, well-rounded surfaces where it may be in contact with the conductors. If used where oils, grease, or other contaminants may be present, the bushing shall be made of a material not deleteriously affected.

445.17 Generator Terminal Housings. Generator terminal housings shall comply with 430.12. Where a horsepower rating is required to determine the required minimum size of the generator terminal housing, the full-load current of the generator shall be compared with comparable motors in Table 430.247 through Table 430.250. The higher horsepower rating of Table 430.247 and Table 430.250 shall be used whenever the generator selection is between two ratings.

Exception: This section shall not apply to generators rated over 600 volts.

445.18 Disconnecting Means and Shutdown of Prime Mover.

• N **(A) Disconnecting Means.** Generators other than cord-and-plug-connected portable shall have one or more disconnecting means. Each disconnecting means shall simultaneously open all associated ungrounded conductors. Each disconnecting means shall be lockable in the open position in accordance with 110.25.

N **(B) Shutdown of Prime Mover.** Generators shall have provisions to shut down the prime mover. The means of shutdown shall comply with all of the following:

(1) Be equipped with provisions to disable all prime mover start control circuits to render the prime mover incapable of starting
(2) Initiate a shutdown mechanism that requires a mechanical reset

The provisions to shut down the prime mover shall be permitted to satisfy the requirements of 445.18(A) where it is capable of being locked in the open position in accordance with 110.25.

Generators with greater than 15 kW rating shall be provided with an additional requirement to shut down the prime mover. This additional shutdown means shall be located outside the equipment room or generator enclosure and shall also meet the requirements of 445.18(B)(1) and (B)(2).

N **(C) Generators Installed in Parallel.** Where a generator is installed in parallel with other generators, the provisions of 445.18(A) shall be capable of isolating the generator output terminals from the paralleling equipment. The disconnecting means shall not be required to be located at the generator.

CHAPTER 22

TRANSFORMERS

INTRODUCTION

This chapter contains requirements from Article 450 — Transformers. Article 450 contains requirements for the installation and use of all transformers other than the special application transformers covered in the eight exceptions to 450.1. The three parts of Article 450 are Part I General, Part II Specific Provisions Applicable to Different Types of Transformers, and Part III Transformer Vaults. The requirements in Article 450 cover transformer location, overcurrent protection, fire protection, disconnecting means, marking of transformers, and construction of transformer vaults. The transformers covered in Article 450 are those used primarily in feeders and branch circuits to supply power for utilization equipment. For the purposes of Article 450, a group of individual transformers connected together to work as a single phase or polyphase transformer and provided with a common nameplate is considered to be a single transformer.

Transformers are a great example of using the *NEC* in a system approach where the specific piece of equipment, the transformer in this case, is just one piece of the overall circuit and the equipment and conductors on the primary and secondary sides of the transformer are covered by other applicable *NEC* articles. For example, the typical power transformer application will involve Article 215 on feeders, Article 240 for overcurrent protection of the primary and secondary conductors, Article 250 for grounding and bonding, and Article 220 for calculating loads which had a directly impact on how to determine the necessary transformer size.

Grounding and bonding of transformer installations is entirely dependent on what type of system is created by the insertion of the transformer into the circuit. Whether the system established through the transformation is separately derived depends on the connections made in the primary (supply) and secondary (output) sides of the transformer. For example, a transformer used to create a small change in voltage (e.g., a buck or a boost transformer) has a shared connection between the supply and output. In that case, there is not complete isolation of the supply and output voltage systems, so a separately derived system has not been created. However, in the case of many power transformers where there is a significant difference between the primary and secondary voltage systems (e.g., 480 volt, 3-phase, 3-wire to 208Y/120 volt, 3 phase, 4-wire), there is no direct connection between any of the circuit conductors on the primary or supply side of the transformer to any of the conductors on the secondary or output side of the transformer. Required grounding and bonding connections that may create a common primary and secondary circuit and equipment connection are not recognized as creating a direct circuit conductor connection. In such cases, where there is circuit conductor isolation between the primary and secondary, the system is grounded and bonded in accordance with the requirements for separately derived systems contained in Article 250.

ARTICLE 450

Transformers and Transformer Vaults (Including Secondary Ties)

PART I. GENERAL PROVISIONS

450.3 Overcurrent Protection. Overcurrent protection of transformers shall comply with 450.3(A), (B), or (C). As used in this section, the word *transformer* shall mean a transformer or polyphase bank of two or more single-phase transformers operating as a unit.

(A) Transformers Over 1000 Volts, Nominal. Overcurrent protection shall be provided in accordance with Table 450.3(A).

(B) Transformers 1000 Volts, Nominal, or Less. Overcurrent protection shall be provided in accordance with Table 450.3(B).

(C) Voltage (Potential) Transformers. Voltage (potential) transformers installed indoors or enclosed shall be protected with primary fuses.

Informational Note: For protection of instrument circuits including voltage transformers, see 408.52.

450.4 Autotransformers 1000 Volts, Nominal, or Less.

(A) Overcurrent Protection. Each autotransformer 1000 volts, nominal, or less shall be protected by an individual overcurrent device installed in series with each ungrounded input conductor. Such overcurrent device shall be rated or set at not more than 125 percent of the rated full-load input current of the autotransformer. Where this calculation does not correspond to a standard rating of a fuse or nonadjustable circuit breaker and the rated input current is 9 amperes or more, the next higher standard rating described in 240.6 shall be permitted. An overcurrent device shall not be installed in series with the shunt winding (the winding common to both the input and the output circuits) of the autotransformer between Points A and B as shown in Figure 450.4(A).

Table 450.3(A) Maximum Rating or Setting of Overcurrent Protection for Transformers Over 1000 Volts (as a Percentage of Transformer-Rated Current)

		Primary Protection over 1000 Volts		Secondary Protection (See Note 2.)		
				Over 1000 Volts		1000 Volts or Less
Location Limitations	Transformer Rated Impedance	Circuit Breaker (See Note 4.)	Fuse Rating	Circuit Breaker (See Note 4.)	Fuse Rating	Circuit Breaker or Fuse Rating
Any location	Not more than 6%	600% (See Note 1.)	300% (See Note 1.)	300% (See Note 1.)	250% (See Note 1.)	125% (See Note 1.)
	More than 6% and not more than 10%	400% (See Note 1.)	300% (See Note 1.)	250% (See Note 1.)	225% (See Note 1.)	125% (See Note 1.)
Supervised locations only (See Note 3.)	Any	300% (See Note 1.)	250% (See Note 1.)	Not required	Not required	Not required
	Not more than 6%	600%	300%	300% (See Note 5.)	250% (See Note 5.)	250% (See Note 5.)
	More than 6% and not more than 10%	400%	300%	250% (See Note 5.)	225% (See Note 5.)	250% (See Note 5.)

Notes:

1. Where the required fuse rating or circuit breaker setting does not correspond to a standard rating or setting, a higher rating or setting that does not exceed the following shall be permitted:

a. The next higher standard rating or setting for fuses and circuit breakers 1000 volts and below, or

b. The next higher commercially available rating or setting for fuses and circuit breakers above 1000 volts.

2. Where secondary overcurrent protection is required, the secondary overcurrent device shall be permitted to consist of not more than six circuit breakers or six sets of fuses grouped in one location. Where multiple overcurrent devices are utilized, the total of all the device ratings shall not exceed the allowed value of a single overcurrent device. If both circuit breakers and fuses are used as the overcurrent device, the total of the device ratings shall not exceed that allowed for fuses.

3. A supervised location is a location where conditions of maintenance and supervision ensure that only qualified persons monitor and service the transformer installation.

4. Electronically actuated fuses that may be set to open at a specific current shall be set in accordance with settings for circuit breakers.

5. A transformer equipped with a coordinated thermal overload protection by the manufacturer shall be permitted to have separate secondary protection omitted.

Table 450.3(B) Maximum Rating or Setting of Overcurrent Protection for Transformers 1000 Volts and Less (as a Percentage of Transformer-Rated Current)

	Primary Protection			Secondary Protection (See Note 2.)	
Protection Method	Currents of 9 Amperes or More	Currents Less Than 9 Amperes	Currents Less Than 2 Amperes	Currents of 9 Amperes or More	Currents Less Than 9 Amperes
Primary only protection	125% (See Note 1.)	167%	300%	Not required	Not required
Primary and secondary protection	250% (See Note 3.)	250% (See Note 3.)	250% (See Note 3.)	125% (See Note 1.)	167%

Notes:

1. Where 125 percent of this current does not correspond to a standard rating of a fuse or nonadjustable circuit breaker, a higher rating that does not exceed the next higher standard rating shall be permitted.

2. Where secondary overcurrent protection is required, the secondary overcurrent device shall be permitted to consist of not more than six circuit breakers or six sets of fuses grouped in one location. Where multiple overcurrent devices are utilized, the total of all the device ratings shall not exceed the allowed value of a single overcurrent device.

3. A transformer equipped with coordinated thermal overload protection by the manufacturer and arranged to interrupt the primary current shall be permitted to have primary overcurrent protection rated or set at a current value that is not more than six times the rated current of the transformer for transformers having not more than 6 percent impedance and not more than four times the rated current of the transformer for transformers having more than 6 percent but not more than 10 percent impedance.

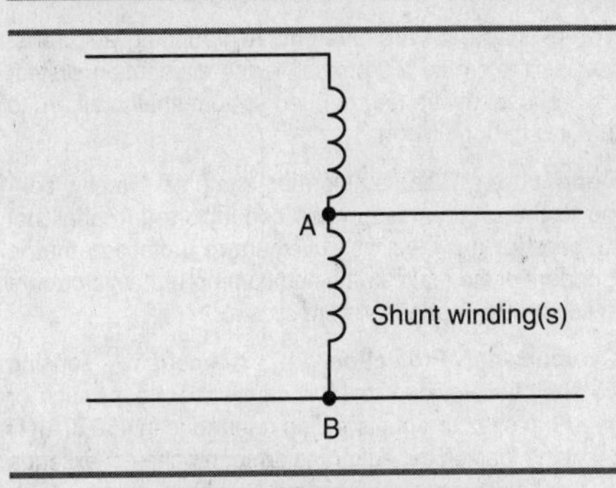

Figure 450.4(A) Autotransformer.

Exception: Where the rated input current of the autotransformer is less than 9 amperes, an overcurrent device rated or set at not more than 167 percent of the input current shall be permitted.

(B) Transformer Field-Connected as an Autotransformer. A transformer field-connected as an autotransformer shall be identified for use at elevated voltage.

Informational Note: For information on permitted uses of autotransformers, see 210.9 and 215.11.

450.5 Grounding Autotransformers. Grounding autotransformers covered in this section are zigzag or T-connected transformers connected to 3-phase, 3-wire ungrounded systems for the purpose of creating a 3-phase, 4-wire distribution system or providing a neutral point for grounding purposes. Such transformers shall have a continuous per-phase current rating and a continuous neutral current rating. Zigzag-connected transformers shall not be installed on the load side of any system grounding connection, including those made in accordance with 250.24(B), 250.30(A)(1), or 250.32(B), Exception No. 1.

(A) Three-Phase, 4-Wire System. A grounding autotransformer used to create a 3-phase, 4-wire distribution system from a 3-phase, 3-wire ungrounded system shall conform to 450.5(A)(1) through (A)(4).

(1) Connections. The transformer shall be directly connected to the ungrounded phase conductors and shall not be switched or provided with overcurrent protection that is independent of the main switch and common-trip overcurrent protection for the 3-phase, 4-wire system.

(2) Overcurrent Protection. An overcurrent sensing device shall be provided that will cause the main switch or common-trip overcurrent protection referred to in 450.5(A)(1) to open if the load on the autotransformer reaches or exceeds 125 percent of its continuous current per-phase or neutral rating. Delayed tripping for temporary overcurrents sensed at the autotransformer overcurrent device shall be permitted for the purpose of allowing proper operation of branch or feeder protective devices on the 4-wire system.

(3) Transformer Fault Sensing. A fault-sensing system that causes the opening of a main switch or common-trip overcurrent device for the 3-phase, 4-wire system shall be provided to guard against single-phasing or internal faults.

Informational Note: This can be accomplished by the use of two subtractive-connected donut-type current transformers installed to sense and signal when an unbalance occurs in the line current to the autotransformer of 50 percent or more of rated current.

(4) Rating. The autotransformer shall have a continuous neutral-current rating that is not less than the maximum possible neutral unbalanced load current of the 4-wire system.

(B) Ground Reference for Fault Protection Devices. A grounding autotransformer used to make available a specified magnitude of ground-fault current for operation of a ground-responsive protective device on a 3-phase, 3-wire ungrounded system shall conform to 450.5(B)(1) and (B)(2).

(1) Rating. The autotransformer shall have a continuous neutral-current rating not less than the specified ground-fault current.

(2) Overcurrent Protection. Overcurrent protection shall comply with (a) and (b).

(a) *Operation and Interrupting Rating.* An overcurrent protective device having an interrupting rating in compliance with 110.9 and that will open simultaneously all ungrounded conductors when it operates shall be applied in the grounding autotransformer branch circuit.

(b) *Ampere Rating.* The overcurrent protection shall be rated or set at a current not exceeding 125 percent of the autotransformer continuous per-phase current rating or 42 percent of the continuous-current rating of any series-connected devices in the autotransformer neutral connection. Delayed tripping for temporary overcurrents to permit the proper operation of ground-responsive tripping devices on the main system shall be permitted but shall not exceed values that would be more than the short-time current rating of the grounding autotransformer or any series connected devices in the neutral connection thereto.

Exception: For high-impedance grounded systems covered in 250.36, where the maximum ground-fault current is designed to be not more than 10 amperes, and where the grounding autotransformer and the grounding impedance are rated for continuous duty, an overcurrent device rated not more than 20 amperes that will simultaneously open all ungrounded conductors shall be permitted to be installed on the line side of the grounding autotransformer.

(C) Ground Reference for Damping Transitory Overvoltages. A grounding autotransformer used to limit transitory overvoltages shall be of suitable rating and connected in accordance with 450.5(A)(1).

450.6 Secondary Ties. As used in this article, a secondary tie is a circuit operating at 1000 volts, nominal, or less between phases that connects two power sources or power supply points, such as the secondaries of two transformers. The tie shall be permitted to consist of one or more conductors per phase or neutral. Conductors connecting the secondaries of transformers in accordance with 450.7 shall not be considered secondary ties.

As used in this section, the word *transformer* means a transformer or a bank of transformers operating as a unit.

(A) Tie Circuits. Tie circuits shall be provided with overcurrent protection at each end as required in Parts I, II, and VIII of Article 240.

Under the conditions described in 450.6(A)(1) and 450.6(A)(2), the overcurrent protection shall be permitted to be in accordance with 450.6(A)(3).

(1) Loads at Transformer Supply Points Only. Where all loads are connected at the transformer supply points at each end of the tie and overcurrent protection is not provided in accordance with Parts I, II, and VIII of Article 240, the rated ampacity of the tie shall not be less than 67 percent of the rated secondary current of the highest rated transformer supplying the secondary tie system.

(2) Loads Connected Between Transformer Supply Points. Where load is connected to the tie at any point between transformer supply points and overcurrent protection is not provided in accordance with Parts I, II, and VIII of Article 240, the rated ampacity of the tie shall not be less than 100 percent of the rated secondary current of the highest rated transformer supplying the secondary tie system.

Exception: Tie circuits comprised of multiple conductors per phase shall be permitted to be sized and protected in accordance with 450.6(A)(4).

(3) Tie Circuit Protection. Under the conditions described in 450.6(A)(1) and (A)(2), both supply ends of each ungrounded tie conductor shall be equipped with a protective device that opens at a predetermined temperature of the tie conductor under short-circuit conditions. This protection shall consist of one of the following: (1) a fusible link cable connector, terminal, or lug, commonly known as a limiter, each being of a size corresponding with that of the conductor and of construction and characteristics according to the operating voltage and the type of insulation on the tie conductors or (2) automatic circuit breakers actuated by devices having comparable time–current characteristics.

(4) Interconnection of Phase Conductors Between Transformer Supply Points. Where the tie consists of more than one conductor per phase or neutral, the conductors of each phase or neutral shall comply with one of the following provisions.

(a) *Interconnected.* The conductors shall be interconnected in order to establish a load supply point, and the protective device specified in 450.6(A)(3) shall be provided in each ungrounded tie conductor at this point on both sides of the interconnection. The means of interconnection shall have an ampacity not less than the load to be served.

(b) *Not Interconnected.* The loads shall be connected to one or more individual conductors of a paralleled conductor tie without interconnecting the conductors of each phase or neutral and without the protection specified in 450.6(A)(3) at load connection points. Where this is done, the tie conductors of each phase or neutral shall have a combined capacity ampacity of not less than 133 percent of the rated secondary current of the highest rated transformer supplying the secondary tie system, the total load of such taps shall not exceed the rated secondary current of the highest rated transformer, and the loads shall be equally divided on each phase and on the individual conductors of each phase as far as practicable.

(5) Tie Circuit Control. Where the operating voltage exceeds 150 volts to ground, secondary ties provided with limiters shall have a switch at each end that, when open, de-energizes the associated tie conductors and limiters. The current rating of the switch shall not be less than the rated current ampacity of the conductors connected to the switch. It shall be capable of interrupting its rated current, and it shall be constructed so that it will not open under the magnetic forces resulting from short-circuit current.

(B) Overcurrent Protection for Secondary Connections. Where secondary ties are used, an overcurrent device rated or set at not more than 250 percent of the rated secondary current of the transformers shall be provided in the secondary connections of each transformer supplying the tie system. In addition, an automatic circuit breaker actuated by a reverse-current relay set to open the circuit at not more than the rated secondary current of the transformer shall be provided in the secondary connection of each transformer.

(C) Grounding. Where the secondary tie system is grounded, each transformer secondary supplying the tie system shall be grounded in accordance with the requirements of 250.30 for separately derived systems.

450.7 Parallel Operation. Transformers shall be permitted to be operated in parallel and switched as a unit, provided the overcurrent protection for each transformer meets the requirements of 450.3(A) for primary and secondary protective devices over 1000 volts, or 450.3(B) for primary and secondary protective devices 1000 volts or less.

450.8 Guarding. Transformers shall be guarded as specified in 450.8(A) through (D).

(A) Mechanical Protection. Appropriate provisions shall be made to minimize the possibility of damage to transformers from external causes where the transformers are exposed to physical damage.

(B) Case or Enclosure. Dry-type transformers shall be provided with a noncombustible moisture-resistant case or enclosure that provides protection against the accidental insertion of foreign objects.

(C) Exposed Energized Parts. Switches or other equipment operating at 1000 volts, nominal, or less and serving only equipment within a transformer enclosure shall be permitted to be installed in the transformer enclosure if accessible to qualified persons only. All energized parts shall be guarded in accordance with 110.27 and 110.34.

(D) Voltage Warning. The operating voltage of exposed live parts of transformer installations shall be indicated by signs or visible markings on the equipment or structures.

450.9 Ventilation. The ventilation shall dispose of the transformer full-load heat losses without creating a temperature rise that is in excess of the transformer rating.

Transformers with ventilating openings shall be installed so that the ventilating openings are not blocked by walls or other obstructions. The required clearances shall be clearly marked on the transformer.

450.10 Grounding.

(A) Dry-Type Transformer Enclosures. Where separate equipment grounding conductors and supply-side bonding jumpers are installed, a terminal bar for all grounding and bonding conductor connections shall be secured inside the transformer enclosure. The terminal bar shall be bonded to the enclosure in accordance with 250.12 and shall not be installed on or over any vented portion of the enclosure.

Exception: Where a dry-type transformer is equipped with wire-type connections (leads), the grounding and bonding connections shall be permitted to be connected together using any of the methods in 250.8 and shall be bonded to the enclosure if of metal.

(B) Other Metal Parts. Where grounded, exposed non–current-carrying metal parts of transformer installations, including fences, guards, and so forth, shall be grounded and bonded under the conditions and in the manner specified for electrical equipment and other exposed metal parts in Parts V, VI, and VII of Article 250.

450.11 Marking.

(B) Source Marking. A transformer shall be permitted to be supplied at the marked secondary voltage, provided that the installation is in accordance with the manufacturer's instructions.

450.13 Accessibility. All transformers and transformer vaults shall be readily accessible to qualified personnel for inspection and maintenance or shall meet the requirements of 450.13(A) or 450.13(B).

(A) Open Installations. Dry-type transformers 1000 volts, nominal, or less, located in the open on walls, columns, or structures, shall not be required to be readily accessible.

(B) Hollow Space Installations. Dry-type transformers 1000 volts, nominal, or less and not exceeding 50 kVA shall be permitted in hollow spaces of buildings not permanently closed in by structure, provided they meet the ventilation requirements of 450.9 and separation from combustible materials requirements of 450.21(A). Transformers so installed shall not be required to be readily accessible.

450.14 Disconnecting Means. Transformers, other than Class 2 or Class 3 transformers, shall have a disconnecting means located either in sight of the transformer or in a remote location. Where located in a remote location, the disconnecting means shall be lockable in accordance with 110.25, and its location shall be field marked on the transformer.

PART II. SPECIFIC PROVISIONS APPLICABLE TO DIFFERENT TYPES OF TRANSFORMERS

450.21 Dry-Type Transformers Installed Indoors.

(A) Not Over 112½ kVA. Dry-type transformers installed indoors and rated 112½ kVA or less shall have a separation of at least 300 mm (12 in.) from combustible material unless separated from the combustible material by a fire-resistant, heat-insulated barrier.

Exception: This rule shall not apply to transformers rated for 1000 volts, nominal, or less that are completely enclosed, except for ventilating openings.

(B) Over 112½ kVA. Individual dry-type transformers of more than 112½ kVA rating shall be installed in a transformer room of fire-resistant construction. Unless specified otherwise in this article, the term *fire resistant* means a construction having a minimum fire rating of 1 hour.

Exception No. 1: Transformers with Class 155 or higher insulation systems and separated from combustible material by a fire-resistant, heat-insulating barrier or by not less than 1.83 m (6 ft) horizontally and 3.7 m (12 ft) vertically.

Exception No. 2: Transformers with Class 155 or higher insulation systems and completely enclosed except for ventilating openings.

(C) Over 35,000 Volts. Dry-type transformers rated over 35,000 volts shall be installed in a vault complying with Part III of this article.

450.22 Dry-Type Transformers Installed Outdoors. Dry-type transformers installed outdoors shall have a weatherproof enclosure.

Transformers exceeding 112½ kVA shall not be located within 300 mm (12 in.) of combustible materials of buildings unless the transformer has Class 155 insulation systems or higher and is completely enclosed except for ventilating openings.

450.23 Less-Flammable Liquid-Insulated Transformers. Transformers insulated with listed less-flammable liquids that have a fire point of not less than 300°C shall be permitted to be installed in accordance with 450.23(A) or 450.23(B).

(A) Indoor Installations. Indoor installations shall be permitted in accordance with one of the following:

(1) In Type I or Type II buildings, in areas where all of the following requirements are met:

a. The transformer is rated 35,000 volts or less.
b. No combustible materials are stored.
c. A liquid confinement area is provided.
d. The installation complies with all the restrictions provided for in the listing of the liquid.
e. With an automatic fire extinguishing system and a liquid confinement area, provided the transformer is rated 35,000 volts or less
f. In accordance with 450.26

(B) Outdoor Installations. Less-flammable liquid-filled transformers shall be permitted to be installed outdoors, attached to, adjacent to, or on the roof of buildings, where installed in accordance with (1) or (2).

(1) For Type I and Type II buildings, the installation shall comply with all the restrictions provided for in the listing of the liquid.
(2) In accordance with 450.27.

450.24 Nonflammable Fluid-Insulated Transformers. Transformers insulated with a dielectric fluid identified as nonflammable shall be permitted to be installed indoors or outdoors. Such transformers installed indoors and rated over 35,000 volts shall be installed in a vault. Such transformers installed indoors shall be furnished with a liquid confinement area and a pressure-relief vent. The transformers shall be furnished with a means for absorbing any gases generated by arcing inside the tank, or the pressure-relief vent shall be connected to a chimney or flue that will carry such gases to an environmentally safe area.

For the purposes of this section, a nonflammable dielectric fluid is one that does not have a flash point or fire point and is not flammable in air.

450.27 Oil-Insulated Transformers Installed Outdoors. Combustible material, combustible buildings, and parts of buildings, fire escapes, and door and window openings shall be safeguarded from fires originating in oil-insulated transformers installed on roofs, attached to or adjacent to a building or combustible material.

In cases where the transformer installation presents a fire hazard, one or more of the following safeguards shall be applied according to the degree of hazard involved:

(1) Space separations
(2) Fire-resistant barriers
(3) Automatic fire suppression systems
(4) Enclosures that confine the oil of a ruptured transformer tank

Oil enclosures shall be permitted to consist of fire-resistant dikes, curbed areas or basins, or trenches filled with coarse, crushed stone. Oil enclosures shall be provided with trapped drains where the exposure and the quantity of oil involved are such that removal of oil is important.

PART III. TRANSFORMER VAULTS

450.41 Location. Vaults shall be located where they can be ventilated to the outside air without using flues or ducts wherever such an arrangement is practicable.

450.42 Walls, Roofs, and Floors. The walls and roofs of vaults shall be constructed of materials that have approved structural strength for the conditions with a minimum fire resistance of 3 hours. The floors of vaults in contact with the earth shall be of concrete that is not less than 100 mm (4 in.)

thick, but, where the vault is constructed with a vacant space or other stories below it, the floor shall have approved structural strength for the load imposed thereon and a minimum fire resistance of 3 hours. For the purposes of this section, studs and wallboard construction shall not be permitted.

Exception: Where transformers are protected with automatic sprinkler, water spray, carbon dioxide, or halon, construction of 1-hour rating shall be permitted.

450.43 Doorways. Vault doorways shall be protected in accordance with 450.43(A), (B), and (C).

(A) Type of Door. Each doorway leading into a vault from the building interior shall be provided with a tight-fitting door that has a minimum fire rating of 3 hours. The authority having jurisdiction shall be permitted to require such a door for an exterior wall opening where conditions warrant.

Exception: Where transformers are protected with automatic sprinkler, water spray, carbon dioxide, or halon, construction of 1-hour rating shall be permitted.

(B) Sills. A door sill or curb that is of an approved height that will confine the oil from the largest transformer within the vault shall be provided, and in no case shall the height be less than 100 mm (4 in.).

(C) Locks. Doors shall be equipped with locks, and doors shall be kept locked, access being allowed only to qualified persons. Personnel doors shall open in the direction of egress and be equipped with listed panic hardware.

450.45 Ventilation Openings. Where required by 450.9, openings for ventilation shall be provided in accordance with 450.45(A) through (F).

(A) Location. Ventilation openings shall be located as far as possible from doors, windows, fire escapes, and combustible material.

(B) Arrangement. A vault ventilated by natural circulation of air shall be permitted to have roughly half of the total area of openings required for ventilation in one or more openings near the floor and the remainder in one or more openings in the roof or in the sidewalls near the roof, or all of the area required for ventilation shall be permitted in one or more openings in or near the roof.

(C) Size. For a vault ventilated by natural circulation of air to an outdoor area, the combined net area of all ventilating openings, after deducting the area occupied by screens, gratings, or louvers, shall not be less than 1900 mm^2 (3 $in.^2$) per kVA of transformer capacity in service, and in no case shall the net area be less than 0.1 m^2 (1 ft^2) for any capacity under 50 kVA.

(D) Covering. Ventilation openings shall be covered with durable gratings, screens, or louvers, according to the treatment required in order to avoid unsafe conditions.

(E) Dampers. All ventilation openings to the indoors shall be provided with automatic closing fire dampers that operate in response to a vault fire. Such dampers shall possess a standard fire rating of not less than 1½ hours.

(F) Ducts. Ventilating ducts shall be constructed of fire-resistant material.

450.46 Drainage. Where practicable, vaults containing more than 100 kVA transformer capacity shall be provided with a drain or other means that will carry off any accumulation of oil or water in the vault unless local conditions make this impracticable. The floor shall be pitched to the drain where provided.

450.47 Water Pipes and Accessories. Any pipe or duct system foreign to the electrical installation shall not enter or pass through a transformer vault. Piping or other facilities provided for vault fire protection, or for transformer cooling, shall not be considered foreign to the electrical installation.

450.48 Storage in Vaults. Materials shall not be stored in transformer vaults.

CHAPTER 23

SWIMMING POOLS, FOUNTAINS, AND SIMILAR INSTALLATIONS

INTRODUCTION

This chapter contains requirements from Article 680 — Swimming Pools, Fountains, and Similar Installations. The requirements of Article 680 cover the installation of electric wiring for (and equipment in or adjacent to) all swimming, wading, and decorative pools, fountains, hot tubs, spas, and hydromassage bathtubs, whether permanently installed or storable, as well as to associated metallic auxiliary equipment such as pumps, filters, and so forth. Article 680 contains requirements on grounding, bonding, wiring methods, conductor clearances, receptacles, lighting, other equipment. Extensive use of ground-fault circuit-interrupter protection and low-voltage equipment is aimed at mitigating the elevated risk of electric shock in environments where the users are partially or completely immersed and have body resistances lower than in dry locations.

More robust requirements on wiring methods and conductor sizes and material address the corrosive effects of chlorine and other treatment agents used to maintain safe water chemistry. Salt water pools have also become very popular in some areas, and the elevated sodium level in the water adversely impacts many of the typical electrical materials used in locations that are noncorrosive. Article 680 requirements covering the impact of corrosive agents on electrical equipment apply only to those portions of the wiring system that are directly exposed to these caustic environments.

Contained in this *Pocket Guide* are four of the seven parts that are included in Article 680: Part I General, Part II Permanently Installed Pools, Part IV Spas and Hot Tubs, and Part V Fountains. The requirements unique to hydromassage bathtubs contained in Part VII of Article 680 are included in the *Pocket Guide to Residential Electrical Installations*. Section 680.2 (not included in this guide) contains definitions that are unique to the bodies of water covered by Article 680. In order to properly apply the requirements of Article 680, it is essential that those designing, performing, or inspecting the electrical installations associated with swimming pools, fountains, hot tubs, and spas be conversant with these definitions.

ARTICLE 680

Swimming Pools, Fountains, and Similar Installations

PART I. GENERAL

680.6 Grounding. Electrical equipment shall be grounded in accordance with Parts V, VI, and VII of Article 250 and connected by wiring methods of Chapter 3, except as modified by this article. The following equipment shall be grounded:

(1) Through-wall lighting assemblies and underwater luminaires, other than those low-voltage lighting products listed for the application without a grounding conductor
(2) All electrical equipment located within 1.5 m (5 ft) of the inside wall of the specified body of water
(3) All electrical equipment associated with the recirculating system of the specified body of water
(4) Junction boxes
(5) Transformer and power supply enclosures
(6) Ground-fault circuit interrupters
(7) Panelboards that are not part of the service equipment and that supply any electrical equipment associated with the specified body of water

N **680.7 Grounding and Bonding Terminals.** Grounding and bonding terminals shall be identified for use in wet and corrosive environments. Field-installed grounding and bonding connections in a damp, wet, or corrosive environment shall be composed of copper, copper alloy, or stainless steel. They shall be listed for direct burial use.

680.8 Cord-and-Plug-Connected Equipment. Fixed or stationary equipment, other than underwater luminaires, for a permanently installed pool shall be permitted to be connected with a flexible cord and plug to facilitate the removal or disconnection for maintenance or repair.

(A) Length. For other than storable pools, the flexible cord shall not exceed 900 mm (3 ft) in length.

(B) Equipment Grounding. The flexible cord shall have a copper equipment grounding conductor sized in accordance with 250.122 but not smaller than 12 AWG. The cord shall terminate in a grounding-type attachment plug.

(C) Construction. The equipment grounding conductors shall be connected to a fixed metal part of the assembly. The removable part shall be mounted on or bonded to the fixed metal part.

680.9 Overhead Conductor Clearances. Overhead conductors shall meet the clearance requirements in this section. Where a minimum clearance from the water level is given, the measurement shall be taken from the maximum water level of the specified body of water.

(A) Power. With respect to service-drop conductors, overhead service conductors, and open overhead wiring, swimming pool and similar installations shall comply with the minimum clearances given in Table 680.9(A) and illustrated in Figure 680.9(A).

(B) Communications Systems. Communications, radio, and television coaxial cables within the scope of Articles 800 through 820 shall be permitted at a height of not less than 3.0 m (10 ft) above swimming and wading pools, diving structures, and observation stands, towers, or platforms.

(C) Network-Powered Broadband Communications Systems. The minimum clearances for overhead network-powered

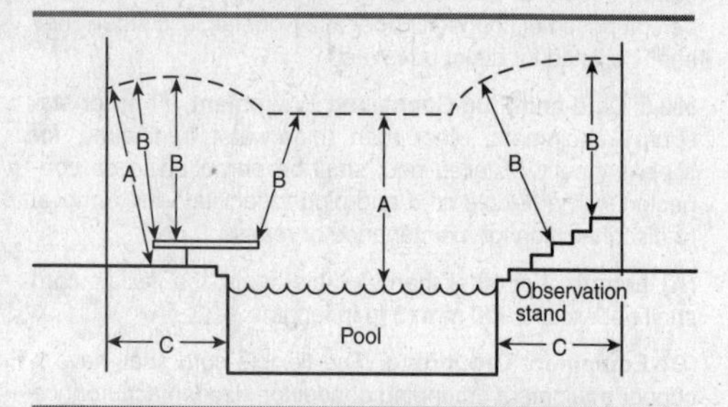

Figure 680.9(A) Clearances from Pool Structures.

broadband communications systems conductors from pools or fountains shall comply with the provisions in Table 680.9(A) for conductors operating at 0 to 750 volts to ground.

680.10 Electric Pool Water Heaters. All electric pool water heaters shall have the heating elements subdivided into loads not exceeding 48 amperes and protected at not over 60 amperes. The ampacity of the branch-circuit conductors and the rating or setting of overcurrent protective devices shall not be less than 125 percent of the total nameplate-rated load.

680.11 Underground Wiring Location. Underground wiring shall be permitted where installed in rigid metal conduit, intermediate metal conduit, rigid polyvinyl chloride conduit, reinforced thermosetting resin conduit, or Type MC cable, suitable for the conditions subject to that location. Underground wiring shall not be permitted under the pool unless this wiring is necessary to supply pool equipment permitted by this article. Minimum cover depths shall be as given in Table 300.5.

•

680.12 Equipment Rooms and Pits. Electrical equipment shall not be installed in rooms or pits that do not have drainage that prevents water accumulation during normal operation or filter maintenance. Equipment shall be suitable for the environment in accordance with 300.6.

680.13 Maintenance Disconnecting Means. One or more means to simultaneously disconnect all ungrounded conductors shall be provided for all utilization equipment other than lighting. Each means shall be readily accessible and within sight from its equipment and shall be located at least 1.5 m (5 ft) horizontally from the inside walls of a pool, spa, fountain, or hot tub unless separated from the open water by a permanently installed barrier that provides a 1.5 m (5 ft) reach path or greater. This horizontal distance shall be measured from the water's edge along the shortest path required to reach the disconnect.

N **680.14 Corrosive Environment.**

(A) General. Areas where pool sanitation chemicals are stored, as well as areas with circulation pumps, automatic chlorinators, filters, open areas under decks adjacent to or abutting the pool structure, and similar locations shall be considered to be a corrosive environment. The air in such areas shall be considered to be laden with acid, chlorine, and bromine vapors, or any combination of acid, chlorine, or bromine vapors, and any liquids or condensation in those areas shall be considered to be laden with acids, chlorine, and bromine vapors, or any combination of acid, chlorine, or bromine vapors.

(B) Wiring Methods. Wiring methods in the areas described in 680.14(A) shall be listed and identified for use in such areas. Rigid metal conduit, intermediate metal conduit, rigid polyvinyl chloride conduit, and reinforced thermosetting resin conduit shall be considered to be resistant to the corrosive environment specified in 680.14(A).

PART II. PERMANENTLY INSTALLED POOLS

680.20 General. Electrical installations at permanently installed pools shall comply with the provisions of Part I and Part II of this article.

680.21 Motors.

(A) Wiring Methods. The wiring to a pool motor shall comply with (A)(1) unless modified for specific circumstances by (A)(2), (A)(3), (A)(4), or (A)(5).

(1) General. Wiring methods installed in the corrosive environment described in 680.14 shall comply with 680.14(B) or shall be type MC cable listed for that location. Wiring methods installed in these locations shall contain an insulated copper equipment grounding conductor sized in accordance with Table 250.122 but not smaller than 12 AWG.

Where installed in noncorrosive environments, branch circuits shall comply with the general requirements in Chapter 3.

(2) Flexible Connections. Where necessary to employ flexible connections at or adjacent to the motor, liquidtight flexible metal or liquidtight flexible nonmetallic conduit with listed fittings shall be permitted.

(3) Cord-and-Plug Connections. Pool-associated motors shall be permitted to employ cord-and-plug connections. The flexible cord shall not exceed 900 mm (3 ft) in length. The flexible cord shall include a copper equipment grounding conductor sized in accordance with 250.122 but not smaller than 12 AWG. The cord shall terminate in a grounding-type attachment plug.

(B) Double Insulated Pool Pumps. A listed cord-and-plug-connected pool pump incorporating an approved system of double insulation that provides a means for grounding only the internal and nonaccessible, non–current-carrying metal parts of the pump shall be connected to any wiring method recognized in Chapter 3 that is suitable for the location. Where the bonding grid is connected to the equipment grounding conductor of the motor circuit in accordance with the second sentence of 680.26(B)(6)(a), the branch-circuit wiring shall comply with 680.21(A).

(C) GFCI Protection. Outlets supplying pool pump motors connected to single-phase, 120-volt through 240-volt branch circuits, whether by receptacle or by direct connection, shall be provided with ground-fault circuit-interrupter protection for personnel.

680.22 Lighting, Receptacles, and Equipment.

(A) Receptacles.

(1) Required Receptacle, Location. Where a permanently installed pool is installed, no fewer than one 125-volt, 15- or 20-ampere receptacle on a general-purpose branch circuit shall be located not less than 1.83 m (6 ft) from, and not more than 6.0 m (20 ft) from, the inside wall of the pool. This receptacle shall be located not more than 2.0 m (6 ft 6 in.) above the floor, platform, or grade level serving the pool.

(2) Circulation and Sanitation System, Location. Receptacles that provide power for water-pump motors or for other loads directly related to the circulation and sanitation system shall be located at least 1.83 m (6 ft) from the inside walls of the pool. These receptacles shall have GFCI protection and be of the grounding type.

(3) Other Receptacles, Location. Other receptacles shall be not less than 1.83 m (6 ft) from the inside walls of a pool.

(4) GFCI Protection. All 15- and 20-ampere, single-phase, 125-volt receptacles located within 6.0 m (20 ft) of the inside walls of a pool shall be protected by a ground-fault circuit interrupter.

(5) Measurements. In determining the dimensions in this section addressing receptacle spacings, the distance to be measured shall be the shortest path the supply cord of an appliance connected to the receptacle would follow without piercing a floor, wall, ceiling, doorway with hinged or sliding door, window opening, or other effective permanent barrier.

(B) Luminaires, Lighting Outlets, and Ceiling-Suspended (Paddle) Fans.

(1) New Outdoor Installation Clearances. In outdoor pool areas, luminaires, lighting outlets, and ceiling-suspended (paddle) fans installed above the pool or the area extending 1.5 m (5 ft) horizontally from the inside walls of the pool shall be installed at a height not less than 3.7 m (12 ft) above the maximum water level of the pool.

(2) Indoor Clearances. For installations in indoor pool areas, the clearances shall be the same as for outdoor areas unless modified as provided in this paragraph. If the branch circuit supplying the equipment is protected by a ground-fault circuit interrupter, the following equipment shall be permitted at a height not less than 2.3 m (7 ft 6 in.) above the maximum pool water level:

(1) Totally enclosed luminaires
(2) Ceiling-suspended (paddle) fans identified for use beneath ceiling structures such as provided on porches or patios

(3) Existing Installations. Existing luminaires and lighting outlets located less than 1.5 m (5 ft) measured horizontally from the inside walls of a pool shall be not less than 1.5 m (5 ft) above the surface of the maximum water level, shall be rigidly attached to the existing structure, and shall be protected by a ground-fault circuit interrupter.

(4) GFCI Protection in Adjacent Areas. Luminaires, lighting outlets, and ceiling-suspended (paddle) fans installed in the area extending between 1.5 m (5 ft) and 3.0 m (10 ft) horizontally from the inside walls of a pool shall be protected by a ground-fault circuit interrupter unless installed not less than 1.5 m (5 ft) above the maximum water level and rigidly attached to the structure adjacent to or enclosing the pool.

(5) Cord-and-Plug-Connected Luminaires. Cord-and-plug-connected luminaires shall comply with the requirements of 680.8 where installed within 4.9 m (16 ft) of any point on the water surface, measured radially.

(6) Low-Voltage Luminaires. Listed low-voltage luminaires not requiring grounding, not exceeding the low-voltage contact limit, and supplied by listed transformers or power supplies that comply with 680.23(A)(2) shall be permitted to be located less than 1.5 m (5 ft) from the inside walls of the pool.

N **(7) Low-Voltage Gas-Fired Luminaires, Decorative Fireplaces, Fire Pits, and Similar Equipment.** Listed low-voltage gas-fired-luminaires, decorative fireplaces, fire pits, and similar equipment using low-voltage ignitors that do not require grounding, and are supplied by listed transformers or power supplies that comply with 680.23(A)(2) with outputs that do not exceed the low-voltage contact limit shall be permitted to be located less than 1.5 m (5 ft) from the inside walls of the pool. Metallic equipment shall be bonded in accordance with the requirements in 680.26(B). Transformers or power supplies supplying this type of equipment shall be installed in accordance with the requirements in 680.24. Metallic gas piping shall be bonded in accordance with the requirements in 250.104(B) and 680.26(B)(7).

(C) Switching Devices. Switching devices shall be located at least 1.5 m (5 ft) horizontally from the inside walls of a pool unless separated from the pool by a solid fence, wall, or other permanent barrier. Alternatively, a switch that is listed as being acceptable for use within 1.5 m (5 ft) shall be permitted.

(D) Other Outlets. Other outlets shall be not less than 3.0 m (10 ft) from the inside walls of the pool. Measurements shall be determined in accordance with 680.22(A)(5).

680.23 Underwater Luminaires. This section covers all luminaires installed below the maximum water level of the pool.

(A) General.

(1) Luminaire Design, Normal Operation. The design of an underwater luminaire supplied from a branch circuit either directly or by way of a transformer or power supply meeting the requirements of this section shall be such that, where the luminaire is properly installed without a ground-fault circuit interrupter, there is no shock hazard with any likely combination of fault conditions during normal use (not relamping).

(2) Transformers and Power Supplies. Transformers and power supplies used for the supply of underwater luminaires, together with the transformer or power supply enclosure, shall be listed, labeled, and identified for swimming pool and spa use. The transformer or power supply shall incorporate either a transformer of the isolated winding type, with an ungrounded secondary that has a grounded metal barrier between the primary and secondary windings, or one that incorporates an approved system of double insulation between the primary and secondary windings.

(3) GFCI Protection, Lamping, Relamping, and Servicing. Ground-fault circuit interrupter protection for personnel shall be installed in the branch circuit supplying luminaires operating at voltages greater than the low-voltage contact limit.

(4) Voltage Limitation. No luminaires shall be installed for operation on supply circuits over 150 volts between conductors.

(5) Location, Wall-Mounted Luminaires. Luminaires mounted in walls shall be installed with the top of the luminaire lens not less than 450 mm (18 in.) below the normal water level of the pool, unless the luminaire is listed and identified for use at lesser depths. No luminaire shall be installed less than 100 mm (4 in.) below the normal water level of the pool.

(6) Bottom-Mounted Luminaires. A luminaire facing upward shall comply with either (1) or (2):

(1) Have the lens guarded to prevent contact by any person
(2) Be listed for use without a guard

(7) Dependence on Submersion. Luminaires that depend on submersion for safe operation shall be inherently protected against the hazards of overheating when not submerged.

(8) Compliance. Compliance with these requirements shall be obtained by the use of a listed underwater luminaire and by installation of a listed ground-fault circuit interrupter in the branch circuit or a listed transformer or power supply for luminaires operating at not more than the low voltage contact limit.

(B) Wet-Niche Luminaires.

(1) Forming Shells. Forming shells shall be installed for the mounting of all wet-niche underwater luminaires and shall be equipped with provisions for conduit entries. Metal parts of the luminaire and forming shell in contact with the pool water shall be of brass or other approved corrosion-resistant metal. All forming shells used with nonmetallic conduit systems, other than those that are part of a listed low-voltage lighting system not requiring grounding, shall include provisions for terminating an 8 AWG copper conductor.

(2) Wiring Extending Directly to the Forming Shell. Conduit shall be installed from the forming shell to a junction box or other enclosure conforming to the requirements in 680.24. Conduit shall be rigid metal, intermediate metal, liquidtight flexible nonmetallic, or rigid nonmetallic.

(a) *Metal Conduit.* Metal conduit shall be approved and shall be of brass or other approved corrosion-resistant metal.

(b) *Nonmetallic Conduit.* Where a nonmetallic conduit is used, an 8 AWG insulated solid or stranded copper bonding jumper shall be installed in this conduit unless a listed low-voltage lighting system not requiring grounding is used. The bonding jumper shall be terminated in the forming shell, junction box or transformer enclosure, or ground-fault circuit-interrupter enclosure. The termination of the 8 AWG bonding jumper in the forming shell shall be covered with, or encapsulated in, a listed potting compound to protect the connection from the possible deteriorating effect of pool water.

(3) Equipment Grounding Provisions for Cords. Other than listed low-voltages lighting systems not requiring grounding wet-niche luminaires that are supplied by a flexible cord or cable shall have all exposed non–current-carrying metal parts grounded by an insulated copper equipment grounding conductor that is an integral part of the cord or cable. This grounding conductor shall be connected to a grounding terminal in the supply junction box, transformer enclosure, or other enclosure. The grounding conductor shall not be smaller than the supply conductors and not smaller than 16 AWG.

(4) Luminaire Grounding Terminations. The end of the flexible-cord jacket and the flexible-cord conductor terminations within a luminaire shall be covered with, or encapsulated in, a suitable potting compound to prevent the entry of water into the luminaire through the cord or its conductors. If present, the grounding connection within a luminaire shall be similarly treated to protect such connection from the deteriorating effect of pool water in the event of water entry into the luminaire.

(5) Luminaire Bonding. The luminaire shall be bonded to, and secured to, the forming shell by a positive locking device that ensures a low-resistance contact and requires a tool to remove the luminaire from the forming shell. Bonding shall not be required for luminaires that are listed for the application and have no non–current-carrying metal parts.

(6) Servicing. All wet-niche luminaires shall be removable from the water for inspection, relamping, or other maintenance. The forming shell location and length of cord in the forming shell shall permit personnel to place the removed luminaire on the deck or other dry location for such maintenance. The luminaire maintenance location shall be accessible without entering or going in the pool water.

(C) Dry-Niche Luminaires.

(1) Construction. A dry-niche luminaire shall have provision for drainage of water. Other than listed low voltage luminaires not requiring grounding, a dry-niche luminaire shall have means for accommodating one equipment grounding conductor for each conduit entry.

(2) Junction Box. A junction box shall not be required but, if used, shall not be required to be elevated or located as specified in 680.24(A)(2) if the luminaire is specifically identified for the purpose.

(D) No-Niche Luminaires. A no-niche luminaire shall meet the construction requirements of 680.23(B)(3) and be installed in accordance with the requirements of 680.23(B). Where connection to a forming shell is specified, the connection shall be to the mounting bracket.

(E) Through-Wall Lighting Assembly. A through-wall lighting assembly shall be equipped with a threaded entry or hub, or a nonmetallic hub, for the purpose of accommodating the termination of the supply conduit. A through-wall lighting assembly shall meet the construction requirements of 680.23(B)(3) and be installed in accordance with the requirements of 680.23. Where connection to a forming shell is specified, the connection shall be to the conduit termination point.

(F) Branch-Circuit Wiring.

(1) Wiring Methods. Where branch-circuit wiring on the supply side of enclosures and junction boxes connected to conduits run to underwater luminaires are installed in corrosive environments as described in 680.14, the wiring method of that portion of the branch circuit shall be as required in 680.14(B) or shall be liquidtight flexible nonmetallic conduit. Wiring methods installed in corrosive environments as described in 680.14 shall contain an insulated copper equipment grounding conductor sized in accordance with Table 250.122, but not smaller than 12 AWG.

Where installed in noncorrosive environments, branch circuits shall comply with the general requirements in Chapter 3.

Exception: Where connecting to transformers or power supplies for pool lights, liquidtight flexible metal conduit shall be permitted. The length shall not exceed 1.8 m (6 ft) for any one length or exceed 3.0 m (10 ft) in total length used.

(2) Equipment Grounding. Other than listed low-voltage luminaires not requiring grounding, all through-wall lighting assemblies, wet-niche, dry-niche, or no-niche luminaires shall be connected to an insulated copper equipment grounding conductor installed with the circuit conductors. The equipment grounding conductor shall be installed without joint or splice except as permitted in (F)(2)(a) and (F)(2)(b). The equipment grounding conductor shall be sized in accordance with Table 250.122 but shall not be smaller than 12 AWG.

Exception: An equipment grounding conductor between the wiring chamber of the secondary winding of a transformer and a junction box shall be sized in accordance with the overcurrent device in this circuit.

(a) If more than one underwater luminaire is supplied by the same branch circuit, the equipment grounding conductor, installed between the junction boxes, transformer enclosures, or other enclosures in the supply circuit to wet-niche luminaires, or between the field-wiring compartments of dry-niche luminaires, shall be permitted to be terminated on grounding terminals.

(b) If the underwater luminaire is supplied from a transformer, ground-fault circuit interrupter, clock-operated switch, or a manual snap switch that is located between the panelboard and a junction box connected to the conduit that extends directly to the underwater luminaire, the equipment grounding conductor shall be permitted to terminate on grounding terminals on the transformer, ground-fault circuit interrupter, clock-operated switch enclosure, or an outlet box used to enclose a snap switch.

(3) Conductors. Conductors on the load side of a ground-fault circuit interrupter or of a transformer, used to comply with the provisions of 680.23(A)(8), shall not occupy raceways, boxes, or enclosures containing other conductors unless one of the following conditions applies:

(1) The other conductors are protected by ground-fault circuit interrupters.
(2) The other conductors are equipment grounding conductors and bonding jumpers as required per 680.23(B)(2)(b).
(3) The other conductors are supply conductors to a feed-through-type ground-fault circuit interrupter.

(4) Ground-fault circuit interrupters shall be permitted in a panelboard that contains circuits protected by other than ground-fault circuit interrupters.

680.24 Junction Boxes and Electrical Enclosures for Transformers or Ground-Fault Circuit Interrupters.

(A) Junction Boxes. A junction box connected to a conduit that extends directly to a forming shell or mounting bracket of a no-niche luminaire shall meet the requirements of this section.

(1) Construction. The junction box shall be listed, labeled, and identified as a swimming pool junction box and shall comply with the following conditions:

(1) Be equipped with threaded entries or hubs or a nonmetallic hub
(2) Be comprised of copper, brass, suitable plastic, or other approved corrosion-resistant material
(3) Be provided with electrical continuity between every connected metal conduit and the grounding terminals by means of copper, brass, or other approved corrosion-resistant metal that is integral with the box

(2) Installation. Where the luminaire operates over the low voltage contact limit, the junction box location shall comply with (A)(2)(a) and (A)(2)(b). Where the luminaire operates at the low voltage contact limit or less, the junction box location shall be permitted to comply with (A)(2)(c).

(a) *Vertical Spacing.* The junction box shall be located not less than 100 mm (4 in.), measured from the inside of the bottom of the box, above the ground level, or pool deck, or not less than 200 mm (8 in.) above the maximum pool water level, whichever provides the greater elevation.

(b) *Horizontal Spacing.* The junction box shall be located not less than 1.2 m (4 ft) from the inside wall of the pool, unless separated from the pool by a solid fence, wall, or other permanent barrier.

(c) *Flush Deck Box.* If used on a lighting system operating at the low voltage contact limit or less, a flush deck box shall be permitted if both of the following conditions are met:

(1) An approved potting compound is used to fill the box to prevent the entrance of moisture.
(2) The flush deck box is located not less than 1.2 m (4 ft) from the inside wall of the pool.

(B) Other Enclosures. An enclosure for a transformer, ground-fault circuit interrupter, or a similar device connected to a conduit that extends directly to a forming shell or mounting bracket of a no-niche luminaire shall meet the requirements of this section.

(1) Construction. The enclosure shall be listed and labeled for the purpose and meet the following requirements:

(1) Equipped with threaded entries or hubs or a nonmetallic hub
(2) Comprised of copper, brass, suitable plastic, or other approved corrosion-resistant material
(3) Provided with an approved seal, such as duct seal at the conduit connection, that prevents circulation of air between the conduit and the enclosures
(4) Provided with electrical continuity between every connected metal conduit and the grounding terminals by means of copper, brass, or other approved corrosion-resistant metal that is integral with the box

(2) Installation.

(a) *Vertical Spacing.* The enclosure shall be located not less than 100 mm (4 in.), measured from the inside of the bottom of the box, above the ground level, or pool deck, or not less than 200 mm (8 in.) above the maximum pool water level, whichever provides the greater elevation.

(b) *Horizontal Spacing.* The enclosure shall be located not less than 1.2 m (4 ft) from the inside wall of the pool, unless separated from the pool by a solid fence, wall, or other permanent barrier.

(C) Protection. Junction boxes and enclosures mounted above the grade of the finished walkway around the pool shall not be located in the walkway unless afforded additional protection, such as by location under diving boards, adjacent to fixed structures, and the like.

(D) Grounding Terminals. Junction boxes, transformer and power-supply enclosures, and ground-fault circuit-interrupter enclosures connected to a conduit that extends directly to a forming shell or mounting bracket of a no-niche luminaire shall be provided with a number of grounding terminals that shall be no fewer than one more than the number of conduit entries.

(E) Strain Relief. The termination of a flexible cord of an underwater luminaire within a junction box, transformer or power-supply enclosure, ground-fault circuit interrupter, or other enclosure shall be provided with a strain relief.

(F) Grounding. The equipment grounding conductor terminals of a junction box, transformer enclosure, or other enclosure in the supply circuit to a wet-niche or no-niche luminaire and the field-wiring chamber of a dry-niche luminaire shall be connected to the equipment grounding terminal of the panelboard. This terminal shall be directly connected to the panelboard enclosure.

680.25 Feeders. These provisions shall apply to any feeder on the supply side of panelboards supplying branch circuits for pool equipment covered in Part II of this article and on the load side of the service equipment or the source of a separately derived system.

(A) Feeders. Where feeders are installed in corrosive environments as described in 680.14, the wiring method of that portion of the feeder shall be as required in 680.14(B) or shall be liquidtight flexible nonmetallic conduit. Wiring methods installed in corrosive environments as described in 680.14 shall contain an insulated copper equipment grounding conductor sized in accordance with Table 250.122, but not smaller than 12 AWG.

Where installed in noncorrosive environments, feeders shall comply with the general requirements in Chapter 3.

(B) Aluminum Conduit. Aluminum conduit shall not be permitted in the pool area where subject to corrosion.

680.26 Equipotential Bonding.

(A) Performance. The equipotential bonding required by this section shall be installed to reduce voltage gradients in the pool area.

(B) Bonded Parts. The parts specified in 680.26(B)(1) through (B)(7) shall be bonded together using solid copper conductors, insulated covered, or bare, not smaller than 8 AWG or with rigid metal conduit of brass or other identified corrosion-resistant metal. Connections to bonded parts shall be made in accordance with 250.8. An 8 AWG or larger solid copper bonding conductor provided to reduce voltage gradients in the pool area shall not be required to be extended or attached to remote panelboards, service equipment, or electrodes.

(1) Conductive Pool Shells. Bonding to conductive pool shells shall be provided as specified in 680.26(B)(1)(a) or (B)(1)(b). Poured concrete, pneumatically applied or sprayed concrete, and concrete block with painted or plastered coatings shall all be considered conductive materials due to water permeability and porosity. Vinyl liners and fiberglass composite shells shall be considered to be nonconductive materials.

(a) *Structural Reinforcing Steel.* Unencapsulated structural reinforcing steel shall be bonded together by steel tie wires or the equivalent. Where structural reinforcing steel is encapsulated in a nonconductive compound, a copper conductor grid shall be installed in accordance with 680.26(B)(1)(b).

(b) *Copper Conductor Grid.* A copper conductor grid shall be provided and shall comply with (b)(1) through (b)(4).

(1) Be constructed of minimum 8 AWG bare solid copper conductors bonded to each other at all points of crossing. The bonding shall be in accordance with 250.8 or other approved means.
(2) Conform to the contour of the pool
(3) Be arranged in a 300-mm (12-in.) by 300-mm (12-in.) network of conductors in a uniformly spaced perpendicular grid pattern with a tolerance of 100 mm (4 in.)

(4) Be secured within or under the pool no more than 150 mm (6 in.) from the outer contour of the pool shell

(2) Perimeter Surfaces. The perimeter surface to be bonded shall be considered to extend for 1 m (3 ft) horizontally beyond the inside walls of the pool and shall include unpaved surfaces and other types of paving. Perimeter surfaces separated from the pool by a permanent wall or building 1.5 m (5 ft) in height or more shall require equipotential bonding only on the pool side of the permanent wall or building. Bonding to perimeter surfaces shall be provided as specified in 680.26(B)(2)(a) or (2)(b) and shall be attached to the pool reinforcing steel or copper conductor grid at a minimum of four (4) points uniformly spaced around the perimeter of the pool. For nonconductive pool shells, bonding at four points shall not be required.

(a) *Structural Reinforcing Steel.* Structural reinforcing steel shall be bonded in accordance with 680.26(B)(1)(a).

(b) *Alternate Means.* Where structural reinforcing steel is not available or is encapsulated in a nonconductive compound, a copper conductor(s) shall be utilized where the following requirements are met:

(1) At least one minimum 8 AWG bare solid copper conductor shall be provided.
(2) The conductors shall follow the contour of the perimeter surface.
(3) Only listed splices shall be permitted.
(4) The required conductor shall be 450 mm to 600 mm (18 in. to 24 in.) from the inside walls of the pool.
(5) The required conductor shall be secured within or under the perimeter surface 100 mm to 150 mm (4 in. to 6 in.) below the subgrade.

(3) Metallic Components. All metallic parts of the pool structure, including reinforcing metal not addressed in 680.26(B)(1)(a), shall be bonded. Where reinforcing steel is encapsulated with a nonconductive compound, the reinforcing steel shall not be required to be bonded.

(4) Underwater Lighting. All metal forming shells and mounting brackets of no-niche luminaires shall be bonded.

Exception: Listed low-voltage lighting systems with nonmetallic forming shells shall not require bonding.

(5) Metal Fittings. All metal fittings within or attached to the pool structure shall be bonded. Isolated parts that are not over 100 mm (4 in.) in any dimension and do not penetrate into the pool structure more than 25 mm (1 in.) shall not require bonding.

(6) Electrical Equipment. Metal parts of electrical equipment associated with the pool water circulating system, including pump motors and metal parts of equipment associated with pool covers, including electric motors, shall be bonded.

Exception: Metal parts of listed equipment incorporating an approved system of double insulation shall not be bonded.

(a) *Double-Insulated Water Pump Motors.* Where a double-insulated water pump motor is installed under the provisions of this rule, a solid 8 AWG copper conductor of sufficient length to make a bonding connection to a replacement motor shall be extended from the bonding grid to an accessible point in the vicinity of the pool pump motor. Where there is no connection between the swimming pool bonding grid and the equipment grounding system for the premises, this bonding conductor shall be connected to the equipment grounding conductor of the motor circuit.

(b) *Pool Water Heaters.* For pool water heaters rated at more than 50 amperes and having specific instructions regarding bonding and grounding, only those parts designated to be bonded shall be bonded and only those parts designated to be grounded shall be grounded.

(7) Fixed Metal Parts. All fixed metal parts shall be bonded including, but not limited to, metal-sheathed cables and raceways, metal piping, metal awnings, metal fences, and metal door and window frames.

Exception No. 1: Those separated from the pool by a permanent barrier that prevents contact by a person shall not be required to be bonded.

Exception No. 2: Those greater than 1.5 m (5 ft) horizontally from the inside walls of the pool shall not be required to be bonded.

Exception No. 3: Those greater than 3.7 m (12 ft) measured vertically above the maximum water level of the pool, or as measured vertically above any observation stands, towers, or platforms, or any diving structures, shall not be required to be bonded.

(C) Pool Water. Where none of the bonded parts is in direct connection with the pool water, the pool water shall be in direct contact with an approved corrosion-resistant conductive surface that exposes not less than 5800 mm^2 (9 $in.^2$) of surface area to the pool water at all times. The conductive surface shall be located where it is not exposed to physical damage or dislodgement during usual pool activities, and it shall be bonded in accordance with 680.26(B).

680.27 Specialized Pool Equipment.

(A) Underwater Audio Equipment. All underwater audio equipment shall be identified.

(1) Speakers. Each speaker shall be mounted in an approved metal forming shell, the front of which is enclosed by a captive metal screen, or equivalent, that is bonded to, and secured to, the forming shell by a positive locking device that ensures a low-resistance contact and requires a tool to open for installation or servicing of the speaker. The forming shell shall be installed in a recess in the wall or floor of the pool.

(2) Wiring Methods. Rigid metal conduit of brass or other identified corrosion-resistant metal, liquidtight flexible nonmetallic conduit (LFNC), rigid polyvinyl chloride conduit, or reinforced thermosetting resin conduit shall extend from the forming shell to a listed junction box or other enclosure as provided in 680.24. Where rigid polyvinyl chloride conduit, reinforced thermosetting resin conduit, or liquidtight flexible nonmetallic conduit is used, an 8 AWG insulated solid or stranded copper bonding jumper shall be installed in this conduit. The bonding jumper shall be terminated in the forming shell and the junction box. The termination of the 8 AWG bonding jumper in the forming shell shall be covered with, or encapsulated in, a listed potting compound to protect such connection from the possible deteriorating effect of pool water.

(3) Forming Shell and Metal Screen. The forming shell and metal screen shall be of brass or other approved corrosion-resistant metal. All forming shells shall include provisions for terminating an 8 AWG copper conductor.

(B) Electrically Operated Pool Covers.

(1) Motors and Controllers. The electric motors, controllers, and wiring shall be located not less than 1.5 m (5 ft) from the inside wall of the pool unless separated from the pool by a wall, cover, or other permanent barrier. Electric motors installed below grade level shall be of the totally enclosed type. The device that controls the operation of the motor for an electrically operated pool cover shall be located such that the operator has full view of the pool.

Exception: Motors that are part of listed systems with ratings not exceeding the low-voltage contact limit that are supplied by listed transformers or power supplies that comply with 680.23(A)(2) shall be permitted to be located less than 1.5 m (5 ft) from the inside walls of the pool.

(2) Protection. The electric motor and controller shall be connected to a branch circuit protected by a ground-fault circuit interrupter.

Exception: Motors that are part of listed systems with ratings not exceeding the low-voltage contact limit that are supplied by listed transformers or power supplies that comply with 680.23(A)(2).

(C) Deck Area Heating. The provisions of this section shall apply to all pool deck areas, including a covered pool, where electrically operated comfort heating units are installed within 6.0 m (20 ft) of the inside wall of the pool.

(1) Unit Heaters. Unit heaters shall be rigidly mounted to the structure and shall be of the totally enclosed or guarded type. Unit heaters shall not be mounted over the pool or within the area extending 1.5 m (5 ft) horizontally from the inside walls of a pool.

(2) Permanently Wired Radiant Heaters. Radiant electric heaters shall be suitably guarded and securely fastened to their mounting device(s). Heaters shall not be installed over a pool or within the area extending 1.5 m (5 ft) horizontally from the inside walls of the pool and shall be mounted at least 3.7 m (12 ft) vertically above the pool deck unless otherwise approved.

(3) Radiant Heating Cables Not Permitted. Radiant heating cables embedded in or below the deck shall not be permitted.

N **680.28 Gas-Fired Water Heater.** Circuits serving gas-fired swimming pool and spa water heaters operating at voltages above the low-voltage contact limit shall be provided with ground-fault circuit-interrupter protection for personnel.

PART IV. SPAS AND HOT TUBS

680.40 General. Electrical installations at spas and hot tubs shall comply with the provisions of Part I and Part IV of this article.

680.41 Emergency Switch for Spas and Hot Tubs. A clearly labeled emergency shutoff or control switch for the purpose of stopping the motor(s) that provides power to the recirculation system and jet system shall be installed at a point readily accessible to the users and not less than 1.5 m (5 ft) away, adjacent to, and within sight of the spa or hot tub. This requirement shall not apply to one-family dwellings.

680.42 Outdoor Installations. A spa or hot tub installed outdoors shall comply with the provisions of Parts I and II of this article, except as permitted in 680.42(A) and (B), that would otherwise apply to pools installed outdoors.

(A) Flexible Connections. Listed packaged spa or hot tub equipment assemblies or self-contained spas or hot tubs utilizing a factory-installed or assembled control panel or panelboard shall be permitted to use flexible connections as covered in 680.42(A)(1) and (A)(2).

(1) Flexible Conduit. Liquidtight flexible metal conduit or liquidtight flexible nonmetallic conduit shall be permitted.

(2) Cord-and-Plug Connections. Cord-and-plug connections with a cord not longer than 4.6 m (15 ft) shall be permitted where protected by a ground-fault circuit interrupter.

(B) Bonding. Bonding by metal-to-metal mounting on a common frame or base shall be permitted. The metal bands or hoops used to secure wooden staves shall not be required to be bonded as required in 680.26.

Equipotential bonding of perimeter surfaces in accordance with 680.26(B)(2) shall not be required to be provided for spas and hot tubs where all of the following conditions apply:

(1) The spa or hot tub shall be listed, labeled, and identified as a self-contained spa for aboveground use.
(2) The spa or hot tub shall not be identified as suitable only for indoor use.
(3) The installation shall be in accordance with the manufacturer's instructions and shall be located on or above grade.
(4) The top rim of the spa or hot tub shall be at least 710 mm (28 in.) above all perimeter surfaces that are within 760 mm (30 in.), measured horizontally from the spa or hot tub. The height of nonconductive external steps for entry to or exit from the self-contained spa shall not be used to reduce or increase this rim height measurement.

(C) Interior Wiring to Outdoor Installations. In the interior of a dwelling unit or in the interior of another building or structure associated with a dwelling unit, any of the wiring methods recognized or permitted in Chapter 3 of this *Code* shall be permitted to be used for the connection to motor disconnecting means and the motor, heating, and control loads that are part of a self-contained spa or hot tub or a packaged spa or hot tub equipment assembly. Wiring to an underwater luminaire shall comply with 680.23 or 680.33.

680.43 Indoor Installations. A spa or hot tub installed indoors shall comply with the provisions of Parts I and II of this article except as modified by this section and shall be connected by the wiring methods of Chapter 3.

Exception No. 1: Listed spa and hot tub packaged units rated 20 amperes or less shall be permitted to be cord-and-plug-connected to facilitate the removal or disconnection of the unit for maintenance and repair.

Exception No. 2: The equipotential bonding requirements for perimeter surfaces in 680.26(B)(2) shall not apply to a listed self-contained spa or hot tub installed above a finished floor.

(A) Receptacles. At least one 125-volt, 15- or 20-ampere receptacle on a general-purpose branch circuit shall be located not less than 1.83 m (6 ft) from, and not exceeding 3.0 m (10 ft) from, the inside wall of the spa or hot tub.

(1) Location. Receptacles shall be located at least 1.83 m (6 ft) measured horizontally from the inside walls of the spa or hot tub.

(2) Protection, General. Receptacles rated 125 volts and 30 amperes or less and located within 3.0 m (10 ft) of the inside walls of a spa or hot tub shall be protected by a ground-fault circuit interrupter.

(3) Protection, Spa or Hot Tub Supply Receptacle. Receptacles that provide power for a spa or hot tub shall be ground-fault circuit-interrupter protected.

(4) Measurements. In determining the dimensions in this section addressing receptacle spacings, the distance to be measured shall be the shortest path the supply cord of an appliance connected to the receptacle would follow without piercing a floor, wall, ceiling, doorway with hinged or sliding door, window opening, or other effective permanent barrier.

(B) Installation of Luminaires, Lighting Outlets, and Ceiling-Suspended (Paddle) Fans.

(1) Elevation. Luminaires, except as covered in 680.43(B)(2), lighting outlets, and ceiling-suspended (paddle) fans located over the spa or hot tub or within 1.5 m (5 ft) from the inside walls of the spa or hot tub shall comply with the clearances specified in (B)(1)(a), (B)(1)(b), and (B)(1)(c) above the maximum water level.

(a) *Without GFCI.* Where no GFCI protection is provided, the mounting height shall be not less than 3.7 m (12 ft).

(b) *With GFCI.* Where GFCI protection is provided, the mounting height shall be permitted to be not less than 2.3 m (7 ft 6 in.).

(c) *Below 2.3 m (7 ft 6 in.).* Luminaires meeting the requirements of item (1) or (2) and protected by a ground-fault circuit interrupter shall be permitted to be installed less than 2.3 m (7 ft 6 in.) over a spa or hot tub:

(1) Recessed luminaires with a glass or plastic lens, nonmetallic or electrically isolated metal trim, and suitable for use in damp locations
(2) Surface-mounted luminaires with a glass or plastic globe, a nonmetallic body, or a metallic body. isolated from contact, and suitable for use in damp locations

(2) Underwater Applications. Underwater luminaires shall comply with the provisions of 680.23 or 680.33.

(C) Switches. Switches shall be located at least 1.5 m (5 ft), measured horizontally, from the inside walls of the spa or hot tub.

(D) Bonding. The following parts shall be bonded together:

(1) All metal fittings within or attached to the spa or hot tub structure.
(2) Metal parts of electrical equipment associated with the spa or hot tub water circulating system, including pump motors, unless part of a listed, labeled, and identified self-contained spa or hot tub.
(3) Metal raceway and metal piping that are within 1.5 m (5 ft) of the inside walls of the spa or hot tub and that are not separated from the spa or hot tub by a permanent barrier.
(4) All metal surfaces that are within 1.5 m (5 ft) of the inside walls of the spa or hot tub and that are not separated from the spa or hot tub area by a permanent barrier.

Exception: Small conductive surfaces not likely to become energized, such as air and water jets and drain fittings, where not connected to metallic piping, towel bars, mirror frames, and similar nonelectrical equipment, shall not be required to be bonded.

(5) Electrical devices and controls that are not associated with the spas or hot tubs and that are located less than 1.5 m (5 ft) from such units; otherwise, they shall be bonded to the spa or hot tub system.

(E) Methods of Bonding. All metal parts associated with the spa or hot tub shall be bonded by any of the following methods:

(1) The interconnection of threaded metal piping and fittings
(2) Metal-to-metal mounting on a common frame or base
(3) The provisions of a solid copper bonding jumper, insulated, covered, or bare, not smaller than 8 AWG

(F) Grounding. The following equipment shall be grounded:

(1) All electrical equipment located within 1.5 m (5 ft) of the inside wall of the spa or hot tub
(2) All electrical equipment associated with the circulating system of the spa or hot tub

(G) Underwater Audio Equipment. Underwater audio equipment shall comply with the provisions of Part II of this article.

680.44 Protection. Except as otherwise provided in this section, the outlet(s) that supplies a self-contained spa or hot tub, a packaged spa or hot tub equipment assembly, or a field-assembled spa or hot tub shall be protected by a ground-fault circuit interrupter.

(A) Listed Units. If so marked, a listed, labeled, and identified self-contained unit or a listed, labeled, and identified packaged equipment assembly that includes integral ground-fault circuit-interrupter protection for all electrical parts within the unit or assembly (pumps, air blowers, heaters, lights, controls, sanitizer generators, wiring, and so forth) shall be permitted without additional GFCI protection.

(B) Other Units. A field-assembled spa or hot tub rated 3 phase or rated over 250 volts or with a heater load of more than 50 amperes shall not require the supply to be protected by a ground-fault circuit interrupter.

PART V. FOUNTAINS

680.50 General. The provisions of Part I and Part V of this article shall apply to all permanently installed fountains as defined in 680.2. Fountains that have water common to a pool shall additionally comply with the requirements in Part II of this article. Part V does not cover self-contained, portable fountains. Portable fountains shall comply with Parts II and III of Article 422.

680.51 Luminaires, Submersible Pumps, and Other Submersible Equipment.

(A) Ground-Fault Circuit Interrupter. Luminaires, submersible pumps, and other submersible equipment, unless listed for operation at low voltage contact limit or less and supplied by a transformer or power supply that complies with 680.23(A)(2), shall be protected by a ground-fault circuit interrupter.

(B) Operating Voltage. No luminaires shall be installed for operation on supply circuits over 150 volts between conductors. Submersible pumps and other submersible equipment shall operate at 300 volts or less between conductors.

(C) Luminaire Lenses. Luminaires shall be installed with the top of the luminaire lens below the normal water level of the fountain unless listed for above-water locations. A luminaire facing upward shall comply with either (1) or (2):

(1) Have the lens guarded to prevent contact by any person
(2) Be listed for use without a guard

(D) Overheating Protection. Electrical equipment that depends on submersion for safe operation shall be protected against overheating by a low-water cutoff or other approved means when not submerged.

(E) Wiring. Equipment shall be equipped with provisions for threaded conduit entries or be provided with a suitable flexible cord. The maximum length of each exposed cord in the fountain shall be limited to 3.0 m (10 ft). Cords extending beyond the fountain perimeter shall be enclosed in approved wiring enclosures. Metal parts of equipment in contact with water shall be of brass or other approved corrosion-resistant metal.

(F) Servicing. All equipment shall be removable from the water for relamping or normal maintenance. Luminaires shall not be permanently embedded into the fountain structure such that the water level must be reduced or the fountain drained for relamping, maintenance, or inspection.

(G) Stability. Equipment shall be inherently stable or be securely fastened in place.

680.52 Junction Boxes and Other Enclosures.

(A) General. Junction boxes and other enclosures used for other than underwater installation shall comply with 680.24.

(B) Underwater Junction Boxes and Other Underwater Enclosures. Junction boxes and other underwater enclosures shall meet the requirements of 680.52(B)(1) and (B)(2).

(1) Construction.

(a) Underwater enclosures shall be equipped with provisions for threaded conduit entries or compression glands or seals for cord entry.

(b) Underwater enclosures shall be submersible and made of copper, brass, or other approved corrosion-resistant material.

(2) Installation. Underwater enclosure installations shall comply with (a) and (b).

(a) Underwater enclosures shall be filled with an approved potting compound to prevent the entry of moisture.

(b) Underwater enclosures shall be firmly attached to the supports or directly to the fountain surface and bonded as required. Where the junction box is supported only by conduits in accordance with 314.23(E) and (F), the conduits shall be of copper, brass, stainless steel, or other approved corrosion-resistant metal. Where the box is fed by nonmetallic conduit, it shall have additional supports and fasteners of copper, brass, or other approved corrosion-resistant material.

680.53 Bonding. All metal piping systems associated with the fountain shall be bonded to the equipment grounding conductor of the branch circuit supplying the fountain.

680.54 Grounding. The following equipment shall be grounded:

(1) Other than listed low-voltage luminaires not requiring grounding, all electrical equipment located within the

fountain or within 1.5 m (5 ft) of the inside wall of the fountain

(2) All electrical equipment associated with the recirculating system of the fountain

(3) Panelboards that are not part of the service equipment and that supply any electrical equipment associated with the fountain

680.55 Methods of Grounding.

(A) Applied Provisions. The provisions of 680.21(A), 680.23(B)(3), 680.23(F)(1) and (F)(2), 680.24(F), and 680.25 shall apply.

(B) Supplied by a Flexible Cord. Electrical equipment that is supplied by a flexible cord shall have all exposed non–current-carrying metal parts grounded by an insulated copper equipment grounding conductor that is an integral part of this cord. The equipment grounding conductor shall be connected to an equipment grounding terminal in the supply junction box, transformer enclosure, power supply enclosure, or other enclosure.

680.56 Cord-and-Plug-Connected Equipment.

(A) Ground-Fault Circuit Interrupter. All electrical equipment, including power-supply cords, shall be protected by ground-fault circuit interrupters.

(B) Cord Type. Flexible cord immersed in or exposed to water shall be of a type for extra-hard usage, as designated in Table 400.4, and shall be a listed type with a "W" suffix.

(C) Sealing. The end of the flexible cord jacket and the flexible cord conductor termination within equipment shall be covered with, or encapsulated in, a suitable potting compound to prevent the entry of water into the equipment through the cord or its conductors. In addition, the ground connection within equipment shall be similarly treated to protect such connections from the deteriorating effect of water that may enter into the equipment.

(D) Terminations. Connections with flexible cord shall be permanent, except that grounding-type attachment plugs and receptacles shall be permitted to facilitate removal or disconnection for maintenance, repair, or storage of fixed or stationary equipment not located in any water-containing part of a fountain.

N

PART VIII. ELECTRICALLY POWERED POOL LIFTS

680.80 General. Electrically powered pool lifts as defined in 680.2 shall comply with Part VIII of this article. They shall not be required to comply with other parts of this article.

680.81 Equipment Approval. Lifts shall be listed, labeled, and identified for swimming pool and spa use.

Exception No. 1: Lifts where the battery is removed for charging at another location and the battery is rated less than or equal to the low-voltage contact limit shall not be required to be listed or labeled.

Exception No. 2: Solar-operated or solar-recharged lifts where the solar panel is attached to the lift and the battery is rated less than or equal to 24 volts shall not be required to be listed or labeled.

Exception No. 3: Lifts that are supplied from a source not exceeding the low-voltage contact limit and supplied by listed transformers or power supplies that comply with 680.23(A)(2) shall not be required to be listed or labeled.

680.82 Protection. Pool lifts connected to premises wiring and operated above the low-voltage contact limit shall be provided with GFCI protection for personnel.

680.84 Switching Devices. Switches and switching devices that are operated above the low-voltage contact limit shall comply with 680.22(C).

CHAPTER 24

INTERACTIVE AND STAND-ALONE PHOTOVOLTAIC (PV) SYSTEMS

INTRODUCTION

This chapter contains requirements from Article 690 — Solar Photovoltaic (PV) Systems, Article 705 — Interconnected Electric Power Production Sources, and Chapter 11 of NFPA 1, *Fire Code*. Article 690 of the *NEC* provides the electrical installation requirements for photovoltaic (PV) electric power production systems that supply electric energy to a premises wiring system. Where the PV system is interconnected with another source of electric power (as is the case with a large majority residential, commercial, institutional and industrial installations), the requirements of Article 705 also have to be used. Article 691 — Large-Scale Photovoltaic (PV) Electric Power covers solar-fueled electric power production facilities, known as solar farms, that are constructed for the primary purpose of providing renewable resource fueled electric power into the transmission and distribution grid of a utility or other large scale user. This article is not covered in this *Pocket Guide*.

As the use of PV systems has seen a rapid expansion across all segments of building construction, so has the need to address the safety concerns of those who have to respond to a building with a PV system in the event of a fire or other emergency event. Access to roofs is an essential part of firefighting strategy, and Chapter 11 of NFPA 1 provides setback and roof area coverage requirements intended to facilitate first responder access. Requirements for quickly reducing power or completely de-energizing conductors and equipment, referred to in Article 690 as "rapid shutdown," are focused exclusively on making rooftops electrically safe for emergency personnel. Recognition that interactive inverter circuit conductors not only export power from the PV system, but can also import fault currents fed by the utility source, is the basis for rules in Article 705 covering the interconnection of sources and limiting the length of "unprotected" conductors on or in buildings and structures.

The requirements in Articles 690 and 705 of the *NEC* have seen significant expansion and modification to address the constantly changing PV technology and the lessons learned from the exponential growth of these systems over the last two decades. Better methods to control and protect PV systems are continuously evolving to keep pace with the demand for this increasingly viable and mainstream power production source.

To provide the user of this *Pocket Guide* with the necessary rules for installing stand-alone and interactive PV systems, installation requirements are provided from the eight parts of Article 690: Part I General, Part II Circuit Requirements, Part III Disconnecting Means, Part IV Wiring Methods, Part V Grounding and Bonding, Part VI Marking, Part VII Connection to Other Sources, and Part VIII Energy Storage Systems. Requirements are also provided from two parts of Article 705 (Part I General, and Part II Interactive Inverters) and from Chapter 11 (Building Services) of NFPA 1. Due to the specialized nature of the terminology used in Articles 690 and 705, several unique definitions that are essential to proper application of the requirements in these two articles have been included. However, users are urged to review all of the definitions contained in these two articles so as to be completely conversant with the terms used in each. Unless specifically modified or indicated otherwise in Articles 690 and 705, the definitions in Article 100 also apply.

ARTICLE 690
Solar Photovoltaic (PV) Systems

PART I. GENERAL

[Article 100.

Photovoltaic (PV) System. The total components and subsystem that, in combination, convert solar energy into electric energy for connection to a utilization load. (CMP-4)]

690.2 Definitions.

N DC-to-DC Converter Output Circuit. Circuit conductors between the dc-to-dc converter source circuit(s) and the inverter or dc utilization equipment.

N DC-to-DC Converter Source Circuit. Circuits between dc-to-dc converters and from dc-to-dc converters to the common connection point(s) of the dc system.

Interactive Inverter Output Circuit. The conductors between the interactive inverter and the service equipment or another electrical power production and distribution network.

Inverter Input Circuit. Conductors connected to the dc input of an inverter.

Inverter Output Circuit. Conductors connected to the ac output of an inverter.

Photovoltaic Output Circuit. Circuit conductors between the PV source circuit(s) and the inverter or dc utilization equipment.

Photovoltaic Source Circuit. Circuits between modules and from modules to the common connection point(s) of the dc system.

N Photovoltaic System DC Circuit Any dc conductor supplied by a PV power source, including PV source circuits, PV output circuits, dc-to-dc converter source circuits, or dc-to-dc converter output circuits.

690.4 General Requirements.

(B) Equipment. Inverters, motor generators, PV modules, PV panels, ac modules, dc combiners, dc-to-dc converters, and charge controllers intended for use in PV systems shall be listed or field labeled for the PV application.

(D) Multiple PV Systems. Multiple PV systems shall be permitted to be installed in or on a single building or structure. Where the PV systems are remotely located from each other, a directory in accordance with 705.10 shall be provided at each PV system disconnecting means.

• N **(E) Locations Not Permitted.** PV system equipment and disconnecting means shall not be installed in bathrooms.

PART II. CIRCUIT REQUIREMENTS

690.7 Maximum Voltage. The maximum voltage of PV system dc circuits shall be the highest voltage between any two circuit conductors or any conductor and ground. PV system dc circuits on or in one- and two-family dwellings shall be permitted to have a maximum voltage of 600 volts or less. PV system dc circuits on or in other types of buildings shall be permitted to have a maximum voltage of 1000 volts or less. Where not located on or in buildings, listed dc PV equipment, rated at a maximum voltage of 1500 volts or less, shall not be required to comply with Parts II and III of Article 490.

(A) Photovoltaic Source and Output Circuits. In a dc PV source circuit or output circuit, the maximum PV system voltage for that circuit shall be calculated in accordance with one of the following methods:

(1) Instructions in listing or labeling of the module: The sum of the PV module–rated open-circuit voltage of the series-connected modules corrected for the lowest expected ambient temperature using the open-circuit voltage temperature coefficients in accordance with the instructions included in the listing or labeling of the module
(2) Crystalline and multicrystalline modules: For crystalline and multicrystalline silicon modules, the sum of

the PV module–rated open-circuit voltage of the series-connected modules corrected for the lowest expected ambient temperature using the correction factor provided in Table 690.7(A)

(3) PV systems of 100 kW or larger: For PV systems with a generating capacity of 100 kW or greater, a documented and stamped PV system design, using an industry standard method and provided by a licensed professional electrical engineer, shall be permitted.

The maximum voltage shall be used to determine the voltage rating of conductors, cables, disconnects, overcurrent devices, and other equipment.

(B) DC-to-DC Converter Source and Output Circuits. In a dc-to-dc converter source and output circuit, the maximum voltage shall be calculated in accordance with 690.7(B)(1) or (B)(2).

(1) Single DC-to-DC Converter. For circuits connected to the output of a single dc-to-dc converter, the maximum voltage shall be the maximum rated voltage output of the dc-to-dc converter.

Table 690.7(A) Voltage Correction Factors for Crystalline and Multicrystalline Silicon Modules

Correction Factors for Ambient Temperatures Below 25°C (77°F). (Multiply the rated open-circuit voltage by the appropriate correction factor shown below.)

Ambient Temperature (°C)	Factor	Ambient Temperature (°F)
24 to 20	1.02	76 to 68
19 to 15	1.04	67 to 59
14 to 10	1.06	58 to 50
9 to 5	1.08	49 to 41
4 to 0	1.10	40 to 32
−1 to −5	1.12	31 to 23
−6 to −10	1.14	22 to 14
−11 to −15	1.16	13 to 5
−16 to −20	1.18	4 to −4
−21 to −25	1.20	−5 to −13
−26 to −30	1.21	−14 to −22
−31 to −35	1.23	−23 to −31
−36 to −40	1.25	−32 to −40

(2) Two or More Series Connected DC-to-DC Converters. For circuits connected to the output of two or more series-connected dc-to-dc converters, the maximum voltage shall be determined in accordance with the instructions included in the listing or labeling of the dc-to-dc converter. If these instructions do not state the rated voltage of series-connected dc-to-dc converters, the maximum voltage shall be the sum of the maximum rated voltage output of the dc-to-dc converters in series.

(C) Bipolar Source and Output Circuits. For 2-wire dc circuits connected to bipolar PV arrays, the maximum voltage shall be the highest voltage between the 2-wire circuit conductors where one conductor of the 2-wire circuit is connected to the functional ground reference (center tap). To prevent overvoltage in the event of a ground-fault or arc-fault, the array shall be isolated from the ground reference and isolated into two 2-wire circuits.

690.8 Circuit Sizing and Current.

(A) Calculation of Maximum Circuit Current. The maximum current for the specific circuit shall be calculated in accordance with 690.8(A)(1) through (A)(6).

(1) Photovoltaic Source Circuit Currents. The maximum current shall be calculated by one of the following methods:

(1) The sum of parallel-connected PV module–rated short-circuit currents multiplied by 125 percent

(2) For PV systems with a generating capacity of 100 kW or greater, a documented and stamped PV system design, using an industry standard method and provided by a licensed professional electrical engineer, shall be permitted. The calculated maximum current value shall be based on the highest 3-hour current average resulting from the simulated local irradiance on the PV array accounting for elevation and orientation. The current value used by this method shall not be less than 70 percent of the value calculated using 690.8(A)(1)(1).

(2) Photovoltaic Output Circuit Currents. The maximum current shall be the sum of parallel source circuit maximum currents as calculated in 690.8(A)(1).

(3) Inverter Output Circuit Current. The maximum current shall be the inverter continuous output current rating.

(4) Stand-Alone Inverter Input Circuit Current. The maximum current shall be the stand-alone continuous inverter input current rating when the inverter is producing rated power at the lowest input voltage.

(5) DC-to-DC Converter Source Circuit Current. The maximum current shall be the dc-to-dc converter continuous output current rating.

(B) Conductor Ampacity. PV system currents shall be considered to be continuous. Circuit conductors shall be sized to carry not less than the larger of 690.8(B)(1) or (B)(2) or where protected by a listed adjustable electronic overcurrent protective device in accordance 690.9(B)(3), not less than the current in 690.8(B)(3).

(1) Before Application of Adjustment and Correction Factors. One hundred twenty-five percent of the maximum currents calculated in 690.8(A) before the application of adjustment and correction factors.

Exception: Circuits containing an assembly, together with its overcurrent device(s), that is listed for continuous operation at 100 percent of its rating shall be permitted to be used at 100 percent of its rating.

(2) After Application of Adjustment and Correction Factors. The maximum currents calculated in 690.8(A) after the application of adjustment and correction factors.

N **(3) Adjustable Electronic Overcurrent Protective Device.** The rating or setting of an adjustable electronic overcurrent protective device installed in accordance with 240.6

(C) Systems with Multiple Direct-Current Voltages. For a PV power source that has multiple output circuit voltages and employs a common-return conductor, the ampacity of the common-return conductor shall not be less than the sum of the ampere ratings of the overcurrent devices of the individual output circuits.

(D) Sizing of Module Interconnection Conductors. Where a single overcurrent device is used to protect a set of two or more parallel-connected module circuits, the ampacity of each of the module interconnection conductors shall not be less than the sum of the rating of the single overcurrent device plus 125 percent of the short-circuit current from the other parallel-connected modules.

690.9 Overcurrent Protection.

(A) Circuits and Equipment. PV system dc circuit and inverter output conductors and equipment shall be protected against overcurrent. Overcurrent protective devices shall not be required for circuits with sufficient ampacity for the highest available current. Circuits connected to current limited supplies (e.g., PV modules, dc-to-dc converters, interactive inverter output circuits) and also connected to sources having higher current availability (e.g., parallel strings of modules, utility power) shall be protected at the higher current source connection.

Exception: An overcurrent device shall not be required for PV modules or PV source circuit or dc-to-dc converters source circuit conductors sized in accordance with 690.8(B) where one of the following applies:

(1) There are no external sources such as parallel-connected source circuits, batteries, or backfeed from inverters.

(2) The short-circuit currents from all sources do not exceed the ampacity of the conductors and the maximum overcurrent protective device size rating specified for the PV module or dc-to-dc converter.

(B) Overcurrent Device Ratings. Overcurrent devices used in PV system dc circuits shall be listed for use in PV systems. Overcurrent devices, where required, shall be rated in accordance with one of the following:

(1) Not less than 125 percent of the maximum currents calculated in 690.8(A).

(2) An assembly, together with its overcurrent device(s), that is listed for continuous operation at 100 percent of its rating shall be permitted to be used at 100 percent of its rating.

(3) Adjustable electronic overcurrent protective devices rated or set in accordance with 240.6.

(C) Photovoltaic Source and Output Circuits. A single overcurrent protective device, where required, shall be permitted to protect the PV modules and conductors of each source circuit or the conductors of each output circuit. Where single overcurrent protection devices are used to protect PV source or output circuits, all overcurrent devices shall be placed in the same polarity for all circuits within a PV system. The overcurrent devices shall be accessible but shall not be required to be readily accessible.

(D) Power Transformers. Overcurrent protection for a transformer with a source(s) on each side shall be provided in accordance with 450.3 by considering first one side of the transformer, then the other side of the transformer, as the primary.

Exception: A power transformer with a current rating on the side connected toward the interactive inverter output, not less than the rated continuous output current of the inverter, shall be permitted without overcurrent protection from the inverter.

690.11 Arc-Fault Circuit Protection (Direct Current). Photovoltaic systems operating at 80 volts dc or greater between any two conductors shall be protected by a listed PV arc-fault circuit interrupter or other system components listed to provide equivalent protection. The system shall detect and interrupt arcing faults resulting from a failure in the intended continuity of a conductor, connection, module, or other system component in the PV system dc circuits.

Exception: For PV systems not installed on or in buildings, PV output circuits and dc-to-dc converter output circuits that are direct buried, installed in metallic raceways, or installed in enclosed metallic cable trays are permitted without arc-fault circuit protection. Detached structures whose sole purpose is to house PV system equipment shall not be considered buildings according to this exception.

690.12 Rapid Shutdown of PV Systems on Buildings. PV system circuits installed on or in buildings shall include a rapid shutdown function to reduce shock hazard for emergency responders in accordance with 690.12(A) through (D).

Exception: Ground mounted PV system circuits that enter buildings, of which the sole purpose is to house PV system equipment, shall not be required to comply with 690.12.

N **(A) Controlled Conductors.** Requirements for controlled conductors shall apply to PV circuits supplied by the PV system.

N **(B) Controlled Limits.** The use of the term *array boundary* in this section is defined as 305 mm (1 ft) from the array in all directions. Controlled conductors outside the array boundary shall comply with 690.12(B)(1) and inside the array boundary shall comply with 690.12(B)(2).

(1) Outside the Array Boundary. Controlled conductors located outside the boundary or more than 1 m (3 ft) from the point of entry inside a building shall be limited to not more than 30 volts within 30 seconds of rapid shutdown initiation. Voltage shall be measured between any two conductors and between any conductor and ground.

(2) Inside the Array Boundary. The PV system shall comply with one of the following:

(1) The PV array shall be listed or field labeled as a rapid shutdown PV array. Such a PV array shall be installed and used in accordance with the instructions included with the rapid shutdown PV array listing or field labeling.
(2) Controlled conductors located inside the boundary or not more than 1 m (3 ft) from the point of penetration of the surface of the building shall be limited to not more than 80 volts within 30 seconds of rapid shutdown initiation. Voltage shall be measured between any two conductors and between any conductor and ground.
(3) PV arrays with no exposed wiring methods, no exposed conductive parts, and installed more than 2.5 m (8 ft) from exposed grounded conductive parts or ground shall not be required to comply with 690.12(B)(2).

The requirement of 690.12(B)(2) shall become effective January 1, 2019.

N **(C) Initiation Device.** The initiation device(s) shall initiate the rapid shutdown function of the PV system. The device "off" position shall indicate that the rapid shutdown function has been initiated for all PV systems connected to that device. For one-family and two-family dwellings, an initiation device(s) shall be located at a readily accessible location outside the building.

The rapid shutdown initiation device(s) shall consist of at least one of the following:

(1) Service disconnecting means
(2) PV system disconnecting means
(3) Readily accessible switch that plainly indicates whether it is in the "off" or "on" position

Where multiple PV systems are installed with rapid shutdown functions on a single service, the initiation device(s) shall consist of not more than six switches or six sets of circuit breakers, or a combination of not more than six switches and sets of circuit breakers, mounted in a single enclosure, or in a group of separate enclosures. These initiation device(s) shall initiate the rapid shutdown of all PV systems with rapid shutdown functions on that service. Where auxiliary initiation devices are installed, these auxiliary devices shall control all PV systems with rapid shutdown functions on that service.

N **(D) Equipment.** Equipment that performs the rapid shutdown functions, other than initiation devices such as listed disconnect switches, circuit breakers, or control switches, shall be listed for providing rapid shutdown protection.

PART III. DISCONNECTING MEANS

690.13 Photovoltaic System Disconnecting Means. Means shall be provided to disconnect the PV system from all wiring systems including power systems, energy storage systems, and utilization equipment and its associated premises wiring.

(A) Location. The PV system disconnecting means shall be installed at a readily accessible location.

(B) Marking. Each PV system disconnecting means shall plainly indicate whether in the open (off) or closed (on) position and be permanently marked "PV SYSTEM DISCONNECT" or equivalent. Additional markings shall be permitted based upon the specific system configuration. For PV system disconnecting means where the line and load terminals may be energized in the open position, the device shall be marked with the following words or equivalent:

WARNING
ELECTRIC SHOCK HAZARD
TERMINALS ON THE LINE AND LOAD SIDES MAY BE
ENERGIZED IN THE OPEN POSITION

The warning sign(s) or label(s) shall comply with 110.21(B).

(C) Suitable for Use. If the PV system is connected to the supply side of the service disconnecting means as permitted in 230.82(6), the PV system disconnecting means shall be listed as suitable for use as service equipment.

(D) Maximum Number of Disconnects. Each PV system disconnecting means shall consist of not more than six switches or six sets of circuit breakers, or a combination of not more than six switches and sets of circuit breakers, mounted in a single enclosure, or in a group of separate enclosures. A single PV system disconnecting means shall be permitted for the combined ac output of one or more inverters or ac modules in an interactive system.

(E) Ratings. The PV system disconnecting means shall have ratings sufficient for the maximum circuit current available short-circuit current, and voltage that is available at the terminals of the PV system disconnect.

N **(F) Type of Disconnect.**

(1) Simultaneous Disconnection. The PV system disconnecting means shall simultaneously disconnect the PV system conductors of the circuit from all conductors of other wiring systems. The PV system disconnecting means shall be an externally operable general-use switch or circuit breaker, or other approved means. A dc PV system disconnecting means shall be marked for use in PV systems or be suitable for backfeed operation.

(2) Devices Marked "Line" and "Load." Devices marked with "line" and "load" shall not be permitted for backfeed or reverse current.

(3) DC-Rated Enclosed Switches, Open-Type Switches, and Low-Voltage Power Circuit Breakers. DC-rated, enclosed switches, open-type switches, and low-voltage power circuit breakers shall be permitted for backfeed operation.

690.15 Disconnection of Photovoltaic Equipment. Isolating devices shall be provided to isolate PV modules, ac PV modules, fuses, dc-to-dc converters inverters, and charge controllers from all conductors that are not solidly grounded. An equipment disconnecting means or a PV system disconnecting means shall be permitted in place of an isolating device. Where the maximum circuit current is greater than 30 amperes for the output circuit of a dc combiner or the input circuit of a charge controller or inverter, an equipment disconnecting means shall be provided for isolation. Where a charge controller or inverter has multiple input circuits, a single equipment disconnecting means shall be permitted to isolate the equipment from the input circuits.

• **(A) Location.** Isolating devices or equipment disconnecting means shall be installed in circuits connected to equipment at a location within the equipment, or within sight and within 3 m (10 ft) of the equipment. An equipment disconnecting means shall be permitted to be remote from the equipment where the equipment disconnecting means can be remotely operated from within 3 m (10 ft) of the equipment.

N **(B) Interrupting Rating.** An equipment disconnecting means shall have an interrupting rating sufficient for the maximum short-circuit current and voltage that is available at the terminals of the equipment. An isolating device shall not be required to have an interrupting rating.

N **(C) Isolating Device.** An isolating device shall not be required to simultaneously disconnect all current-carrying conductors of a circuit. The isolating device shall be one of the following:

(1) A connector meeting the requirements of 690.33 and listed and identified for use with specific equipment
(2) A finger safe fuse holder
(3) An isolating switch that requires a tool to open
(4) An isolating device listed for the intended application

An isolating device shall be rated to open the maximum circuit current under load or be marked "Do Not Disconnect Under Load" or "Not for Current Interrupting."

N **(D) Equipment Disconnecting Means.** An equipment disconnecting means shall simultaneously disconnect all current-carrying conductors that are not solidly grounded of the circuit to which it is connected. An equipment disconnecting means shall be externally operable without exposing the operator to contact with energized parts, shall indicate whether in the open (off) or closed (on) position, and shall be lockable in accordance with 110.25. An equipment disconnecting means shall be one of the following devices:

(1) A manually operable switch or circuit breaker
(2) A connector meeting the requirements of 690.33(E)(1)
(3) A load break fused pull out switch
(4) A remote controlled circuit breaker that is operable locally and opens automatically when control power is interrupted

For equipment disconnecting means, other than those complying with 690.33, where the line and load terminals can be energized in the open position, the device shall be marked in accordance with the warning in 690.13(B).

•

PART IV. WIRING METHODS

690.31 Methods Permitted.

(A) Wiring Systems. All raceway and cable wiring methods included in this *Code*, other wiring systems and fittings specifically listed for use on PV arrays, and wiring as part of a listed system shall be permitted. Where wiring devices with integral enclosures are used, sufficient length of cable shall be provided to facilitate replacement.

Where PV source and output circuits operating at voltages greater than 30 volts are installed in readily accessible locations, circuit conductors shall be guarded or installed in Type MC cable or in raceway. For ambient temperatures exceeding 30°C (86°F), conductor ampacities shall be corrected in accordance with Table 690.31(A).

• **(B) Identification and Grouping.** PV source circuits and PV output circuits shall not be contained in the same raceway, cable tray, cable, outlet box, junction box, or similar fitting as conductors, feeders, branch circuits of other non-PV systems, or inverter output circuits, unless the conductors of the different systems are separated by a partition. PV system circuit conductors shall be identified and grouped as required by 690.31(B)(1) through (2). The means of identification shall be permitted by separate color coding, marking tape, tagging, or other approved means.

(1) Identification. PV system circuit conductors shall be identified at all accessible points of termination, connection, and splices.

N **Table 690.31(A) Correction Factors**

Ambient Temperature (°C)	Temperature Rating of Conductor				Ambient Temperature (°F)
	60°C (140°F)	75°C (167°F)	90°C (194°F)	105°C (221°F)	
30	1.00	1.00	1.00	1.00	86
31–35	0.91	0.94	0.96	0.97	87–95
36–40	0.82	0.88	0.91	0.93	96–104
41–45	0.71	0.82	0.87	0.89	105–113
46–50	0.58	0.75	0.82	0.86	114–122
51–55	0.41	0.67	0.76	0.82	123–131
56–60	—	0.58	0.71	0.77	132–140
61–70	—	0.33	0.58	0.68	141–158
71–80	—	—	0.41	0.58	159–176

The means of identification shall be permitted by separate color coding, marking tape, tagging, or other approved means. Only solidly grounded PV system circuit conductors, in accordance with 690.41(A)(5), shall be marked in accordance with 200.6.

Exception: Where the identification of the conductors is evident by spacing or arrangement, further identification shall not be required.

(2) Grouping. Where the conductors of more than one PV system occupy the same junction box or raceway with a removable cover(s), the ac and dc conductors of each system shall be grouped separately by cable ties or similar means at least once and shall then be grouped at intervals not to exceed 1.8 m (6 ft).

Exception: The requirement for grouping shall not apply if the circuit enters from a cable or raceway unique to the circuit that makes the grouping obvious.

(C) Single-Conductor Cable.

(1) **General.** Single-conductor cable Type USE-2 and single-conductor cable listed and identified as photovoltaic (PV) wire shall be permitted in exposed outdoor locations in PV source circuits within the PV array. PV wire shall be installed in accordance with 338.10(B)(4)(b) and 334.30.

(2) **Cable Tray.** PV source circuits and PV output circuits using single-conductor cable listed and identified as photovoltaic (PV) wire of all sizes, with or without a cable tray marking/rating, shall be permitted in cable trays installed in outdoor locations, provided that the cables are supported at intervals not to exceed 300 mm (12 in.) and secured at intervals not to exceed 1.4 m (4½ ft).

(D) Multiconductor Cable. Jacketed multiconductor cable assemblies listed and identified for the application shall be permitted in outdoor locations. The cable shall be secured at intervals not exceeding 1.8 m (6 ft).

(E) Flexible Cords and Cables Connected to Tracking PV Arrays. Flexible cords and flexible cables, where connected to moving parts of tracking PV arrays, shall comply with Article 400 and shall be of a type identified as a hard service cord or portable power cable; they shall be suitable for extra-hard usage, listed for outdoor use, water resistant, and sunlight resistant. Allowable ampacities shall be in accordance with 400.5. Stranded copper PV wire shall be permitted to be connected to moving parts of tracking PV arrays in accordance with the minimum number of strands specified in Table 690.31(E).

(F) Small-Conductor Cables. Single-conductor cables listed for outdoor use that are sunlight resistant and moisture resistant in sizes 16 AWG and 18 AWG shall be permitted for module interconnections where such cables meet the ampacity requirements of 400.5. Section 310.15 shall be used to determine the cable ampacity adjustment and correction factors.

(G) Photovoltaic System Direct Current Circuits on or in a Building. Where PV system dc circuits run inside a building, they shall be contained in metal raceways, Type MC metal-clad cable that complies with 250.118(10), or metal enclosures from the point of penetration of the surface of the building to the first readily accessible disconnecting means. The disconnecting means shall comply with 690.13(B) and (C) and 690.15(A) and (B). The wiring methods shall comply with the additional installation requirements in 690.31(G)(1) through (4).

(1) Embedded in Building Surfaces. Where circuits are embedded in built-up, laminate, or membrane roofing materials in roof areas not covered by PV modules and associated equipment, the location of circuits shall be clearly marked using a marking protocol that is approved as being suitable for continuous exposure to sunlight and weather.

(2) Flexible Wiring Methods. Where flexible metal conduit (FMC) smaller than metric designator 21 (trade size ¾) or Type MC cable smaller than 25 mm (1 in.) in diameter containing PV power circuit conductors is installed across ceilings or floor joists, the raceway or cable shall be protected by substantial guard strips that are at least as high as the raceway or cable. Where run exposed, other than within 1.8 m (6 ft) of their connection to equipment, these wiring methods shall closely follow the building surface or be protected from physical damage by an approved means.

(3) Marking and Labeling Required. The following wiring methods and enclosures that contain PV system dc circuit conductors shall be marked with the wording WARNING: PHOTOVOLTAIC POWER SOURCE by means of permanently affixed labels or other approved permanent marking:

(1) Exposed raceways, cable trays, and other wiring methods
(2) Covers or enclosures of pull boxes and junction boxes
(3) Conduit bodies in which any of the available conduit openings are unused

(4) Marking and Labeling Methods and Locations. The labels or markings shall be visible after installation. The labels shall be reflective, and all letters shall be capitalized and shall be a minimum height of 9.5 mm (3/8 in.) in white on a red background. PV system dc circuit labels shall appear on every section of the wiring system that is separated by enclosures, walls, partitions, ceilings, or floors. Spacing between labels or markings, or between a label and a marking, shall not be more than 3 m (10 ft). Labels required by this section shall be suitable for the environment where they are installed.

(H) Flexible, Fine-Stranded Cables. Flexible, fine-stranded cables shall be terminated only with terminals, lugs, devices, or connectors in accordance with 110.14.

(I) Bipolar Photovoltaic Systems. Where the sum, without consideration of polarity, of the voltages of the two monopole subarrays exceeds the rating of the conductors and connected equipment, monopole subarrays in a bipolar PV system shall be physically separated, and the electrical output circuits from each monopole subarray shall be installed in separate raceways until connected to the inverter. The disconnecting means and overcurrent protective devices for each monopole subarray output shall be in separate enclosures. All conductors from each separate monopole subarray shall be routed in the same raceway. Solidly grounded bipolar PV systems shall be clearly marked with a permanent, legible warning notice indicating that the disconnection of the grounded conductor(s) may result in overvoltage on the equipment.

Exception: Listed switchgear rated for the maximum voltage between circuits and containing a physical barrier separating the disconnecting means for each monopole subarray shall be permitted to be used instead of disconnecting means in separate enclosures.

• **690.32 Component Interconnections.** Fittings and connectors that are intended to be concealed at the time of on-site assembly, where listed for such use, shall be permitted for on-site interconnection of modules or other array components. Such fittings and connectors shall be equal to the wiring method employed in insulation, temperature rise, and fault-current withstand, and shall be capable of resisting the effects of the environment in which they are used.

690.33 Connectors. Connectors, other than those covered by 690.32, shall comply with 690.33(A) through (E).

(A) Configuration. The connectors shall be polarized and shall have a configuration that is noninterchangeable with receptacles in other electrical systems on the premises.

(B) Guarding. The connectors shall be constructed and installed so as to guard against inadvertent contact with live parts by persons.

(C) Type. The connectors shall be of the latching or locking type. Connectors that are readily accessible and that are used in circuits operating at over 30 volts dc or 15 volts ac shall require a tool for opening.

(D) Grounding Member. The grounding member shall be the first to make and the last to break contact with the mating connector.

(E) Interruption of Circuit. Connectors shall be either (1) or (2):

(1) Be rated for interrupting current without hazard to the operator.
(2) Be a type that requires the use of a tool to open and marked “Do Not Disconnect Under Load” or “Not for Current Interrupting.”

690.34 Access to Boxes. Junction, pull, and outlet boxes located behind modules or panels shall be so installed that the wiring contained in them can be rendered accessible directly or by displacement of a module(s) or panel(s) secured by removable fasteners and connected by a flexible wiring system.

PART V. GROUNDING AND BONDING

690.41 System Grounding.

(A) PV System Grounding Configurations. One or more of the following system grounding configurations shall be employed:

(1) 2-wire PV arrays with one functional grounded conductor
(2) Bipolar PV arrays according to 690.7(C) with a functional ground reference (center tap)
(3) PV arrays not isolated from the grounded inverter output circuit
(4) Ungrounded PV arrays
(5) Solidly grounded PV arrays as permitted in 690.41(B) Exception
(6) PV systems that use other methods that accomplish equivalent system protection in accordance with 250.4(A) with equipment listed and identified for the use

(B) Ground-Fault Protection. DC PV arrays shall be provided with dc ground-fault protection meeting the requirements of 690.41(B)(1) and (2) to reduce fire hazards.

Exception: PV arrays with not more than two PV source circuits and with all PV system dc circuits not on or in buildings shall be permitted without ground-fault protection where solidly grounded.

(1) Ground-Fault Detection. The ground fault protective device or system shall detect ground fault(s) in the PV array dc current–carrying conductors and components, including any functional grounded conductors, and be listed for providing PV ground-fault protection.

(2) Isolating Faulted Circuits. The faulted circuits shall be isolated by one of the following methods:

(1) The current-carrying conductors of the faulted circuit shall be automatically disconnected.
(2) The inverter or charge controller fed by the faulted circuit shall automatically cease to supply power to output circuits and isolate the PV system dc circuits from the ground reference in a functional grounded system.

•

690.42 Point of System Grounding Connection. Systems with a ground-fault protective device in accordance with 690.41(B) shall have any current-carrying conductor-to-ground connection made by the ground-fault protective device. For solidly grounded PV systems, the dc circuit grounding connection shall be made at any single point on the PV output circuit.

•

690.43 Equipment Grounding and Bonding. Exposed non–current-carrying metal parts of PV module frames, electrical equipment, and conductor enclosures of PV systems shall be grounded in accordance with 250.134 or 250.136(A), regardless of voltage. Equipment grounding conductors and devices shall comply with 690.43(A) through (C).

(A) Photovoltaic Module Mounting Systems and Devices. Devices and systems used for mounting PV modules that are also used for bonding module frames shall be listed, labeled, and identified for bonding PV modules. Devices that mount adjacent PV modules shall be permitted to bond adjacent PV modules.

(B) Equipment Secured to Grounded Metal Supports. Devices listed, labeled, and identified for bonding and grounding the metal parts of PV systems shall be permitted to bond the equipment to grounded metal supports. Metallic support structures shall have identified bonding jumpers connected between separate metallic sections or shall be identified for equipment bonding and shall be connected to the equipment grounding conductor.

(C) With Circuit Conductors. Equipment grounding conductors for the PV array and support structure (where installed) shall be contained within the same raceway, cable, or otherwise run with the PV array circuit conductors when those circuit conductors leave the vicinity of the PV array.

•

690.45 Size of Equipment Grounding Conductors. Equipment grounding conductors for PV source and PV output circuits shall be sized in accordance with 250.122. Where no overcurrent protective device is used in the circuit, an assumed overcurrent device rated in accordance with 690.9(B) shall be used when applying Table 250.122. Increases in equipment grounding conductor size to address voltage drop considerations shall not be required. An equipment grounding conductor shall not be smaller than 14 AWG.

690.46 Array Equipment Grounding Conductors. For PV modules, equipment grounding conductors smaller than 6 AWG shall comply with 250.120(C).

•

690.47 Grounding Electrode System.

(A) Buildings or Structures Supporting a PV Array. A building or structure supporting a PV array shall have a grounding electrode system installed in accordance with Part III of Article 250.

PV array equipment grounding conductors shall be connected to the grounding electrode system of the building or structure supporting the PV array in accordance with Part VII of Article 250. This connection shall be in addition to any other equipment grounding conductor requirements in 690.43(C). The PV array equipment grounding conductors shall be sized in accordance with 690.45.

For PV systems that are not solidly grounded, the equipment grounding conductor for the output of the PV system, connected to associated distribution equipment, shall be permitted to be the connection to ground for ground-fault protection and equipment grounding of the PV array.

For solidly grounded PV systems, as permitted in 690.41(A)(5), the grounded conductor shall be connected to a grounding electrode system by means of a grounding electrode conductor sized in accordance with 250.166.

N **(B) Additional Auxiliary Electrodes for Array Grounding.** Grounding electrodes shall be permitted to be installed in accordance with 250.52 and 250.54 at the location of ground- and roof-mounted PV arrays. The electrodes shall be permitted to be connected directly to the array frame(s) or structure. The grounding electrode conductor shall be sized according to 250.66. The structure of a ground-mounted PV array shall be permitted to be considered a grounding electrode if it meets the requirements of 250.52. Roof mounted PV arrays shall be permitted to use the metal frame of a building or structure if the requirements of 250.52(A)(2) are met.

690.50 Equipment Bonding Jumpers. Equipment bonding jumpers, if used, shall comply with 250.120(C).

PART VI. MARKING

690.53 Direct-Current Photovoltaic Power Source. A permanent label for the dc PV power source indicating the information specified in (1) through (3) shall be provided by the installer at dc PV system disconnecting means and at each dc equipment disconnecting means required by 690.15. Where a disconnecting means has more than one dc PV power source, the values in 690.53(1) through (3) shall be specified for each source.

(1) Maximum voltage
(2) Maximum circuit current
(3) Maximum rated output current of the charge controller or dc-to-dc converter (if installed)

690.54 Interactive System Point of Interconnection. All interactive system(s) points of interconnection with other sources shall be marked at an accessible location at the disconnecting means as a power source and with the rated ac output current and the nominal operating ac voltage.

690.55 Photovoltaic Systems Connected to Energy Storage Systems. The PV system output circuit conductors shall be marked to indicate the polarity where connected to energy storage systems.

690.56 Identification of Power Sources.

(A) Facilities with Stand-Alone Systems. Any structure or building with a PV power system that is not connected to a utility service source and is a stand-alone system shall have a permanent plaque or directory installed on the exterior of the building or structure at a readily visible location. The plaque or directory shall indicate the location of system disconnecting means and that the structure contains a stand-alone electrical power system.

(B) Facilities with Utility Services and Photovoltaic Systems. Plaques or directories shall be installed in accordance with 705.10.

(C) Buildings with Rapid Shutdown. Buildings with PV systems shall have permanent labels as described in 690.56(C)(1) through (C)(3).

N **(1) Rapid Shutdown Type.** The type of PV system rapid shutdown shall be labeled as described in 690.56(C)(1)(a) or (1)(b):

(a) For PV systems that shut down the array and conductors leaving the array:

SOLAR PV SYSTEM IS EQUIPPED
WITH RAPID SHUTDOWN.
TURN RAPID SHUTDOWN SWITCH TO
THE "OFF" POSITION TO SHUT DOWN PV
SYSTEM AND REDUCE SHOCK HAZARD IN ARRAY.

The title "SOLAR PV SYSTEM IS EQUIPPED WITH RAPID SHUTDOWN" shall utilize capitalized characters with a minimum height of 9.5 mm (3⁄8 in.) in black on yellow background, and the remaining characters shall be capitalized with a minimum height of 4.8 mm (3⁄16 in.) in black on white background. *[See Figure 690.56(C)(1)(a).]*

(b) For PV systems that only shut down conductors leaving the array:

SOLAR PV SYSTEM IS EQUIPPED
WITH RAPID SHUTDOWN
TURN RAPID SHUTDOWN SWITCH TO
THE "OFF" POSITION TO SHUT DOWN
CONDUCTORS OUTSIDE THE ARRAY.
CONDUCTORS IN ARRAY REMAIN
ENERGIZED IN SUNLIGHT.

The title "SOLAR PV SYSTEM IS EQUIPPED WITH RAPID SHUTDOWN" shall utilize capitalized characters with a minimum height of 9.5 mm (3⁄8 in.) in white on red background, and the remaining characters shall be capitalized with a minimum height of 4.8 mm (3⁄16 in.) in black on white background. *[See Figure 690.56(C)(1)(b).]*

The labels in 690.56(C)(1)(a) and (b) shall include a simple diagram of a building with a roof. The diagram shall have sections in red to signify sections of the PV system that are not shut down when the rapid shutdown switch is operated.

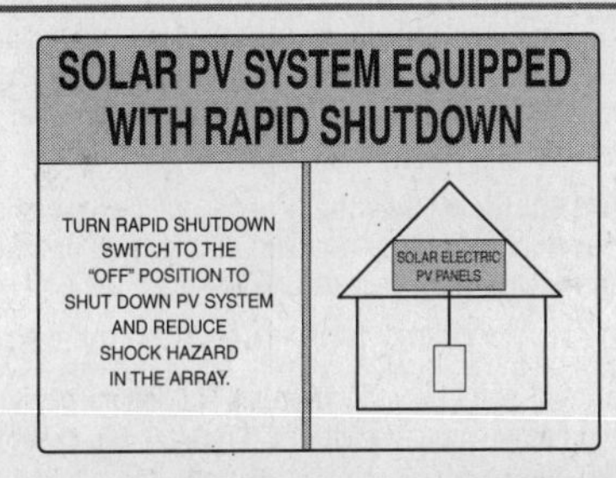

N Figure 690.56(C)(1)(a) Label for PV Systems That Shut Down the Array and the Conductors Leaving the Array.

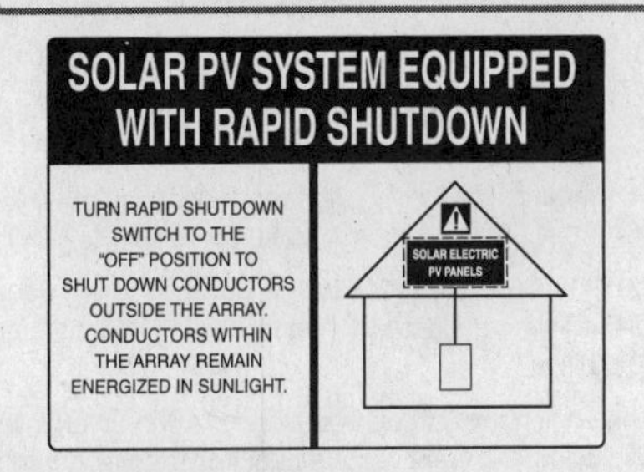

N Figure 690.56(C)(1)(b) Label for PV Systems That Shut Down the Conductors Leaving the Array Only.

The rapid shutdown label in 690.56(C)(1) shall be located on or no more than 1 m (3 ft) from the service disconnecting means to which the PV systems are connected and shall indicate the location of all identified rapid shutdown switches if not at the same location.

N **(2) Buildings with More Than One Rapid Shutdown Type.** For buildings that have PV systems with both rapid shutdown types or a PV system with a rapid shutdown type and a PV system with no rapid shutdown, a detailed plan view diagram of the roof shall be provided showing each different PV system and a dotted line around areas that remain energized after the rapid shutdown switch is operated.

N **(3) Rapid Shutdown Switch.** A rapid shutdown switch shall have a label located on or no more than 1 m (3 ft) from the switch that includes the following wording:

RAPID SHUTDOWN SWITCH FOR SOLAR PV SYSTEM

The label shall be reflective, with all letters capitalized and having a minimum height of 9.5 mm (⅜ in.), in white on red background.

PART VII. CONNECTION TO OTHER SOURCES

N **690.59 Connection to Other Sources.** PV systems connected to other sources shall be installed in accordance with Parts I and II of Article 705.

•

PART VIII. ENERGY STORAGE SYSTEMS

690.71 General. An energy storage system connected to a PV system shall be installed in accordance with Article 706.

690.72 Self-Regulated PV Charge Control. The PV source circuit shall be considered to comply with the requirements of 706.23 if:

(1) The PV source circuit is matched to the voltage rating and charge current requirements of the interconnected battery cells and,

(2) The maximum charging current multiplied by 1 hour is less than 3 percent of the rated battery capacity

expressed in ampere-hours or as recommended by the battery manufacturer

ARTICLE 705
Interconnected Electric Power Production Sources

PART I. GENERAL

705.2 Definitions.

Interactive Inverter Output Circuit. The conductors between the interactive inverter and the service equipment or another electric power production source, such as a utility, for electrical production and distribution network.

705.6 Equipment Approval. All equipment shall be approved for the intended use. Interactive inverters for interconnection to systems interactive equipment intended to operate in parallel with the electric power system including, but not limited to, interactive inverters, engine generators, energy storage equipment, and wind turbines shall be listed and or field labeled for the intended use of interconnection service.

705.10 Directory. A permanent plaque or directory denoting the location of all electric power source disconnecting means on or in the premises shall be installed at each service equipment location and at the location(s) of the system disconnect(s) for all electric power production sources capable of being interconnected. The marking shall comply with 110.21(B).

Exception: Installations with large numbers of power production sources shall be permitted to be designated by groups.

705.12 Point of Connection. The output of an interconnected electric power source shall be connected as specified in 705.12(A) or (B).

(A) Supply Side. An electric power production source shall be permitted to be connected to the supply side of the service disconnecting means as permitted in 230.82(6). The sum of the ratings of all overcurrent devices connected to power production sources shall not exceed the rating of the service.

(B) Load Side. The output of an interconnected electric power source shall be permitted to be connected to the load side of the service disconnecting means of the other source(s) at any distribution equipment on the premises. Where distribution equipment, including switchgear, switchboards, or panelboards, is fed simultaneously by a primary source(s) of electricity and one or more other power source(s), and where this distribution equipment is capable of supplying multiple branch circuits or feeders, or both, the interconnecting provisions for other power sources shall comply with 705.12(B)(1) through (B)(5).

(1) Dedicated Overcurrent and Disconnect. Each source interconnection of one or more power sources installed in one system shall be made at a dedicated circuit breaker or fusible disconnecting means.

(2) Bus or Conductor Ampere Rating. One hundred twenty-five percent of the power source output circuit current shall be used in ampacity calculations for the following:

(1) *Feeders*. Where the power source output connection is made to a feeder at a location other than the opposite end of the feeder from the primary source overcurrent device, that portion of the feeder on the load side of the power source output connection shall be protected by one of the following:
   a. The feeder ampacity shall be not less than the sum of the primary source overcurrent device and 125 percent of the power source output circuit current.
   b. An overcurrent device on the load side of the power source connection shall be rated not greater than the ampacity of the feeder.

(2) *Taps*. In systems where power source output connections are made at feeders, any taps shall be sized based on the sum of 125 percent of the power source(s) output circuit current and the rating of the overcurrent device protecting the feeder conductors as calculated in 240.21(B).

(3) *Busbars*. One of the methods that follows shall be used to determine the ratings of busbars in panelboards.

(a) The sum of 125 percent of the power source(s) output circuit current and the rating of the overcurrent device protecting the busbar shall not exceed the ampacity of the busbar.

(b) Where two sources, one a primary power source and the other another power source, are located at opposite ends of a busbar that contains loads, the sum of 125 percent of the power source(s) output circuit current and the rating of the overcurrent device protecting the busbar shall not exceed 120 percent of the ampacity of the busbar. The busbar shall be sized for the loads connected in accordance with Article 220. A permanent warning label shall be applied to the distribution equipment adjacent to the back-fed breaker from the power source that displays the following or equivalent wording:

WARNING:
POWER SOURCE OUTPUT CONNECTION —
DO NOT RELOCATE THIS OVERCURRENT DEVICE.

The warning sign(s) or label(s) shall comply with 110.21(B).

(c) The sum of the ampere ratings of all overcurrent devices on panelboards, both load and supply devices, excluding the rating of the overcurrent device protecting the busbar, shall not exceed the ampacity of the busbar. The rating of the overcurrent device protecting the busbar shall not exceed the rating of the busbar. Permanent warning labels shall be applied to distribution equipment displaying the following or equivalent wording:

WARNING:
THIS EQUIPMENT FED BY MULTIPLE SOURCES.
TOTAL RATING OF ALL OVERCURRENT DEVICES
EXCLUDING MAIN SUPPLY OVERCURRENT DEVICE
SHALL NOT EXCEED AMPACITY OF BUSBAR.

The warning sign(s) or label(s) shall comply with 110.21(B).

(d) A connection at either end, but not both ends, of a center-fed panelboard in dwellings shall be permitted where the sum of 125 percent of the power source(s) output circuit current and the rating of the overcurrent device protecting the busbar does not exceed 120 percent of the current rating of the busbar.

(e) Connections shall be permitted on multiple-ampacity busbars where designed under engineering supervision that includes available fault current and busbar load calculations.

(3) Marking. Equipment containing overcurrent devices in circuits supplying power to a busbar or conductor supplied from multiple sources shall be marked to indicate the presence of all sources.

(4) Suitable for Backfeed. Circuit breakers, if backfed, shall be suitable for such operation.

(5) Fastening. Listed plug-in-type circuit breakers backfed from electric power sources that are listed and identified as interactive shall be permitted to omit the additional fastener normally required by 408.36(D) for such applications.

705.16 Interrupting and Short-Circuit Current Rating. Consideration shall be given to the contribution of fault currents from all interconnected power sources for the interrupting and short-circuit current ratings of equipment on interactive systems.

705.20 Disconnecting Means, Sources. Means shall be provided to disconnect all ungrounded conductors of an electric power production source(s) from all other conductors.

705.21 Disconnecting Means, Equipment. Means shall be provided to disconnect power production equipment, such as interactive inverters or transformers associated with a power production source, from all ungrounded conductors of all sources of supply. Equipment intended to be operated and maintained as an integral part of a power production source exceeding 1000 volts shall not be required to have a disconnecting means.

705.22 Disconnect Device. The disconnecting means for ungrounded conductors shall consist of a manual or power operated switch(es) or circuit breaker(s) that complies with the following:

(1) Located where readily accessible
(2) Externally operable without exposing the operator to contact with live parts and, if power operated, of a type that is opened by hand in the event of a power-supply failure
(3) Plainly indicate whether in the open (off) or closed (on) position
(4) Have ratings sufficient for the maximum circuit current, available short-circuit current, and voltage that is available at the terminals
(5) Where the line and load terminals are capable of being energized in the open position, marked in accordance with the warning in 690.13(B)
(6) Simultaneously disconnect all ungrounded conductors of the circuit
(7) Be lockable in the open (off) position in accordance with 110.25

N **705.23 Interactive System Disconnecting Means.** A readily accessible means shall be provided to disconnect the interactive system from all wiring systems including power systems, energy storage systems, and utilization equipment and its associated premises wiring.

705.30 Overcurrent Protection. Conductors shall be protected in accordance with Article 240. Equipment and conductors connected to more than one electrical source shall have a sufficient number of overcurrent devices located so as to provide protection from all sources.

(A) Solar Photovoltaic Systems. Solar photovoltaic systems shall be protected in accordance with Article 690.

(D) Interactive Inverters. Interactive inverters shall be protected in accordance with 705.65.

705.31 Location of Overcurrent Protection. Overcurrent protection for electric power production source conductors, connected to the supply side of the service disconnecting means in accordance with 705.12(A), shall be located within 3 m (10 ft) of the point where the electric power production source conductors are connected to the service.

Exception: Where the overcurrent protection for the power production source is located more than 3 m (10 ft) from the point of connection for the electric power production source to the service, cable limiters or current-limited circuit breakers for each ungrounded conductor shall be installed at the point where the electric power production conductors are connected to the service.

705.32 Ground-Fault Protection. Where ground-fault protection is used, the output of an interactive system shall be connected to the supply side of the ground-fault protection.

Exception: Connection shall be permitted to be made to the load side of ground-fault protection, if there is ground-fault protection for equipment from all ground-fault current sources.

705.50 Grounding. Interconnected electric power production sources shall be grounded in accordance with Article 250.

Exception: For direct-current systems connected through an inverter directly to a grounded service, other methods that accomplish equivalent system protection and that utilize equipment listed and identified for the use shall be permitted.

PART II. INTERACTIVE INVERTERS

705.60 Circuit Sizing and Current.

(A) Calculation of Maximum Circuit Current. The maximum current for the specific circuit shall be calculated in accordance with 705.60(A)(1) and (A)(2).

(1) Inverter Input Circuit Currents. The maximum current shall be the maximum rated input current of the inverter.

(2) Inverter Output Circuit Current. The maximum current shall be the inverter continuous output current rating.

(B) Ampacity and Overcurrent Device Ratings. Inverter system currents shall be considered to be continuous. The circuit conductors and overcurrent devices shall be sized to carry not less than 125 percent of the maximum currents as calculated in 705.60(A). The rating or setting of overcurrent devices shall be permitted in accordance with 240.4(B) and (C).

Exception: Circuits containing an assembly together with its overcurrent device(s) that is listed for continuous operation at 100 percent of its rating shall be permitted to be utilized at 100 percent of its rating.

705.65 Overcurrent Protection.

(A) Circuits and Equipment. Inverter input circuits, inverter output circuits, and storage battery circuit conductors and equipment shall be protected in accordance with the requirements of Article 240. Circuits connected to more than one electrical source shall have overcurrent devices located so as to provide overcurrent protection from all sources.

Exception: An overcurrent device shall not be required for circuit conductors sized in accordance with 705.60(B) and located where one of the following applies:

(1) There are no external sources such as parallel-connected source circuits, batteries, or backfeed from inverters.
(2) The short-circuit currents from all sources do not exceed the ampacity of the conductors.

N **(C) Conductor Ampacity.** Power source output circuit conductors that are connected to a feeder, if smaller than the feeder conductors, shall be sized to carry not less than the larger of the current as calculated in 705.60(B) or as calculated in accordance with 240.21(B) based on the over-current device protecting the feeder.

705.70 Interactive Inverters Mounted in Not Readily Accessible Locations. Interactive inverters shall be permitted to be mounted on roofs or other exterior areas that are not readily accessible. These installations shall comply with (1) through (4):

(1) A dc disconnecting means shall be mounted within sight of or in the inverter.
(2) An ac disconnecting means shall be mounted within sight of or in the inverter.

(3) An additional ac disconnecting means for the inverter shall comply with 705.22.
(4) A plaque shall be installed in accordance with 705.10.

705.80 Utility-Interactive Power Systems Employing Energy Storage. Utility-interactive power systems employing energy storage shall also be marked with the maximum operating voltage, including any equalization voltage, and the polarity of the grounded circuit conductor.

705.95 Ampacity of Neutral Conductor. The ampacity of the neutral conductors shall comply with either (A) or (B).

(A) Neutral Conductor for Single Phase, 2-Wire Inverter Output. If a single-phase, 2-wire inverter output is connected to the neutral and one ungrounded conductor (only) of a 3-wire system or of a 3-phase, 4-wire, wye-connected system, the maximum load connected between the neutral and any one ungrounded conductor plus the inverter output rating shall not exceed the ampacity of the neutral conductor.

(B) Neutral Conductor for Instrumentation, Voltage, Detection or Phase Detection. A conductor used solely for instrumentation, voltage detection, or phase detection and connected to a single-phase or 3-phase interactive inverter, shall be permitted to be sized at less than the ampacity of the other current-carrying conductors and shall be sized equal to or larger than the equipment grounding conductor.

NFPA 1
CHAPTER 11
BUILDING SERVICES

11.12 Photovoltaic Systems.

11.12.1 Photovoltaic systems shall be in accordance with Section 11.12 and NFPA 70.

11.12.2 Building-Mounted Photovoltaic Installations.

11.12.2.1* Marking. Photovoltaic systems shall be permanently marked as specified in this subsection.

11.12.2.1.1 Main Service Disconnect Marking. A label shall be permanently affixed to the main service disconnect panel serving alternating current (ac) and direct current (dc) photovoltaic systems. The label shall be red with white capital letters at least ¾ in. (19 mm) in height and in a nonserif font, to read: "WARNING: PHOTOVOLTAIC POWER SOURCE." The materials used for the label shall be reflective, weather resistant, and suitable for the environment.

11.12.2.1.2 Circuit Disconnecting Means Marking. A permanent label shall be affixed adjacent to the circuit breaker controlling the inverter or other photovoltaic system electrical controller serving ac and dc photovoltaic systems. The label shall have contrasting color with capital letters at least ⅜ in. (10 mm) in height and in a nonserif font, to read: "PHOTOVOLTAIC DISCONNECT." The label shall be constructed of durable adhesive material or other approved material.

11.12.2.1.3* Conduit, Raceway, Enclosure, Cable Assembly, and Junction Box Markings. Marking shall be required on all interior and exterior dc conduits, raceways, enclosures, cable assemblies, and junction boxes.

11.12.2.1.3.1 Marking Locations. Marking shall be placed on all dc conduits, raceways, enclosures, and cable assemblies every 10 ft (3048 mm), at turns, and above and below penetrations. Marking shall be placed on all dc combiner and junction boxes.

11.12.2.1.3.2* Marking Content and Format. Marking for dc conduits, raceways, enclosures, cable assemblies, and junction boxes shall be red with white lettering with minimum 3⁄8 in. (10 mm) capital letters in a nonserif font, to read: "WARNING: PHOTOVOLTAIC POWER SOURCE." Marking shall be reflective, weather resistant, and suitable for the environment.

11.12.2.1.4 Secondary Power Source Markings. Where photovoltaic systems are interconnected to battery systems, generator backup systems, or other secondary power systems, additional signage acceptable to the AHJ shall be required indicating the location of the secondary power source shutoff switch.

11.12.2.1.5 Installer Information. Signage, acceptable to the AHJ, shall be installed adjacent to the main disconnect indicating the name and emergency telephone number of the installing contractor.

11.12.2.1.6* Inverter Marking. Markings shall not be required for inverters.

11.12.2.2 Access, Pathways, and Smoke Ventilation.

11.12.2.2.1 General. Access and spacing requirements shall be required to provide emergency access to the roof, provide pathways to specific areas of the roof, provide for smoke ventilation opportunity areas, and to provide emergency egress from the roof.

11.12.2.2.1.1 Exceptions. The AHJ shall be permitted to grant exceptions where access, pathway, or ventilation requirements are reduced due to any of the following circumstances:

(1) Proximity and type of adjacent exposures
(2) Alternative access opportunities, as from adjoining roofs
(3) Ground level access to the roof
(4) Adequate ventilation opportunities beneath photovoltaic module arrays
(5) Adequate ventilation opportunities afforded by module set back from other rooftop equipment
(6) Automatic ventilation devices
(7) New technologies, methods, or other innovations that ensure adequate fire department access, pathways, and ventilation opportunities

11.12.2.2.1.2 Pitch. Designation of ridge, hip, and valley shall not apply to roofs with 2-in-12 or less pitch.

11.12.2.2.1.3 Roof Access Points. Roof access points shall be defined as areas where fire department ladders are not placed over openings (windows or doors), are located at strong points of building construction, and are in locations where they will not conflict with overhead obstructions (tree limbs, wires, or signs).

11.12.2.2.2 One- and Two-Family Dwellings and Townhouses. Photovoltaic systems installed in one- and two-family dwellings and townhouses shall be in accordance with this section.

11.12.2.2.2.1 Access and Pathways.

11.12.2.2.2.1.1 Hip Roof Layouts. Photovoltaic modules shall be located in a manner that provides a 3 ft (914 mm) wide clear access pathway from the eave to the ridge of each roof slope where the photovoltaic modules are located. The access pathway shall be located at a structurally strong location of the building, such as a bearing wall.

Exception: The requirement of 11.12.2.2.2.1.1 shall not apply where adjoining roof planes provide a 3 ft (914 mm) wide clear access pathway.

11.12.2.2.2.1.2 Single Ridge Layouts. Photovoltaic modules shall be located in a manner that provides two 3 ft (914 mm) wide access pathways from the eave to the ridge on each roof slope where the modules are located.

11.12.2.2.2.1.3 Hip and Valley Layouts. Photovoltaic modules shall be located no closer than 1½ ft (457 mm) to a hip or valley if modules are to be placed on both sides of the hip or valley. Where modules are located on only one side of a hip or valley of equal length, the photovoltaic modules shall be allowed to be placed directly adjacent to the hip or valley.

11.12.2.2.2.2 Ridge Setback. Photovoltaic modules shall be located not less than 3 ft (914 mm) below the ridge.

11.12.2.2.3 Buildings Other Than One- and Two-Family Dwellings and Townhouses. Photovoltaic energy systems installed in any building other than one- and two-family dwellings and townhouses shall be in accordance with this section. Where the AHJ determines that the roof configuration is similar to a one- and two-family dwelling or townhouse, the AHJ shall allow the requirements of 11.12.2.2.2.

11.12.2.2.3.1 Access. A minimum 4 ft (1219 mm) wide clear perimeter shall be provided around the edges of the roof for buildings with a length or width of 250 ft (76.2 m) or less along either axis. A minimum 6 ft (1829 mm) wide clear perimeter shall be provided around the edges of the roof for buildings having length or width greater than 250 ft (76.2 m) along either axis.

11.12.2.2.3.2 Pathways. Pathways shall be established as follows:

(1) Pathways shall be over areas capable of supporting the live load of fire fighters accessing the roof.
(2) Centerline axis pathways shall be provided in both axes of the roof.
(3) Centerline axis pathways shall run where the roof structure is capable of supporting the live load of fire fighters accessing the roof.
(4) Pathways shall be in a straight line not less than 4 ft (1219 mm) clear to skylights, ventilation hatches, and roof standpipes.
(5) Pathways shall provide not less than 4 ft (1219 mm) clear around roof access hatches with at least one not less than 4 ft (1219 mm) clear pathway to the parapet or roof edge.

11.12.2.2.3.3 Smoke Ventilation. Ability for fire department smoke ventilation shall be provided in accordance with this section.

11.12.2.2.3.3.1 Maximum Array. Arrays of photovoltaic modules shall be no greater than 150 ft (45.7 m) × 150 ft (45.7 m) in distance in either axis.

11.12.2.2.3.3.2 Ventilation Options. Ventilation options between array sections shall be one of the following:

(1) A pathway 8 ft (2438 mm) or greater in width
(2) A pathway 4 ft (1219 mm) or greater in width and bordering on existing roof skylights or ventilation hatches
(3) A pathway 4 ft (1219 mm) or greater in width and bordering 4 ft (1219 mm) × 8 ft (2438 mm) venting cutouts options every 20 ft (6096 mm) on alternating sides of the pathway

11.12.2.2.4 Location of Direct Current (DC) Conductors.

11.12.2.2.4.1 Exterior-mounted dc conduits, wiring systems, and raceways for photovoltaic circuits shall be located as close as possible to the ridge, hip, or valley and from the hip or valley as directly as possible to an outside wall to reduce trip hazards and maximize ventilation opportunities.

11.12.2.2.4.2 Conduit runs between subarrays and to dc combiner boxes shall be designed to take the shortest path from the array to the dc combiner box.

11.12.2.2.4.3 DC combiner boxes shall be located so that conduit runs are minimized in the pathways between arrays.

11.12.2.2.4.4 DC wiring shall be run in metallic conduit or raceways where located within enclosed spaces in a building.

11.12.2.2.4.4.1 Where dc wiring is run perpendicular or parallel to load-bearing members, a minimum 10 in. (254 mm) space below roof decking or sheathing shall be maintained.

11.12.3 Ground-Mounted Photovoltaic System Installations. Ground-mounted photovoltaic systems shall be installed in accordance with 11.12.3.1 through 11.12.3.3.

11.12.3.1* Clearances. A clear area of 10 ft (3048 mm) around ground-mounted photovoltaic installations shall be provided.

11.12.3.2* Noncombustible Base. A gravel base or other noncombustible base acceptable to the AHJ shall be installed and maintained under and around the installation.

11.12.3.3* Security Barriers. Fencing, skirting, or other suitable security barriers shall be installed when required by the AHJ.

CHAPTER 25

EMERGENCY SYSTEMS

INTRODUCTION

This chapter contains requirements from Article 700 — Emergency Systems. Article 700 contains requirements that cover the installation, operation, and maintenance of emergency systems. These systems consist of circuits and equipment intended to supply, distribute, and control electricity for illumination or power, or both, to required facilities when the normal electrical supply or system is interrupted. The equipment supplied by the emergency system is typically used to provide a safe level of lighting and power for building evacuation. Fuel supplies and operational times are required to allow for 1.5 hours of operation. The *NEC* does not preclude longer operational times where it may be necessary or desirable to maintain building operations and not discharge the occupants, as in the case of a storm shelter or other place of refuge.

The requirements of Article 700 impact all types of occupancies including mandatory application to certain segments of the essential electrical system (ESS) in hospitals. Standby systems installed in a dwelling unit are not legally required by codes and therefore do not fall under the scope of Article 700. These optional standby systems installed at the discretion of the property owner to provide power during normal source outages are covered by the requirements contained in Chapter 26 of this *Pocket Guide*.

Article 700 is divided into six parts: Part I General, Part II Circuit Wiring, Part III Sources of Power, Part IV Emergency System Circuits for Lighting and Power, Part V Control — Emergency Lighting Circuits, and Part VI Overcurrent Protection. All of these requirements are focused on providing a reliable and resilient system to avoid building occupant panic in the event that the normal power source is compromised.

Requirements in other codes and standards such as *NFPA 101®, Life Safety Code®*, and *NFPA 110, Standard for Emergency and Standby Power Systems,* drive the requirements for when and where emergency power supplies are required and for the ongoing maintenance of such systems to ensure they remain fully operational.

ARTICLE 700
Emergency Systems

PART I. GENERAL

700.2 Definitions.

N **Branch Circuit Emergency Lighting Transfer Switch.** A device connected on the load side of a branch circuit overcurrent protective device that transfers only emergency lighting loads from the normal supply to an emergency supply.

Emergency Systems. Those systems legally required and classed as emergency by municipal, state, federal, or other codes, or by any governmental agency having jurisdiction. These systems are intended to automatically supply illumination, power, or both, to designated areas and equipment in the event of failure of the normal supply or in the event of accident to elements of a system intended to supply, distribute, and control power and illumination essential for safety to human life.

Informational Note: Emergency systems are generally installed in places of assembly where artificial illumination is required for safe exiting and for panic control in buildings subject to occupancy by large numbers of persons, such as hotels, theaters, sports arenas, health care facilities, and similar institutions. Emergency systems may also provide power for such functions as ventilation where essential to maintain life, fire detection and alarm systems, elevators, fire pumps, public safety communications systems, industrial processes where current interruption would produce serious life safety or health hazards, and similar functions.

700.3 Tests and Maintenance.

(A) Conduct or Witness Test. The authority having jurisdiction shall conduct or witness a test of the complete system upon installation and periodically afterward.

N **(F) Temporary Source of Power for Maintenance or Repair of the Alternate Source of Power.** If the emergency system relies on a single alternate source of power, which will be disabled for maintenance or repair, the emergency system shall include permanent switching means to connect a portable or temporary alternate source of power, which shall be available for the duration of the maintenance or repair. The permanent switching means to connect a portable or temporary alternate source of power shall comply with the following:

(1) Connection to the portable or temporary alternate source of power shall not require modification of the permanent system wiring.
(2) Transfer of power between the normal power source and the emergency power source shall be in accordance with 700.12.
(3) The connection point for the portable or temporary alternate source shall be marked with the phase rotation and system bonding requirements.
(4) Mechanical or electrical interlocking shall prevent inadvertent interconnection of power sources.
(5) The switching means shall include a contact point that shall annunciate at a location remote from the generator or at another facility monitoring system to indicate that the permanent emergency source is disconnected from the emergency system.

It shall be permissible to utilize manual switching to switch from the permanent source of power to the portable or temporary alternate source of power and to utilize the switching means for connection of a load bank.

Informational Note: There are many possible methods to achieve the requirements of 700.3(F). See Figure 700.3(F) for one example.

Exception: The permanent switching means to connect a portable or temporary alternate source of power, for the duration of the maintenance or repair, shall not be required where any of the following conditions exists:

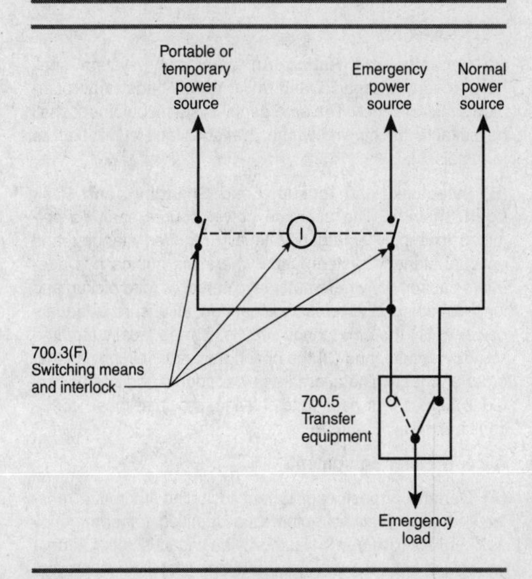

N **Figure 700.3(F)**

(1) *All processes that rely on the emergency system source are capable of being disabled during maintenance or repair of the emergency source of power.*
(2) *The building or structure is unoccupied and fire suppression systems are fully functional and do not require an alternate power source.*
(3) *Other temporary means can be substituted for the emergency system.*
(4) *A permanent alternate emergency source, such as, but not limited to, a second on-site standby generator or separate electric utility service connection, capable of supporting the emergency system, exists.*

700.4 Capacity.

(A) Capacity and Rating. An emergency system shall have adequate capacity and rating for all loads to be operated simultaneously. The emergency system equipment shall be suitable for the maximum available fault current at its terminals.

(B) Selective Load Pickup, Load Shedding, and Peak Load Shaving. The alternate power source shall be permitted to supply emergency, legally required standby, and optional standby system loads where the source has adequate capacity or where automatic selective load pickup and load shedding is provided as needed to ensure adequate power to (1) the emergency circuits, (2) the legally required standby circuits, and (3) the optional standby circuits, in that order of priority. The alternate power source shall be permitted to be used for peak load shaving, provided these conditions are met.

700.5 Transfer Equipment.

(A) General. Transfer equipment, including automatic transfer switches, shall be automatic, identified for emergency use, and approved by the authority having jurisdiction. Transfer equipment shall be designed and installed to prevent the inadvertent interconnection of normal and emergency sources of supply in any operation of the transfer equipment. Transfer equipment and electric power production systems installed to permit operation in parallel with the normal source shall meet the requirements of Article 705.

(B) Bypass Isolation Switches. Means shall be permitted to bypass and isolate the transfer equipment. Where bypass isolation switches are used, inadvertent parallel operation shall be avoided.

(C) Automatic Transfer Switches. Automatic transfer switches shall be electrically operated and mechanically held. Automatic transfer switches shall be listed for emergency system use.

(D) Use. Transfer equipment shall supply only emergency loads.

N **(E) Documentation.** The short-circuit current rating of the transfer equipment, based on the specific overcurrent protective device type and settings protecting the transfer equipment, shall be field marked on the exterior of the transfer equipment.

700.6 Signals. Audible and visual signal devices shall be provided, where practicable, for the purpose described in 700.6(A) through (D).

(A) Malfunction. To indicate malfunction of the emergency source.

(B) Carrying Load. To indicate that the battery is carrying load.

(C) Not Functioning. To indicate that the battery charger is not functioning.

(D) Ground Fault. To indicate a ground fault in solidly grounded wye emergency systems of more than 150 volts to ground and circuit-protective devices rated 1000 amperes or more. The sensor for the ground-fault signal devices shall be located at, or ahead of, the main system disconnecting means for the emergency source, and the maximum setting of the signal devices shall be for a ground-fault current of 1200 amperes. Instructions on the course of action to be taken in event of indicated ground fault shall be located at or near the sensor location.

• For systems with multiple emergency sources connected to a paralleling bus, the ground fault sensor shall be permitted to be at an alternative location.

700.7 Signs.

(A) Emergency Sources. A sign shall be placed at the service-entrance equipment, indicating type and location of each on-site emergency power source.

Exception: A sign shall not be required for individual unit equipment as specified in 700.12(F).

(B) Grounding. Where removal of a grounding or bonding connection in normal power source equipment interrupts the grounding electrode conductor connection to the alternate power source(s) grounded conductor, a warning sign shall be installed at the normal power source equipment stating:

WARNING
SHOCK HAZARD EXISTS IF GROUNDING
ELECTRODE CONDUCTOR OR BONDING JUMPER
CONNECTION IN THIS EQUIPMENT IS REMOVED
WHILE ALTERNATE SOURCE(S) IS ENERGIZED.

The warning sign(s) or label(s) shall comply with 110.21(B).

700.8 Surge Protection. A listed SPD shall be installed in or on all emergency systems switchboards and panelboards.

PART II. CIRCUIT WIRING

700.10 Wiring, Emergency System.

(A) Identification. Emergency circuits shall be permanently marked so they will be readily identified as a component of an emergency circuit or system by the following methods:

(1) All boxes and enclosures (including transfer switches, generators, and power panels) for emergency circuits shall be permanently marked as a component of an emergency circuit or system.
(2) Where boxes or enclosures are not encountered, exposed cable or raceway systems shall be permanently marked to be identified as a component of an emergency circuit or system, at intervals not to exceed 7.6 m (25 ft).

Receptacles supplied from the emergency system shall have a distinctive color or marking on the receptacle cover plates or the receptacles.

(B) Wiring. Wiring of two or more emergency circuits supplied from the same source shall be permitted in the same raceway, cable, box, or cabinet. Wiring from an emergency source or emergency source distribution overcurrent protection to emergency loads shall be kept entirely independent of all other wiring and equipment, unless otherwise permitted in 700.10(B)(1) through (5):

(1) Wiring from the normal power source located in transfer equipment enclosures
(2) Wiring supplied from two sources in exit or emergency luminaires
(3) Wiring from two sources in a listed load control relay supplying exit or emergency luminaires, or in a common junction box, attached to exit or emergency luminaires
(4) Wiring within a common junction box attached to unit equipment, containing only the branch circuit supplying the unit equipment and the emergency circuit supplied by the unit equipment
(5) Wiring from an emergency source to supply emergency and other (nonemergency) loads in accordance with 700.10(B)(5)a., b., c., and d. as follows:

a. Separate vertical switchgear sections or separate vertical switchboard sections, with or without a common bus, or individual disconnects mounted in separate enclosures shall be used to separate emergency loads from all other loads.
b. The common bus of separate sections of the switchgear, separate sections of the switchboard, or the individual enclosures shall be either of the following:

(i) Supplied by single or multiple feeders without overcurrent protection at the source
(ii) Supplied by single or multiple feeders with overcurrent protection, provided that the overcurrent protection that is common to an emergency system and any non-emergency system(s) is selectively coordinated with the next downstream overcurrent protective device in the nonemergency system(s)

c. Emergency circuits shall not originate from the same vertical switchgear section, vertical switchboard section, panelboard enclosure, or individual disconnect enclosure as other circuits.
d. It shall be permissible to utilize single or multiple feeders to supply distribution equipment between an emergency source and the point where the emergency loads are separated from all other loads.

(C) Wiring Design and Location. Emergency wiring circuits shall be designed and located so as to minimize the hazards that might cause failure due to flooding, fire, icing, vandalism, and other adverse conditions.

(D) Fire Protection. Emergency systems shall meet the additional requirements in (D)(1) through (D)(3) in the following occupancies:

(1) Assembly occupancies for not less than 1000 persons
(2) Buildings above 23 m (75 ft) in height
(3) Health care occupancies where persons are not capable of self preservation
(4) Educational occupancies with more than 300 occupants

(1) Feeder-Circuit Wiring. Feeder-circuit wiring shall meet one of the following conditions:

(1) The cable or raceway is installed in spaces or areas that are fully protected by an approved automatic fire suppression system.
(2) The cable or raceway is protected by a listed electrical circuit protective system with a minimum 2-hour fire rating.
(3) The cable or raceway is a listed fire-resistive cable system.
(4) The cable or raceway is protected by a listed fire-rated assembly that has a minimum fire rating of 2 hours and contains only emergency circuits
(5) The cable or raceway is encased in a minimum of 50 mm (2 in.) of concrete.

(2) Feeder-Circuit Equipment. Equipment for feeder circuits (including transfer switches, transformers, and panelboards) shall be located either in spaces fully protected by approved automatic fire suppression systems (including sprinklers, carbon dioxide systems) or in spaces with a 2-hour fire resistance rating.

(3) Generator Control Wiring. Control conductors installed between the transfer equipment and the emergency generator shall be kept entirely independent of all other wiring and shall meet the conditions of 700.10(D)(1). The integrity of the generator control wiring shall be continuously monitored.

Loss of integrity of the remote start circuit(s) shall initiate visual and audible annunciation of generator malfunction at the generator local and remote annunciator(s) and start the generator(s).

PART III. SOURCES OF POWER

700.12 General Requirements. Current supply shall be such that, in the event of failure of the normal supply to, or within, the building or group of buildings concerned, emergency lighting, emergency power, or both shall be available within the time required for the application but not to exceed 10 seconds. The supply system for emergency purposes, in addition to the normal services to the building and meeting the general requirements of this section, shall be one or more of the types of systems described in 700.12(A) through (E). Unit equipment in accordance with 700.12(F) shall satisfy the applicable requirements of this article.

In selecting an emergency source of power, consideration shall be given to the occupancy and the type of service to be rendered, whether of minimum duration, as for evacuation of a theater, or longer duration, as for supplying emergency power and lighting due to an indefinite period of current failure from trouble either inside or outside the building.

Equipment shall be designed and located so as to minimize the hazards that might cause complete failure due to flooding, fires, icing, and vandalism.

Equipment for sources of power as described in 700.12(A) through (E) shall be installed either in spaces fully protected by approved automatic fire suppression systems (sprinklers, carbon dioxide systems, and so forth) or in spaces with a 1-hour fire rating where located within the following:

(1) Assembly occupancies for more than 1000 persons
(2) Buildings above 23 m (75 ft) in height with any of the following occupancy classes — assembly, educational, residential, detention and correctional, business, and mercantile
(3) Health care occupancies where persons are not capable of self-preservation
(4) Educational occupancies with more than 300 occupants

(A) Storage Battery. Storage batteries shall be of suitable rating and capacity to supply and maintain the total load for a minimum period of 1½ hours, without the voltage applied to the load falling below 87½ percent of normal. Automotive-type batteries shall not be used.

An automatic battery charging means shall be provided.

(B) Generator Set.

(1) Prime Mover-Driven. For a generator set driven by a prime mover acceptable to the authority having jurisdiction and sized in accordance with 700.4, means shall be provided for automatically starting the prime mover on failure of the normal service and for automatic transfer and operation of all required electrical circuits. A time-delay feature permitting a 15-minute setting shall be provided to avoid retransfer in case of short-time reestablishment of the normal source.

(2) Internal Combustion Engines as Prime Movers. Where internal combustion engines are used as the prime mover, an on-site fuel supply shall be provided with an on-premises fuel supply sufficient for not less than 2 hours' full-demand operation of the system. Where power is needed for the operation of the fuel transfer pumps to deliver fuel to a generator set day tank, this pump shall be connected to the emergency power system.

(3) Dual Supplies. Prime movers shall not be solely dependent on a public utility gas system for their fuel supply or municipal water supply for their cooling systems. Means shall be provided for automatically transferring from one fuel supply to another where dual fuel supplies are used.

Exception: Where acceptable to the authority having jurisdiction, the use of other than on-site fuels shall be permitted where there is a low probability of a simultaneous failure of both the off-site fuel delivery system and power from the outside electrical utility company.

(4) Battery Power and Dampers. Where a storage battery is used for control or signal power or as the means of starting the prime mover, it shall be suitable for the purpose and shall be equipped with an automatic charging means independent of the generator set. Where the battery charger is required for the operation of the generator set, it shall be connected to the emergency system. Where power is required for the operation of dampers used to ventilate the generator set, the dampers shall be connected to the emergency system.

(5) Auxiliary Power Supply. Generator sets that require more than 10 seconds to develop power shall be permitted if an auxiliary power supply energizes the emergency system until the generator can pick up the load.

(6) Outdoor Generator Sets. Where an outdoor housed generator set is equipped with a readily accessible disconnecting means in accordance with 445.18, and the disconnecting means is located within sight of the building or structure supplied, an additional disconnecting means shall not be required where ungrounded conductors serve or pass through the building or structure. Where the generator supply conductors terminate at a disconnecting means in or on a building or structure, the disconnecting means shall meet the requirements of 225.36.

Exception: For installations under single management, where conditions of maintenance and supervision ensure that only qualified persons will monitor and service the installation and where documented safe switching procedures are established and maintained for disconnection, the generator set disconnecting means shall not be required to be located within sight of the building or structure served.

(C) Uninterruptible Power Supplies. Uninterruptible power supplies used to provide power for emergency systems shall comply with the applicable provisions of 700.12(A) and (B).

(D) Separate Service. Where approved by the authority having jurisdiction as suitable for use as an emergency source of power, an additional service shall be permitted. This service shall be in accordance with the applicable provisions of Article 230 and the following additional requirements:

(1) Separate overhead service conductors, service drops, underground service conductors, or service laterals shall be installed.
(2) The service conductors for the separate service shall be installed sufficiently remote electrically and physically

from any other service conductors to minimize the possibility of simultaneous interruption of supply.

(E) Fuel Cell System. Fuel cell systems used as a source of power for emergency systems shall be of suitable rating and capacity to supply and maintain the total load for not less than 2 hours of full-demand operation.

Installation of a fuel cell system shall meet the requirements of Parts II through VIII of Article 692.

Where a single fuel cell system serves as the normal supply for the building or group of buildings concerned, it shall not serve as the sole source of power for the emergency standby system.

(F) Unit Equipment.

(1) Components of Unit Equipment. Individual unit equipment for emergency illumination shall consist of the following:

(1) A rechargeable battery
(2) A battery charging means
(3) Provisions for one or more lamps mounted on the equipment, or shall be permitted to have terminals for remote lamps, or both
(4) A relaying device arranged to energize the lamps automatically upon failure of the supply to the unit equipment

(2) Installation of Unit Equipment. Unit equipment shall be installed in accordance with 700.12(F)(2)(1) through (6).

(1) The batteries shall be of suitable rating and capacity to supply and maintain the total lamp load associated with the unit in accordance with (a) or (b):

(a) For a period of at least 1½ hours without the voltage falling below 87½ percent of normal battery voltage

(b) The unit equipment shall supply and maintain not less than 60 percent of the initial emergency illumination for a period of at least 1½ hours

(2) Unit equipment shall be permanently fixed (i.e., not portable) in place and shall have all wiring to each unit installed in accordance with the requirements of any of the wiring methods in Chapter 3. Flexible cord-and-plug connection shall be permitted, provided that the cord does not exceed 900 mm (3 ft) in length.

(3) The branch circuit feeding the unit equipment shall be the same branch circuit as that serving the normal lighting in the area and connected ahead of any local switches.

Exception: In a separate and uninterrupted area supplied by a minimum of three normal lighting circuits that are not part of a multiwire branch circuit, a separate branch circuit for unit equipment shall be permitted if it originates from the same panelboard as that of the normal lighting circuits and is provided with a lock-on feature.

(4) The branch circuit that feeds unit equipment shall be clearly identified at the distribution panel.
(5) Emergency luminaires that obtain power from a unit equipment and are not part of the unit equipment shall be wired to the unit equipment as required by 700.10 and by one of the wiring methods of Chapter 3.
(6) Remote heads providing lighting for the exterior of an exit door shall be permitted to be supplied by the unit equipment serving the area immediately inside the exit door.

PART IV. EMERGENCY SYSTEM CIRCUITS FOR LIGHTING AND POWER

700.15 Loads on Emergency Branch Circuits. No appliances and no lamps, other than those specified as required for emergency use, shall be supplied by emergency lighting circuits.

700.16 Emergency Illumination. Emergency illumination shall include means of egress lighting, illuminated exit signs, and all other luminaires specified as necessary to provide required illumination.

Emergency lighting systems shall be designed and installed so that the failure of any individual lighting element, such as the burning out of a lamp, cannot leave in total darkness any space that requires emergency illumination.

Where high-intensity discharge lighting such as high- and low-pressure sodium, mercury vapor, and metal halide is used as the sole source of normal illumination, the emergency lighting system shall be required to operate until normal illumination has been restored.

Where an emergency system is installed, emergency illumination shall be provided in the area of the disconnecting means required by 225.31 and 230.70, as applicable, where the disconnecting means are installed indoors.

Exception: Alternative means that ensure that the emergency lighting illumination level is maintained shall be permitted.

700.17 Branch Circuits for Emergency Lighting. Branch circuits that supply emergency lighting shall be installed to provide service from a source complying with 700.12 when the normal supply for lighting is interrupted. Such installations shall provide either of the following:

(1) An emergency lighting supply, independent of the normal lighting supply, with provisions for automatically transferring the emergency lights upon the event of failure of the normal lighting branch circuit
(2) Two or more branch circuits supplied from separate and complete systems with independent power sources. One of the two power sources and systems shall be part of the emergency system, and the other shall be permitted to be part of the normal power source and system. Each system shall provide sufficient power for emergency lighting purposes.

    Unless both systems are used for regular lighting purposes and are both kept lighted, means shall be provided for automatically energizing either system upon failure of the other. Either or both systems shall be permitted to be a part of the general lighting of the protected occupancy if circuits supplying lights for emergency illumination are installed in accordance with other sections of this article.

700.18 Circuits for Emergency Power. For branch circuits that supply equipment classed as emergency, there shall be an emergency supply source to which the load will be transferred automatically upon the failure of the normal supply.

700.19 Multiwire Branch Circuits. The branch circuit serving emergency lighting and power circuits shall not be part of a multiwire branch circuit.

PART V. CONTROL — EMERGENCY LIGHTING CIRCUITS

700.20 Switch Requirements. The switch or switches installed in emergency lighting circuits shall be arranged so that only authorized persons have control of emergency lighting.

Exception No. 1: Where two or more single-throw switches are connected in parallel to control a single circuit, at least one of these switches shall be accessible only to authorized persons.

Exception No. 2: Additional switches that act only to put emergency lights into operation but not disconnect them shall be permissible.

Switches connected in series or 3- and 4-way switches shall not be used.

700.21 Switch Location. All manual switches for controlling emergency circuits shall be in locations convenient to authorized persons responsible for their actuation. In facilities covered by Articles 518 and 520, a switch for controlling emergency lighting systems shall be located in the lobby or at a place conveniently accessible thereto.

In no case shall a control switch for emergency lighting be placed in a motion-picture projection booth or on a stage or platform.

Exception: Where multiple switches are provided, one such switch shall be permitted in such locations where arranged so that it can only energize the circuit but cannot de-energize the circuit.

700.22 Exterior Lights. Those lights on the exterior of a building that are not required for illumination when there is sufficient daylight shall be permitted to be controlled by an automatic light-actuated device.

700.23 Dimmer and Relay Systems. A dimmer or relay system containing more than one dimmer or relay and listed for use in emergency systems shall be permitted to be used as a control device for energizing emergency lighting circuits. Upon failure of normal power, the dimmer or relay system shall be permitted to selectively energize only those branch circuits required to provide minimum emergency illumination. All branch circuits supplied by the dimmer or relay system cabinet shall comply with the wiring methods of Article 700.

700.24 Directly Controlled Luminaires. Where emergency illumination is provided by one or more directly controlled luminaires that respond to an external control input to bypass normal control upon loss of normal power, such luminaires and external bypass controls shall be individually listed for use in emergency systems.

N **700.25 Branch Circuit Emergency Lighting Transfer Switch.** Emergency lighting loads supplied by branch circuits rated at not greater than 20 amperes shall be permitted to be transferred from the normal branch circuit to an emergency branch circuit using a listed branch circuit emergency lighting transfer switch. The mechanically held requirement of 700.5(C) shall not apply to listed branch circuit emergency lighting transfer switches.

700.26 Automatic Load Control Relay. If an emergency lighting load is automatically energized upon loss of the normal supply, a listed automatic load control relay shall be permitted to energize the load. The load control relay shall not be used as transfer equipment.

PART VI. OVERCURRENT PROTECTION

700.30 Accessibility. The branch-circuit overcurrent devices in emergency circuits shall be accessible to authorized persons only.

700.31 Ground-Fault Protection of Equipment. The alternate source for emergency systems shall not be required to provide ground-fault protection of equipment with automatic disconnecting means. Ground-fault indication at the emergency source shall be provided in accordance with 700.6(D) if ground-fault protection of equipment with automatic disconnecting means is not provided.

700.32 Selective Coordination. Emergency system(s) overcurrent devices shall be selectively coordinated with all supply-side overcurrent protective devices.

Selective coordination shall be selected by a licensed professional engineer or other qualified persons engaged primarily in the design, installation, or maintenance of electrical systems. The selection shall be documented and made available to those authorized to design, install, inspect, maintain, and operate the system.

Exception: Selective coordination shall not be required between two overcurrent devices located in series if no loads are connected in parallel with the downstream device.

CHAPTER 26

OPTIONAL STANDBY SYSTEMS

INTRODUCTION

This chapter contains requirements from Article 702 — Optional Standby Systems. The requirements of Article 702 cover the installation and operation of optional standby systems which are used to provide power for occupant comfort, business continuity, or both and are installed at the owner's discretion rather than by a codified or AHJ mandate to provide for occupant or emergency responder safety, which is the function of the systems covered in Articles 700 — Emergency Systems and 701 — Legally Required Standby Systems. Article 702 comprises Part I General, and Part II Wiring. The requirements of Article 702 cover the capacity and rating of the system and alternate source, transfer equipment, conductor installation and routing, alternate power source connections to the premises wiring system, and disconnecting means for standby generators located outdoors. Generators used to supply optional standby loads can also be used to supply emergency loads provided the generator either has capacity to fully supply all loads connected to it, or equipment is installed to shed loads in order that the emergency loads are given priority operational status. Additionally, a generator that is used to supply emergency loads must be able to provide the necessary level and quality of power to the emergency loads within ten seconds of loss of normal (typically utility) power. This requirement does not apply to generators that are used to supply only optional standby loads.

Article 702 requirements are organized into two parts: Part I — General, and Part II — Wiring. In contrast to the requirements in Article 700, it is not necessary to provide operational time, system segregation, and wiring resiliency requirements for optional standby systems. This is because the impact of an interruption of its operation may be inconvenient, but it does not place the occupant into the same level of peril and potential for mass panic that, for instance, could occur at a large assembly occupancy or high-rise building.

ARTICLE 702
Optional Standby Systems

PART I. GENERAL

702.1 Scope. The provisions of this article apply to the installation and operation of optional standby systems.

The systems covered by this article consist of those that are permanently installed in their entirety, including prime movers, and those that are arranged for a connection to a premises wiring system from a portable alternate power supply.

702.2 Definition.

Optional Standby Systems. Those systems intended to supply power to public or private facilities or property where life safety does not depend on the performance of the system. These systems are intended to supply on-site generated power to selected loads either automatically or manually.

Informational Note: Optional standby systems are typically installed to provide an alternate source of electric power for such facilities as industrial and commercial buildings, farms, and residences and to serve loads such as heating and refrigeration systems, data processing and communications systems, and industrial processes that, when stopped during any power outage, could cause discomfort, serious interruption of the process, damage to the product or process, or the like.

702.4 Capacity and Rating.

(A) Available Short-Circuit Current. Optional standby system equipment shall be suitable for the maximum available short-circuit current at its terminals.

(B) System Capacity. The calculations of load on the standby source shall be made in accordance with Article 220 or by another approved method.

(1) Manual Transfer Equipment. Where manual transfer equipment is used, an optional standby system shall have adequate capacity and rating for the supply of all equipment intended to be operated at one time. The user of the optional standby system shall be permitted to select the load connected to the system.

(2) Automatic Transfer Equipment. Where automatic transfer equipment is used, an optional standby system shall comply with (2)(a) or (2)(b).

(a) *Full Load.* The standby source shall be capable of supplying the full load that is transferred by the automatic transfer equipment.

(b) *Load Management.* Where a system is employed that will automatically manage the connected load, the standby source shall have a capacity sufficient to supply the maximum load that will be connected by the load management system.

702.5 Transfer Equipment. Transfer equipment shall be suitable for the intended use and designed and installed so as to prevent the inadvertent interconnection of normal and alternate sources of supply in any operation of the transfer equipment. Transfer equipment and electric power production systems installed to permit operation in parallel with the normal source shall meet the requirements of Article 705.

Transfer equipment, located on the load side of branch circuit protection, shall be permitted to contain supplemental overcurrent protection having an interrupting rating sufficient for the available fault current that the generator can deliver. The supplementary overcurrent protection devices shall be part of a listed transfer equipment.

Transfer equipment shall be required for all standby systems subject to the provisions of this article and for which an electric utility supply is either the normal or standby source.

Exception: Temporary connection of a portable generator without transfer equipment shall be permitted where conditions of maintenance and supervision ensure that only qualified persons service the installation and where the normal supply is physically isolated by a lockable disconnecting means or by disconnection of the normal supply conductors.

The short-circuit current rating of the transfer equipment, based on the specific overcurrent protective device type and settings protecting the transfer equipment, shall be field marked on the exterior of the transfer equipment.

702.6 Signals. Audible and visual signal devices shall be provided, where practicable, for the following purposes specified in 702.6(A) and (B).

(A) Malfunction. To indicate malfunction of the optional standby source.

(B) Carrying Load. To indicate that the optional standby source is carrying load.

Exception: Signals shall not be required for portable standby power sources.

702.7 Signs.

(A) Standby. A sign shall be placed at the service-entrance equipment that indicates the type and location of each on-site optional standby power source. A sign shall not be required for individual unit equipment for standby illumination.

(B) Grounding. Where removal of a grounding or bonding connection in normal power source equipment interrupts the grounding electrode conductor connection to the alternate power source(s) grounded conductor, a warning sign shall be installed at the normal power source equipment stating:

WARNING
SHOCK HAZARD EXISTS IF GROUNDING ELECTRODE CONDUCTOR OR BONDING JUMPER CONNECTION IN THIS EQUIPMENT IS REMOVED WHILE ALTERNATE SOURCE(S) IS ENERGIZED.

The warning sign(s) or label(s) shall comply with 110.21(B).

(C) Power Inlet. Where a power inlet is used for a temporary connection to a portable generator, a warning sign shall be placed near the inlet to indicate the type of derived system that the system is capable of based on the wiring of the transfer equipment. The sign shall display one of the following warnings:

WARNING:
FOR CONNECTION OF A SEPARATELY DERIVED (BONDED NEUTRAL) SYSTEM ONLY

or

WARNING:
FOR CONNECTION OF A NONSEPARATELY DERIVED
(FLOATING NEUTRAL) SYSTEM ONLY

PART II. WIRING

702.10 Wiring Optional Standby Systems. The optional standby system wiring shall be permitted to occupy the same raceways, cables, boxes, and cabinets with other general wiring.

702.11 Portable Generator Grounding.

(A) Separately Derived System. Where a portable optional standby source is used as a separately derived system, it shall be grounded to a grounding electrode in accordance with 250.30.

(B) Nonseparately Derived System. Where a portable optional standby source is used as a nonseparately derived system, the equipment grounding conductor shall be bonded to the system grounding electrode.

702.12 Outdoor Generator Sets.

(A) Portable Generators Greater Than 15 kW and Permanently Installed Generators. Where an outdoor housed generator set is equipped with a readily accessible disconnecting means in accordance with 445.18, and the disconnecting means is located within sight of the building or structure supplied, an additional disconnecting means shall not be required where ungrounded conductors serve or pass through the building or structure. Where the generator supply conductors terminate at a disconnecting means in or on a building or structure, the disconnecting means shall meet the requirements of 225.36.

(B) Portable Generators 15 kW or Less. Where a portable generator, rated 15 kW or less, is installed using a flanged inlet or other cord- and plug-type connection, a disconnecting means shall not be required where ungrounded conductors serve or pass through a building or structure.

N **(C) Power Inlets Rated at 100 Amperes or Greater, for Portable Generators.** Equipment containing power inlets for the connection of a generator source shall be listed for the intended use. Systems with power inlets shall be equipped with an interlocked disconnecting means.

Exception No. 1: If the inlet device is rated as a disconnecting means

Exception No. 2: Supervised industrial installations where permanent space is identified for the portable generator located within line of sight of the power inlets shall not be required to have interlocked disconnecting means nor inlets rated as disconnects.

CHAPTER 27

CONDUIT AND TUBING FILL TABLES

INTRODUCTION

This chapter contains tables and notes from Chapter 9 — Tables and from Annex C — Conduit and Tubing Fill Tables for Conductors and Fixture Wires of the Same Size. These tables are used to determine the maximum number of conductors permitted in raceways (conduit and tubing). Tables 4 and 5 of Chapter 9 are used to determine the minimum size cylindrical raceway based on the same size conductors being installed or on combinations of different size conductors being installed. Annex C is permitted to be used where all the conductors are the same size. It should be noted that in some cases the number of conductors specified for a given insulation type in Annex C will be less than that yielded through calculation using Tables 4 and 5. This is because Table 5 has been revised in recent years to reflect smaller nominal dimensions for several conductor insulation types. Table 8 — Conductor Properties (in Chapter 9) provides useful information such as size (AWG or kcmil), circular mils, overall area, and the resistance of copper and aluminum conductors. Where bare equipment grounding conductors are used, the dimensions from Table 8 can be used in calculating the minimum required raceway size.

Metal and nonmetallic wireways contained conductor fill requirements that cap the fill at 20 percent of the wireway cross-section. Using the wireway size and the conductor dimensions contained in Table 5, the wireway fill for all the same size conductors or for a mixture of conductor sizes can be calculated. Additionally, listed wireway will specify the maximum size conductor that can be installed. While 20 percent fill may seem to be quite restrictive, the advantage that metal wireway enjoys in comparison to cylindrical raceways like EMT, RMC, and Rigid PVC Conduit is the number of conductors that can be installed without having to apply ampacity adjustment factors. A metal wireway can contain up to 30 current-carrying conductors without having to reduce the conductor ampacity. Ampacity adjustment for cylindrical raceways is required where there are more than 3 current-carrying conductors.

CHAPTER 9 TABLES

Table 1 Percent of Cross Section of Conduit and Tubing for Conductors and Cables

Number of Conductors and/or Cables	Cross-Sectional Area (%)
1	53
2	31
Over 2	40

Informational Note No. 1: Table 1 is based on common conditions of proper cabling and alignment of conductors where the length of the pull and the number of bends are within reasonable limits. It should be recognized that, for certain conditions, a larger size conduit or a lesser conduit fill should be considered.

Informational Note No. 2: When pulling three conductors or cables into a raceway, if the ratio of the raceway (inside diameter) to the conductor or cable (outside diameter) is between 2.8 and 3.2, jamming can occur. While jamming can occur when pulling four or more conductors or cables into a raceway, the probability is very low.

Notes to Tables

(1) See Informative Annex C for the maximum number of conductors and fixture wires, all of the same size (total cross-sectional area including insulation) permitted in trade sizes of the applicable conduit or tubing.

(2) Table 1 applies only to complete conduit or tubing systems and is not intended to apply to sections of conduit or tubing used to protect exposed wiring from physical damage.

(3) Equipment grounding or bonding conductors, where installed, shall be included when calculating conduit or tubing fill. The actual dimensions of the equipment grounding or bonding conductor (insulated or bare) shall be used in the calculation.

(4) Where conduit or tubing nipples having a maximum length not to exceed 600 mm (24 in.) are installed between boxes, cabinets, and similar enclosures, the nipples shall be permitted to be filled to 60 percent of their total cross-sectional area, and 310.15(B)(3)(a) adjustment factors need not apply to this condition.

(5) For conductors not included in Chapter 9, such as multiconductor cables and optical fiber cables, the actual dimensions shall be used.
(6) For combinations of conductors of different sizes, use actual dimensions or Table 5 and Table 5A for dimensions of conductors and Table 4 for the applicable conduit or tubing dimensions.
(7) When calculating the maximum number of conductors or cables permitted in a conduit or tubing, all of the same size (total cross-sectional area including insulation), the next higher whole number shall be used to determine the maximum number of conductors permitted when the calculation results in a decimal greater than or equal to 0.8. When calculating the size for conduit or tubing permitted for a single conductor, one conductor shall be permitted when the calculation results in a decimal greater than or equal to 0.8.
(8) Where bare conductors are permitted by other sections of this *Code*, the dimensions for bare conductors in Table 8 shall be permitted.
(9) A multiconductor cable, optical fiber cable, or flexible cord of two or more conductors shall be treated as a single conductor for calculating percentage conduit or tubing fill area. For cables that have elliptical cross sections, the cross-sectional area calculation shall be based on using the major diameter of the ellipse as a circle diameter. Assemblies of single insulated conductors without an overall covering shall not be considered a cable when determining conduit or tubing fill area. The conduit or tubing fill for the assemblies shall be calculated based upon the individual conductors.
(10) The values for approximate conductor diameter and area shown in Table 5 are based on worst-case scenario and indicate round concentric-lay-stranded conductors. Solid and round concentric-lay-stranded conductor values are grouped together for the purpose of Table 5. Round compact-stranded conductor values are shown in Table 5A. If the actual values of the conductor diameter and area are known, they shall be permitted to be used.

Table 2 Radius of Conduit and Tubing Bends

Conduit or Tubing Size		One Shot and Full Shoe Benders		Other Bends	
Metric Designator	Trade Size	mm	in.	mm	in.
16	½	101.6	4	101.6	4
21	¾	114.3	4½	127	5
27	1	146.05	5¾	152.4	6
35	1¼	184.15	7¼	203.2	8
41	1½	209.55	8¼	254	10
53	2	241.3	9½	304.8	12
63	2½	266.7	10½	381	15
78	3	330.2	13	457.2	18
91	3½	381	15	533.4	21
103	4	406.4	16	609.6	24
129	5	609.6	24	762	30
155	6	762	30	914.4	36

Table 4 Dimensions and Percent Area of Conduit and Tubing (Areas of Conduit or Tubing for the Combinations of Wires Permitted in Table 1, Chapter 9)

Metric Designator	Trade Size	Over 2 Wires 40%		60%		1 Wire 53%		2 Wires 31%		Nominal Internal Diameter		Total Area 100%	
		mm^2	$in.^2$	mm^2	$in.^2$	mm^2	$in.^2$	mm^2	$in.^2$	mm	in.	mm^2	$in.^2$
Article 358 — Electrical Metallic Tubing (EMT)													
16	½	78	0.122	118	0.182	104	0.161	61	0.094	15.8	0.622	196	0.304
21	¾	137	0.213	206	0.320	182	0.283	106	0.165	20.9	0.824	343	0.533
27	1	222	0.346	333	0.519	295	0.458	172	0.268	26.6	1.049	556	0.864
35	1¼	387	0.598	581	0.897	513	0.793	300	0.464	35.1	1.380	968	1.496
41	1½	526	0.814	788	1.221	696	1.079	407	0.631	40.9	1.610	1314	2.036
53	2	866	1.342	1299	2.013	1147	1.778	671	1.040	52.5	2.067	2165	3.356
63	2½	1513	2.343	2270	3.515	2005	3.105	1173	1.816	69.4	2.731	3783	5.858
78	3	2280	3.538	3421	5.307	3022	4.688	1767	2.742	85.2	3.356	5701	8.846
91	3½	2980	4.618	4471	6.927	3949	6.119	2310	3.579	97.4	3.834	7451	11.545
103	4	3808	5.901	5712	8.852	5046	7.819	2951	4.573	110.1	4.334	9521	14.753

Article 362 — Electrical Nonmetallic Tubing (ENT)

16	½	73	0.114	110	0.171	97	0.151	57	0.088	15.3	0.602	184	0.285
21	¾	131	0.203	197	0.305	174	0.269	102	0.157	20.4	0.804	328	0.508
27	1	215	0.333	322	0.499	284	0.441	166	0.258	26.1	1.029	537	0.832
35	1¼	375	0.581	562	0.872	497	0.770	291	0.450	34.5	1.36	937	1.453
41	1½	512	0.794	769	1.191	679	1.052	397	0.616	40.4	1.59	1281	1.986
53	2	849	1.316	1274	1.975	1125	1.744	658	1.020	52	2.047	2123	3.291
63	2½	—	—	—	—	—	—	—	—	—	—	—	—
78	3	—	—	—	—	—	—	—	—	—	—	—	—
91	3½	—	—	—	—	—	—	—	—	—	—	—	—

Article 348 — Flexible Metal Conduit (FMC)

12	⅜	30	0.046	44	0.069	39	0.061	23	0.036	9.7	0.384	74	0.116
16	½	81	0.127	122	0.190	108	0.168	63	0.098	16.1	0.635	204	0.317
21	¾	137	0.213	206	0.320	182	0.283	106	0.165	20.9	0.824	343	0.533
27	1	211	0.327	316	0.490	279	0.433	163	0.253	25.9	1.020	527	0.817
35	1¼	330	0.511	495	0.766	437	0.677	256	0.396	32.4	1.275	824	1.277
41	1½	480	0.743	720	1.115	636	0.985	372	0.576	39.1	1.538	1201	1.858
53	2	843	1.307	1264	1.961	1117	1.732	653	1.013	51.8	2.040	2107	3.269
63	2½	1267	1.963	1900	2.945	1678	2.602	982	1.522	63.5	2.500	3167	4.909
78	3	1824	2.827	2736	4.241	2417	3.746	1414	2.191	76.2	3.000	4560	7.069

(continued)

Table 4 Dimensions and Percent Area of Conduit and Tubing (Areas of Conduit or Tubing for the Combinations of Wires Permitted in Table 1, Chapter 9) ***Continued***

Metric Designator	Trade Size	Over 2 Wires 40%		60%		1 Wire 53%		2 Wires 31%		Nominal Internal Diameter		Total Area 100%	
		mm²	in.²	mm²	in.²	mm²	in.²	mm²	in.²	mm	in.	mm²	in.²
91	3½	2483	3.848	3724	5.773	3290	5.099	1924	2.983	88.9	3.500	6207	9.621
103	4	3243	5.027	4864	7.540	4297	6.660	2513	3.896	101.6	4.000	8107	12.566
Article 342 — Intermediate Metal Conduit (IMC)													
12	⅜	—	—	—	—	—	—	—	—	—	—	—	—
16	½	89	0.137	133	0.205	117	0.181	69	0.106	16.8	0.660	222	0.342
21	¾	151	0.235	226	0.352	200	0.311	117	0.182	21.9	0.864	377	0.586
27	1	248	0.384	372	0.575	329	0.508	192	0.297	28.1	1.105	620	0.959
35	1¼	425	0.659	638	0.988	564	0.873	330	0.510	36.8	1.448	1064	1.647
41	1½	573	0.890	859	1.335	759	1.179	444	0.690	42.7	1.683	1432	2.225
53	2	937	1.452	1405	2.178	1241	1.924	726	1.125	54.6	2.150	2341	3.630
63	2½	1323	2.054	1985	3.081	1753	2.722	1026	1.592	64.9	2.557	3308	5.135
78	3	2046	3.169	3069	4.753	2711	4.199	1586	2.456	80.7	3.176	5115	7.922
91	3½	2729	4.234	4093	6.351	3616	5.610	2115	3.281	93.2	3.671	6822	10.584
103	4	3490	5.452	5235	8.179	4624	7.224	2705	4.226	105.4	4.166	8725	13.631

Article 356 — Liquidtight Flexible Nonmetallic Conduit (LFNC-A*)

12	⅜	50	0.077	75	0.115	66	0.102	39	0.060	12.6	0.495	125	0.192
16	½	80	0.125	121	0.187	107	0.165	62	0.097	16.0	0.630	201	0.312
21	¾	139	0.214	208	0.321	184	0.283	107	0.166	21.0	0.825	346	0.535
27	1	221	0.342	331	0.513	292	0.453	171	0.265	26.5	1.043	552	0.854
35	1¼	387	0.601	581	0.901	513	0.796	300	0.466	35.1	1.383	968	1.502
41	1½	520	0.807	781	1.211	690	1.070	403	0.626	40.7	1.603	1301	2.018
53	2	863	1.337	1294	2.006	1143	1.772	669	1.036	52.4	2.063	2157	3.343

*Corresponds to 356.2(1).

Article 356 — Liquidtight Flexible Nonmetallic Conduit (LFNC-B*)

12	⅜	49	0.077	74	0.115	65	0.102	38	0.059	12.5	0.494	123	0.192
16	½	81	0.125	122	0.188	108	0.166	63	0.097	16.1	0.632	204	0.314
21	¾	140	0.216	210	0.325	185	0.287	108	0.168	21.1	0.830	350	0.541
27	1	226	0.349	338	0.524	299	0.462	175	0.270	26.8	1.054	564	0.873
35	1¼	394	0.611	591	0.917	522	0.810	305	0.474	35.4	1.395	984	1.528
41	1½	510	0.792	765	1.188	676	1.050	395	0.614	40.3	1.588	1276	1.981
53	2	836	1.298	1255	1.948	1108	1.720	648	1.006	51.6	2.033	2091	3.246

*Corresponds to 356.2(2).

(continued)

Table 4 Dimensions and Percent Area of Conduit and Tubing (Areas of Conduit or Tubing for the Combinations of Wires Permitted in Table 1, Chapter 9) ***Continued***

Metric Designator	Trade Size	Over 2 Wires 40%		60%		1 Wire 53%		2 Wires 31%		Nominal Internal Diameter		Total Area 100%	
		mm²	in.²	mm²	in.²	mm²	in.²	mm²	in.²	mm	in.	mm²	in.²
N **Article 356 — Liquidtight Flexible Nonmetallic Conduit (LFNC-C*)**													
12	⅜	47.7	0.074	71.5	0.111	63.2	0.098	36.9	0.057	12.3	0.485	119.19	0.185
16	½	77.9	0.121	116.9	0.181	103.2	0.160	60.4	0.094	15.7	0.620	194.778	0.302
21	¾	134.6	0.209	201.9	0.313	178.4	0.276	104.3	0.162	20.7	0.815	336.568	0.522
27	1	215.0	0.333	322.5	0.500	284.9	0.442	166.6	0.258	26.2	1.030	537.566	0.833
35	1¼	380.4	0.590	570.6	0.884	504.1	0.781	294.8	0.457	34.8	1.370	951.039	1.474
41	1½	509.2	0.789	763.8	1.184	674.7	1.046	394.6	0.612	40.3	1.585	1272.963	1.973
53	2	847.6	1.314	1271.4	1.971	1123.1	1.741	656.9	1.018	51.9	2.045	2119.063	3.285

*Corresponds to 356.2(3).

Metric Designator	Trade Size	Over 2 Wires 40% mm²	Over 2 Wires 40% in.²	60% mm²	60% in.²	1 Wire 53% mm²	1 Wire 53% in.²	2 Wires 31% mm²	2 Wires 31% in.²	Nominal Internal Diameter mm	Nominal Internal Diameter in.	Total Area 100% mm²	Total Area 100% in.²
Article 350 — Liquidtight Flexible Metal Conduit (LFMC)													
12	⅜	49	0.077	74	0.115	65	0.102	38	0.059	12.5	0.494	123	0.192
16	½	81	0.125	122	0.188	108	0.166	63	0.097	16.1	0.632	204	0.314

21	¾	140	0.216	210	0.325	185	0.287	108	0.168	21.1	0.830	350	0.541
27	1	226	0.349	338	0.524	299	0.462	175	0.270	26.8	1.054	564	0.873
35	1¼	394	0.611	591	0.917	522	0.810	305	0.474	35.4	1.395	984	1.528
41	1½	510	0.792	765	1.188	676	1.050	395	0.614	40.3	1.588	1276	1.981
53	2	836	1.298	1255	1.948	1108	1.720	648	1.006	51.6	2.033	2091	3.246
63	2½	1259	1.953	1888	2.929	1668	2.587	976	1.513	63.3	2.493	3147	4.881
78	3	1931	2.990	2896	4.485	2559	3.962	1497	2.317	78.4	3.085	4827	7.475
91	1½	2511	3.893	3766	5.839	3327	5.158	1946	3.017	89.4	3.520	6277	9.731
103	4	3275	5.077	4912	7.615	4339	6.727	2538	3.935	102.1	4.020	8187	12.692
129	5	—	—	—	—	—	—	—	—	—	—	—	—
155	6	—	—	—	—	—	—	—	—	—	—	—	—
Article 344 — Rigid Metal Conduit (RMC)													
12	⅜	—	—	—	—	—	—	—	—	—	—	—	—
16	½	81	0.125	122	0.188	108	0.166	63	0.097	16.1	0.632	204	0.314
21	¾	141	0.220	212	0.329	187	0.291	109	0.170	21.2	0.836	353	0.549
27	1	229	0.355	344	0.532	303	0.470	177	0.275	27.0	1.063	573	0.887
35	1¼	394	0.610	591	0.916	522	0.809	305	0.473	35.4	1.394	984	1.526
41	1½	533	0.829	800	1.243	707	1.098	413	0.642	41.2	1.624	1333	2.071
53	2	879	1.363	1319	2.045	1165	1.806	681	1.056	52.9	2.083	2198	3.408
63	2½	1255	1.946	1882	2.919	1663	2.579	972	1.508	63.2	2.489	3137	4.866
78	3	1936	3.000	2904	4.499	2565	3.974	1500	2.325	78.5	3.090	4840	7.499

(continued)

Table 4 Dimensions and Percent Area of Conduit and Tubing (Areas of Conduit or Tubing for the Combinations of Wires Permitted in Table 1, Chapter 9) ***Continued***

Metric Designator	Trade Size	Over 2 Wires 40%		60%		1 Wire 53%		2 Wires 31%		Nominal Internal Diameter		Total Area 100%	
		mm²	in.²	mm²	in.²	mm²	in.²	mm²	in.²	mm	in.	mm²	in.²
91	3½	2584	4.004	3877	6.006	3424	5.305	2003	3.103	90.7	3.570	6461	10.010
103	4	3326	5.153	4990	7.729	4408	6.828	2578	3.994	102.9	4.050	8316	12.882
129	5	5220	8.085	7830	12.127	6916	10.713	4045	6.266	128.9	5.073	13050	20.212
155	6	7528	11.663	11292	17.495	9975	15.454	5834	9.039	154.8	6.093	18821	29.158
Article 352 — Rigid PVC Conduit (PVC), Schedule 80													
12	⅜	—	—	—	—	—	—	—	—	—	—	—	—
16	½	56	0.087	85	0.130	75	0.115	44	0.067	13.4	0.526	141	0.217
21	¾	105	0.164	158	0.246	139	0.217	82	0.127	18.3	0.722	263	0.409
27	1	178	0.275	267	0.413	236	0.365	138	0.213	23.8	0.936	445	0.688
35	1¼	320	0.495	480	0.742	424	0.656	248	0.383	31.9	1.255	799	1.237
41	1½	442	0.684	663	1.027	585	0.907	342	0.530	37.5	1.476	1104	1.711
53	2	742	1.150	1113	1.725	983	1.523	575	0.891	48.6	1.913	1855	2.874
63	2½	1064	1.647	1596	2.471	1410	2.183	825	1.277	58.2	2.290	2660	4.119

78	3	1660	2.577	2491	3.865	2200	3.414	1287	1.997	72.7	2.864	4151	6.442
91	3½	2243	3.475	3365	5.213	2972	4.605	1738	2.693	84.5	3.326	5608	8.688
103	4	2907	4.503	4361	6.755	3852	5.967	2253	3.490	96.2	3.786	7268	11.258
129	5	4607	7.142	6911	10.713	6105	9.463	3571	5.535	121.1	4.768	11518	17.855
155	6	6605	10.239	9908	15.359	8752	13.567	5119	7.935	145.0	5.709	16513	25.598
Articles 352 and 353 — Rigid PVC Conduit (PVC), Schedule 40, and HDPE Conduit (HDPE)													
12	⅜	—	—	—	—	—	—	—	—	—	—	—	—
16	½	74	0.114	110	0.171	97	0.151	57	0.088	15.3	0.602	184	0.285
21	¾	131	0.203	196	0.305	173	0.269	101	0.157	20.4	0.804	327	0.508
27	1	214	0.333	321	0.499	284	0.441	166	0.258	26.1	1.029	535	0.832
35	1¼	374	0.581	561	0.872	495	0.770	290	0.450	34.5	1.360	935	1.453
41	1½	513	0.794	769	1.191	679	1.052	397	0.616	40.4	1.590	1282	1.986
53	2	849	1.316	1274	1.975	1126	1.744	658	1.020	52.0	2.047	2124	3.291
63	2½	1212	1.878	1817	2.817	1605	2.488	939	1.455	62.1	2.445	3029	4.695
78	3	1877	2.907	2816	4.361	2487	3.852	1455	2.253	77.3	3.042	4693	7.268
91	3½	2511	3.895	3766	5.842	3327	5.161	1946	3.018	89.4	3.521	6277	9.737
103	4	3237	5.022	4855	7.532	4288	6.654	2508	3.892	101.5	3.998	8091	12.554
129	5	5099	7.904	7649	11.856	6756	10.473	3952	6.126	127.4	5.016	12748	19.761
155	6	7373	11.427	11060	17.140	9770	15.141	5714	8.856	153.2	6.031	18433	28.567

Table 5 Dimensions of Insulated Conductors and Fixture Wires

Type	Size (AWG or kcmil)	Approximate Area		Approximate Diameter	
		mm²	in.²	mm	in.
Type: FFH-2, RFH-1, RFH-2, RFHH-2, RHH*, RHW*, RHW-2*, RHH, RHW, RHW-2, SF-1, SF-2, SFF-1, SFF-2, TF, TFF, THHW, THW, THW-2, TW, XF, XFF					
RFH-2, FFH-2, RFHH-2	18	9.355	0.0145	3.454	0.136
	16	11.10	0.0172	3.759	0.148
RHH, RHW, RHW-2	14	18.90	0.0293	4.902	0.193
	12	22.77	0.0353	5.385	0.212
	10	28.19	0.0437	5.994	0.236
	8	53.87	0.0835	8.280	0.326
	6	67.16	0.1041	9.246	0.364
	4	86.00	0.1333	10.46	0.412
	3	98.13	0.1521	11.18	0.440
	2	112.9	0.1750	11.99	0.472
	1	171.6	0.2660	14.78	0.582
	1/0	196.1	0.3039	15.80	0.622
	2/0	226.1	0.3505	16.97	0.668

	3/0	262.7	0.4072	18.29	0.720
	4/0	306.7	0.4754	19.76	0.778
	250	405.9	0.6291	22.73	0.895
	300	457.3	0.7088	24.13	0.950
	350	507.7	0.7870	25.43	1.001
	400	556.5	0.8626	26.62	1.048
	500	650.5	1.0082	28.78	1.133
	600	782.9	1.2135	31.57	1.243
	700	874.9	1.3561	33.38	1.314
	750	920.8	1.4272	34.24	1.348
	800	965.0	1.4957	35.05	1.380
	900	1057	1.6377	36.68	1.444
	1000	1143	1.7719	38.15	1.502
	1250	1515	2.3479	43.92	1.729
	1500	1738	2.6938	47.04	1.852
	1750	1959	3.0357	49.94	1.966
	2000	2175	3.3719	52.63	2.072

(continued)

Table 5 Dimensions of Insulated Conductors and Fixture Wires ***Continued***

Type	Size (AWG or kcmil)	Approximate Area		Approximate Diameter	
		mm²	in.²	mm	in.
Type: FFH-2, RFH-1, RFH-2, RFHH-2, RHH*, RHW*, RHW-2*, RHH, RHW, RHW-2, SF-1, SF-2, SFF-1, SFF-2, TF, TFF, THHW, THW, THW-2, TW, XF, XFF ***Continued***					
SF-2, SFF-2	18	7.419	0.0115	3.073	0.121
	16	8.968	0.0139	3.378	0.133
	14	11.10	0.0172	3.759	0.148
SF-1, SFF-1	18	4.194	0.0065	2.311	0.091
RFH-1, TF, TFF, XF, XFF	18	5.161	0.0088	2.692	0.106
TF, TFF, XF, XFF	16	7.032	0.0109	2.997	0.118
TW, XF, XFF, THHW, THW, THW-2	14	8.968	0.0139	3.378	0.133
TW, THHW, THW, THW-2	12	11.68	0.0181	3.861	0.152
	10	15.68	0.0243	4.470	0.176
	8	28.19	0.0437	5.994	0.236

RHH*, RHW*, RHW-2*	14	13.48	0.0209	4.140	0.163
RHH*, RHW*, RHW-2*, XF, XFF	12	16.77	0.0260	4.623	0.182
Type: RHH*, RHW*, RHW-2*, THHN, THHW, THW, THW-2, TFN, TFFN, THWN, THWN-2, XF, XFF					
RHH*, RHW*, RHW-2*, XF, XFF	10	21.48	0.0333	5.232	0.206
RHH*, RHW*, RHW-2*	8	35.87	0.0556	6.756	0.266
TW, THW, THHW, THW-2, RHH*, RHW*, RHW-2*	6	46.84	0.0726	7.722	0.304
	4	62.77	0.0973	8.941	0.652
	3	73.16	0.1134	9.652	0.380
	2	86.00	0.1333	10.46	0.412
	1	122.6	0.1901	12.50	0.492
	1/0	143.4	0.2223	13.51	0.532
	2/0	169.3	0.2624	14.68	0.578
	3/0	201.1	0.3117	16.00	0.630
	4/0	239.9	0.3718	17.48	0.688
	250	296.5	0.4596	19.43	0.765
	300	340.7	0.5281	20.83	0.820

(continued)

Table 5 Dimensions of Insulated Conductors and Fixture Wires ***Continued***

Type	Size (AWG or kcmil)	Approximate Area		Approximate Diameter	
		mm²	in.²	mm	in.
Type: RHH*, RHW*, RHW-2*, THHN, THHW, THW, THW-2, TFN, TFFN, THWN, THWN-2, XF, XFF ***Continued***					
	350	384.4	0.5958	22.12	0.871
	400	427.0	0.6619	23.32	0.918
	500	509.7	0.7901	25.48	1.003
	600	627.7	0.9729	28.27	1.113
	700	710.3	1.1010	30.07	1.184
	750	751.7	1.1652	30.94	1.218
	800	791.7	1.2272	31.75	1.250
	900	874.9	1.3561	33.38	1.314
	1000	953.8	1.4784	34.85	1.372
	1250	1200	1.8602	39.09	1.539
	1500	1400	2.1695	42.21	1.662
	1750	1598	2.4773	45.11	1.776
	2000	1795	2.7818	47.80	1.882

TFN, TFFN	18	3.548	0.0055	2.134	0.084
	16	4.645	0.0072	2.438	0.096
THHN, THWN, THWN-2	14	6.258	0.0097	2.819	0.111
	12	8.581	0.0133	3.302	0.130
	10	13.61	0.0211	4.166	0.164
	8	23.61	0.0366	5.486	0.216
	6	32.71	0.0507	6.452	0.254
	4	53.16	0.0824	8.230	0.324
	3	62.77	0.0973	8.941	0.352
	2	74.71	0.1158	9.754	0.384
	1	100.8	0.1562	11.33	0.446
	1/0	119.7	0.1855	12.34	0.486
	2/0	143.4	0.2223	13.51	0.532
	3/0	172.8	0.2679	14.83	0.584
	4/0	208.8	0.3237	16.31	0.642
	250	256.1	0.3970	18.06	0.711
	300	297.3	0.4608	19.46	0.766

(continued)

Table 5 Dimensions of Insulated Conductors and Fixture Wires ***Continued***

Type	Size (AWG or kcmil)	Approximate Area		Approximate Diameter	
		mm²	in.²	mm	in.
Type: FEP, FEPB, PAF, PAFF, PF PFA, PFAH, PFF, PGF, PGFF, PTF, PTFF, TFE, THHN, THWN, TWHN-2, Z, ZF, ZFF, ZHF					
THHN, THWN, THWN-2	350	338.2	0.5242	20.75	0.817
	400	378.3	0.5863	21.95	0.864
	500	456.3	0.7073	24.10	0.949
	600	559.7	0.8676	26.70	1.051
	700	637.9	0.9887	28.50	1.122
	750	677.2	1.0496	29.36	1.156
	800	715.2	1.1085	30.18	1.188
	900	794.3	1.2311	31.80	1.252
	1000	869.5	1.3478	33.27	1.310
PF, PGFF, PGF, PFF, PTF, PAF, PTFF, PAFF	18	3.742	0.0058	2.184	0.086
	16	4.839	0.0075	2.489	0.098
PF, PGFF, PGF, PFF, PTF, PAF, PTFF, PAFF, TFE, FEP, PFA, FEPB, PFAH	14	6.452	0.0100	2.870	0.113

TFE, FEP, PFA, FEPB, PFAH	12	8.839	0.0137	3.353	0.132
	10	12.32	0.0191	3.962	0.156
	8	21.48	0.0333	5.232	0.206
	6	30.19	0.0468	6.198	0.244
	4	43.23	0.0670	7.417	0.292
	3	51.87	0.0804	8.128	0.320
	2	62.77	0.0973	8.941	0.352
TFE, PFAH, PFA	1	90.26	0.1399	10.72	0.422
TFE, PFA, PFAH, Z	1/0	108.1	0.1676	11.73	0.462
	2/0	130.8	0.2027	12.90	0.508
	3/0	158.9	0.2463	14.22	0.560
	4/0	193.5	0.3000	15.70	0.618
ZF, ZFF, ZHF	18	2.903	0.0045	1.930	0.076
	16	3.935	0.0061	2.235	0.088
Z, ZF, ZFF, ZHF	14	5.355	0.0083	2.616	0.103
Z	12	7.548	0.0117	3.099	0.122
	10	12.32	0.0191	3.962	0.156
	8	19.48	0.0302	4.978	0.196

(continued)

Table 5 Dimensions of Insulated Conductors and Fixture Wires ***Continued***

Type	Size (AWG or kcmil)	Approximate Area		Approximate Diameter	
		mm²	in.²	mm	in.
Type: FEP, FEPB, PAF, PAFF, PF PFA, PFAH, PFF, PGF, PGFF, PTF, PTFF, TFE, THHN, THWN, TWHN-2, Z, ZF, ZFF, ZHF ***Continued***					
	6	27.74	0.0430	5.944	0.234
	4	40.32	0.0625	7.163	0.282
	3	55.16	0.0855	8.382	0.330
	2	66.39	0.1029	9.195	0.362
	1	81.87	0.1269	10.21	0.402
Type: KF-1, KF-2, KFF-1, KFF-2, XHH, XHHW, XHHW-2, ZW					
XHHW, ZW, XHHW-2, XHH	14	8.968	0.0139	3.378	0.133
	12	11.68	0.0181	3.861	0.152
	10	15.68	0.0243	4.470	0.176
	8	28.19	0.0437	5.994	0.236
	6	38.06	0.0590	6.960	0.274
	4	52.52	0.0814	8.179	0.322
	3	62.06	0.0962	8.890	0.350
	2	73.94	0.1146	9.703	0.382

XHHW, XHHW-2, XHH	1	98.97	0.1534	11.23	0.442
	1/0	117.7	0.1825	12.24	0.482
	2/0	141.3	0.2190	13.41	0.528
	3/0	170.5	0.2642	14.73	0.580
	4/0	206.3	0.3197	16.21	0.638
	250	251.9	0.3904	17.91	0.705
	300	292.6	0.4536	19.30	0.760
	350	333.3	0.5166	20.60	0.811
	400	373.0	0.5782	21.79	0.858
	500	450.6	0.6984	23.95	0.943
	600	561.9	0.8709	26.75	1.053
	700	640.2	0.9923	28.55	1.124
	750	679.5	1.0532	29.41	1.158
	800	717.5	1.1122	30.23	1.190
	900	796.8	1.2351	31.85	1.254
	1000	872.2	1.3519	33.32	1.312
	1250	1108.0	1.7180	37.57	1.479
	1500	1300.0	2.0156	40.69	1.602
	1750	1492.0	2.3127	43.59	1.716
	2000	1682.0	2.6073	46.28	1.822

(continued)

Table 5 Dimensions of Insulated Conductors and Fixture Wires ***Continued***

Type	Size (AWG or kcmil)	Approximate Area		Approximate Diameter	
		mm²	in.²	mm	in.
Type: KF-1, KF-2, KFF-1, KFF-2, XHH, XHHW, XHHW-2, ZW ***Continued***					
KF-2, KFF-2	18	2.000	0.003	1.575	0.062
	16	2.839	0.0043	1.88	0.074
	14	4.129	0.0064	2.286	0.090
	12	6.000	0.0092	2.743	0.108
	10	8.968	0.0139	3.378	0.133
KF-1, KFF-1	18	1.677	0.0026	1.448	0.057
	16	2.387	0.0037	1.753	0.069
	14	3.548	0.0055	2.134	0.084
	12	5.355	0.0083	2.616	0.103
	10	8.194	0.0127	3.226	0.127

*Types RHH, RHW, and RHW-2 without outer covering.

Table 5A Compact Copper and Aluminum Building Wire Nominal Dimensions* and Areas

Size (AWG or kcmil)	Bare Conductor Diameter		Types RHH**, RHW**, or USE Approximate Diameter		Types RHH**, RHW**, or USE Approximate Area		Types THW and THHW Approximate Diameter		Types THW and THHW Approximate Area		Type THHN Approximate Diameter		Type THHN Approximate Area		Type XHHW Approximate Diameter		Type XHHW Approximate Area		Size (AWG or kcmil)
	mm	in.	mm	in.	mm²	in.²	mm	in.	mm²	in.²	mm	in.	mm²	in.²	mm	in.	mm²	in.²	
8	3.404	0.134	6.604	0.260	34.25	0.0531	6.477	0.255	32.90	0.0510	—	—	—	—	5.690	0.224	25.42	0.0394	8
6	4.293	0.169	7.493	0.295	44.10	0.0683	7.366	0.290	42.58	0.0660	6.096	0.240	29.16	0.0452	6.604	0.260	34.19	0.0530	6
4	5.410	0.213	8.509	0.335	56.84	0.0881	8.509	0.335	56.84	0.0881	7.747	0.305	47.10	0.0730	7.747	0.305	47.10	0.0730	4
2	6.807	0.268	9.906	0.390	77.03	0.1194	9.906	0.390	77.03	0.1194	9.144	0.360	65.61	0.1017	9.144	0.360	65.61	0.1017	2
1	7.595	0.299	11.81	0.465	109.5	0.1698	11.81	0.465	109.5	0.1698	10.54	0.415	87.23	0.1352	10.54	0.415	87.23	0.1352	1
1/0	8.534	0.336	12.70	0.500	126.6	0.1963	12.70	0.500	126.6	0.1963	11.43	0.450	102.6	0.1590	11.43	0.450	102.6	0.1590	1/0
2/0	9.550	0.376	13.72	0.540	147.8	0.2290	13.84	0.545	150.5	0.2332	12.57	0.495	124.1	0.1924	12.45	0.490	121.6	0.1885	2/0
3/0	10.74	0.423	14.99	0.590	176.3	0.2733	14.99	0.590	176.3	0.2733	13.72	0.540	147.7	0.2290	13.72	0.540	147.7	0.2290	3/0
4/0	12.07	0.475	16.26	0.640	207.6	0.3217	16.38	0.645	210.8	0.3267	15.11	0.595	179.4	0.2780	14.99	0.590	176.3	0.2733	4/0

(continued)

Table 5A Compact Copper and Aluminum Building Wire Nominal Dimensions* and Areas *Continued*

Size (AWG or kcmil)	Bare Conductor		Types RHH**, RHW**, or USE				Types THW and THHW				Type THHN				Type XHHW				Size (AWG or kcmil)
	Diameter		Approximate Diameter		Approximate Area		Approximate Diameter		Approximate Area		Approximate Diameter		Approximate Area		Approximate Diameter		Approximate Area		
	mm	in.	mm	in.	mm^2	$in.^2$	mm	in.	mm^2	$in.^2$	mm	in.	mm^2	$in.^2$	mm	in.	mm^2	$in.^2$	
250	13.21	0.520	18.16	0.715	259.0	0.4015	18.42	0.725	266.3	0.4128	17.02	0.670	227.4	0.3525	16.76	0.660	220.7	0.3421	250
300	14.48	0.570	19.43	0.765	296.5	0.4596	19.69	0.775	304.3	0.4717	18.29	0.720	262.6	0.4071	18.16	0.715	259.0	0.4015	300
350	15.65	0.616	20.57	0.810	332.3	0.5153	20.83	0.820	340.7	0.5281	19.56	0.770	300.4	0.4656	19.30	0.760	292.6	0.4536	350
400	16.74	0.659	21.72	0.855	370.5	0.5741	21.97	0.865	379.1	0.5876	20.70	0.815	336.5	0.5216	20.32	0.800	324.3	0.5026	400
500	18.69	0.736	23.62	0.930	438.2	0.6793	23.88	0.940	447.7	0.6939	22.48	0.885	396.8	0.6151	22.35	0.880	392.4	0.6082	500
600	20.65	0.813	26.29	1.035	542.8	0.8413	26.67	1.050	558.6	0.8659	25.02	0.985	491.6	0.7620	24.89	0.980	486.6	0.7542	600
700	22.28	0.877	27.94	1.100	613.1	0.9503	28.19	1.110	624.3	0.9676	26.67	1.050	558.6	0.8659	26.67	1.050	558.6	0.8659	700
750	23.06	0.908	28.83	1.135	652.8	1.0118	29.21	1.150	670.1	1.0386	27.31	1.075	585.5	0.9076	27.69	1.090	602.0	0.9331	750
900	25.37	0.999	31.50	1.240	779.3	1.2076	31.09	1.224	759.1	1.1766	30.33	1.194	722.5	1.1196	29.69	1.169	692.3	1.0733	900
1000	26.92	1.060	32.64	1.285	836.6	1.2968	32.64	1.285	836.6	1.2968	31.88	1.255	798.1	1.2370	31.24	1.230	766.6	1.1882	1000

*Dimensions are from industry sources.

**Types RHH and RHW without outer coverings.

Table 8 Conductor Properties

Size (AWG or kcmil)	Area		Conductors							Direct-Current Resistance at 75°C (167°F)					
			Stranding			Overall				Copper					
				Diameter		Diameter		Area		Uncoated		Coated		Aluminum	
	mm²	Circular mils	Qty	mm	in.	mm	in.	mm²	in.²	ohm/km	ohm/kFT	ohm/km	ohm/kFT	ohm/km	ohm/kFT
18	0.823	1620	1	—	—	1.02	0.040	0.823	0.001	25.5	7.77	26.5	8.08	42.0	12.8
18	0.823	1620	7	0.39	0.015	1.16	0.046	1.06	0.002	26.1	7.95	27.7	8.45	42.8	13.1
16	1.31	2580	1	—	—	1.29	0.051	1.31	0.002	16.0	4.89	16.7	5.08	26.4	8.05
16	1.31	2580	7	0.49	0.019	1.46	0.058	1.68	0.003	16.4	4.99	17.3	5.29	26.9	8.21
14	2.08	4110	1	—	—	1.63	0.064	2.08	0.003	10.1	3.07	10.4	3.19	16.6	5.06
14	2.08	4110	7	0.62	0.024	1.85	0.073	2.68	0.004	10.3	3.14	10.7	3.26	16.9	5.17
12	3.31	6530	1	—	—	2.05	0.081	3.31	0.005	6.34	1.93	6.57	2.01	10.45	3.18
12	3.31	6530	7	0.78	0.030	2.32	0.092	4.25	0.006	6.50	1.98	6.73	2.05	10.69	3.25

(continued)

Table 8 Conductor Properties *Continued*

Size (AWG or kcmil)	Area		Conductors							Direct-Current Resistance at 75°C (167°F)					
			Stranding			Overall				Copper				Aluminum	
				Diameter		Diameter		Area		Uncoated		Coated			
	mm²	Circular mils	Qty	mm	in.	mm	in.	mm²	in.²	ohm/km	ohm/kFT	ohm/km	ohm/kFT	ohm/km	ohm/kFT
10	5.261	10380	1	—	—	2.588	0.102	5.26	0.008	3.984	1.21	4.148	1.26	6.561	2.00
10	5.261	10380	7	0.98	0.038	2.95	0.116	6.76	0.011	4.070	1.24	4.226	1.29	6.679	2.04
8	8.367	16510	1	—	—	3.264	0.128	8.37	0.013	2.506	0.764	2.579	0.786	4.125	1.26
8	8.367	16510	7	1.23	0.049	3.71	0.146	10.76	0.017	2.551	0.778	2.653	0.809	4.204	1.28
6	13.30	26240	7	1.56	0.061	4.67	0.184	17.09	0.027	1.608	0.491	1.671	0.510	2.652	0.808
4	21.15	41740	7	1.96	0.077	5.89	0.232	27.19	0.042	1.010	0.308	1.053	0.321	1.666	0.508
3	26.67	52620	7	2.20	0.087	6.60	0.260	34.28	0.053	0.802	0.245	0.833	0.254	1.320	0.403
2	33.62	66360	7	2.47	0.097	7.42	0.292	43.23	0.067	0.634	0.194	0.661	0.201	1.045	0.319
1	42.41	83690	19	1.69	0.066	8.43	0.332	55.80	0.087	0.505	0.154	0.524	0.160	0.829	0.253

1/0	53.49	105600	19	1.89	0.074	9.45	0.372	70.41	0.109	0.399	0.122	0.415	0.127	0.660	0.201
2/0	67.43	133100	19	2.13	0.084	10.62	0.418	88.74	0.137	0.3170	0.0967	0.329	0.101	0.523	0.159
3/0	85.01	167800	19	2.39	0.094	11.94	0.470	111.9	0.173	0.2512	0.0766	0.2610	0.0797	0.413	0.126
4/0	107.2	211600	19	2.68	0.106	13.41	0.528	141.1	0.219	0.1996	0.0608	0.2050	0.0626	0.328	0.100
250	127	—	37	2.09	0.082	14.61	0.575	168	0.260	0.1687	0.0515	0.1753	0.0535	0.2778	0.0847
300	152	—	37	2.29	0.090	16.00	0.630	201	0.312	0.1409	0.0429	0.1463	0.0446	0.2318	0.0707
350	177	—	37	2.47	0.097	17.30	0.681	235	0.364	0.1205	0.0367	0.1252	0.0382	0.1984	0.0605
400	203	—	37	2.64	0.104	18.49	0.728	268	0.416	0.1053	0.0321	0.1084	0.0331	0.1737	0.0529
500	253	—	37	2.95	0.116	20.65	0.813	336	0.519	0.0845	0.0258	0.0869	0.0265	0.1391	0.0424
600	304	—	61	2.52	0.099	22.68	0.893	404	0.626	0.0704	0.0214	0.0732	0.0223	0.1159	0.0353
700	355	—	61	2.72	0.107	24.49	0.964	471	0.730	0.0603	0.0184	0.0622	0.0189	0.0994	0.0303
750	390	—	61	2.82	0.111	25.35	0.998	505	0.782	0.0563	0.0171	0.0579	0.0176	0.0927	0.0282
800	405	—	61	2.91	0.114	26.16	1.030	538	0.834	0.0528	0.0161	0.0544	0.0166	0.0868	0.0265
900	456	—	61	3.09	0.122	27.79	1.094	606	0.940	0.0470	0.0143	0.0481	0.0147	0.0770	0.0235
1000	507	—	61	3.25	0.128	29.26	1.152	673	1.042	0.0423	0.0129	0.0434	0.0132	0.0695	0.0212
1250	633	—	91	2.98	0.117	32.74	1.289	842	1.305	0.0338	0.0103	0.0347	0.0106	0.0554	0.0169

(continued)

Table 8 Conductor Properties ***Continued***

Size (AWG or kcmil)	Conductors: Area mm²	Conductors: Area Circular mils	Stranding: Qty	Stranding: Diameter mm	Stranding: Diameter in.	Overall: Diameter mm	Overall: Diameter in.	Overall: Area mm²	Overall: Area in.²	DC Resistance at 75°C (167°F): Copper Uncoated ohm/km	Copper Uncoated ohm/kFT	Copper Coated ohm/km	Copper Coated ohm/kFT	Aluminum ohm/km	Aluminum ohm/kFT
1500	760	—	91	3.26	0.128	35.86	1.412	1011	1.566	0.02814	0.00858	0.02814	0.00883	0.0464	0.0141
1750	887	—	127	2.98	0.117	38.76	1.526	1180	1.829	0.02410	0.00735	0.02410	0.00756	0.0397	0.0121
2000	1013	—	127	3.19	0.126	41.45	1.632	1349	2.092	0.02109	0.00643	0.02109	0.00662	0.0348	0.0106

Notes:

1. These resistance values are valid **only** for the parameters as given. Using conductors having coated strands, different stranding type, and, especially, other temperatures changes the resistance.
2. Equation for temperature change: $R_2 = R_1 [1 + \alpha(T_2 - 75)]$, where $\alpha_{cu} = 0.00323$, $\alpha_{AL} = 0.00330$ at 75°C.
3. Conductors with compact and compressed stranding have about 9 percent and 3 percent, respectively, smaller bare conductor diameters than those shown. See Table 5A for actual compact cable dimensions.
4. The IACS conductivities used: bare copper = 100%, aluminum = 61%.
5. Class B stranding is listed as well as solid for some sizes. Its overall diameter and area are those of its circumscribing circle.

Table 10 Conductor Stranding

Conductor Size		Number of Strands		
		Copper		Aluminum
AWG or kcmil	mm²	Class B[a]	Class C	Class B[a]
24–30	0.20–0.05	[b]	—	—
22	0.32	7	—	—
20	0.52	10	—	—
18	0.82	16	—	—
16	1.3	26	—	—
14–2	2.1–33.6	7	19	7[c]
1–4/0	42.4–107	19	37	19
250–500	127–253	37	61	37
600–1000	304–508	61	91	61
1250–1500	635–759	91	127	91
1750–2000	886–1016	127	271	127

[a]Conductors with a lesser number of strands shall be permitted based on an evaluation for connectability and bending.

[b]Number of strands vary.

[c]Aluminum 14 AWG (2.1 mm²) is not available.

ANNEX C

Conduit and Tubing Fill Tables for Conductors and Fixture Wires of the Same Size

*Where this table is used in conjunction with Tables C.1 through C.13, the conductors installed must be of the compact type.

Table C.1 Maximum Number of Conductors or Fixture Wires in Electrical Metallic Tubing (EMT) (Based on Chapter 9: Table 1, Table 4, and Table 5)

Type	Conductor Size (AWG/ kcmil)	Trade Size (Metric Designator)												
		⅜ (12)	½ (16)	¾ (21)	1 (27)	1¼ (35)	1½ (41)	2 (53)	2½ (63)	3 (78)	3½ (91)	4 (103)	5 (129)	6 (155)
CONDUCTORS														
RHH, RHW, RHW-2	14	—	4	7	11	20	27	46	80	120	157	201	—	—
	12	—	3	6	9	17	23	38	66	100	131	167	—	—
	10	—	2	5	8	13	18	30	53	81	105	135	—	—
	8	—	1	2	4	7	9	16	28	42	55	70	—	—
	6	—	1	1	3	5	8	13	22	34	44	56	—	—
	4	—	1	1	2	4	6	10	17	26	34	44	—	—
	3	—	1	1	1	4	5	9	15	23	30	38	—	—
	2	—	1	1	1	3	4	7	13	20	26	33	—	—
	1	—	0	1	1	1	3	5	9	13	17	22	—	—
	1/0	—	0	1	1	1	2	4	7	11	15	19	—	—
	2/0	—	0	1	1	1	2	4	6	10	13	17	—	—
	3/0	—	0	0	1	1	1	3	5	8	11	14	—	—
	4/0	—	0	0	1	1	1	3	5	7	9	12	—	—
	250	—	0	0	0	1	1	1	3	5	7	9	—	—
	300	—	0	0	0	1	1	1	3	5	6	8	—	—
	350	—	0	0	0	1	1	1	3	4	6	7	—	—
	400	—	0	0	0	1	1	1	2	4	5	7	—	—
	500	—	0	0	0	0	1	1	2	3	4	6	—	—

(continued)

Table C.1 Maximum Number of Conductors or Fixture Wires in Electrical Metallic Tubing (EMT) (Based on *NEC* Chapter 9: Table 1, Table 4, and Table 5) ***Continued***

Type	Conductor Size (AWG/ kcmil)	Trade Size (Metric Designator)												
		⅜ (12)	½ (16)	¾ (21)	1 (27)	1¼ (35)	1½ (41)	2 (53)	2½ (63)	3 (78)	3½ (91)	4 (103)	5 (129)	6 (155)
		CONDUCTORS												
	600	—	0	0	0	0	1	1	1	3	4	5	—	—
	700	—	0	0	0	0	0	1	1	2	3	4	—	—
	750	—	0	0	0	0	0	1	1	2	3	4	—	—
	800	—	0	0	0	0	0	1	1	2	3	4	—	—
	900	—	0	0	0	0	0	1	1	1	3	3	—	—
	1000	—	0	0	0	0	0	1	1	1	2	3	—	—
	1250	—	0	0	0	0	0	0	1	1	1	2	—	—
	1500	—	0	0	0	0	0	0	1	1	1	1	—	—
	1750	—	0	0	0	0	0	0	1	1	1	1	—	—
	2000	—	0	0	0	0	0	0	1	1	1	1	—	—
TW, THHW, THW, THW-2	14	—	8	15	25	43	58	96	168	254	332	424	—	—
	12	—	6	11	19	33	45	74	129	195	255	326	—	—
	10	—	5	8	14	24	33	55	96	145	190	243	—	—
	8	—	2	5	8	13	18	30	53	81	105	135	—	—

RHH*, RHW*, RHW-2*	14	—	6	10	16	28	39	64	112	169	221	282	—	—
	12	—	4	8	13	23	31	51	90	136	177	227	—	—
	10	—	3	6	10	18	24	40	70	106	138	177	—	—
	8	—	1	4	6	10	14	24	42	63	83	106	—	—
TW, THW, THHW, THW-2, RHH*, RHW*, RHW-2*	6	—	1	3	4	8	11	18	32	48	63	81	—	—
	4	—	1	1	3	6	8	13	24	36	47	60	—	—
	3	—	1	1	3	5	7	12	20	31	40	52	—	—
	2	—	1	1	2	4	6	10	17	26	34	44	—	—
	1	—	1	1	1	3	4	7	12	18	24	31	—	—
	1/0	—	0	1	1	2	3	6	10	16	20	26	—	—
	2/0	—	0	1	1	1	3	5	9	13	17	22	—	—
	3/0	—	0	1	1	1	2	4	7	11	15	19	—	—
	4/0	—	0	0	1	1	1	3	6	9	12	16	—	—
	250	—	0	0	1	1	1	3	5	7	10	13	—	—
	300	—	0	0	1	1	1	2	4	6	8	11	—	—
	350	—	0	0	0	1	1	1	4	6	7	10	—	—
	400	—	0	0	0	1	1	1	3	5	7	9	—	—
	500	—	0	0	0	1	1	1	3	4	6	7	—	—
	600	—	0	0	0	1	1	1	2	3	4	6	—	—
	700	—	0	0	0	0	1	1	1	3	4	5	—	—
	750	—	0	0	0	0	1	1	1	3	4	5	—	—
	800	—	0	0	0	0	1	1	1	3	3	5	—	—
	900	—	0	0	0	0	0	1	1	2	3	4	—	—

(continued)

Table C.1 Maximum Number of Conductors or Fixture Wires in Electrical Metallic Tubing (EMT) (Based on *NEC* Chapter 9: Table 1, Table 4, and Table 5) ***Continued***

Type	Conductor Size (AWG/kcmil)	Trade Size (Metric Designator)												
		3/8 (12)	1/2 (16)	3/4 (21)	1 (27)	1¼ (35)	1½ (41)	2 (53)	2½ (63)	3 (78)	3½ (91)	4 (103)	5 (129)	6 (155)
CONDUCTORS														
	1000	—	0	0	0	0	0	1	1	2	3	4	—	—
	1250	—	0	0	0	0	0	1	1	1	2	3	—	—
	1500	—	0	0	0	0	0	1	1	1	1	2	—	—
	1750	—	0	0	0	0	0	0	1	1	1	2	—	—
	2000	—	0	0	0	0	0	0	1	1	1	1	—	—
THHN, THWN, THWN-2	14	—	12	22	35	61	84	138	241	364	476	608	—	—
	12	—	9	16	26	45	61	101	176	266	347	443	—	—
	10	—	5	10	16	28	38	63	111	167	219	279	—	—
	8	—	3	6	9	16	22	36	64	96	126	161	—	—
	6	—	2	4	7	12	16	26	46	69	91	116	—	—
	4	—	1	2	4	7	10	16	28	43	56	71	—	—
	3	—	1	1	3	6	8	13	24	36	47	60	—	—
	2	—	1	1	3	5	7	11	20	30	40	51	—	—
	1	—	1	1	1	4	5	8	15	22	29	37	—	—
	1/0	—	1	1	1	3	4	7	12	19	25	32	—	—
	2/0	—	0	1	1	2	3	6	10	16	20	26	—	—
	3/0	—	0	1	1	1	3	5	8	13	17	22	—	—
	4/0	—	0	1	1	1	2	4	7	11	14	18	—	—

	250	—	0	0	1	1	1	3	6	9	11	15	—	—
	300	—	0	0	1	1	1	3	5	7	10	13	—	—
	350	—	0	0	1	1	1	2	4	6	9	11	—	—
	400	—	0	0	0	1	1	1	4	6	8	10	—	—
	500	—	0	0	0	1	1	1	3	5	6	8	—	—
	600	—	0	0	0	1	1	1	2	4	5	7	—	—
	700	—	0	0	0	1	1	1	2	3	4	6	—	—
	750	—	0	0	0	0	1	1	1	3	4	5	—	—
	800	—	0	0	0	0	1	1	1	3	4	5	—	—
	900	—	0	0	0	0	1	1	1	3	3	4	—	—
	1000	—	0	0	0	0	1	1	1	2	3	4	—	—
FEP, FEPB, PFA, PFAH, TFE	14	—	12	21	34	60	81	134	234	354	462	590	—	—
	12	—	9	15	25	43	59	98	171	258	337	430	—	—
	10	—	6	11	18	31	42	70	122	185	241	309	—	—
	8	—	3	6	10	18	24	40	70	106	138	177	—	—
	6	—	2	4	7	12	17	28	50	75	98	126	—	—
	4	—	1	3	5	9	12	20	35	53	69	88	—	—
	3	—	1	2	4	7	10	16	29	44	57	73	—	—
	2	—	1	1	3	6	8	13	24	36	47	60	—	—
PFA, PFAH, TFE	1	—	1	1	2	4	6	9	16	25	33	42	—	—
PFA, PFAH, TFE, Z	1/0	—	1	1	1	3	5	8	14	21	27	35	—	—
	2/0	—	0	1	1	3	4	6	11	17	22	29	—	—
	3/0	—	0	1	1	2	3	5	9	14	18	24	—	—
	4/0	—	0	1	1	1	2	4	8	11	15	19	—	—

(continued)

Table C.1 Maximum Number of Conductors or Fixture Wires in Electrical Metallic Tubing (EMT) (Based on *NEC* Chapter 9: Table 1, Table 4, and Table 5) ***Continued***

Type	Conductor Size (AWG/ kcmil)	Trade Size (Metric Designator)												
		⅜ (12)	½ (16)	¾ (21)	1 (27)	1¼ (35)	1½ (41)	2 (53)	2½ (63)	3 (78)	3½ (91)	4 (103)	5 (129)	6 (155)
CONDUCTORS														
Z	14	—	14	25	41	72	98	161	282	426	556	711	—	—
	12	—	10	18	29	51	69	114	200	302	394	504	—	—
	10	—	6	11	18	31	42	70	122	185	241	309	—	—
	8	—	4	7	11	20	27	44	77	117	153	195	—	—
	6	—	3	5	8	14	19	31	54	82	107	137	—	—
	4	—	1	3	5	9	13	21	37	56	74	94	—	—
	3	—	1	2	4	7	9	15	27	41	54	69	—	—
	2	—	1	1	3	6	8	13	22	34	45	57	—	—
	1	—	1	1	2	4	6	10	18	28	36	46	—	—
XHHW, ZW, XHHW-2, XHH	14	—	8	15	25	43	58	96	168	254	332	424	—	—
	12	—	6	11	19	33	45	74	129	195	255	326	—	—
	10	—	5	8	14	24	33	55	96	145	190	243	—	—
	8	—	2	5	8	13	18	30	53	81	105	135	—	—
	6	—	1	3	6	10	14	22	39	60	78	100	—	—
	4	—	1	2	4	7	10	16	28	43	56	72	—	—
	3	—	1	1	3	6	8	14	24	36	48	61	—	—
	2	—	1	1	3	5	7	11	20	31	40	51	—	—

XHHW, XHHW-2, XHH	1	—	1	1	1	4	5	8	15	23	30	38	—	—
	1/0	—	1	1	1	3	4	7	13	19	25	32	—	—
	2/0	—	0	1	1	2	3	6	10	16	21	27	—	—
	3/0	—	0	1	1	1	3	5	9	13	17	22	—	—
	4/0	—	0	1	1	1	2	4	7	11	14	18	—	—
	250	—	0	0	1	1	1	3	6	9	12	15	—	—
	300	—	0	0	1	1	1	3	5	8	10	13	—	—
	350	—	0	0	1	1	1	2	4	7	9	11	—	—
	400	—	0	0	0	1	1	1	4	6	8	10	—	—
	500	—	0	0	0	1	1	1	3	5	6	8	—	—
	600	—	0	0	0	1	1	1	2	4	5	6	—	—
	700	—	0	0	0	0	1	1	2	3	4	6	—	—
	750	—	0	0	0	0	1	1	1	3	4	5	—	—
	800	—	0	0	0	0	1	1	1	3	4	5	—	—
	900	—	0	0	0	0	1	1	1	3	3	4	—	—
	1000	—	0	0	0	0	0	1	1	2	3	4	—	—
	1250	—	0	0	0	0	0	1	1	1	2	3	—	—
	1500	—	0	0	0	0	0	1	1	1	1	3	—	—
	1750	—	0	0	0	0	0	0	1	1	1	2	—	—
	2000	—	0	0	0	0	0	0	1	1	1	1	—	—

(continued)

Table C.1 Maximum Number of Conductors or Fixture Wires in Electrical Metallic Tubing (EMT) (Based on *NEC* Chapter 9: Table 1, Table 4, and Table 5) *Continued*

Type	Conductor Size (AWG/ kcmil)	Trade Size (Metric Designator)												
		3/8 (12)	1/2 (16)	3/4 (21)	1 (27)	1 1/4 (35)	1 1/2 (41)	2 (53)	2 1/2 (63)	3 (78)	3 1/2 (91)	4 (103)	5 (129)	6 (155)
		FIXTURE WIRES												
RFH-2, FFH-2, RFHH-2	18	—	8	14	24	41	56	92	161	244	318	407	—	—
	16	—	7	12	20	34	47	78	136	205	268	343	—	—
SF-2, SFF-2	18	—	10	18	30	52	71	116	203	307	401	513	—	—
	16	—	8	15	25	43	58	96	168	254	332	424	—	—
	14	—	7	12	20	34	47	78	136	205	268	343	—	—
SF-1, SFF-1	18	—	18	33	53	92	125	206	360	544	710	908	—	—
RFH-1, TF, TFF, XF, XFF	18	—	14	24	39	68	92	152	266	402	524	670	—	—
	16	—	11	19	31	55	74	123	215	324	423	541	—	—
XF, XFF	14	—	8	15	25	43	58	96	168	254	332	424	—	—
TFN, TFFN	18	—	22	38	63	109	148	244	426	643	839	1073	—	—
	16	—	17	29	48	83	113	186	325	491	641	819	—	—
PF, PFF, PGF, PGFF, PAF, PTF, PTFF, PAFF	18	—	21	36	59	103	140	231	404	610	796	1017	—	—
	16	—	16	28	46	79	108	179	312	471	615	787	—	—
	14	—	12	21	34	60	81	134	234	354	462	590	—	—

ZF, ZFF, ZHF	18	—	27	47	77	133	181	298	520	786	1026	1311	—	—
	16	—	20	35	56	98	133	220	384	580	757	967	—	—
	14	—	14	25	41	72	98	161	282	426	556	711	—	—
KF-2, KFF-2	18	—	40	71	115	199	271	447	781	1179	1539	1967	—	—
	16	—	28	49	80	139	189	312	545	823	1074	1372	—	—
	14	—	19	33	54	93	127	209	366	553	721	922	—	—
	12	—	13	23	37	65	88	146	254	384	502	641	—	—
	10	—	8	15	25	43	58	96	168	254	332	424	—	—
KF-1, KFF-1	18	—	46	82	133	230	313	516	901	1361	1776	2269	—	—
	16	—	33	57	93	161	220	363	633	956	1248	1595	—	—
	14	—	22	38	63	109	148	244	426	643	839	1073	—	—
	12	—	14	25	41	72	98	161	282	426	556	711	—	—
	10	—	9	16	27	47	64	105	184	278	363	464	—	—
XF, XFF	12	—	4	8	13	23	31	51	90	136	177	227	—	—
	10	—	3	6	10	18	24	40	70	106	138	177	—	—

Notes:

1. This table is for concentric stranded conductors only. For compact stranded conductors, Table C.1(A) should be used.
2. Two-hour fire-rated RHH cable has ceramifiable insulation, which has much larger diameters than other RHH wires. Consult manufacturer's conduit fill tables.

*Types RHH, RHW, and RHW-2 without outer covering.

Table C.1(A) Maximum Number of Conductors or Fixture Wires in Electrical Metallic Tubing (EMT) (Based on Chapter 9: Table 1, Table 4, and Table 5A)

Type	Conductor Size (AWG/ kcmil)	Trade Size (Metric Designator)												
		⅜ (12)	½ (16)	¾ (21)	1 (27)	1¼ (35)	1½ (41)	2 (53)	2½ (63)	3 (78)	3½ (91)	4 (103)	5 (129)	6 (155)
COMPACT CONDUCTORS														
THW, THW-2, THHW	8	—	2	4	6	11	16	26	46	69	90	115	—	—
	6	—	1	3	5	9	12	20	35	53	70	89	—	—
	4	—	1	2	4	6	9	15	26	40	52	67	—	—
	2	—	1	1	3	5	7	11	19	29	38	49	—	—
	1	—	1	1	1	3	4	8	13	21	27	34	—	—
	1/0	—	1	1	1	3	4	7	12	18	23	30	—	—
	2/0	—	0	1	1	2	3	5	10	15	20	25	—	—
	3/0	—	0	1	1	1	3	5	8	13	17	21	—	—
	4/0	—	0	1	1	1	2	4	7	11	14	18	—	—
	250	—	0	0	1	1	1	3	5	8	11	14	—	—
	300	—	0	0	1	1	1	3	5	7	9	12	—	—
	350	—	0	0	1	1	1	2	4	6	8	11	—	—
	400	—	0	0	0	1	1	1	4	6	8	10	—	—
	500	—	0	0	0	1	1	1	3	5	6	8	—	—

	600	—	0	0	0	1	1	1	2	4	5	7	—	—
	700	—	0	0	0	1	1	1	2	3	4	6	—	—
	750	—	0	0	0	0	1	1	1	3	4	5	—	—
	900	—	0	0	0	0	1	1	1	3	4	5	—	—
	1000	—	0	0	0	0	1	1	1	2	3	4	—	—
THHN, THWN, THWN-2	8	—	—	—	—	—	—	—	—	—	—	—	—	—
	6	—	2	4	7	13	18	29	52	78	102	130	—	—
	4	—	1	3	4	8	11	18	32	48	63	81	—	—
	2	—	1	1	3	6	8	13	23	34	45	58	—	—
	1	—	1	1	2	4	6	10	17	26	34	43	—	—
	1/0	—	1	1	1	3	5	8	14	22	29	37	—	—
	2/0	—	1	1	1	3	4	7	12	18	24	30	—	—
	3/0	—	0	1	1	2	3	6	10	15	20	25	—	—
	4/0	—	0	1	1	1	3	5	8	12	16	21	—	—
	250	—	0	1	1	1	1	4	6	10	13	16	—	—
	300	—	0	0	1	1	1	3	5	8	11	14	—	—
	350	—	0	0	1	1	1	3	5	7	10	12	—	—
	400	—	0	0	1	1	1	2	4	6	9	11	—	—
	500	—	0	0	0	1	1	1	4	5	7	9	—	—

(continued)

Table C.1(A) Maximum Number of Conductors or Fixture Wires in Electrical Metallic Tubing (EMT) (Based on Chapter 9: Table 1, Table 4, and Table 5A) ***Continued***

Type	Conductor Size (AWG/kcmil)	Trade Size (Metric Designator)												
		⅜ (12)	½ (16)	¾ (21)	1 (27)	1¼ (35)	1½ (41)	2 (53)	2½ (63)	3 (78)	3½ (91)	4 (103)	5 (129)	6 (155)
COMPACT CONDUCTORS														
	600	—	0	0	0	1	1	1	3	4	6	7	—	—
	700	—	0	0	0	1	1	1	2	4	5	7	—	—
	750	—	0	0	0	1	1	1	2	4	5	6	—	—
	900	—	0	0	0	0	1	1	1	3	4	5	—	—
	1000	—	0	0	0	0	1	1	1	3	3	4	—	—
XHHW, XHHW-2	8	—	3	5	8	15	20	34	59	90	117	149	—	—
	6	—	1	4	6	11	15	25	44	66	87	111	—	—
	4	—	1	3	4	8	11	18	32	48	63	81	—	—
	2	—	1	1	3	6	8	13	23	34	45	58	—	—
	1	—	1	1	2	4	6	10	17	26	34	43	—	—

	1/0	—	1	1	1	3	5	8	14	22	29	37	—	—
	2/0	—	1	1	1	3	4	7	12	18	24	31	—	—
	3/0	—	0	1	1	2	3	6	10	15	20	25	—	—
	4/0	—	0	1	1	1	3	5	8	13	17	21	—	—
	250	—	0	1	1	1	2	4	7	10	13	17	—	—
	300	—	0	0	1	1	1	3	6	9	11	14	—	—
	350	—	0	0	1	1	1	3	5	8	10	13	—	—
	400	—	0	0	1	1	1	2	4	7	9	11	—	—
	500	—	0	0	0	1	1	1	4	6	7	9	—	—
	600	—	0	0	0	1	1	1	3	4	6	8	—	—
	700	—	0	0	0	1	1	1	2	4	5	7	—	—
	750	—	0	0	0	1	1	1	2	3	5	6	—	—
	900	—	0	0	0	0	1	1	1	3	4	5	—	—
	1000	—	0	0	0	0	1	1	1	3	4	5	—	—

Definition: *Compact stranding* is the result of a manufacturing process where the stranded conductor is compressed to the extent that the interstices (voids between strand wires) are virtually eliminated.

Table C.2 Maximum Number of Conductors or Fixture Wires in Electrical Nonmetallic Tubing (ENT) (Based on Chapter 9: Table 1, Table 4, and Table 5)

Type	Conductor Size (AWG/ kcmil)	Trade Size (Metric Designator)												
		⅜ (12)	½ (16)	¾ (21)	1 (27)	1¼ (35)	1½ (41)	2 (53)	2½ (63)	3 (78)	3½ (91)	4 (103)	5 (129)	6 (155)
CONDUCTORS														
RHH, RHW, RHW-2	14	—	4	7	11	20	27	45	—	—	—	—	—	—
	12	—	3	5	9	16	22	37	—	—	—	—	—	—
	10	—	2	4	7	13	18	30	—	—	—	—	—	—
	8	—	1	2	4	7	9	15	—	—	—	—	—	—
	6	—	1	1	3	5	7	12	—	—	—	—	—	—
	4	—	1	1	2	4	6	10	—	—	—	—	—	—
	3	—	1	1	1	4	5	8	—	—	—	—	—	—
	2	—	1	1	1	3	4	7	—	—	—	—	—	—
	1	—	0	1	1	1	3	5	—	—	—	—	—	—
	1/0	—	0	1	1	1	2	4	—	—	—	—	—	—
	2/0	—	0	0	1	1	1	3	—	—	—	—	—	—
	3/0	—	0	0	1	1	1	3	—	—	—	—	—	—
	4/0	—	0	0	1	1	1	2	—	—	—	—	—	—
	250	—	0	0	0	1	1	1	—	—	—	—	—	—
	300	—	0	0	0	1	1	1	—	—	—	—	—	—
	350	—	0	0	0	1	1	1	—	—	—	—	—	—

	400	—	0	0	0	1	1	1	—	—	—	—	—	—
	500	—	0	0	0	0	1	1	—	—	—	—	—	—
	600	—	0	0	0	0	1	1	—	—	—	—	—	—
	700	—	0	0	0	0	0	1	—	—	—	—	—	—
	750	—	0	0	0	0	0	1	—	—	—	—	—	—
	800	—	0	0	0	0	0	1	—	—	—	—	—	—
	900	—	0	0	0	0	0	1	—	—	—	—	—	—
	1000	—	0	0	0	0	0	1	—	—	—	—	—	—
	1250	—	0	0	0	0	0	0	—	—	—	—	—	—
	1500	—	0	0	0	0	0	0	—	—	—	—	—	—
	1750	—	0	0	0	0	0	0	—	—	—	—	—	—
	2000	—	0	0	0	0	0	0	—	—	—	—	—	—
TW, THHW, THW, THW-2	14	—	8	14	24	42	57	94	—	—	—	—	—	—
	12	—	6	11	18	32	44	72	—	—	—	—	—	—
	10	—	4	8	13	24	32	54	—	—	—	—	—	—
	8	—	2	4	7	13	18	30	—	—	—	—	—	—
RHH*, RHW*, RHW-2*	14	—	5	9	16	28	38	63	—	—	—	—	—	—
	12	—	4	8	13	22	30	50	—	—	—	—	—	—
	10	—	3	6	10	17	24	39	—	—	—	—	—	—
	8	—	1	3	6	10	14	23	—	—	—	—	—	—

(continued)

Table C.2 Maximum Number of Conductors or Fixture Wires in Electrical Nonmetallic Tubing (ENT) (Based on *NEC* Chapter 9: Table 1, Table 4, and Table 5) *Continued*

Type	Conductor Size (AWG/ kcmil)	Trade Size (Metric Designator)												
		⅜ (12)	½ (16)	¾ (21)	1 (27)	1¼ (35)	1½ (41)	2 (53)	2½ (63)	3 (78)	3½ (91)	4 (103)	5 (129)	6 (155)
CONDUCTORS														
TW, THW, THHW, THW-2, RHH*, RHW*, RHW-2*	6	—	1	2	4	8	11	18	—	—	—	—	—	—
	4	—	1	1	3	6	8	13	—	—	—	—	—	—
	3	—	1	1	3	5	7	11	—	—	—	—	—	—
	2	—	1	1	2	4	6	10	—	—	—	—	—	—
	1	—	0	1	1	3	4	7	—	—	—	—	—	—
	1/0	—	0	1	1	2	3	6	—	—	—	—	—	—
	2/0	—	0	1	1	1	3	5	—	—	—	—	—	—
	3/0	—	0	1	1	1	2	4	—	—	—	—	—	—
	4/0	—	0	0	1	1	1	3	—	—	—	—	—	—
	250	—	0	0	1	1	1	3	—	—	—	—	—	—
	300	—	0	0	1	1	1	2	—	—	—	—	—	—
	350	—	0	0	0	1	1	1	—	—	—	—	—	—
	400	—	0	0	0	1	1	1	—	—	—	—	—	—
	500	—	0	0	0	1	1	1	—	—	—	—	—	—

	600	—	0	0	0	0	1	1	—	—	—	—	—	—
	700	—	0	0	0	0	1	1	—	—	—	—	—	—
	750	—	0	0	0	0	1	1	—	—	—	—	—	—
	800	—	0	0	0	0	1	1	—	—	—	—	—	—
	900	—	0	0	0	0	0	1	—	—	—	—	—	—
	1000	—	0	0	0	0	0	1	—	—	—	—	—	—
	1250	—	0	0	0	0	0	1	—	—	—	—	—	—
	1500	—	0	0	0	0	0	1	—	—	—	—	—	—
	1750	—	0	0	0	0	0	0	—	—	—	—	—	—
	2000	—	0	0	0	0	0	0	—	—	—	—	—	—
THHN, THWN, THWN-2	14	—	11	21	34	60	82	135	—	—	—	—	—	—
	12	—	8	15	25	43	59	99	—	—	—	—	—	—
	10	—	5	9	15	27	37	62	—	—	—	—	—	—
	8	—	3	5	9	16	21	36	—	—	—	—	—	—
	6	—	1	4	6	11	15	26	—	—	—	—	—	—
	4	—	1	2	4	7	9	16	—	—	—	—	—	—
	3	—	1	1	3	6	8	13	—	—	—	—	—	—
	2	—	1	1	3	5	7	11	—	—	—	—	—	—
	1	—	1	1	1	3	5	8	—	—	—	—	—	—
	1/0	—	1	1	1	3	4	7	—	—	—	—	—	—
	2/0	—	0	1	1	2	3	6	—	—	—	—	—	—
	3/0	—	0	1	1	1	3	5	—	—	—	—	—	—
	4/0	—	0	1	1	1	2	4	—	—	—	—	—	—

(continued)

Table C.2 Maximum Number of Conductors or Fixture Wires in Electrical Nonmetallic Tubing (ENT) (Based on *NEC* Chapter 9: Table 1, Table 4, and Table 5) *Continued*

Type	Conductor Size (AWG/kcmil)	Trade Size (Metric Designator)												
		⅜ (12)	½ (16)	¾ (21)	1 (27)	1¼ (35)	1½ (41)	2 (53)	2½ (63)	3 (78)	3½ (91)	4 (103)	5 (129)	6 (155)
CONDUCTORS														
	250	—	0	0	1	1	1	3	—	—	—	—	—	—
	300	—	0	0	1	1	1	3	—	—	—	—	—	—
	350	—	0	0	1	1	1	2	—	—	—	—	—	—
	400	—	0	0	0	1	1	1	—	—	—	—	—	—
	500	—	0	0	0	1	1	1	—	—	—	—	—	—
	600	—	0	0	0	1	1	1	—	—	—	—	—	—
	700	—	0	0	0	0	1	1	—	—	—	—	—	—
	750	—	0	0	0	0	1	1	—	—	—	—	—	—
	800	—	0	0	0	0	1	1	—	—	—	—	—	—
	900	—	0	0	0	0	1	1	—	—	—	—	—	—
	1000	—	0	0	0	0	0	1	—	—	—	—	—	—
FEP, FEPB, PFA, PFAH, TFE	14	—	11	20	33	58	79	131	—	—	—	—	—	—
	12	—	8	15	24	42	58	96	—	—	—	—	—	—
	10	—	6	10	17	30	41	69	—	—	—	—	—	—
	8	—	3	6	10	17	24	39	—	—	—	—	—	—

	6	—	2	4	7	12	17	28	—	—	—	—	—	—
	4	—	1	3	5	8	12	19	—	—	—	—	—	—
	3	—	1	2	4	7	10	16	—	—	—	—	—	—
	2	—	1	1	3	6	8	13	—	—	—	—	—	—
PFA, PFAH, TFE	1	—	1	1	2	4	5	9	—	—	—	—	—	—
PFA, PFAH, TFE, Z	1/0	—	1	1	1	3	4	8	—	—	—	—	—	—
	2/0	—	0	1	1	3	4	6	—	—	—	—	—	—
	3/0	—	0	1	1	2	3	5	—	—	—	—	—	—
	4/0	—	0	1	1	1	2	4	—	—	—	—	—	—
Z	14	—	13	24	40	70	95	158	—	—	—	—	—	—
	12	—	9	17	28	49	68	112	—	—	—	—	—	—
	10	—	6	10	17	30	41	69	—	—	—	—	—	—
	8	—	3	6	11	19	26	43	—	—	—	—	—	—
	6	—	2	4	7	13	18	30	—	—	—	—	—	—
	4	—	1	3	5	9	12	21	—	—	—	—	—	—
	3	—	1	2	4	6	9	15	—	—	—	—	—	—
	2	—	1	1	3	5	7	12	—	—	—	—	—	—
	1	—	1	1	2	4	6	10	—	—	—	—	—	—
XHHW, ZW, XHHW-2, XHH	14	—	8	14	24	42	57	94	—	—	—	—	—	—
	12	—	6	11	18	32	44	72	—	—	—	—	—	—
	10	—	4	8	13	24	32	54	—	—	—	—	—	—
	8	—	2	4	7	13	18	30	—	—	—	—	—	—
	6	—	1	3	5	10	13	22	—	—	—	—	—	—

(continued)

Table C.2 Maximum Number of Conductors or Fixture Wires in Electrical Nonmetallic Tubing (ENT) (Based on *NEC* Chapter 9: Table 1, Table 4, and Table 5) *Continued*

Type	Conductor Size (AWG/kcmil)	Trade Size (Metric Designator)												
		⅜ (12)	½ (16)	¾ (21)	1 (27)	1¼ (35)	1½ (41)	2 (53)	2½ (63)	3 (78)	3½ (91)	4 (103)	5 (129)	6 (155)
CONDUCTORS														
	4	—	1	2	4	7	9	16	—	—	—	—	—	—
	3	—	1	1	3	6	8	13	—	—	—	—	—	—
	2	—	1	1	3	5	7	11	—	—	—	—	—	—
XHHW, XHHW-2, XHH	1	—	1	1	1	3	5	8	—	—	—	—	—	—
	1/0	—	0	1	1	3	4	7	—	—	—	—	—	—
	2/0	—	0	1	1	2	3	6	—	—	—	—	—	—
	3/0	—	0	1	1	1	3	5	—	—	—	—	—	—
	4/0	—	0	1	1	1	2	4	—	—	—	—	—	—
	250	—	0	0	1	1	1	3	—	—	—	—	—	—
	300	—	0	0	1	1	1	3	—	—	—	—	—	—
	350	—	0	0	1	1	1	2	—	—	—	—	—	—
	400	—	0	0	0	1	1	1	—	—	—	—	—	—
	500	—	0	0	0	1	1	1	—	—	—	—	—	—

	600	—	0	0	0	1	1	1	—	—	—	—	—	—
	700	—	0	0	0	0	1	1	—	—	—	—	—	—
	750	—	0	0	0	0	1	1	—	—	—	—	—	—
	800	—	0	0	0	0	1	1	—	—	—	—	—	—
	900	—	0	0	0	0	1	1	—	—	—	—	—	—
	1000	—	0	0	0	0	0	1	—	—	—	—	—	—
	1250	—	0	0	0	0	0	1	—	—	—	—	—	—
	1500	—	0	0	0	0	0	1	—	—	—	—	—	—
	1750	—	0	0	0	0	0	0	—	—	—	—	—	—
	2000	—	0	0	0	0	0	0	—	—	—	—	—	—
FIXTURE WIRES														
RFH-2, FFH-2, RFHH-2	18	—	8	14	23	40	54	90	—	—	—	—	—	—
	16	—	6	12	19	33	46	76	—	—	—	—	—	—
SF-2, SFF-2	18	—	10	17	29	50	69	114	—	—	—	—	—	—
	16	—	8	14	24	42	57	94	—	—	—	—	—	—
	14	—	6	12	19	33	46	76	—	—	—	—	—	—
SF-1, SFF-1	18	—	17	31	51	89	122	202	—	—	—	—	—	—
RFH-1, TF, TFF, XF, XFF	18	—	13	23	38	66	90	149	—	—	—	—	—	—
	16	—	10	18	30	53	73	120	—	—	—	—	—	—
XF, XFF	14	—	8	14	24	42	57	94	—	—	—	—	—	—

(continued)

Table C.2 Maximum Number of Conductors or Fixture Wires in Electrical Nonmetallic Tubing (ENT) (Based on *NEC* Chapter 9: Table 1, Table 4, and Table 5) *Continued*

Type	Conductor Size (AWG/ kcmil)	Trade Size (Metric Designator)												
		⅜ (12)	½ (16)	¾ (21)	1 (27)	1¼ (35)	1½ (41)	2 (53)	2½ (63)	3 (78)	3½ (91)	4 (103)	5 (129)	6 (155)
FIXTURE WIRES														
TFN, TFFN	18	—	20	37	60	105	144	239	—	—	—	—	—	—
	16	—	16	28	46	80	110	183	—	—	—	—	—	—
PF, PFF, PGF, PGFF, PAF, PTF, PTFF, PAFF	18	—	19	35	57	100	137	227	—	—	—	—	—	—
	16	—	15	27	44	77	106	175	—	—	—	—	—	—
	14	—	11	20	33	58	79	131	—	—	—	—	—	—
ZF, ZFF, ZHF	18	—	25	45	74	129	176	292	—	—	—	—	—	—
	16	—	18	33	54	95	130	216	—	—	—	—	—	—
	14	—	13	24	40	70	95	158	—	—	—	—	—	—
KF-2, KFF-2	18	—	38	67	111	193	265	439	—	—	—	—	—	—
	16	—	26	47	77	135	184	306	—	—	—	—	—	—
	14	—	18	31	52	91	124	205	—	—	—	—	—	—
	12	—	12	22	36	63	86	143	—	—	—	—	—	—
	10	—	8	14	24	42	57	94	—	—	—	—	—	—

KF-1, KFF-1	18	—	44	78	128	223	305	506	—	—	—	—	—	—
	16	—	31	55	90	157	214	355	—	—	—	—	—	—
	14	—	20	37	60	105	144	239	—	—	—	—	—	—
	12	—	13	24	40	70	95	158	—	—	—	—	—	—
	10	—	9	16	26	45	62	103	—	—	—	—	—	—
XF, XFF	12	—	4	8	13	22	30	50	—	—	—	—	—	—
	10	—	3	6	10	17	24	39	—	—	—	—	—	—

Notes:

1. This table is for concentric stranded conductors only. For compact stranded conductors, Table C.2(A) should be used.
2. Two-hour fire-rated RHH cable has ceramifiable insulation, which has much larger diameters than other RHH wires. Consult manufacturer's conduit fill tables.

*Types RHH, RHW, and RHW-2 without outer covering.

Table C.3 Maximum Number of Conductors or Fixture Wires in Flexible Metal Conduit (FMC) (Based on Chapter 9: Table 1, Table 4, and Table 5)

Type	Conductor Size (AWG/kcmil)	Trade Size (Metric Designator)												
		⅜ (12)	½ (16)	¾ (21)	1 (27)	1¼ (35)	1½ (41)	2 (53)	2½ (63)	3 (78)	3½ (91)	4 (103)	5 (129)	6 (155)
CONDUCTORS														
RHH, RHW, RHW-2	14	1	4	7	11	17	25	44	67	96	131	171	—	—
	12	1	3	6	9	14	21	37	55	80	109	142	—	—
	10	1	3	5	7	11	17	30	45	64	88	115	—	—
	8	0	1	2	4	6	9	15	23	34	46	60	—	—
	6	0	1	1	3	5	7	12	19	27	37	48	—	—
	4	0	1	1	2	4	5	10	14	21	29	37	—	—
	3	0	1	1	1	3	5	8	13	18	25	33	—	—
	2	0	1	1	1	3	4	7	11	16	22	28	—	—
	1	0	0	1	1	1	2	5	7	10	14	19	—	—
	1/0	0	0	1	1	1	2	4	6	9	12	16	—	—
	2/0	0	0	1	1	1	1	3	5	8	11	14	—	—
	3/0	0	0	0	1	1	1	3	5	7	9	12	—	—
	4/0	0	0	0	1	1	1	2	4	6	8	10	—	—

	250	0	0	0	0	1	1	1	3	4	6	8	—	—
	300	0	0	0	0	1	1	1	2	4	5	7	—	—
	350	0	0	0	0	1	1	1	2	3	5	6	—	—
	400	0	0	0	0	0	1	1	1	3	4	6	—	—
	500	0	0	0	0	0	1	1	1	3	4	5	—	—
	600	0	0	0	0	0	1	1	1	2	3	4	—	—
	700	0	0	0	0	0	0	1	1	1	3	3	—	—
	750	0	0	0	0	0	0	1	1	1	2	3	—	—
	800	0	0	0	0	0	0	1	1	1	2	3	—	—
	900	0	0	0	0	0	0	1	1	1	2	3	—	—
	1000	0	0	0	0	0	0	1	1	1	1	3	—	—
	1250	0	0	0	0	0	0	0	1	1	1	1	—	—
	1500	0	0	0	0	0	0	0	1	1	1	1	—	—
	1750	0	0	0	0	0	0	0	1	1	1	1	—	—
	2000	0	0	0	0	0	0	0	0	1	1	1	—	—
TW, THHW, THW, THW-2	14	3	9	15	23	36	53	94	141	203	277	361	—	—
	12	2	7	11	18	28	41	72	108	156	212	277	—	—
	10	1	5	8	13	21	30	54	81	116	158	207	—	—
	8	1	3	5	7	11	17	30	45	64	88	115	—	—
RHH*, RHW*, RHW-2*	14	1	6	10	15	24	35	62	94	135	184	240	—	—
	12	1	5	8	12	19	28	50	75	108	148	193	—	—
	10	1	4	6	10	15	22	39	59	85	115	151	—	—
	8	1	1	4	6	9	13	23	35	51	69	90	—	—

(continued)

Table C.3 Maximum Number of Conductors or Fixture Wires in Flexible Metal Conduit (FMC) (Based on *NEC* Chapter 9: Table 1, Table 4, and Table 5) ***Continued***

Type	Conductor Size (AWG/kcmil)	Trade Size (Metric Designator)												
		⅜ (12)	½ (16)	¾ (21)	1 (27)	1¼ (35)	1½ (41)	2 (53)	2½ (63)	3 (78)	3½ (91)	4 (103)	5 (129)	6 (155)
		CONDUCTORS												
TW, THW, THHW, THW-2, RHH*, RHW*, RHW-2*	6	1	1	3	4	7	10	18	27	39	53	69	—	—
	4	0	1	1	3	5	7	13	20	29	39	51	—	—
	3	0	1	1	3	4	6	11	17	25	34	44	—	—
	2	0	1	1	2	4	5	10	14	21	29	37	—	—
	1	0	1	1	1	2	4	7	10	15	20	26	—	—
	1/0	0	0	1	1	1	3	6	9	12	17	22	—	—
	2/0	0	0	1	1	1	3	5	7	10	14	19	—	—
	3/0	0	0	1	1	1	2	4	6	9	12	16	—	—
	4/0	0	0	0	1	1	1	3	5	7	10	13	—	—
	250	0	0	0	1	1	1	3	4	6	8	11	—	—
	300	0	0	0	1	1	1	2	3	5	7	9	—	—
	350	0	0	0	0	1	1	1	3	4	6	8	—	—
	400	0	0	0	0	1	1	1	3	4	6	7	—	—
	500	0	0	0	0	1	1	1	2	3	5	6	—	—

	600	0	0	0	0	0	1	1	1	3	4	5	—	—
	700	0	0	0	0	0	1	1	1	2	3	4	—	—
	750	0	0	0	0	0	1	1	1	2	3	4	—	—
	800	0	0	0	0	0	1	1	1	1	3	4	—	—
	900	0	0	0	0	0	0	1	1	1	3	3	—	—
	1000	0	0	0	0	0	0	1	1	1	2	3	—	—
	1250	0	0	0	0	0	0	1	1	1	1	2	—	—
	1500	0	0	0	0	0	0	0	1	1	1	1	—	—
	1750	0	0	0	0	0	0	0	1	1	1	1	—	—
	2000	0	0	0	0	0	0	0	1	1	1	1	—	—
THHN, THWN, THWN-2	14	4	13	22	33	52	76	135	202	291	396	518	—	—
	12	3	9	16	24	38	56	98	147	212	289	378	—	—
	10	1	6	10	15	24	35	62	93	134	182	238	—	—
	8	1	3	6	9	14	20	35	53	77	105	137	—	—
	6	1	2	4	6	10	14	25	38	55	76	99	—	—
	4	0	1	2	4	6	9	16	24	34	46	61	—	—
	3	0	1	1	3	5	7	13	20	29	39	51	—	—
	2	0	1	1	3	4	6	11	17	24	33	43	—	—
	1	0	1	1	1	3	4	8	12	18	24	32	—	—
	1/0	0	1	1	1	2	4	7	10	15	20	27	—	—
	2/0	0	0	1	1	1	3	6	9	12	17	22	—	—
	3/0	0	0	1	1	1	2	5	7	10	14	18	—	—
	4/0	0	0	1	1	1	1	4	6	8	12	15	—	—

(continued)

Table C.3 Maximum Number of Conductors or Fixture Wires in Flexible Metal Conduit (FMC) (Based on *NEC* Chapter 9: Table 1, Table 4, and Table 5) *Continued*

Type	Conductor Size (AWG/kcmil)	Trade Size (Metric Designator)												
		⅜ (12)	½ (16)	¾ (21)	1 (27)	1¼ (35)	1½ (41)	2 (53)	2½ (63)	3 (78)	3½ (91)	4 (103)	5 (129)	6 (155)
CONDUCTORS														
	250	0	0	0	1	1	1	3	5	7	9	12	—	—
	300	0	0	0	1	1	1	3	4	6	8	11	—	—
	350	0	0	0	1	1	1	2	3	5	7	9	—	—
	400	0	0	0	0	1	1	1	3	5	6	8	—	—
	500	0	0	0	0	1	1	1	2	4	5	7	—	—
	600	0	0	0	0	0	1	1	1	3	4	5	—	—
	700	0	0	0	0	0	1	1	1	3	4	5	—	—
	750	0	0	0	0	0	1	1	1	2	3	4	—	—
	800	0	0	0	0	0	1	1	1	2	3	4	—	—
	900	0	0	0	0	0	0	1	1	1	3	4	—	—
	1000	0	0	0	0	0	0	1	1	1	3	3	—	—
FEP, FEPB, PFA, PFAH, TFE	14	4	12	21	32	51	74	130	196	282	385	502	—	—
	12	3	9	15	24	37	54	95	143	206	281	367	—	—
	10	2	6	11	17	26	39	68	103	148	201	263	—	—
	8	1	4	6	10	15	22	39	59	85	115	151	—	—

	6	1	2	4	7	11	16	28	42	60	82	107	—	—
	4	1	1	3	5	7	11	19	29	42	57	75	—	—
	3	0	1	2	4	6	9	16	24	35	48	62	—	—
	2	0	1	1	3	5	7	13	20	29	39	51	—	—
PFA, PFAH, TFE	1	0	1	1	2	3	5	9	14	20	27	36	—	—
PFA, PFAH, TFE, Z	1/0	0	1	1	1	3	4	8	11	17	23	30	—	—
	2/0	0	1	1	1	2	3	6	9	14	19	24	—	—
	3/0	0	0	1	1	1	3	5	8	11	15	20	—	—
	4/0	0	0	1	1	1	2	4	6	9	13	16	—	—
Z	14	5	15	25	39	61	89	157	236	340	463	605	—	—
	12	4	11	18	28	43	63	111	168	241	329	429	—	—
	10	2	6	11	17	26	39	68	103	148	201	263	—	—
	8	1	4	7	11	17	24	43	65	93	127	166	—	—
	6	1	3	5	7	12	17	30	45	65	89	117	—	—
	4	1	1	3	5	8	12	21	31	45	61	80	—	—
	3	0	1	2	4	6	8	15	23	33	45	58	—	—
	2	0	1	1	3	5	7	12	19	27	37	49	—	—
	1	0	1	1	2	4	6	10	15	22	30	39	—	—
XHHW, ZW, XHHW-2, XHH	14	3	9	15	23	36	53	94	141	203	277	361	—	—
	12	2	7	11	18	28	41	72	108	156	212	277	—	—
	10	1	5	8	13	21	30	54	81	116	158	207	—	—
	8	1	3	5	7	11	17	30	45	64	88	115	—	—
	6	1	1	3	5	8	12	22	33	48	65	85	—	—

(continued)

Table C.3 Maximum Number of Conductors or Fixture Wires in Flexible Metal Conduit (FMC) (Based on *NEC* Chapter 9: Table 1, Table 4, and Table 5) ***Continued***

Type	Conductor Size (AWG/kcmil)	Trade Size (Metric Designator)												
		⅜ (12)	½ (16)	¾ (21)	1 (27)	1¼ (35)	1½ (41)	2 (53)	2½ (63)	3 (78)	3½ (91)	4 (103)	5 (129)	6 (155)
		CONDUCTORS												
	4	0	1	2	4	6	9	16	24	34	47	61	—	—
	3	0	1	1	3	5	7	13	20	29	40	52	—	—
	2	0	1	1	3	4	6	11	17	24	33	44	—	—
XHH, XHHW, XHHW-2	1	0	1	1	1	3	5	8	13	18	25	32	—	—
	1/0	0	1	1	1	2	4	7	10	15	21	27	—	—
	2/0	0	0	1	1	2	3	6	9	13	17	23	—	—
	3/0	0	0	1	1	1	3	5	7	10	14	19	—	—
	4/0	0	0	1	1	1	2	4	6	9	12	15	—	—
	250	0	0	0	1	1	1	3	5	7	10	13	—	—
	300	0	0	0	1	1	1	3	4	6	8	11	—	—
	350	0	0	0	1	1	1	2	4	5	7	9	—	—
	400	0	0	0	0	1	1	1	3	5	6	8	—	—
	500	0	0	0	0	1	1	1	3	4	5	7	—	—

	600	0	0	0	0	0	1	1	1	3	4	5	—	—
	700	0	0	0	0	0	1	1	1	3	4	5	—	—
	750	0	0	0	0	0	1	1	1	2	3	4	—	—
	800	0	0	0	0	0	1	1	1	2	3	4	—	—
	900	0	0	0	0	0	0	1	1	1	3	4	—	—
	1000	0	0	0	0	0	0	1	1	1	3	3	—	—
	1250	0	0	0	0	0	0	1	1	1	1	3	—	—
	1500	0	0	0	0	0	0	1	1	1	1	2	—	—
	1750	0	0	0	0	0	0	0	1	1	1	1	—	—
	2000	0	0	0	0	0	0	0	1	1	1	1	—	—
FIXTURE WIRES														
RFH-2, FFH-2, RFHH-2	18	3	8	14	22	35	51	90	135	195	265	346	—	—
	16	2	7	12	19	29	43	76	114	164	223	292	—	—
SF-2, SFF-2	18	4	11	18	28	44	64	113	170	246	334	437	—	—
	16	3	9	15	23	36	53	94	141	203	277	361	—	—
	14	2	7	12	19	29	43	76	114	164	223	292	—	—
SF-1, SFF-1	18	7	19	33	50	78	114	201	302	435	592	773	—	—
RFH-1, TF, TFF, XF, XFF	18	5	14	24	37	58	84	148	223	321	437	571	—	—
	16	4	11	19	30	47	68	120	180	259	353	461	—	—
XF, XFF	14	3	9	15	23	36	53	94	141	203	277	361	—	—

(continued)

Table C.3 Maximum Number of Conductors or Fixture Wires in Flexible Metal Conduit (FMC) (Based on *NEC* Chapter 9: Table 1, Table 4, and Table 5) *Continued*

Type	Conductor Size (AWG/kcmil)	Trade Size (Metric Designator)												
		⅜ (12)	½ (16)	¾ (21)	1 (27)	1¼ (35)	1½ (41)	2 (53)	2½ (63)	3 (78)	3½ (91)	4 (103)	5 (129)	6 (155)
FIXTURE WIRES														
TFN, TFFN	18	8	23	38	59	93	135	237	357	514	699	914	—	—
	16	6	17	29	45	71	103	181	272	392	534	698	—	—
PF, PFF, PGF, PGFF, PAF, PTF, PTFF, PAFF	18	8	22	36	56	88	128	225	338	487	663	866	—	—
	16	6	17	28	43	68	99	174	262	377	513	670	—	—
	14	4	12	21	32	51	74	130	196	282	385	502	—	—
ZF, ZFF, ZHF	18	10	28	47	72	113	165	290	436	628	855	1117	—	—
	16	7	20	35	53	83	122	214	322	463	631	824	—	—
	14	5	15	25	39	61	89	157	236	340	463	605	—	—
KF-2, KFF-2	18	15	42	71	109	170	247	436	654	942	1282	1675	—	—
	16	10	29	49	76	118	173	304	456	657	895	1169	—	—
	14	7	20	33	51	80	116	204	307	442	601	785	—	—
	12	5	13	23	35	55	80	142	213	307	418	546	—	—
	10	3	9	15	23	36	53	94	141	203	277	361	—	—

KF-1, KFF-1	18	18	48	82	125	196	286	503	755	1087	1480	1933	—	—
	16	12	34	57	88	138	201	353	530	764	1040	1358	—	—
	14	8	23	38	59	93	135	237	357	514	699	914	—	—
	12	5	15	25	39	61	89	157	236	340	463	605	—	—
	10	3	10	16	25	40	58	103	154	222	303	395	—	—
XF, XFF	12	1	5	8	12	19	28	50	75	108	148	193	—	—
	10	1	4	6	10	15	22	39	59	85	115	151	—	—

Notes:

1. This table is for concentric stranded conductors only. For compact stranded conductors, Table C.3(A) should be used.
2. Two-hour fire-rated RHH cable has ceramifiable insulation, which has much larger diameters than other RHH wires. Consult manufacturer's conduit fill tables.

*Types RHH, RHW, and RHW-2 without outer covering.

Table C.4 Maximum Number of Conductors or Fixture Wires in Intermediate Metal Conduit (IMC) (Based on Chapter 9: Table 1, Table 4, and Table 5)

Type	Conductor Size (AWG/kcmil)	Trade Size (Metric Designator)												
		⅜ (12)	½ (16)	¾ (21)	1 (27)	1¼ (35)	1½ (41)	2 (53)	2½ (63)	3 (78)	3½ (91)	4 (103)	5 (129)	6 (155)
CONDUCTORS														
RHH, RHW, RHW-2	14	—	4	8	13	22	30	49	70	108	144	186	—	—
	12	—	4	6	11	18	25	41	58	89	120	154	—	—
	10	—	3	5	8	15	20	33	47	72	97	124	—	—
	8	—	1	3	4	8	10	17	24	38	50	65	—	—
	6	—	1	1	3	6	8	14	19	30	40	52	—	—
	4	—	1	1	3	5	6	11	15	23	31	41	—	—
	3	—	1	1	2	4	6	9	13	21	28	36	—	—
	2	—	1	1	1	3	5	8	11	18	24	31	—	—
	1	—	0	1	1	2	3	5	7	12	16	20	—	—
	1/0	—	0	1	1	1	3	4	6	10	14	18	—	—
	2/0	—	0	1	1	1	2	4	6	9	12	15	—	—
	3/0	—	0	0	1	1	1	3	5	7	10	13	—	—
	4/0	—	0	0	1	1	1	3	4	6	9	11	—	—

	250	—	0	0	1	1	1	1	3	5	6	8	—	—
	300	—	0	0	0	1	1	1	3	4	6	7	—	—
	350	—	0	0	0	1	1	1	2	4	5	7	—	—
	400	—	0	0	0	1	1	1	2	3	5	6	—	—
	500	—	0	0	0	1	1	1	1	3	4	5	—	—
	600	—	0	0	0	0	1	1	1	2	3	4	—	—
	700	—	0	0	0	0	1	1	1	2	3	4	—	—
	750	—	0	0	0	0	1	1	1	1	3	4	—	—
	800	—	0	0	0	0	0	1	1	1	3	3	—	—
	900	—	0	0	0	0	0	1	1	1	2	3	—	—
	1000	—	0	0	0	0	0	1	1	1	2	3	—	—
	1250	—	0	0	0	0	0	1	1	1	1	1	—	—
	1500	—	0	0	0	0	0	0	1	1	1	1	—	—
	1750	—	0	0	0	0	0	0	1	1	1	1	—	—
	2000	—	0	0	0	0	0	0	1	1	1	1	—	—
TW, THHW, THW, THW-2	14	—	10	17	27	47	64	104	147	228	304	392	—	—
	12	—	7	13	21	36	49	80	113	175	234	301	—	—
	10	—	5	9	15	27	36	59	84	130	174	224	—	—
	8	—	3	5	8	15	20	33	47	72	97	124	—	—
RHH*, RHW*, RHW-2*	14	—	6	11	18	31	42	69	98	151	202	261	—	—
	12	—	5	9	14	25	34	56	79	122	163	209	—	—
	10	—	4	7	11	19	26	43	61	95	127	163	—	—
	8	—	2	4	7	12	16	26	37	57	76	98	—	—

(continued)

Table C.4 Maximum Number of Conductors or Fixture Wires in Intermediate Metal Conduit (IMC) (Based on Chapter 9: Table 1, Table 4, and Table 5) ***Continued***

Type	Conductor Size (AWG/kcmil)	Trade Size (Metric Designator)												
		⅜ (12)	½ (16)	¾ (21)	1 (27)	1¼ (35)	1½ (41)	2 (53)	2½ (63)	3 (78)	3½ (91)	4 (103)	5 (129)	6 (155)
CONDUCTORS														
TW, THW, THHW, THW-2, RHH*, RHW*, RHW-2*	6	—	1	3	5	9	12	20	28	43	58	75	—	—
	4	—	1	2	4	6	9	15	21	32	43	56	—	—
	3	—	1	1	3	6	8	13	18	28	37	48	—	—
	2	—	1	1	3	5	6	11	15	23	31	41	—	—
	1	—	1	1	1	3	4	7	11	16	22	28	—	—
	1/0	—	1	1	1	3	4	6	9	14	19	24	—	—
	2/0	—	0	1	1	2	3	5	8	12	16	20	—	—
	3/0	—	0	1	1	1	3	4	6	10	13	17	—	—
	4/0	—	0	1	1	1	2	4	5	8	11	14	—	—
	250	—	0	0	1	1	1	3	4	7	9	12	—	—
	300	—	0	0	1	1	1	2	4	6	8	10	—	—
	350	—	0	0	1	1	1	2	3	5	7	9	—	—
	400	—	0	0	0	1	1	1	3	4	6	8	—	—
	500	—	0	0	0	1	1	1	2	4	5	7	—	—

	600	—	0	0	0	1	1	1	1	3	4	5	—	—
	700	—	0	0	0	0	1	1	1	3	4	5	—	—
	750	—	0	0	0	0	1	1	1	2	3	4	—	—
	800	—	0	0	0	0	1	1	1	2	3	4	—	—
	900	—	0	0	0	0	1	1	1	2	3	4	—	—
	1000	—	0	0	0	0	0	1	1	1	3	3	—	—
	1250	—	0	0	0	0	0	1	1	1	1	3	—	—
	1500	—	0	0	0	0	0	1	1	1	1	2	—	—
	1750	—	0	0	0	0	0	0	1	1	1	1	—	—
	2000	—	0	0	0	0	0	0	1	1	1	1	—	—
THHN, THWN, THWN-2	14	—	14	24	39	68	91	149	211	326	436	562	—	—
	12	—	10	17	29	49	67	109	154	238	318	410	—	—
	10	—	6	11	18	31	42	69	97	150	200	258	—	—
	8	—	3	6	10	18	24	39	56	86	115	149	—	—
	6	—	2	4	7	13	17	28	40	62	83	107	—	—
	4	—	1	3	4	8	11	17	25	38	51	66	—	—
	3	—	1	2	4	6	9	15	21	32	43	56	—	—
	2	—	1	1	3	5	7	12	17	27	36	47	—	—
	1	—	1	1	2	4	5	9	13	20	27	35	—	—
	1/0	—	1	1	1	3	4	8	11	17	23	29	—	—
	2/0	—	1	1	1	3	4	6	9	14	19	24	—	—
	3/0	—	0	1	1	2	3	5	7	12	16	20	—	—
	4/0	—	0	1	1	1	2	4	6	9	13	17	—	—

(continued)

Table C.4 Maximum Number of Conductors or Fixture Wires in Intermediate Metal Conduit (IMC) (Based on Chapter 9: Table 1, Table 4, and Table 5) ***Continued***

Type	Conductor Size (AWG/ kcmil)	Trade Size (Metric Designator)												
		⅜ (12)	½ (16)	¾ (21)	1 (27)	1¼ (35)	1½ (41)	2 (53)	2½ (63)	3 (78)	3½ (91)	4 (103)	5 (129)	6 (155)
					CONDUCTORS									
	250	—	0	0	1	1	1	3	5	8	10	13	—	—
	300	—	0	0	1	1	1	3	4	7	9	12	—	—
	350	—	0	0	1	1	1	2	4	6	8	10	—	—
	400	—	0	0	1	1	1	2	3	5	7	9	—	—
	500	—	0	0	0	1	1	1	3	4	6	7	—	—
	600	—	0	0	0	1	1	1	2	3	5	6	—	—
	700	—	0	0	0	1	1	1	1	3	4	5	—	—
	750	—	0	0	0	1	1	1	1	3	4	5	—	—
	800	—	0	0	0	0	1	1	1	3	4	5	—	—
	900	—	0	0	0	0	1	1	1	2	3	4	—	—
	1000	—	0	0	0	0	1	1	1	2	3	4	—	—
FEP, FEPB, PFA, PFAH, TFE	14	—	13	23	38	66	89	145	205	317	423	545	—	—
	12	—	10	17	28	48	65	106	150	231	309	398	—	—
	10	—	7	12	20	34	46	76	107	166	221	285	—	—
	8	—	4	7	11	19	26	43	61	95	127	163	—	—

	6	—	3	5	8	14	19	31	44	67	90	116	—	—
	4	—	1	3	5	10	13	21	30	47	63	81	—	—
	3	—	1	3	4	8	11	18	25	39	52	68	—	—
	2	—	1	2	4	6	9	15	21	32	43	56	—	—
PFA, PFAH, TFE	1	—	1	1	2	4	6	10	14	22	30	39	—	—
PFA, PFAH, TFE, Z	1/0	—	1	1	1	4	5	8	12	19	25	32	—	—
	2/0	—	1	1	1	3	4	7	10	15	21	27	—	—
	3/0	—	0	1	1	2	3	6	8	13	17	22	—	—
	4/0	—	0	1	1	1	3	5	7	10	14	18	—	—
Z	14	—	16	28	46	79	107	175	247	381	510	657	—	—
	12	—	11	20	32	56	76	124	175	271	362	466	—	—
	10	—	7	12	20	34	46	76	107	166	221	285	—	—
	8	—	4	7	12	22	29	48	68	105	140	180	—	—
	6	—	3	5	9	15	20	33	47	73	98	127	—	—
	4	—	1	3	6	10	14	23	33	50	67	87	—	—
	3	—	1	2	4	7	10	17	24	37	49	63	—	—
	2	—	1	1	3	6	8	14	20	30	41	53	—	—
	1	—	1	1	3	5	7	11	16	25	33	43	—	—
XHHW, ZW, XHHW-2, XHH	14	—	10	17	27	47	64	104	147	228	304	392	—	—
	12	—	7	13	21	36	49	80	113	175	234	301	—	—
	10	—	5	9	15	27	36	59	84	130	174	224	—	—
	8	—	3	5	8	15	20	33	47	72	97	124	—	—
	6	—	1	4	6	11	15	24	35	53	71	92	—	—

(continued)

Table C.4 Maximum Number of Conductors or Fixture Wires in Intermediate Metal Conduit (IMC) (Based on Chapter 9: Table 1, Table 4, and Table 5) ***Continued***

Type	Conductor Size (AWG/ kcmil)	Trade Size (Metric Designator)												
		⅜ (12)	½ (16)	¾ (21)	1 (27)	1¼ (35)	1½ (41)	2 (53)	2½ (63)	3 (78)	3½ (91)	4 (103)	5 (129)	6 (155)
		CONDUCTORS												
	4	—	1	3	4	8	11	18	25	39	52	67	—	—
	3	—	1	2	4	7	9	15	21	33	44	56	—	—
	2	—	1	1	3	5	7	12	18	27	37	47	—	—
XHHW, XHHW-2, XHH	1	—	1	1	2	4	6	9	13	20	27	35	—	—
	1/0	—	1	1	1	3	5	8	11	17	23	30	—	—
	2/0	—	1	1	1	3	4	6	9	14	19	25	—	—
	3/0	—	0	1	1	2	3	5	7	12	16	20	—	—
	4/0	—	0	1	1	1	2	4	6	10	13	17	—	—
	250	—	0	0	1	1	1	3	5	8	11	14	—	—
	300	—	0	0	1	1	1	3	4	7	9	12	—	—
	350	—	0	0	1	1	1	3	4	6	8	10	—	—
	400	—	0	0	1	1	1	2	3	5	7	9	—	—
	500	—	0	0	0	1	1	1	3	4	6	8	—	—

	600	—	0	0	0	1	1	1	2	3	5	6	—	—
	700	—	0	0	0	1	1	1	1	3	4	5	—	—
	750	—	0	0	0	1	1	1	1	3	4	5	—	—
	800	—	0	0	0	0	1	1	1	3	4	5	—	—
	900	—	0	0	0	0	1	1	1	2	3	4	—	—
	1000	—	0	0	0	0	1	1	1	2	3	4	—	—
	1250	—	0	0	0	0	0	1	1	1	2	3	—	—
	1500	—	0	0	0	0	0	1	1	1	1	2	—	—
	1750	—	0	0	0	0	0	1	1	1	1	2	—	—
	2000	—	0	0	0	0	0	0	1	1	1	1	—	—
FIXTURE WIRES														
RFH-2, FFH-2, RFHH-2	18	—	9	16	26	45	61	100	141	218	292	376	—	—
	16	—	8	13	22	38	51	84	119	184	246	317	—	—
SF-2, SFF-2	18	—	12	20	33	57	77	126	178	275	368	474	—	—
	16	—	10	17	27	47	64	104	147	228	304	392	—	—
	14	—	8	13	22	38	51	84	119	184	246	317	—	—
SF-1, SFF-1	18	—	21	36	59	101	137	223	316	487	651	839	—	—
RFH-1, TF, TFF, XF, XFF	18	—	15	26	43	75	101	165	233	360	481	619	—	—
	16	—	12	21	35	60	81	133	188	290	388	500	—	—
XF, XFF	14	—	10	17	27	47	64	104	147	228	304	392	—	—
TFN, TFFN	18	—	25	42	69	119	162	264	373	576	769	991	—	—
	16	—	19	32	53	91	123	201	285	440	588	757	—	—

(continued)

Table C.4 Maximum Number of Conductors or Fixture Wires in Intermediate Metal Conduit (IMC) (Based on Chapter 9: Table 1, Table 4, and Table 5) ***Continued***

Type	Conductor Size (AWG/ kcmil)	Trade Size (Metric Designator)												
		3/8 (12)	1/2 (16)	3/4 (21)	1 (27)	1¼ (35)	1½ (41)	2 (53)	2½ (63)	3 (78)	3½ (91)	4 (103)	5 (129)	6 (155)
FIXTURE WIRES														
PF, PFF, PGF, PGFF, PAF, PTF, PTFF, PAFF	18	—	23	40	66	113	153	250	354	546	730	940	—	—
	16	—	18	31	51	88	118	193	274	422	564	727	—	—
	14	—	13	23	38	66	89	145	205	317	423	545	—	—
ZF, ZFF, ZHF	18	—	30	52	85	146	197	322	456	704	941	1211	—	—
	16	—	22	38	63	108	146	238	336	519	694	894	—	—
	14	—	16	28	46	79	107	175	247	381	510	657	—	—
KF-2, KFF-2	18	—	45	78	128	219	296	484	684	1056	1411	1817	—	—
	16	—	32	54	89	153	207	337	477	737	984	1268	—	—
	14	—	21	36	60	103	139	227	321	495	661	852	—	—
	12	—	15	25	41	71	96	158	223	344	460	592	—	—
	10	—	10	17	27	47	64	104	147	228	304	392	—	—

KF-1, KFF-1	18	—	52	90	147	253	342	558	790	1218	1628	2097	—	—
	16	—	37	63	103	178	240	392	555	856	1144	1473	—	—
	14	—	25	42	69	119	162	264	373	576	769	991	—	—
	12	—	16	28	46	79	107	175	247	381	510	657	—	—
	10	—	10	18	30	52	70	114	161	249	333	429	—	—
XF, XFF	12	—	5	9	14	25	34	56	79	122	163	209	—	—
	10	—	4	7	11	19	26	43	61	95	127	163	—	—

Notes:

1. This table is for concentric stranded conductors only. For compact stranded conductors, Table C.4(A) should be used.

2. Two-hour fire-rated RHH cable has ceramifiable insulation, which has much larger diameters than other RHH wires. Consult manufacturer's conduit fill tables.

*Types RHH, RHW, and RHW-2 without outer covering.

Table C.5 Maximum Number of Conductors or Fixture Wires in Liquidtight Flexible Nonmetallic Conduit (Type LFNC-A) (Based on Chapter 9: Table 1, Table 4, and Table 5)

Type	Conductor Size (AWG/ kcmil)	Trade Size (Metric Designator)												
		⅜ (12)	½ (16)	¾ (21)	1 (27)	1¼ (35)	1½ (41)	2 (53)	2½ (63)	3 (78)	3½ (91)	4 (103)	5 (129)	6 (155)
CONDUCTORS														
RHH, RHW, RHW-2	14	2	4	7	11	20	27	45	—	—	—	—	—	—
	12	1	3	6	9	17	23	38	—	—	—	—	—	—
	10	1	3	5	8	13	18	30	—	—	—	—	—	—
	8	1	1	2	4	7	9	16	—	—	—	—	—	—
	6	1	1	1	3	5	7	13	—	—	—	—	—	—
	4	0	1	1	2	4	6	10	—	—	—	—	—	—
	3	0	1	1	1	4	5	8	—	—	—	—	—	—
	2	0	1	1	1	3	4	7	—	—	—	—	—	—
	1	0	0	1	1	1	3	5	—	—	—	—	—	—
	1/0	0	0	1	1	1	2	4	—	—	—	—	—	—
	2/0	0	0	1	1	1	1	4	—	—	—	—	—	—
	3/0	0	0	0	1	1	1	3	—	—	—	—	—	—
	4/0	0	0	0	1	1	1	3	—	—	—	—	—	—
	250	0	0	0	0	1	1	1	—	—	—	—	—	—
	300	0	0	0	0	1	1	1	—	—	—	—	—	—
	350	0	0	0	0	1	1	1	—	—	—	—	—	—

	400	0	0	0	0	1	1	1	—	—	—	—	—	—
	500	0	0	0	0	0	1	1	—	—	—	—	—	—
	600	0	0	0	0	0	1	1	—	—	—	—	—	—
	700	0	0	0	0	0	0	1	—	—	—	—	—	—
	750	0	0	0	0	0	0	1	—	—	—	—	—	—
	800	0	0	0	0	0	0	1	—	—	—	—	—	—
	900	0	0	0	0	0	0	1	—	—	—	—	—	—
	1000	0	0	0	0	0	0	1	—	—	—	—	—	—
	1250	0	0	0	0	0	0	0	—	—	—	—	—	—
	1500	0	0	0	0	0	0	0	—	—	—	—	—	—
	1750	0	0	0	0	0	0	0	—	—	—	—	—	—
	2000	0	0	0	0	0	0	0	—	—	—	—	—	—
TW, THHW, THW, THW-2	14	5	9	15	24	43	58	96	—	—	—	—	—	—
	12	4	7	12	19	33	44	74	—	—	—	—	—	—
	10	3	5	9	14	24	33	55	—	—	—	—	—	—
	8	1	3	5	8	13	18	30	—	—	—	—	—	—
RHH*, RHW*, RHW-2*	14	3	6	10	16	28	38	64	—	—	—	—	—	—
	12	3	5	8	13	23	31	51	—	—	—	—	—	—
	10	1	3	6	10	18	24	40	—	—	—	—	—	—
	8	1	1	4	6	11	14	24	—	—	—	—	—	—

(continued)

Table C.5 Maximum Number of Conductors or Fixture Wires in Liquidtight Flexible Nonmetallic Conduit (Type LFNC-A) (Based on Chapter 9: Table 1, Table 4, and Table 5) ***Continued***

Type	Conductor Size (AWG/kcmil)	Trade Size (Metric Designator)												
		⅜ (12)	½ (16)	¾ (21)	1 (27)	1¼ (35)	1½ (41)	2 (53)	2½ (63)	3 (78)	3½ (91)	4 (103)	5 (129)	6 (155)
CONDUCTORS														
TW, THW, THHW, THW-2, RHH*, RHW*, RHW-2*	6	1	1	3	4	8	11	18	—	—	—	—	—	—
	4	1	1	1	3	6	8	13	—	—	—	—	—	—
	3	1	1	1	3	5	7	11	—	—	—	—	—	—
	2	0	1	1	2	4	6	10	—	—	—	—	—	—
	1	0	1	1	1	3	4	7	—	—	—	—	—	—
	1/0	0	0	1	1	2	3	6	—	—	—	—	—	—
	2/0	0	0	1	1	1	3	5	—	—	—	—	—	—
	3/0	0	0	1	1	1	2	4	—	—	—	—	—	—
	4/0	0	0	0	1	1	1	3	—	—	—	—	—	—
	250	0	0	0	1	1	1	3	—	—	—	—	—	—
	300	0	0	0	1	1	1	2	—	—	—	—	—	—
	350	0	0	0	0	1	1	1	—	—	—	—	—	—
	400	0	0	0	0	1	1	1	—	—	—	—	—	—
	500	0	0	0	0	1	1	1	—	—	—	—	—	—
	600	0	0	0	0	1	1	1	—	—	—	—	—	—
	700	0	0	0	0	0	1	1	—	—	—	—	—	—

	750	0	0	0	0	0	1	1	—	—	—	—	—	—
	800	0	0	0	0	0	1	1	—	—	—	—	—	—
	900	0	0	0	0	0	0	1	—	—	—	—	—	—
	1000	0	0	0	0	0	0	1	—	—	—	—	—	—
	1250	0	0	0	0	0	0	1	—	—	—	—	—	—
	1500	0	0	0	0	0	0	1	—	—	—	—	—	—
	1750	0	0	0	0	0	0	0	—	—	—	—	—	—
	2000	0	0	0	0	0	0	0	—	—	—	—	—	—
THHN, THWN, THWN-2	14	8	13	22	35	62	83	138	—	—	—	—	—	—
	12	5	9	16	25	45	60	100	—	—	—	—	—	—
	10	3	6	10	16	28	38	63	—	—	—	—	—	—
	8	1	3	6	9	16	22	36	—	—	—	—	—	—
	6	1	2	4	6	12	16	26	—	—	—	—	—	—
	4	1	1	2	4	7	9	16	—	—	—	—	—	—
	3	1	1	1	3	6	8	13	—	—	—	—	—	—
	2	1	1	1	3	5	7	11	—	—	—	—	—	—
	1	0	1	1	1	4	5	8	—	—	—	—	—	—
	1/0	0	1	1	1	3	4	7	—	—	—	—	—	—
	2/0	0	0	1	1	2	3	6	—	—	—	—	—	—
	3/0	0	0	1	1	1	3	5	—	—	—	—	—	—
	4/0	0	0	1	1	1	2	4	—	—	—	—	—	—
	250	0	0	0	1	1	1	3	—	—	—	—	—	—
	300	0	0	0	1	1	1	3	—	—	—	—	—	—
	350	0	0	0	1	1	1	2	—	—	—	—	—	—

(continued)

Table C.5 Maximum Number of Conductors or Fixture Wires in Liquidtight Flexible Nonmetallic Conduit (Type LFNC-A) (Based on Chapter 9: Table 1, Table 4, and Table 5) ***Continued***

Type	Conductor Size (AWG/ kcmil)	Trade Size (Metric Designator)												
		⅜ (12)	½ (16)	¾ (21)	1 (27)	1¼ (35)	1½ (41)	2 (53)	2½ (63)	3 (78)	3½ (91)	4 (103)	5 (129)	6 (155)
		CONDUCTORS												
	400	0	0	0	0	1	1	1	—	—	—	—	—	—
	500	0	0	0	0	1	1	1	—	—	—	—	—	—
	600	0	0	0	0	1	1	1	—	—	—	—	—	—
	700	0	0	0	0	1	1	1	—	—	—	—	—	—
	750	0	0	0	0	0	1	1	—	—	—	—	—	—
	800	0	0	0	0	0	1	1	—	—	—	—	—	—
	900	0	0	0	0	0	1	1	—	—	—	—	—	—
	1000	0	0	0	0	0	0	1	—	—	—	—	—	—
FEP, FEPB, PFA, PFAH, TFE	14	7	12	21	34	60	80	133	—	—	—	—	—	—
	12	5	9	15	25	44	59	97	—	—	—	—	—	—
	10	4	6	11	18	31	42	70	—	—	—	—	—	—
	8	1	3	6	10	18	24	40	—	—	—	—	—	—
	6	1	2	4	7	13	17	28	—	—	—	—	—	—
	4	1	1	3	5	9	12	20	—	—	—	—	—	—
	3	1	1	2	4	7	10	16	—	—	—	—	—	—
	2	1	1	1	3	6	8	13	—	—	—	—	—	—

PFA, PFAH, TFE	1	0	1	1	2	4	5	9	—	—	—	—	—	—
PFA, PFAH, TFE, Z	1/0	0	1	1	1	3	5	8	—	—	—	—	—	—
	2/0	0	1	1	1	3	4	6	—	—	—	—	—	—
	3/0	0	0	1	1	2	3	5	—	—	—	—	—	—
	4/0	0	0	1	1	1	2	4	—	—	—	—	—	—
Z	14	9	15	25	41	72	97	161	—	—	—	—	—	—
	12	6	10	18	29	51	69	114	—	—	—	—	—	—
	10	4	6	11	18	31	42	70	—	—	—	—	—	—
	8	2	4	7	11	20	26	44	—	—	—	—	—	—
	6	1	3	5	8	14	18	31	—	—	—	—	—	—
	4	1	1	3	5	9	13	21	—	—	—	—	—	—
	3	1	1	2	4	7	9	15	—	—	—	—	—	—
	2	1	1	1	3	6	8	13	—	—	—	—	—	—
	1	1	1	1	2	4	6	10	—	—	—	—	—	—
XHHW, ZW, XHHW-2, XHH	14	5	9	15	24	43	58	96	—	—	—	—	—	—
	12	4	7	12	19	33	44	74	—	—	—	—	—	—
	10	3	5	9	14	24	33	55	—	—	—	—	—	—
	8	1	3	5	8	13	18	30	—	—	—	—	—	—
	6	1	1	3	5	10	13	22	—	—	—	—	—	—
	4	1	1	2	4	7	10	16	—	—	—	—	—	—
	3	1	1	1	3	6	8	14	—	—	—	—	—	—
	2	1	1	1	3	5	7	11	—	—	—	—	—	—

(continued)

Table C.5 Maximum Number of Conductors or Fixture Wires in Liquidtight Flexible Nonmetallic Conduit (Type LFNC-A) (Based on Chapter 9: Table 1, Table 4, and Table 5) ***Continued***

Type	Conductor Size (AWG/kcmil)	Trade Size (Metric Designator)												
		⅜ (12)	½ (16)	¾ (21)	1 (27)	1¼ (35)	1½ (41)	2 (53)	2½ (63)	3 (78)	3½ (91)	4 (103)	5 (129)	6 (155)
CONDUCTORS														
XHHW, XHHW-2, XHH	1	0	1	1	1	4	5	8	—	—	—	—	—	—
	1/0	0	1	1	1	3	4	7	—	—	—	—	—	—
	2/0	0	0	1	1	2	3	6	—	—	—	—	—	—
	3/0	0	0	1	1	1	3	5	—	—	—	—	—	—
	4/0	0	0	1	1	1	2	4	—	—	—	—	—	—
	250	0	0	0	1	1	1	3	—	—	—	—	—	—
	300	0	0	0	1	1	1	3	—	—	—	—	—	—
	350	0	0	0	1	1	1	2	—	—	—	—	—	—
	400	0	0	0	0	1	1	1	—	—	—	—	—	—
	500	0	0	0	0	1	1	1	—	—	—	—	—	—
	600	0	0	0	0	1	1	1	—	—	—	—	—	—
	700	0	0	0	0	1	1	1	—	—	—	—	—	—
	750	0	0	0	0	0	1	1	—	—	—	—	—	—
	800	0	0	0	0	0	1	1	—	—	—	—	—	—
	900	0	0	0	0	0	1	1	—	—	—	—	—	—
	1000	0	0	0	0	0	0	1	—	—	—	—	—	—
	1250	0	0	0	0	0	0	1	—	—	—	—	—	—
	1500	0	0	0	0	0	0	1	—	—	—	—	—	—

	1750	0	0	0	0	0	0	0	—	—	—	—	—	—
	2000	0	0	0	0	0	0	0	—	—	—	—	—	—
FIXTURE WIRES														
RFH-2, FFH-2, RFHH-2	18	5	8	14	23	41	55	92	—	—	—	—	—	—
	16	4	7	12	20	35	47	77	—	—	—	—	—	—
SF-2, SFF-2	18	6	11	18	29	52	70	116	—	—	—	—	—	—
	16	5	9	15	24	43	58	96	—	—	—	—	—	—
	14	4	7	12	20	35	47	77	—	—	—	—	—	—
SF-1, SFF-1	18	12	19	33	52	92	124	205	—	—	—	—	—	—
RFH-1, TF, TFF, XF, XFF	18	8	14	24	39	68	91	152	—	—	—	—	—	—
	16	7	11	19	31	55	74	122	—	—	—	—	—	—
XF, XFF	14	5	9	15	24	43	58	96	—	—	—	—	—	—
TFN, TFFN	18	14	22	39	62	109	146	243	—	—	—	—	—	—
	16	10	17	29	47	83	112	185	—	—	—	—	—	—
PF, PFF, PGF, PGFF, PAF, PTF, PTFF, PAFF	18	13	21	37	59	103	139	230	—	—	—	—	—	—
	16	10	16	28	45	80	107	178	—	—	—	—	—	—
	14	7	12	21	34	60	80	133	—	—	—	—	—	—
ZF, ZFF, ZHF	18	17	27	47	76	133	179	297	—	—	—	—	—	—
	16	12	20	35	56	98	132	219	—	—	—	—	—	—
	14	9	15	25	41	72	97	161	—	—	—	—	—	—
KF-2, KFF-2	18	25	41	71	114	200	269	445	—	—	—	—	—	—
	16	18	29	49	79	139	187	311	—	—	—	—	—	—

(continued)

Table C.5 Maximum Number of Conductors or Fixture Wires in Liquidtight Flexible Nonmetallic Conduit (Type LFNC-A) (Based on Chapter 9: Table 1, Table 4, and Table 5) ***Continued***

Type	Conductor Size (AWG/ kcmil)	Trade Size (Metric Designator)												
		⅜ (12)	½ (16)	¾ (21)	1 (27)	1¼ (35)	1½ (41)	2 (53)	2½ (63)	3 (78)	3½ (91)	4 (103)	5 (129)	6 (155)
		FIXTURE WIRES												
	14	12	19	33	53	94	126	209	—	—	—	—	—	—
	12	8	13	23	37	65	87	145	—	—	—	—	—	—
	10	5	9	15	24	43	58	96	—	—	—	—	—	—
KF-1, KFF-1	18	29	48	82	131	231	310	514	—	—	—	—	—	—
	16	20	33	58	92	162	218	361	—	—	—	—	—	—
	14	14	22	39	62	109	146	243	—	—	—	—	—	—
	12	9	15	25	41	72	97	161	—	—	—	—	—	—
	10	6	10	17	27	47	63	105	—	—	—	—	—	—
XF, XFF	12	3	5	8	13	23	31	51	—	—	—	—	—	—
	10	1	3	6	10	18	24	40	—	—	—	—	—	—

Notes:

1. This table is for concentric stranded conductors only. For compact stranded conductors, Table C.6(A) should be used.

2. Two-hour fire-rated RHH cable has ceramifiable insulation, which has much larger diameters than other RHH wires. Consult manufacturer's conduit fill tables.

*Types RHH, RHW, and RHW-2 without outer covering.

Table C.6 Maximum Number of Conductors or Fixture Wires in Liquidtight Flexible Nonmetallic Conduit (Type LFNC-B*) (Based on Chapter 9: Table 1, Table 4, and Table 5)

Type	Conductor Size (AWG/kcmil)	Trade Size (Metric Designator)												
		⅜ (12)	½ (16)	¾ (21)	1 (27)	1¼ (35)	1½ (41)	2 (53)	2½ (63)	3 (78)	3½ (91)	4 (103)	5 (129)	6 (155)
CONDUCTORS														
RHH, RHW, RHW-2	14	2	4	7	12	21	27	44	—	—	—	—	—	—
	12	1	3	6	10	17	22	36	—	—	—	—	—	—
	10	1	3	5	8	14	18	29	—	—	—	—	—	—
	8	1	1	2	4	7	9	15	—	—	—	—	—	—
	6	1	1	1	3	6	7	12	—	—	—	—	—	—
	4	0	1	1	2	4	6	9	—	—	—	—	—	—
	3	0	1	1	1	4	5	8	—	—	—	—	—	—
	2	0	1	1	1	3	4	7	—	—	—	—	—	—
	1	0	0	1	1	1	3	5	—	—	—	—	—	—
	1/0	0	0	1	1	1	2	4	—	—	—	—	—	—
	2/0	0	0	1	1	1	1	3	—	—	—	—	—	—

(continued)

Table C.6 Maximum Number of Conductors or Fixture Wires in Liquidtight Flexible Nonmetallic Conduit (Type LFNC-B*) (Based on Chapter 9: Table 1, Table 4, and Table 5) ***Continued***

Type	Conductor Size (AWG/ kcmil)	Trade Size (Metric Designator)												
		⅜ (12)	½ (16)	¾ (21)	1 (27)	1¼ (35)	1½ (41)	2 (53)	2½ (63)	3 (78)	3½ (91)	4 (103)	5 (129)	6 (155)
						CONDUCTORS								
	3/0	0	0	0	1	1	1	3	—	—	—	—	—	—
	4/0	0	0	0	1	1	1	2	—	—	—	—	—	—
	250	0	0	0	0	1	1	1	—	—	—	—	—	—
	300	0	0	0	0	1	1	1	—	—	—	—	—	—
	350	0	0	0	0	1	1	1	—	—	—	—	—	—
	400	0	0	0	0	1	1	1	—	—	—	—	—	—
	500	0	0	0	0	1	1	1	—	—	—	—	—	—
	600	0	0	0	0	0	1	1	—	—	—	—	—	—
	700	0	0	0	0	0	0	1	—	—	—	—	—	—
	750	0	0	0	0	0	0	1	—	—	—	—	—	—
	800	0	0	0	0	0	0	1	—	—	—	—	—	—
	900	0	0	0	0	0	0	1	—	—	—	—	—	—
	1000	0	0	0	0	0	0	1	—	—	—	—	—	—
	1250	0	0	0	0	0	0	0	—	—	—	—	—	—
	1500	0	0	0	0	0	0	0	—	—	—	—	—	—
	1750	0	0	0	0	0	0	0	—	—	—	—	—	—
	2000	0	0	0	0	0	0	0	—	—	—	—	—	—

TW, THHW, THW, THW-2	14	5	9	15	25	44	57	93	—	—	—	—	—	—
	12	4	7	12	19	33	43	71	—	—	—	—	—	—
	10	3	5	9	14	25	32	53	—	—	—	—	—	—
	8	1	3	5	8	14	18	29	—	—	—	—	—	—
RHH*, RHW*, RHW-2*	14	3	6	10	16	29	38	62	—	—	—	—	—	—
	12	3	5	8	13	23	30	50	—	—	—	—	—	—
	10	1	3	6	10	18	23	39	—	—	—	—	—	—
	8	1	1	4	6	11	14	23	—	—	—	—	—	—
TW, THW, THHW, THW-2, RHH*, RHW*, RHW-2*	6	1	1	3	5	8	11	18	—	—	—	—	—	—
	4	1	1	1	3	6	8	13	—	—	—	—	—	—
	3	1	1	1	3	5	7	11	—	—	—	—	—	—
	2	0	1	1	2	4	6	9	—	—	—	—	—	—
	1	0	1	1	1	3	4	7	—	—	—	—	—	—
	1/0	0	0	1	1	2	3	6	—	—	—	—	—	—
	2/0	0	0	1	1	2	3	5	—	—	—	—	—	—
	3/0	0	0	1	1	1	2	4	—	—	—	—	—	—
	4/0	0	0	0	1	1	1	3	—	—	—	—	—	—
	250	0	0	0	1	1	1	3	—	—	—	—	—	—
	300	0	0	0	1	1	1	2	—	—	—	—	—	—
	350	0	0	0	0	1	1	1	—	—	—	—	—	—
	400	0	0	0	0	1	1	1	—	—	—	—	—	—
	500	0	0	0	0	1	1	1	—	—	—	—	—	—

(continued)

Table C.6 Maximum Number of Conductors or Fixture Wires in Liquidtight Flexible Nonmetallic Conduit (Type LFNC-B*) (Based on Chapter 9: Table 1, Table 4, and Table 5) ***Continued***

Type	Conductor Size (AWG/kcmil)	Trade Size (Metric Designator)												
		⅜ (12)	½ (16)	¾ (21)	1 (27)	1¼ (35)	1½ (41)	2 (53)	2½ (63)	3 (78)	3½ (91)	4 (103)	5 (129)	6 (155)
CONDUCTORS														
	600	0	0	0	0	1	1	1	—	—	—	—	—	—
	700	0	0	0	0	0	1	1	—	—	—	—	—	—
	750	0	0	0	0	0	1	1	—	—	—	—	—	—
	800	0	0	0	0	0	1	1	—	—	—	—	—	—
	900	0	0	0	0	0	0	1	—	—	—	—	—	—
	1000	0	0	0	0	0	0	1	—	—	—	—	—	—
	1250	0	0	0	0	0	0	1	—	—	—	—	—	—
	1500	0	0	0	0	0	0	0	—	—	—	—	—	—
	1750	0	0	0	0	0	0	0	—	—	—	—	—	—
	2000	0	0	0	0	0	0	0	—	—	—	—	—	—
THHN, THWN, THWN-2	14	8	13	22	36	63	81	134	—	—	—	—	—	—
	12	5	9	16	26	46	59	97	—	—	—	—	—	—
	10	3	6	10	16	29	37	61	—	—	—	—	—	—
	8	1	3	6	9	16	21	35	—	—	—	—	—	—
	6	1	2	4	7	12	15	25	—	—	—	—	—	—
	4	1	1	2	4	7	9	15	—	—	—	—	—	—
	3	1	1	1	3	6	8	13	—	—	—	—	—	—
	2	1	1	1	3	5	7	11	—	—	—	—	—	—
	1	0	1	1	1	4	5	8	—	—	—	—	—	—

	1/0	0	1	1	1	3	4	7	—	—	—	—	—	—
	2/0	0	0	1	1	2	3	6	—	—	—	—	—	—
	3/0	0	0	1	1	1	3	5	—	—	—	—	—	—
	4/0	0	0	1	1	1	2	4	—	—	—	—	—	—
	250	0	0	0	1	1	1	3	—	—	—	—	—	—
	300	0	0	0	1	1	1	3	—	—	—	—	—	—
	350	0	0	0	1	1	1	2	—	—	—	—	—	—
	400	0	0	0	0	1	1	1	—	—	—	—	—	—
	500	0	0	0	0	1	1	1	—	—	—	—	—	—
	600	0	0	0	0	1	1	1	—	—	—	—	—	—
	700	0	0	0	0	1	1	1	—	—	—	—	—	—
	750	0	0	0	0	0	1	1	—	—	—	—	—	—
	800	0	0	0	0	0	1	1	—	—	—	—	—	—
	900	0	0	0	0	0	1	1	—	—	—	—	—	—
	1000	0	0	0	0	0	0	1	—	—	—	—	—	—
FEP, FEPB, PFA, PFAH, TFE	14	7	12	21	35	61	79	130	—	—	—	—	—	—
	12	5	9	15	25	44	58	94	—	—	—	—	—	—
	10	4	6	11	18	32	41	68	—	—	—	—	—	—
	8	1	3	6	10	18	23	39	—	—	—	—	—	—
	6	1	2	4	7	13	17	27	—	—	—	—	—	—
	4	1	1	3	5	9	12	19	—	—	—	—	—	—
	3	1	1	2	4	7	10	16	—	—	—	—	—	—
	2	1	1	1	3	6	8	13	—	—	—	—	—	—

(continued)

Table C.6 Maximum Number of Conductors or Fixture Wires in Liquidtight Flexible Nonmetallic Conduit (Type LFNC-B*) (Based on Chapter 9: Table 1, Table 4, and Table 5) ***Continued***

Type	Conductor Size (AWG/kcmil)	Trade Size (Metric Designator)												
		3/8 (12)	1/2 (16)	3/4 (21)	1 (27)	1 1/4 (35)	1 1/2 (41)	2 (53)	2 1/2 (63)	3 (78)	3 1/2 (91)	4 (103)	5 (129)	6 (155)
CONDUCTORS														
PFA, PFAH, TFE	1	0	1	1	2	4	5	9	—	—	—	—	—	—
PFA, PFAH, TFE, Z	1/0	0	1	1	1	3	4	7	—	—	—	—	—	—
	2/0	0	1	1	1	3	4	6	—	—	—	—	—	—
	3/0	0	0	1	1	2	3	5	—	—	—	—	—	—
	4/0	0	0	1	1	1	2	4	—	—	—	—	—	—
Z	14	9	15	26	42	73	95	156	—	—	—	—	—	—
	12	6	10	18	30	52	67	111	—	—	—	—	—	—
	10	4	6	11	18	32	41	68	—	—	—	—	—	—
	8	2	4	7	11	20	26	43	—	—	—	—	—	—
	6	1	3	5	8	14	18	30	—	—	—	—	—	—
	4	1	1	3	5	9	12	20	—	—	—	—	—	—
	3	1	1	2	4	7	9	15	—	—	—	—	—	—
	2	1	1	1	3	6	7	12	—	—	—	—	—	—
	1	1	1	1	2	5	6	10	—	—	—	—	—	—

XHHW, ZW, XHHW-2, XHH	14	5	9	15	25	44	57	93	—	—	—	—	—	—
	12	4	7	12	19	33	43	71	—	—	—	—	—	—
	10	3	5	9	14	25	32	53	—	—	—	—	—	—
	8	1	3	5	8	14	18	29	—	—	—	—	—	—
	6	1	1	3	6	10	13	22	—	—	—	—	—	—
	4	1	1	2	4	7	9	16	—	—	—	—	—	—
	3	1	1	1	3	6	8	13	—	—	—	—	—	—
	2	1	1	1	3	5	7	11	—	—	—	—	—	—
XHHW, XHHW-2, XHH	1	0	1	1	1	4	5	8	—	—	—	—	—	—
	1/0	0	1	1	1	3	4	7	—	—	—	—	—	—
	2/0	0	0	1	1	2	3	6	—	—	—	—	—	—
	3/0	0	0	1	1	1	3	5	—	—	—	—	—	—
	4/0	0	0	1	1	1	2	4	—	—	—	—	—	—
	250	0	0	0	1	1	1	3	—	—	—	—	—	—
	300	0	0	0	1	1	1	3	—	—	—	—	—	—
	350	0	0	0	1	1	1	2	—	—	—	—	—	—
	400	0	0	0	1	1	1	1	—	—	—	—	—	—
	500	0	0	0	0	1	1	1	—	—	—	—	—	—
	600	0	0	0	0	1	1	1	—	—	—	—	—	—
	700	0	0	0	0	1	1	1	—	—	—	—	—	—
	750	0	0	0	0	0	1	1	—	—	—	—	—	—
	800	0	0	0	0	0	1	1	—	—	—	—	—	—
	900	0	0	0	0	0	1	1	—	—	—	—	—	—

(continued)

Table C.6 Maximum Number of Conductors or Fixture Wires in Liquidtight Flexible Nonmetallic Conduit (Type LFNC-B*) (Based on Chapter 9: Table 1, Table 4, and Table 5) ***Continued***

Type	Conductor Size (AWG/ kcmil)	Trade Size (Metric Designator)												
		3/8 (12)	1/2 (16)	3/4 (21)	1 (27)	1¼ (35)	1½ (41)	2 (53)	2½ (63)	3 (78)	3½ (91)	4 (103)	5 (129)	6 (155)
				CONDUCTORS										
	1000	0	0	0	0	0	0	1	—	—	—	—	—	—
	1250	0	0	0	0	0	0	1	—	—	—	—	—	—
	1500	0	0	0	0	0	0	1	—	—	—	—	—	—
	1750	0	0	0	0	0	0	0	—	—	—	—	—	—
	2000	0	0	0	0	0	0	0	—	—	—	—	—	—
				FIXTURE WIRES										
RFH-2, FFH-2, RFHH-2	18	5	8	15	24	42	54	89	—	—	—	—	—	—
	16	4	7	12	20	35	46	75	—	—	—	—	—	—
SF-2, SFF-2	18	6	11	19	30	53	69	113	—	—	—	—	—	—
	16	5	9	15	25	44	57	93	—	—	—	—	—	—
	14	4	7	12	20	35	46	75	—	—	—	—	—	—
SF-1, SFF-1	18	12	19	33	53	94	122	199	—	—	—	—	—	—
RFH-1, TF, TFF, XF, XFF	18	8	14	24	39	69	90	147	—	—	—	—	—	—
	16	7	11	20	32	56	72	119	—	—	—	—	—	—
XF, XFF	14	5	9	15	25	44	57	93	—	—	—	—	—	—
TFN, TFFN	18	14	23	39	63	111	144	236	—	—	—	—	—	—
	16	10	17	30	48	85	110	180	—	—	—	—	—	—

PF, PFF, PGF, PGFF, PAF, PTF, PTFF, PAFF	18	13	21	37	60	105	136	224	—	—	—	—	—	—
	16	10	16	29	46	81	105	173	—	—	—	—	—	—
	14	7	12	21	35	61	79	130	—	—	—	—	—	—
ZF, ZFF, ZHF	18	17	28	48	77	136	176	288	—	—	—	—	—	—
	16	12	20	35	57	100	130	213	—	—	—	—	—	—
	14	9	15	26	42	73	95	156	—	—	—	—	—	—
KF-2, KFF-2	18	25	42	72	116	203	264	433	—	—	—	—	—	—
	16	18	29	50	81	142	184	302	—	—	—	—	—	—
	14	12	19	34	54	95	124	203	—	—	—	—	—	—
	12	8	13	23	38	66	86	141	—	—	—	—	—	—
	10	5	9	15	25	44	57	93	—	—	—	—	—	—
KF-1, KFF-1	18	29	48	83	134	235	304	499	—	—	—	—	—	—
	16	20	34	58	94	165	214	351	—	—	—	—	—	—
	14	14	23	39	63	111	144	236	—	—	—	—	—	—
	12	9	15	26	42	73	95	156	—	—	—	—	—	—
	10	6	10	17	27	48	62	102	—	—	—	—	—	—
XF, XFF	12	3	5	8	13	23	30	50	—	—	—	—	—	—
	10	1	3	6	10	18	23	39	—	—	—	—	—	—

Notes:

1. This table is for concentric stranded conductors only. For compact stranded conductors, Table C.5(A) should be used.

2. Two-hour fire-rated RHH cable has ceramifiable insulation, which has much larger diameters than other RHH wires. Consult manufacturer's conduit fill tables.

*Types RHH, RHW, and RHW-2 without outer covering.

N

Table C.7 Maximum Number of Conductors of Fixture Wires in Liquidtight Flexible Nonmetallic Conduit (Type LFNC-C) (Based on Chapter 9: Table 1, Table 4, and Table 5)

Type	Conductor Size (AWG/ kcmil)	Trade Size (Metric Designator)												
		⅜ (12)	½ (16)	¾ (21)	1 (27)	1¼ (35)	1½ (41)	2 (53)	2½ (63)	3 (78)	3½ (91)	4 (103)	5 (129)	6 (155)
CONDUCTORS														
RHH, RHW, RHW-2	14	2	4	7	11	20	27	45	—	—	—	—	—	—
	12	1	3	6	9	16	22	37	—	—	—	—	—	—
	10	1	2	4	7	13	18	30	—	—	—	—	—	—
	8	1	1	2	4	7	9	15	—	—	—	—	—	—
	6	0	1	1	3	5	7	12	—	—	—	—	—	—
	4	0	1	1	2	4	6	10	—	—	—	—	—	—
	3	0	1	1	1	4	5	8	—	—	—	—	—	—
	2	0	0	1	1	3	4	7	—	—	—	—	—	—
	1	0	0	1	1	1	3	5	—	—	—	—	—	—
	1/0	0	0	0	1	1	2	4	—	—	—	—	—	—
	2/0	0	0	0	1	1	1	3	—	—	—	—	—	—
	3/0	0	0	0	1	1	1	3	—	—	—	—	—	—
	4/0	0	0	0	0	1	1	2	—	—	—	—	—	—

	250	0	0	0	0	1	1	1	—	—	—	—	—	—
	300	0	0	0	0	1	1	1	—	—	—	—	—	—
	350	0	0	0	0	0	1	1	—	—	—	—	—	—
	400	0	0	0	0	0	1	1	—	—	—	—	—	—
	500	0	0	0	0	0	1	1	—	—	—	—	—	—
	600	0	0	0	0	0	0	1	—	—	—	—	—	—
	700	0	0	0	0	0	0	1	—	—	—	—	—	—
	750	0	0	0	0	0	0	1	—	—	—	—	—	—
	800	0	0	0	0	0	0	1	—	—	—	—	—	—
	900	0	0	0	0	0	0	1	—	—	—	—	—	—
	1000	0	0	0	0	0	0	0	—	—	—	—	—	—
	1250	0	0	0	0	0	0	0	—	—	—	—	—	—
	1500	0	0	0	0	0	0	0	—	—	—	—	—	—
	1750	0	0	0	0	0	0	0	—	—	—	—	—	—
	2000	0	0	0	0	0	0	0	—	—	—	—	—	—
TW, THHW, THW, THW-2	14	5	8	15	24	42	56	94	—	—	—	—	—	—
	12	4	6	11	18	32	43	72	—	—	—	—	—	—
	10	3	5	8	13	24	32	54	—	—	—	—	—	—
	8	1	2	4	7	13	18	30	—	—	—	—	—	—

(continued)

Table C.7 Maximum Number of Conductors of Fixture Wires in Liquidtight Flexible Nonmetallic Conduit (Type LFNC-C) (Based on Chapter 9: Table 1, Table 4, and Table 5) ***Continued***

Type	Conductor Size (AWG/ kcmil)	Trade Size (Metric Designator)												
		3/8 (12)	1/2 (16)	3/4 (21)	1 (27)	1¼ (35)	1½ (41)	2 (53)	2½ (63)	3 (78)	3½ (91)	4 (103)	5 (129)	6 (155)
					CONDUCTORS									
RHH*, RHW*, RHW-2*	14	2	5	10	16	28	37	63	—	—	—	—	—	—
	12	2	4	8	13	22	30	50	—	—	—	—	—	—
	10	1	3	6	10	17	23	39	—	—	—	—	—	—
	8	1	1	3	6	10	14	23	—	—	—	—	—	—
TW, THW, THHW, THW-2, RHH*, RHW*, RHW-2	6	1	1	3	4	8	11	18	—	—	—	—	—	—
	4	1	1	1	3	6	8	13	—	—	—	—	—	—
	3	0	1	1	3	5	7	11	—	—	—	—	—	—
	2	0	1	1	2	4	6	10	—	—	—	—	—	—
	1	0	0	1	1	3	4	7	—	—	—	—	—	—
	1/0	0	0	1	1	2	3	6	—	—	—	—	—	—
	2/0	0	0	1	1	1	3	5	—	—	—	—	—	—
	3/0	0	0	0	1	1	2	4	—	—	—	—	—	—
	4/0	0	0	0	1	1	1	3	—	—	—	—	—	—

	250	0	0	0	0	1	1	3	—	—	—	—	—	—
	300	0	0	0	0	1	1	2	—	—	—	—	—	—
	350	0	0	0	0	1	1	1	—	—	—	—	—	—
	400	0	0	0	0	1	1	1	—	—	—	—	—	—
	500	0	0	0	0	0	1	1	—	—	—	—	—	—
	600	0	0	0	0	0	1	1	—	—	—	—	—	—
	700	0	0	0	0	0	0	1	—	—	—	—	—	—
	750	0	0	0	0	0	0	1	—	—	—	—	—	—
	800	0	0	0	0	0	0	1	—	—	—	—	—	—
	900	0	0	0	0	0	0	1	—	—	—	—	—	—
	1000	0	0	0	0	0	0	1	—	—	—	—	—	—
	1250	0	0	0	0	0	0	0	—	—	—	—	—	—
	1500	0	0	0	0	0	0	0	—	—	—	—	—	—
	1750	0	0	0	0	0	0	0	—	—	—	—	—	—
	2000	0	0	0	0	0	0	0	—	—	—	—	—	—
THHW, THWN, THWN-2	14	7	12	21	34	61	81	135	—	—	—	—	—	—
	12	5	9	15	25	44	59	98	—	—	—	—	—	—
	10	3	5	10	15	28	37	62	—	—	—	—	—	—
	8	1	3	5	9	16	21	36	—	—	—	—	—	—
	6	1	2	4	6	11	15	26	—	—	—	—	—	—

(continued)

Table C.7 Maximum Number of Conductors of Fixture Wires in Liquidtight Flexible Nonmetallic Conduit (Type LFNC-C) (Based on Chapter 9: Table 1, Table 4, and Table 5) ***Continued***

Type	Conductor Size (AWG/kcmil)	Trade Size (Metric Designator)												
		⅜ (12)	½ (16)	¾ (21)	1 (27)	1¼ (35)	1½ (41)	2 (53)	2½ (63)	3 (78)	3½ (91)	4 (103)	5 (129)	6 (155)
CONDUCTORS														
	4	1	1	2	4	7	9	16	—	—	—	—	—	—
	3	0	1	1	3	6	8	13	—	—	—	—	—	—
	2	0	1	1	3	5	7	11	—	—	—	—	—	—
	1	0	1	1	1	3	5	8	—	—	—	—	—	—
	1/0	0	1	1	1	3	4	7	—	—	—	—	—	—
	2/0	0	0	1	1	2	3	6	—	—	—	—	—	—
	3/0	0	0	1	1	1	3	5	—	—	—	—	—	—
	4/0	0	0	1	1	1	2	4	—	—	—	—	—	—
	250	0	0	0	1	1	1	3	—	—	—	—	—	—
	300	0	0	0	1	1	1	3	—	—	—	—	—	—
	350	0	0	0	1	1	1	2	—	—	—	—	—	—
	400	0	0	0	0	1	1	1	—	—	—	—	—	—
	500	0	0	0	0	1	1	1	—	—	—	—	—	—

	600	0	0	0	0	1	1	1	—	—	—	—	—	—
	700	0	0	0	0	0	1	1	—	—	—	—	—	—
	750	0	0	0	0	0	1	1	—	—	—	—	—	—
	800	0	0	0	0	0	1	1	—	—	—	—	—	—
	900	0	0	0	0	0	1	1	—	—	—	—	—	—
	1000	0	0	0	0	0	0	1	—	—	—	—	—	—
FEP, FEPB, PFA, PFAH, TFE	14	7	12	21	33	59	79	131	—	—	—	—	—	—
	12	5	9	15	24	43	57	96	—	—	—	—	—	—
	10	4	6	11	17	31	41	68	—	—	—	—	—	—
	8	1	3	6	10	17	23	39	—	—	—	—	—	—
	6	1	2	4	7	12	17	28	—	—	—	—	—	—
	4	1	1	3	5	9	11	19	—	—	—	—	—	—
	3	1	1	2	4	7	10	16	—	—	—	—	—	—
	2	1	1	1	3	6	8	13	—	—	—	—	—	—
PFA, PFAH, TFE	1	0	1	1	2	4	5	9	—	—	—	—	—	—
PFA, PFAH, TFE, Z	1/0	0	1	1	1	3	4	8	—	—	—	—	—	—
	2/0	0	0	1	1	3	4	6	—	—	—	—	—	—
	3/0	0	0	1	1	2	3	5	—	—	—	—	—	—
	4/0	0	0	1	1	1	2	4	—	—	—	—	—	—

(continued)

Table C.7 Maximum Number of Conductors of Fixture Wires in Liquidtight Flexible Nonmetallic Conduit (Type LFNC-C) (Based on Chapter 9: Table 1, Table 4, and Table 5) ***Continued***

Type	Conductor Size (AWG/ kcmil)	Trade Size (Metric Designator)												
		3/8 (12)	1/2 (16)	3/4 (21)	1 (27)	1¼ (35)	1½ (41)	2 (53)	2½ (63)	3 (78)	3½ (91)	4 (103)	5 (129)	6 (155)
		CONDUCTORS												
Z	14	9	14	25	40	71	95	158	—	—	—	—	—	—
	12	6	10	18	28	50	67	112	—	—	—	—	—	—
	10	4	6	11	17	31	41	68	—	—	—	—	—	—
	8	2	4	7	11	19	26	43	—	—	—	—	—	—
	6	1	3	5	7	13	18	30	—	—	—	—	—	—
	4	1	1	3	5	9	12	21	—	—	—	—	—	—
	3	1	1	2	4	7	9	15	—	—	—	—	—	—
	2	1	1	1	3	5	7	12	—	—	—	—	—	—
	1	0	1	1	2	4	6	10	—	—	—	—	—	—
XHHW, ZW, XHHW-2, XHH	14	5	8	15	24	42	56	94	—	—	—	—	—	—
	12	4	6	11	18	32	43	72	—	—	—	—	—	—
	10	3	5	8	13	24	32	54	—	—	—	—	—	—
	8	1	2	4	7	13	18	30	—	—	—	—	—	—
	6	1	1	3	5	10	13	22	—	—	—	—	—	—

	4	1	1	2	4	7	9	16	—	—	—	—	—	—
	3	1	1	1	3	6	8	13	—	—	—	—	—	—
	2	1	1	1	3	5	7	11	—	—	—	—	—	—
XHHW, XHHW-2, XHH	1	0	1	1	1	4	5	8	—	—	—	—	—	—
	1/0	0	1	1	1	3	4	7	—	—	—	—	—	—
	2/0	0	0	1	1	2	3	6	—	—	—	—	—	—
	3/0	0	0	1	1	2	3	5	—	—	—	—	—	—
	4/0	0	0	1	1	1	2	4	—	—	—	—	—	—
	250	0	0	0	1	1	1	3	—	—	—	—	—	—
	300	0	0	0	1	1	1	3	—	—	—	—	—	—
	350	0	0	0	1	1	1	2	—	—	—	—	—	—
	400	0	0	0	0	1	1	1	—	—	—	—	—	—
	500	0	0	0	0	1	1	1	—	—	—	—	—	—
	600	0	0	0	0	1	1	1	—	—	—	—	—	—
	700	0	0	0	0	0	1	1	—	—	—	—	—	—
	750	0	0	0	0	0	1	1	—	—	—	—	—	—
	800	0	0	0	0	0	1	1	—	—	—	—	—	—
	900	0	0	0	0	0	1	1	—	—	—	—	—	—

(continued)

Table C.7 Maximum Number of Conductors of Fixture Wires in Liquidtight Flexible Nonmetallic Conduit (Type LFNC-C) (Based on Chapter 9: Table 1, Table 4, and Table 5) ***Continued***

Type	Conductor Size (AWG/ kcmil)	Trade Size (Metric Designator)												
		⅜ (12)	½ (16)	¾ (21)	1 (27)	1¼ (35)	1½ (41)	2 (53)	2½ (63)	3 (78)	3½ (91)	4 (103)	5 (129)	6 (155)
		CONDUCTORS												
	1000	0	0	0	0	0	0	1	—	—	—	—	—	—
	1250	0	0	0	0	0	0	1	—	—	—	—	—	—
	1500	0	0	0	0	0	0	1	—	—	—	—	—	—
	1750	0	0	0	0	0	0	0	—	—	—	—	—	—
	2000	0	0	0	0	0	0	0	—	—	—	—	—	—
		FIXTURE WIRES												
RFH-2, FFH-2, RFHH-2	18	5	8	14	23	40	54	90	—	—	—	—	—	—
	16	4	7	12	19	34	46	76	—	—	—	—	—	—
SF-2, SFF-2	18	6	10	18	29	51	68	114	—	—	—	—	—	—
	16	5	8	15	24	42	56	94	—	—	—	—	—	—
	14	4	7	12	19	34	46	76	—	—	—	—	—	—
SF-1, SFF-1	18	11	18	32	51	90	121	202	—	—	—	—	—	—

RFH-1, TF, TFF, XF, XFF	18	8	13	23	38	67	89	149	—	—	—	—	—	—
	16	6	11	19	30	54	72	120	—	—	—	—	—	—
XF, XFF	14	5	8	15	24	42	56	94	—	—	—	—	—	—
TFN, TFFN	18	13	22	38	60	107	143	239	—	—	—	—	—	—
	16	10	17	29	46	82	109	182	—	—	—	—	—	—
PF, PFF, PGF, PGFF, PAF, PTF, PTFF, PAFF	18	12	21	36	57	101	136	226	—	—	—	—	—	—
	16	10	16	28	44	78	105	175	—	—	—	—	—	—
	14	7	12	21	33	59	79	131	—	—	—	—	—	—
ZF, ZFF, ZHF	18	16	27	46	74	131	175	292	—	—	—	—	—	—
	16	12	20	34	54	96	129	215	—	—	—	—	—	—
	14	9	14	25	40	71	95	131	—	—	—	—	—	—
KF-2, KFF-2	18	24	40	69	111	196	263	438	—	—	—	—	—	—
	16	17	28	48	77	137	183	305	—	—	—	—	—	—
	14	11	19	32	52	92	123	205	—	—	—	—	—	—
	12	8	13	22	36	64	85	142	—	—	—	—	—	—
	10	5	8	15	24	42	56	94	—	—	—	—	—	—

(continued)

Table C.7 Maximum Number of Conductors of Fixture Wires in Liquidtight Flexible Nonmetallic Conduit (Type LFNC-C) (Based on Chapter 9: Table 1, Table 4, and Table 5) ***Continued***

Type	Conductor Size (AWG/ kcmil)	Trade Size (Metric Designator)												
		⅜ (12)	½ (16)	¾ (21)	1 (27)	1¼ (35)	1½ (41)	2 (53)	2½ (63)	3 (78)	3½ (91)	4 (103)	5 (129)	6 (155)
FIXTURE WIRES														
KF-1, KFF-1	18	28	46	80	128	227	303	505	—	—	—	—	—	—
	16	20	32	56	90	159	213	355	—	—	—	—	—	—
	14	13	22	38	60	107	143	239	—	—	—	—	—	—
	12	9	14	25	40	71	95	158	—	—	—	—	—	—
	10	6	9	16	26	46	62	103	—	—	—	—	—	—
XF, XFF	12	3	4	8	13	22	30	50	—	—	—	—	—	—
	10	1	3	6	10	17	23	39	—	—	—	—	—	—

Notes:

1. This table is for concentric stranded conductors only. For compact stranded conductors, Table C.5(A) should be used.
2. Two-hour fire-rated RHH cable has ceramifiable insulation, which has larger diameters than other RHH wires.

Consult manufacturer's conduit fill tables.

*Types RHH, RHW, and RHW-2 without outer covering.

Table C.8 Maximum Number of Conductors or Fixture Wires in Liquidtight Flexible Metal Conduit (LFMC) (Based on *NEC* Chapter 9: Table 1, Table 4, and Table 5)

Type	Conductor Size (AWG/kcmil)	Trade Size (Metric Designator)												
		⅜ (12)	½ (16)	¾ (21)	1 (27)	1¼ (35)	1½ (41)	2 (53)	2½ (63)	3 (78)	3½ (91)	4 (103)	5 (129)	6 (155)
						CONDUCTORS								
RHH, RHW, RHW-2	14	2	4	7	12	21	27	44	66	102	133	173	—	—
	12	1	3	6	10	17	22	36	55	84	110	144	—	—
	10	1	3	5	8	14	18	29	44	68	89	116	—	—
	8	1	1	2	4	7	9	15	23	36	46	61	—	—
	6	1	1	1	3	6	7	12	18	28	37	48	—	—
	4	0	1	1	2	4	6	9	14	22	29	38	—	—
	3	0	1	1	1	4	5	8	13	19	25	33	—	—
	2	0	1	1	1	3	4	7	11	17	22	29	—	—
	1	0	0	1	1	1	3	5	7	11	14	19	—	—
	1/0	0	0	1	1	1	2	4	6	10	13	16	—	—
	2/0	0	0	1	1	1	1	3	5	8	11	14	—	—
	3/0	0	0	0	1	1	1	3	4	7	9	12	—	—
	4/0	0	0	0	1	1	1	2	4	6	8	10	—	—
	250	0	0	0	0	1	1	1	3	4	6	8	—	—
	300	0	0	0	0	1	1	1	2	4	5	7	—	—
	350	0	0	0	0	1	1	1	2	3	5	6	—	—

(continued)

Table C.8 Maximum Number of Conductors or Fixture Wires in Liquidtight Flexible Metal Conduit (LFMC) (Based on *NEC* Chapter 9: Table 1, Table 4, and Table 5) ***Continued***

Type	Conductor Size (AWG/ kcmil)	Trade Size (Metric Designator)												
		⅜ (12)	½ (16)	¾ (21)	1 (27)	1¼ (35)	1½ (41)	2 (53)	2½ (63)	3 (78)	3½ (91)	4 (103)	5 (129)	6 (155)
CONDUCTORS														
	400	0	0	0	0	1	1	1	1	3	4	6	—	—
	500	0	0	0	0	1	1	1	1	3	4	5	—	—
	600	0	0	0	0	0	1	1	1	2	3	4	—	—
	700	0	0	0	0	0	0	1	1	1	3	3	—	—
	750	0	0	0	0	0	0	1	1	1	2	3	—	—
	800	0	0	0	0	0	0	1	1	1	2	3	—	—
	900	0	0	0	0	0	0	1	1	1	2	3	—	—
	1000	0	0	0	0	0	0	1	1	1	1	3	—	—
	1250	0	0	0	0	0	0	0	1	1	1	1	—	—
	1500	0	0	0	0	0	0	0	1	1	1	1	—	—
	1750	0	0	0	0	0	0	0	1	1	1	1	—	—
	2000	0	0	0	0	0	0	0	0	1	1	1	—	—
TW, THHW, THW, THW-2	14	5	9	15	25	44	57	93	140	215	280	365	—	—
	12	4	7	12	19	33	43	71	108	165	215	280	—	—
	10	3	5	9	14	25	32	53	80	123	160	209	—	—
	8	1	3	5	8	14	18	29	44	68	89	116	—	—

RHH*, RHW*, RHW-2*	14	3	6	10	16	29	38	62	93	143	186	243	—	—
	12	3	5	8	13	23	30	50	75	115	149	195	—	—
	10	1	3	6	10	18	23	39	58	89	117	152	—	—
	8	1	1	4	6	11	14	23	35	53	70	91	—	—
TW, THW, THHW, THW-2, RHH*, RHW*, RHW-2*	6	1	1	3	5	8	11	18	27	41	53	70	—	—
	4	1	1	1	3	6	8	13	20	30	40	52	—	—
	3	1	1	1	3	5	7	11	17	26	34	44	—	—
	2	0	1	1	2	4	6	9	14	22	29	38	—	—
	1	0	1	1	1	3	4	7	10	15	20	26	—	—
	1/0	0	0	1	1	2	3	6	8	13	17	23	—	—
	2/0	0	0	1	1	2	3	5	7	11	15	19	—	—
	3/0	0	0	1	1	1	2	4	6	9	12	16	—	—
	4/0	0	0	0	1	1	1	3	5	8	10	13	—	—
	250	0	0	0	1	1	1	3	4	6	8	11	—	—
	300	0	0	0	1	1	1	2	3	5	7	9	—	—
	350	0	0	0	0	1	1	1	3	5	6	8	—	—
	400	0	0	0	0	1	1	1	3	4	6	7	—	—
	500	0	0	0	0	1	1	1	2	3	5	6	—	—
	600	0	0	0	0	1	1	1	1	3	4	5	—	—
	700	0	0	0	0	0	1	1	1	2	3	4	—	—
	750	0	0	0	0	0	1	1	1	2	3	4	—	—
	800	0	0	0	0	0	1	1	1	2	3	4	—	—
	900	0	0	0	0	0	0	1	1	1	3	3	—	—
	1000	0	0	0	0	0	0	1	1	1	2	3	—	—
	1250	0	0	0	0	0	0	1	1	1	1	2	—	—

(continued)

Table C.8 Maximum Number of Conductors or Fixture Wires in Liquidtight Flexible Metal Conduit (LFMC) (Based on *NEC* Chapter 9: Table 1, Table 4, and Table 5) ***Continued***

Type	Conductor Size (AWG/kcmil)	Trade Size (Metric Designator)												
		⅜ (12)	½ (16)	¾ (21)	1 (27)	1¼ (35)	1½ (41)	2 (53)	2½ (63)	3 (78)	3½ (91)	4 (103)	5 (129)	6 (155)
CONDUCTORS														
	1500	0	0	0	0	0	0	0	1	1	1	2	—	—
	1750	0	0	0	0	0	0	0	1	1	1	1	—	—
	2000	0	0	0	0	0	0	0	1	1	1	1	—	—
THHN, THWN, THWN-2	14	8	13	22	36	63	81	134	201	308	401	523	—	—
	12	5	9	16	26	46	59	97	146	225	292	381	—	—
	10	3	6	10	16	29	37	61	92	141	184	240	—	—
	8	1	3	6	9	16	21	35	53	81	106	138	—	—
	6	1	2	4	7	12	15	25	38	59	76	100	—	—
	4	1	1	2	4	7	9	15	23	36	47	61	—	—
	3	1	1	1	3	6	8	13	20	30	40	52	—	—
	2	1	1	1	3	5	7	11	17	26	33	44	—	—
	1	0	1	1	1	4	5	8	12	19	25	32	—	—
	1/0	0	1	1	1	3	4	7	10	16	21	27	—	—
	2/0	0	0	1	1	2	3	6	8	13	17	23	—	—
	3/0	0	0	1	1	1	3	5	7	11	14	19	—	—
	4/0	0	0	1	1	1	2	4	6	9	12	15	—	—

	250	0	0	0	1	1	1	3	5	7	10	12	—	—
	300	0	0	0	1	1	1	3	4	6	8	11	—	—
	350	0	0	0	1	1	1	2	3	5	7	9	—	—
	400	0	0	0	0	1	1	1	3	5	6	8	—	—
	500	0	0	0	0	1	1	1	2	4	5	7	—	—
	600	0	0	0	0	1	1	1	1	3	4	6	—	—
	700	0	0	0	0	1	1	1	1	3	4	5	—	—
	750	0	0	0	0	0	1	1	1	3	3	5	—	—
	800	0	0	0	0	0	1	1	1	2	3	4	—	—
	900	0	0	0	0	0	1	1	1	2	3	4	—	—
	1000	0	0	0	0	0	0	1	1	1	3	3	—	—
FEP, FEPB, PFA, PFAH, TFE	14	7	12	21	35	61	79	130	195	299	389	507	—	—
	12	5	9	15	25	44	58	94	142	218	284	370	—	—
	10	4	6	11	18	32	41	68	102	156	203	266	—	—
	8	1	3	6	10	18	23	39	58	89	117	152	—	—
	6	1	2	4	7	13	17	27	41	64	83	108	—	—
	4	1	1	3	5	9	12	19	29	44	58	75	—	—
	3	1	1	2	4	7	10	16	24	37	48	63	—	—
	2	1	1	1	3	6	8	13	20	30	40	52	—	—
PFA, PFAH, TFE	1	0	1	1	2	4	5	9	14	21	28	36	—	—
PFA, PFAH, TFE, Z	1/0	0	1	1	1	3	4	7	11	18	23	30	—	—
	2/0	0	1	1	1	3	4	6	9	14	19	25	—	—
	3/0	0	0	1	1	2	3	5	8	12	16	20	—	—
	4/0	0	0	1	1	1	2	4	6	10	13	17	—	—

(continued)

Table C.8 Maximum Number of Conductors or Fixture Wires in Liquidtight Flexible Metal Conduit (LFMC) (Based on *NEC* Chapter 9: Table 1, Table 4, and Table 5) ***Continued***

Type	Conductor Size (AWG/kcmil)	Trade Size (Metric Designator)												
		⅜ (12)	½ (16)	¾ (21)	1 (27)	1¼ (35)	1½ (41)	2 (53)	2½ (63)	3 (78)	3½ (91)	4 (103)	5 (129)	6 (155)
		CONDUCTORS												
Z	14	9	15	26	42	73	95	156	235	360	469	611	—	—
	12	6	10	18	30	52	67	111	167	255	332	434	—	—
	10	4	6	11	18	32	41	68	102	156	203	266	—	—
	8	2	4	7	11	20	26	43	64	99	129	168	—	—
	6	1	3	5	8	14	18	30	45	69	90	118	—	—
	4	1	1	3	5	9	12	20	31	48	62	81	—	—
	3	1	1	2	4	7	9	15	23	35	45	59	—	—
	2	1	1	1	3	6	7	12	19	29	38	49	—	—
	1	1	1	1	2	5	6	10	15	23	30	40	—	—
XHHW, ZW, XHHW-2, XHH	14	5	9	15	25	44	57	93	140	215	280	365	—	—
	12	4	7	12	19	33	43	71	108	165	215	280	—	—
	10	3	5	9	14	25	32	53	80	123	160	209	—	—
	8	1	3	5	8	14	18	29	44	68	89	116	—	—
	6	1	1	3	6	10	13	22	33	50	66	86	—	—
	4	1	1	2	4	7	9	16	24	36	48	62	—	—
	3	1	1	1	3	6	8	13	20	31	40	52	—	—
	2	1	1	1	3	5	7	11	17	26	34	44	—	—

XHHW, XHHW-2, XHH	1	0	1	1	1	4	5	8	12	19	25	33	—	—
	1/0	0	1	1	1	3	4	7	10	16	21	28	—	—
	2/0	0	0	1	1	2	3	6	9	13	17	23	—	—
	3/0	0	0	1	1	1	3	5	7	11	14	19	—	—
	4/0	0	0	1	1	1	2	4	6	9	12	16	—	—
	250	0	0	0	1	1	1	3	5	7	10	13	—	—
	300	0	0	0	1	1	1	3	4	6	8	11	—	—
	350	0	0	0	1	1	1	2	3	5	7	10	—	—
	400	0	0	0	1	1	1	1	3	5	6	8	—	—
	500	0	0	0	0	1	1	1	2	4	5	7	—	—
	600	0	0	0	0	1	1	1	1	3	4	6	—	—
	700	0	0	0	0	1	1	1	1	3	4	5	—	—
	750	0	0	0	0	0	1	1	1	3	3	5	—	—
	800	0	0	0	0	0	1	1	1	2	3	4	—	—
	900	0	0	0	0	0	1	1	1	2	3	4	—	—
	1000	0	0	0	0	0	0	1	1	1	3	3	—	—
	1250	0	0	0	0	0	0	1	1	1	1	3	—	—
	1500	0	0	0	0	0	0	1	1	1	1	2	—	—
	1750	0	0	0	0	0	0	0	1	1	1	1	—	—
	2000	0	0	0	0	0	0	0	1	1	1	1	—	—

(continued)

Table C.8 Maximum Number of Conductors or Fixture Wires in Liquidtight Flexible Metal Conduit (LFMC) (Based on *NEC* Chapter 9: Table 1, Table 4, and Table 5) ***Continued***

Type	Conductor Size (AWG/ kcmil)	Trade Size (Metric Designator)												
		3/8 (12)	1/2 (16)	3/4 (21)	1 (27)	1-1/4 (35)	1 1/2 (41)	2 (53)	2 1/2 (63)	3 (78)	3 1/2 (91)	4 (103)	5 (129)	6 (155)
				FIXTURE WIRES										
RFH-2, FFH-2, RFHH-2	18	5	8	15	24	42	54	89	134	206	268	350	—	—
	16	4	7	12	20	35	46	75	113	174	226	295	—	—
SF-2, SFF-2	18	6	11	19	30	53	69	113	169	260	338	441	—	—
	16	5	9	15	25	44	57	93	140	215	280	365	—	—
	14	4	7	12	20	35	46	75	113	174	226	295	—	—
SF-1, SFF-1	18	12	19	33	53	94	122	199	300	460	599	781	—	—
RFH-1, TF, TFF, XF, XFF	18	8	14	24	39	69	90	147	222	339	442	577	—	—
	16	7	11	20	32	56	72	119	179	274	357	465	—	—
XF, XFF	14	5	9	15	25	44	57	93	140	215	280	365	—	—
TFN, TFFN	18	14	23	39	63	111	144	236	355	543	707	923	—	—
	16	10	17	30	48	85	110	180	271	415	540	705	—	—
PF, PFF, PGF, PGFF, PAF, PTF, PTFF, PAFF	18	13	21	37	60	105	136	224	336	515	671	875	—	—
	16	10	16	29	46	81	105	173	260	398	519	677	—	—
	14	7	12	21	35	61	79	130	195	299	389	507	—	—

ZF, ZFF, ZHF	18	17	28	48	77	136	176	288	434	664	865	1128	—	—
	16	12	20	35	57	100	130	213	320	490	638	832	—	—
	14	9	15	26	42	73	95	156	235	360	469	611	—	—
KF-2, KFF-2	18	25	42	72	116	203	264	433	651	996	1297	1692	—	—
	16	18	29	50	81	142	184	302	454	695	905	1180	—	—
	14	12	19	34	54	95	124	203	305	467	608	793	—	—
	12	8	13	23	38	66	86	141	212	325	423	552	—	—
	10	5	9	15	25	44	57	93	140	215	280	365	—	—
KF-1, KFF-1	18	29	48	83	134	235	304	499	751	1150	1497	1952	—	—
	16	20	34	58	94	165	214	351	527	808	1052	1372	—	—
	14	14	23	39	63	111	144	236	355	543	707	923	—	—
	12	9	15	26	42	73	95	156	235	360	469	611	—	—
	10	6	10	17	27	48	62	102	153	235	306	399	—	—
XF, XFF	12	3	5	8	13	23	30	50	75	115	149	195	—	—
	10	1	3	6	10	18	23	39	58	89	117	152	—	—

Notes:

1. This table is for concentric stranded conductors only. For compact stranded conductors, Table C.7(A) should be used.

2. Two-hour fire-rated RHH cable has ceramifiable insulation, which has much larger diameters than other RHH wires. Consult manufacturer's conduit fill tables.

*Types RHH, RHW, and RHW-2 without outer covering.

Table C.9 Maximum Number of Conductors or Fixture Wires in Rigid Metal Conduit (RMC) (Based on Chapter 9: Table 1, Table 4, and Table 5)

Type	Conductor Size (AWG/kcmil)	Trade Size (Metric Designator)												
		⅜ (12)	½ (16)	¾ (21)	1 (27)	1¼ (35)	1½ (41)	2 (53)	2½ (63)	3 (78)	3½ (91)	4 (103)	5 (129)	6 (155)
CONDUCTORS														
RHH, RHW, RHW-2	14	—	4	7	12	21	28	46	66	102	136	176	276	398
	12	—	3	6	10	17	23	38	55	85	113	146	229	330
	10	—	3	5	8	14	19	31	44	68	91	118	185	267
	8	—	1	2	4	7	10	16	23	36	48	61	97	139
	6	—	1	1	3	6	8	13	18	29	38	49	77	112
	4	—	1	1	2	4	6	10	14	22	30	38	60	87
	3	—	1	1	2	4	5	9	12	19	26	34	53	76
	2	—	1	1	1	3	4	7	11	17	23	29	46	66
	1	—	0	1	1	1	3	5	7	11	15	19	30	44
	1/0	—	0	1	1	1	2	4	6	10	13	17	26	38
	2/0	—	0	1	1	1	2	4	5	8	11	14	23	33

3/0	—	0	0	1	1	1	3	4	7	10	12	20	28
4/0	—	0	0	1	1	1	3	4	6	8	11	17	24
250	—	0	0	0	1	1	1	3	4	6	8	13	18
300	—	0	0	0	1	1	1	2	4	5	7	11	16
350	—	0	0	0	1	1	1	2	4	5	6	10	15
400	—	0	0	0	1	1	1	1	3	4	6	9	13
500	—	0	0	0	1	1	1	1	3	4	5	8	11
600	—	0	0	0	0	1	1	1	2	3	4	6	9
700	—	0	0	0	0	1	1	1	1	3	3	6	8
750	—	0	0	0	0	0	1	1	1	3	3	5	8
800	—	0	0	0	0	0	1	1	1	2	3	5	7
900	—	0	0	0	0	0	1	1	1	2	3	5	7
1000	—	0	0	0	0	0	1	1	1	1	3	4	6
1250	—	0	0	0	0	0	0	1	1	1	1	3	5
1500	—	0	0	0	0	0	0	1	1	1	1	3	4
1750	—	0	0	0	0	0	0	1	1	1	1	2	4
2000	—	0	0	0	0	0	0	0	1	1	1	2	3

(continued)

Table C.9 Maximum Number of Conductors or Fixture Wires in Rigid Metal Conduit (RMC) (Based on Chapter 9: Table 1, Table 4, and Table 5) ***Continued***

Type	Conductor Size (AWG/kcmil)	Trade Size (Metric Designator)												
		⅜ (12)	½ (16)	¾ (21)	1 (27)	1¼ (35)	1½ (41)	2 (53)	2½ (63)	3 (78)	3½ (91)	4 (103)	5 (129)	6 (155)
CONDUCTORS														
TW, THHW, THW, THW-2	14	—	9	15	25	44	59	98	140	215	288	370	581	839
	12	—	7	12	19	33	45	75	107	165	221	284	446	644
	10	—	5	9	14	25	34	56	80	123	164	212	332	480
	8	—	3	5	8	14	19	31	44	68	91	118	185	267
RHH*, RHW*, RHW-2*	14	—	6	10	17	29	39	65	93	143	191	246	387	558
	12	—	5	8	13	23	32	52	75	115	154	198	311	448
	10	—	3	6	10	18	25	41	58	90	120	154	242	350
	8	—	1	4	6	11	15	24	35	54	72	92	145	209
TW, THW, THHW, THW-2, RHH*, RHW*, RHW-2*	6	—	1	3	5	8	11	18	27	41	55	71	111	160
	4	—	1	1	3	6	8	14	20	31	41	53	83	120
	3	—	1	1	3	5	7	12	17	26	35	45	71	103
	2	—	1	1	2	4	6	10	14	22	30	38	60	87
	1	—	1	1	1	3	4	7	10	15	21	27	42	61

1/0	—	0	1	1	2	3	6	8	13	18	23	36	52
2/0	—	0	1	1	2	3	5	7	11	15	19	31	44
3/0	—	0	1	1	1	2	4	6	9	13	16	26	37
4/0	—	0	0	1	1	1	3	5	8	10	14	21	31
250	—	0	0	1	1	1	3	4	6	8	11	17	25
300	—	0	0	1	1	1	2	3	5	7	9	15	22
350	—	0	0	0	1	1	1	3	5	6	8	13	19
400	—	0	0	0	1	1	1	3	4	6	7	12	17
500	—	0	0	0	1	1	1	2	3	5	6	10	14
600	—	0	0	0	1	1	1	1	3	4	5	8	12
700	—	0	0	0	0	1	1	1	2	3	4	7	10
750	—	0	0	0	0	1	1	1	2	3	4	7	10
800	—	0	0	0	0	1	1	1	2	3	4	6	9
900	—	0	0	0	0	1	1	1	1	3	3	6	8
1000	—	0	0	0	0	0	1	1	1	2	3	5	8
1250	—	0	0	0	0	0	1	1	1	1	2	4	6
1500	—	0	0	0	0	0	1	1	1	1	2	3	5
1750	—	0	0	0	0	0	0	1	1	1	1	3	4
2000	—	0	0	0	0	0	0	1	1	1	1	3	4

(continued)

Table C.9 Maximum Number of Conductors or Fixture Wires in Rigid Metal Conduit (RMC) (Based on Chapter 9: Table 1, Table 4, and Table 5) ***Continued***

Type	Conductor Size (AWG/kcmil)	Trade Size (Metric Designator)												
		⅜ (12)	½ (16)	¾ (21)	1 (27)	1¼ (35)	1½ (41)	2 (53)	2½ (63)	3 (78)	3½ (91)	4 (103)	5 (129)	6 (155)
CONDUCTORS														
THHN, THWN, THWN-2	14	—	13	22	36	63	85	140	200	309	412	531	833	1202
	12	—	9	16	26	46	62	102	146	225	301	387	608	877
	10	—	6	10	17	29	39	64	92	142	189	244	383	552
	8	—	3	6	9	16	22	37	53	82	109	140	221	318
	6	—	2	4	7	12	16	27	38	59	79	101	159	230
	4	—	1	2	4	7	10	16	23	36	48	62	98	141
	3	—	1	1	3	6	8	14	20	31	41	53	83	120
	2	—	1	1	3	5	7	11	17	26	34	44	70	100
	1	—	1	1	1	4	5	8	12	19	25	33	51	74
THHN, THWN, THWN-2	1/0	—	1	1	1	3	4	7	10	16	21	27	43	63
	2/0	—	0	1	1	2	3	6	8	13	18	23	36	52
	3/0	—	0	1	1	1	3	5	7	11	15	19	30	43
	4/0	—	0	1	1	1	2	4	6	9	12	16	25	36

	250	—	0	0	1	1	1	3	5	7	10	13	20	29
	300	—	0	0	1	1	1	3	4	6	8	11	17	25
	350	—	0	0	1	1	1	2	3	5	7	10	15	22
	400	—	0	0	1	1	1	2	3	5	7	8	13	20
	500	—	0	0	0	1	1	1	2	4	5	7	11	16
	600	—	0	0	0	1	1	1	1	3	4	6	9	13
	700	—	0	0	0	1	1	1	1	3	4	5	8	11
	750	—	0	0	0	0	1	1	1	3	4	5	7	11
	800	—	0	0	0	0	1	1	1	2	3	4	7	10
	900	—	0	0	0	0	1	1	1	2	3	4	6	9
	1000	—	0	0	0	0	1	1	1	1	3	4	6	8
FEP, FEPB, PFA, PFAH, TFE	14	—	12	22	35	61	83	136	194	300	400	515	808	1166
	12	—	9	16	26	44	60	99	142	219	292	376	590	851
	10	—	6	11	18	32	43	71	102	157	209	269	423	610
	8	—	3	6	10	18	25	41	58	90	120	154	242	350
	6	—	2	4	7	13	17	29	41	64	85	110	172	249
	4	—	1	3	5	9	12	20	29	44	59	77	120	174
	3	—	1	2	4	7	10	17	24	37	50	64	100	145
	2	—	1	1	3	6	8	14	20	31	41	53	83	120

(continued)

Table C.9 Maximum Number of Conductors or Fixture Wires in Rigid Metal Conduit (RMC) (Based on Chapter 9: Table 1, Table 4, and Table 5) ***Continued***

Type	Conductor Size (AWG/kcmil)	Trade Size (Metric Designator)												
		3/8 (12)	1/2 (16)	3/4 (21)	1 (27)	1¼ (35)	1½ (41)	2 (53)	2½ (63)	3 (78)	3½ (91)	4 (103)	5 (129)	6 (155)
CONDUCTORS														
PFA, PFAH,	1	—	1	1	2	4	6	9	14	21	28	37	57	83
TFE		—												
PFA, PFAH,	1/0	—	1	1	1	3	5	8	11	18	24	30	48	69
TFE, Z	2/0	—	1	1	1	3	4	6	9	14	19	25	40	57
	3/0	—	0	1	1	2	3	5	8	12	16	21	33	47
	4/0	—	0	1	1	1	2	4	6	10	13	17	27	39
Z	14	—	15	26	42	73	100	164	234	361	482	621	974	1405
	12	—	10	18	30	52	71	116	166	256	342	440	691	997
	10	—	6	11	18	32	43	71	102	157	209	269	423	610
	8	—	4	7	11	20	27	45	64	99	132	170	267	386
	6	—	3	5	8	14	19	31	45	69	93	120	188	271

	4	—	1	3	5	9	13	22	31	48	64	82	129	186
	3	—	1	2	4	7	9	16	22	35	47	60	94	136
	2	—	1	1	3	6	8	13	19	29	39	50	78	113
	1	—	1	1	2	5	6	10	15	23	31	40	63	92
XHHW, ZW, XHHW-2, XHH	14	—	9	15	25	44	59	98	140	215	288	370	581	839
	12	—	7	12	19	33	45	75	107	165	221	284	446	644
	10	—	5	9	14	25	34	56	80	123	164	212	332	480
	8	—	3	5	8	14	19	31	44	68	91	118	185	267
	6	—	1	3	6	10	14	23	33	51	68	87	137	197
	4	—	1	2	4	7	10	16	24	37	49	63	99	143
	3	—	1	1	3	6	8	14	20	31	41	53	84	121
	2	—	1	1	3	5	7	12	17	26	35	45	70	101
	1	—	1	1	1	4	5	9	12	19	26	33	52	76
XHHW, XHHW-2, XHH	1/0	—	1	1	1	3	4	7	10	16	22	28	44	64
	2/0	—	0	1	1	2	3	6	9	13	18	23	37	53
	3/0	—	0	1	1	1	3	5	7	11	15	19	30	44
	4/0	—	0	1	1	1	2	4	6	9	12	16	25	36

(continued)

Table C.9 Maximum Number of Conductors or Fixture Wires in Rigid Metal Conduit (RMC) (Based on Chapter 9: Table 1, Table 4, and Table 5) ***Continued***

Type	Conductor Size (AWG/kcmil)	Trade Size (Metric Designator)												
		3/8 (12)	1/2 (16)	3/4 (21)	1 (27)	1 1/4 (35)	1 1/2 (41)	2 (53)	2 1/2 (63)	3 (78)	3 1/2 (91)	4 (103)	5 (129)	6 (155)
CONDUCTORS														
	250	—	0	0	1	1	1	3	5	7	10	13	20	30
	300	—	0	0	1	1	1	3	4	6	9	11	18	25
	350	—	0	0	1	1	1	2	3	6	7	10	15	22
	400	—	0	0	1	1	1	2	3	5	7	9	14	20
	500	—	0	0	0	1	1	1	2	4	5	7	11	16
	600	—	0	0	0	1	1	1	1	3	4	6	9	13
	700	—	0	0	0	1	1	1	1	3	4	5	8	11
	750	—	0	0	0	0	1	1	1	3	4	5	7	11
	800	—	0	0	0	0	1	1	1	2	3	4	7	10
	900	—	0	0	0	0	1	1	1	2	3	4	6	9
	1000	—	0	0	0	0	1	1	1	1	3	4	6	8
	1250	—	0	0	0	0	0	1	1	1	2	3	4	6
	1500	—	0	0	0	0	0	1	1	1	1	2	4	5

	1750	—	0	0	0	0	0	0	1	1	1	1	3	5
	2000	—	0	0	0	0	0	0	1	1	1	1	3	4
FIXTURE WIRES													
RFH-2, FFH-2, RFHH-2	18	—	8	15	24	42	57	94	134	207	276	355	557	804
	16	—	7	12	20	35	48	79	113	174	232	299	470	678
SF-2, SFF-2	18	—	11	19	31	53	72	118	169	261	348	448	703	1014
	16	—	9	15	25	44	59	98	140	215	288	370	581	839
	14	—	7	12	20	35	48	79	113	174	232	299	470	678
SF-1, SFF-1	18	—	19	33	54	94	127	209	299	461	616	792	1244	1794
RFH-1, TF, TFF, XF, XFF	18	—	14	25	40	69	94	155	221	341	455	585	918	1325
	16	—	11	20	32	56	76	125	178	275	367	472	741	1070
XF, XFF	14	—	9	15	25	44	59	98	140	215	288	370	581	839
TFN, TFFN	18	—	23	40	64	111	150	248	354	545	728	937	1470	2120
	16	—	17	30	49	84	115	189	270	416	556	715	1123	1620
PF, PFF, PGF, PGFF, PAF, PTF, PTFF, PAFF	18	—	21	38	61	105	143	235	335	517	690	888	1394	2011
	16	—	16	29	47	81	110	181	259	400	534	687	1078	1555
	14	—	12	22	35	61	83	136	194	300	400	515	808	1166

(continued)

Table C.9 Maximum Number of Conductors or Fixture Wires in Rigid Metal Conduit (RMC) (Based on Chapter 9: Table 1, Table 4, and Table 5) ***Continued***

Type	Conductor Size (AWG/kcmil)	Trade Size (Metric Designator)												
		⅜ (12)	½ (16)	¾ (21)	1 (27)	1¼ (35)	1½ (41)	2 (53)	2½ (63)	3 (78)	3½ (91)	4 (103)	5 (129)	6 (155)
CONDUCTORS														
ZF, ZFF, ZHF	18	—	28	49	79	135	184	303	432	666	889	1145	1796	2592
	16	—	20	36	58	100	136	223	319	491	656	844	1325	1912
	14	—	15	26	42	73	100	164	234	361	482	621	974	1405
KF-2, KFF-2	18	—	42	73	118	203	276	454	648	1000	1334	1717	2695	3887
	16	—	29	51	82	142	192	317	452	697	931	1198	1880	2712
	14	—	19	34	55	95	129	213	304	468	625	805	1263	1822
	12	—	13	24	38	66	90	148	211	326	435	560	878	1267
	10	—	9	15	25	44	59	98	140	215	288	370	581	839
KF-1, KFF-1	18	—	48	84	136	234	318	524	748	1153	1540	1982	3109	4486
	16	—	34	59	96	165	224	368	526	810	1082	1392	2185	3152
	14	—	23	40	64	111	150	248	354	545	728	937	1470	2120
	12	—	15	26	42	73	100	164	234	361	482	621	974	1405
	10	—	10	17	28	48	65	107	153	236	315	405	636	918

XF, XFF	12	—	5	8	13	23	32	52	75	115	154	198	311	448
	10	—	3	6	10	18	25	41	58	90	120	154	242	350

Notes:

1. This table is for concentric stranded conductors only. For compact stranded conductors, Table C.8(A) should be used.

2. Two-hour fire-rated RHH cable has ceramifiable insulation, which has much larger diameters than other RHH wires. Consult manufacturer's conduit fill tables.

*Types RHH, RHW, and RHW-2 without outer covering.

Table C.9(A) Maximum Number of Conductors or Fixture Wires in Rigid Metal Conduit (RMC) (Based on Chapter 9: Table 1, Table 4, and Table 5A)

Type	Conductor Size (AWG/kcmil)	Trade Size (Metric Designator)												
		⅜ (12)	½ (16)	¾ (21)	1 (27)	1¼ (35)	1½ (41)	2 (53)	2½ (63)	3 (78)	3½ (91)	4 (103)	5 (129)	6 (155)
COMPACT CONDUCTORS														
THW, THW-2, THHW	8	—	2	4	7	12	16	26	38	59	78	101	158	228
	6	—	1	3	5	9	12	20	29	45	60	78	122	176
	4	—	1	2	4	7	9	15	22	34	45	58	91	132
	2	—	1	1	3	5	7	11	16	25	33	43	67	97
	1	—	1	1	1	3	5	8	11	17	23	30	47	68
	1/0	—	1	1	1	3	4	7	10	15	20	26	41	59
	2/0	—	0	1	1	2	3	6	8	13	17	22	34	50
	3/0	—	0	1	1	1	3	5	7	11	14	19	29	42
	4/0	—	0	1	1	1	2	4	6	9	12	15	24	35
	250	—	0	0	1	1	1	3	4	7	9	12	19	28
	300	—	0	0	1	1	1	3	4	6	8	11	17	24
	350	—	0	0	1	1	1	2	3	5	7	9	15	22
	400	—	0	0	1	1	1	1	3	5	7	8	13	20
	500	—	0	0	0	1	1	1	3	4	5	7	11	17

	600	—	0	0	0	1	1	1	1	3	4	6	9	13
	700	—	0	0	0	1	1	1	1	3	4	5	8	12
	750	—	0	0	0	0	1	1	1	3	4	5	7	11
	900	—	0	0	0	0	1	1	1	2	3	4	7	10
	1000	—	0	0	0	0	1	1	1	1	3	4	6	9
THHN, THWN, THWN-2	8	—	—	—	—	—	—	—	—	—	—	—	—	—
	6	—	2	5	8	13	18	30	43	66	88	114	179	258
	4	—	1	3	5	8	11	18	26	41	55	70	110	159
	2	—	1	1	3	6	8	13	19	29	39	50	79	114
	1	—	1	1	2	4	6	10	14	22	29	38	59	86
	1/0	—	1	1	1	4	5	8	12	19	25	32	51	73
	2/0	—	1	1	1	3	4	7	10	15	21	26	42	60
	3/0	—	0	1	1	2	3	6	8	13	17	22	35	51
	4/0	—	0	1	1	1	3	5	7	10	14	18	29	42
	250	—	0	1	1	1	2	4	5	8	11	14	23	33
	300	—	0	0	1	1	1	3	4	7	10	12	20	28
	350	—	0	0	1	1	1	3	4	6	8	11	17	25
	400	—	0	0	1	1	1	2	3	5	7	10	15	22
	500	—	0	0	0	1	1	1	3	5	6	8	13	19

(continued)

Table C.9(A) Maximum Number of Conductors or Fixture Wires in Rigid Metal Conduit (RMC) (Based on Chapter 9: Table 1, Table 4, and Table 5A) ***Continued***

Type	Conductor Size (AWG/kcmil)	Trade Size (Metric Designator)												
		⅜ (12)	½ (16)	¾ (21)	1 (27)	1¼ (35)	1½ (41)	2 (53)	2½ (63)	3 (78)	3½ (91)	4 (103)	5 (129)	6 (155)
COMPACT CONDUCTORS														
	600	—	0	0	0	1	1	1	2	4	5	6	10	15
	700	—	0	0	0	1	1	1	1	3	4	6	9	13
	750	—	0	0	0	1	1	1	1	3	4	5	9	13
	900	—	0	0	0	0	1	1	1	2	3	4	7	10
	1000	—	0	0	0	0	1	1	1	2	3	4	6	9
XHHW, XHHW-2	8	—	3	5	9	15	21	34	49	76	101	130	205	296
	6	—	2	4	6	11	15	25	36	56	75	97	152	220
	4	—	1	3	5	8	11	18	26	41	55	70	110	159
	2	—	1	1	3	6	8	13	19	29	39	50	79	114
	1	—	1	1	2	4	6	10	14	22	29	38	59	86
	1/0	—	1	1	1	4	5	8	12	19	25	32	51	73
	2/0	—	1	1	1	3	4	7	10	16	21	27	43	62
	3/0	—	0	1	1	2	3	6	8	13	17	22	35	51
	4/0	—	0	1	1	1	3	5	7	11	14	19	29	42

250	—	0	1	1	1	2	4	5	8	11	15	23	34
300	—	0	0	1	1	1	3	5	7	10	13	20	29
350	—	0	0	1	1	1	3	4	6	9	11	18	25
400	—	0	0	1	1	1	2	4	6	8	10	16	23
500	—	0	0	0	1	1	1	3	5	6	8	13	19
600	—	0	0	0	1	1	1	2	4	5	7	10	15
700	—	0	0	0	1	1	1	1	3	4	6	9	13
750	—	0	0	0	1	1	1	1	3	4	5	8	12
900	—	0	0	0	0	1	1	1	2	3	5	7	11
1000	—	0	0	0	0	1	1	1	2	3	4	7	10

Definition: *Compact stranding* is the result of a manufacturing process where the stranded conductor is compressed to the extent that the interstices (voids between strand wires) are virtually eliminated.

Table C.10 Maximum Number of Conductors or Fixture Wires in Rigid PVC Conduit, Schedule 80 (Based on Chapter 9: Table 1, Table 4, and Table 5)

Type	Conductor Size (AWG/kcmil)	Trade Size (Metric Designator)												
		⅜ (12)	½ (16)	¾ (21)	1 (27)	1¼ (35)	1½ (41)	2 (53)	2½ (63)	3 (78)	3½ (91)	4 (103)	5 (129)	6 (155)
CONDUCTORS														
RHH, RHW, RHW-2	14	—	3	5	9	17	23	39	56	88	118	153	243	349
	12	—	2	4	7	14	19	32	46	73	98	127	202	290
	10	—	1	3	6	11	15	26	37	59	79	103	163	234
	8	—	1	1	3	6	8	13	19	31	41	54	85	122
	6	—	1	1	2	4	6	11	16	24	33	43	68	98
	4	—	1	1	1	3	5	8	12	19	26	33	53	77
	3	—	0	1	1	3	4	7	11	17	23	29	47	67
	2	—	0	1	1	3	4	6	9	14	20	25	41	58
	1	—	0	1	1	1	2	4	6	9	13	17	27	38
	1/0	—	0	0	1	1	1	3	5	8	11	15	23	33
	2/0	—	0	0	1	1	1	3	4	7	10	13	20	29

	3/0	—	0	0	1	1	1	3	4	6	8	11	17	25
	4/0	—	0	0	0	1	1	2	3	5	7	9	15	21
	250	—	0	0	0	1	1	1	2	4	5	7	11	16
	300	—	0	0	0	1	1	1	2	3	5	6	10	14
	350	—	0	0	0	1	1	1	1	3	4	5	9	13
	400	—	0	0	0	0	1	1	1	3	4	5	8	12
	500	—	0	0	0	0	1	1	1	2	3	4	7	10
	600	—	0	0	0	0	0	1	1	1	3	3	6	8
	700	—	0	0	0	0	0	1	1	1	2	3	5	7
	750	—	0	0	0	0	0	1	1	1	2	3	5	7
	800	—	0	0	0	0	0	1	1	1	2	3	4	7
	900	—	0	0	0	0	0	1	1	1	1	2	4	6
	1000	—	0	0	0	0	0	1	1	1	1	2	4	5
	1250	—	0	0	0	0	0	0	1	1	1	1	3	4
	1500	—	0	0	0	0	0	0	1	1	1	1	2	4
	1750	—	0	0	0	0	0	0	0	1	1	1	2	3
	2000	—	0	0	0	0	0	0	0	1	1	1	1	3

(continued)

Table C.10 Maximum Number of Conductors or Fixture Wires in Rigid PVC Conduit, Schedule 80 (Based on Chapter 9: Table 1, Table 4, and Table 5) ***Continued***

Type	Conductor Size (AWG/kcmil)	Trade Size (Metric Designator)												
		⅜ (12)	½ (16)	¾ (21)	1 (27)	1¼ (35)	1½ (41)	2 (53)	2½ (63)	3 (78)	3½ (91)	4 (103)	5 (129)	6 (155)
CONDUCTORS														
TW, THHW, THW, THW-2	14	—	6	11	19	35	49	82	118	185	250	324	514	736
	12	—	4	9	15	27	38	63	91	142	192	248	394	565
	10	—	3	6	11	20	28	47	68	106	143	185	294	421
	8	—	1	3	6	11	15	26	37	59	79	103	163	234
RHH*, RHW-2*, RHW*	14	—	4	8	13	23	32	55	79	123	166	215	341	490
	12	—	3	6	10	19	26	44	63	99	133	173	274	394
	10	—	2	5	8	15	20	34	49	77	104	135	214	307
	8	—	1	3	5	9	12	20	29	46	62	81	128	184
TW, THW, THHW, THW-2, RHH*, RHW*, RHW-2*	6	—	1	1	3	7	9	16	22	35	48	62	98	141
	4	—	1	1	3	5	7	12	17	26	35	46	73	105
	3	—	1	1	2	4	6	10	14	22	30	39	63	90
	2	—	1	1	1	3	5	8	12	19	26	33	53	77
	1	—	0	1	1	2	3	6	8	13	18	23	37	54

1/0	—	0	1	1	1	3	5	7	11	15	20	32	46
2/0	—	0	1	1	1	2	4	6	10	13	17	27	39
3/0	—	0	0	1	1	1	3	5	8	11	14	23	33
4/0	—	0	0	1	1	1	3	4	7	9	12	19	27
250	—	0	0	0	1	1	2	3	5	7	9	15	22
300	—	0	0	0	1	1	1	3	5	6	8	13	19
350	—	0	0	0	1	1	1	2	4	6	7	12	17
400	—	0	0	0	1	1	1	2	4	5	7	10	15
500	—	0	0	0	1	1	1	1	3	4	5	9	13
600	—	0	0	0	0	1	1	1	2	3	4	7	10
700	—	0	0	0	0	1	1	1	2	3	4	6	9
750	—	0	0	0	0	0	1	1	1	3	4	6	8
800	—	0	0	0	0	0	1	1	1	3	3	6	8
900	—	0	0	0	0	0	1	1	1	2	3	5	7
1000	—	0	0	0	0	0	1	1	1	2	3	5	7
1250	—	0	0	0	0	0	1	1	1	1	2	4	5
1500	—	0	0	0	0	0	0	1	1	1	1	3	4
1750	—	0	0	0	0	0	0	1	1	1	1	3	4
2000	—	0	0	0	0	0	0	0	1	1	1	2	3

(continued)

Table C.10 Maximum Number of Conductors or Fixture Wires in Rigid PVC Conduit, Schedule 80 (Based on Chapter 9: Table 1, Table 4, and Table 5) ***Continued***

Type	Conductor Size (AWG/kcmil)	Trade Size (Metric Designator)												
		⅜ (12)	½ (16)	¾ (21)	1 (27)	1¼ (35)	1½ (41)	2 (53)	2½ (63)	3 (78)	3½ (91)	4 (103)	5 (129)	6 (155)
CONDUCTORS														
THHN, THWN, THWN-2	14	—	9	17	28	51	70	118	170	265	358	464	736	1055
	12	—	6	12	20	37	51	86	124	193	261	338	537	770
	10	—	4	7	13	23	32	54	78	122	164	213	338	485
	8	—	2	4	7	13	18	31	45	70	95	123	195	279
	6	—	1	3	5	9	13	22	32	51	68	89	141	202
	4	—	1	1	3	6	8	14	20	31	42	54	86	124
	3	—	1	1	3	5	7	12	17	26	35	46	73	105
	2	—	1	1	2	4	6	10	14	22	30	39	61	88
	1	—	0	1	1	3	4	7	10	16	22	29	45	65
THHN, THWN, THWN-2	1/0	—	0	1	1	2	3	6	9	14	18	24	38	55
	2/0	—	0	1	1	1	3	5	7	11	15	20	32	46
	3/0	—	0	1	1	1	2	4	6	9	13	17	26	38
	4/0	—	0	0	1	1	1	3	5	8	10	14	22	31

	250	—	0	0	1	1	1	3	4	6	8	11	18	25
	300	—	0	0	0	1	1	2	3	5	7	9	15	22
	350	—	0	0	0	1	1	1	3	5	6	8	13	19
	400	—	0	0	0	1	1	1	3	4	6	7	12	17
	500	—	0	0	0	1	1	1	2	3	5	6	10	14
	600	—	0	0	0	0	1	1	1	3	4	5	8	12
	700	—	0	0	0	0	1	1	1	2	3	4	7	10
	750	—	0	0	0	0	1	1	1	2	3	4	7	9
	800	—	0	0	0	0	1	1	1	2	3	4	6	9
	900	—	0	0	0	0	0	1	1	1	3	3	6	8
	1000	—	0	0	0	0	0	1	1	1	2	3	5	7
FEP, FEPB, PFA, PFAH, TFE	14	—	8	16	27	49	68	115	164	257	347	450	714	1024
	12	—	6	12	20	36	50	84	120	188	253	328	521	747
	10	—	4	8	14	26	36	60	86	135	182	235	374	536
	8	—	2	5	8	15	20	34	49	77	104	135	214	307
	6	—	1	3	6	10	14	24	35	55	74	96	152	218
	4	—	1	2	4	7	10	17	24	38	52	67	106	153
	3	—	1	1	3	6	8	14	20	32	43	56	89	127
	2	—	1	1	3	5	7	12	17	26	35	46	73	105

(continued)

Table C.10 Maximum Number of Conductors or Fixture Wires in Rigid PVC Conduit, Schedule 80 (Based on Chapter 9: Table 1, Table 4, and Table 5) ***Continued***

Type	Conductor Size (AWG/kcmil)	Trade Size (Metric Designator)												
		⅜ (12)	½ (16)	¾ (21)	1 (27)	1¼ (35)	1½ (41)	2 (53)	2½ (63)	3 (78)	3½ (91)	4 (103)	5 (129)	6 (155)
		CONDUCTORS												
PFA, PFAH, TFE	1	—	1	1	1	3	5	8	11	18	25	32	51	73
PFA, PFAH, TFE, Z	1/0	—	0	1	1	3	4	7	10	15	20	27	42	61
	2/0	—	0	1	1	2	3	5	8	12	17	22	35	50
	3/0	—	0	1	1	1	2	4	6	10	14	18	29	41
	4/0	—	0	0	1	1	1	4	5	8	11	15	24	34
Z	14	—	10	19	33	59	82	138	198	310	418	542	860	1233
	12	—	7	14	23	42	58	98	141	220	297	385	610	875
	10	—	4	8	14	26	36	60	86	135	182	235	374	536
	8	—	3	5	9	16	22	38	54	85	115	149	236	339
	6	—	1	4	6	11	16	26	38	60	81	104	166	238

	4	—	1	2	4	8	11	18	26	41	55	72	114	164
	3	—	1	1	3	5	8	13	19	30	40	52	83	119
	2	—	1	1	2	5	6	11	16	25	33	43	69	99
	1	—	1	1	1	4	5	9	13	20	27	35	56	80
XHHW, ZW, XHHW-2, XHH	14	—	6	11	19	35	49	82	118	185	250	324	514	736
	12	—	4	9	15	27	38	63	91	142	192	248	394	565
	10	—	3	6	11	20	28	47	68	106	143	185	294	421
	8	—	1	3	6	11	15	26	37	59	79	103	163	234
	6	—	1	2	4	8	11	19	28	43	59	76	121	173
	4	—	1	1	3	6	8	14	20	31	42	55	87	125
	3	—	1	1	3	5	7	12	17	26	36	47	74	106
	2	—	1	1	2	4	6	10	14	22	30	39	62	89
XHHW, XHHW-2, XHH	1	—	0	1	1	3	4	7	10	16	22	29	46	66
	1/0	—	0	1	1	2	3	6	9	14	19	24	39	56
	2/0	—	0	1	1	1	3	5	7	11	16	20	32	46
	3/0	—	0	1	1	1	2	4	6	9	13	17	27	38
	4/0	—	0	0	1	1	1	3	5	8	11	14	22	32

(continued)

Table C.10 Maximum Number of Conductors or Fixture Wires in Rigid PVC Conduit, Schedule 80 (Based on Chapter 9: Table 1, Table 4, and Table 5) ***Continued***

Type	Conductor Size (AWG/kcmil)	Trade Size (Metric Designator)												
		⅜ (12)	½ (16)	¾ (21)	1 (27)	1¼ (35)	1½ (41)	2 (53)	2½ (63)	3 (78)	3½ (91)	4 (103)	5 (129)	6 (155)
					CONDUCTORS									
	250	—	0	0	1	1	1	3	4	6	9	11	18	26
	300	—	0	0	1	1	1	2	3	5	7	10	15	22
	350	—	0	0	0	1	1	1	3	5	6	8	14	20
	400	—	0	0	0	1	1	1	3	4	6	7	12	17
	500	—	0	0	0	1	1	1	2	3	5	6	10	14
	600	—	0	0	0	0	1	1	1	3	4	5	8	11
	700	—	0	0	0	0	1	1	1	2	3	4	7	10
	750	—	0	0	0	0	1	1	1	2	3	4	6	9
	800	—	0	0	0	0	1	1	1	1	3	4	6	9
	900	—	0	0	0	0	0	1	1	1	3	3	5	8
	1000	—	0	0	0	0	0	1	1	1	2	3	5	7
	1250	—	0	0	0	0	0	1	1	1	1	2	4	6
	1500	—	0	0	0	0	0	0	1	1	1	1	3	5
	1750	—	0	0	0	0	0	0	1	1	1	1	3	4
	2000	—	0	0	0	0	0	0	1	1	1	1	2	4

FIXTURE WIRES														
RFH-2, FFH-2, RFHH-2	18	—	6	11	19	34	47	79	113	177	239	310	492	706
	16	—	5	9	16	28	39	67	95	150	202	262	415	595
SF-2, SFF-2	18	—	7	14	24	43	59	100	143	224	302	391	621	890
	16	—	6	11	19	35	49	82	118	185	250	324	514	736
	14	—	5	9	16	28	39	67	95	150	202	262	415	595
SF-1, SFF-1	18	—	13	25	42	76	105	177	253	396	534	692	1098	1575
RFH-1, TF, TFF, XF, XFF	18	—	10	18	31	56	77	130	187	293	395	511	811	1163
	16	—	8	15	25	45	62	105	151	236	319	413	655	939
XF, XFF	14	—	6	11	19	35	49	82	118	185	250	324	514	736
TFN, TFFN	18	—	15	29	50	90	124	209	299	468	632	818	1298	1861
	16	—	12	22	38	68	95	159	229	358	482	625	992	1422
PF, PFF, PGF, PGFF, PAF, PTF, PTFF, PAFF	18	—	15	28	47	85	118	198	284	444	599	776	1231	1765
	16	—	11	22	36	66	91	153	219	343	463	600	952	1365
	14	—	8	16	27	49	68	115	164	257	347	450	714	1024
ZF, ZFF, ZHF	18	—	19	36	61	110	152	255	366	572	772	1000	1587	2275
	16	—	14	27	45	81	112	188	270	422	569	738	1171	1678
	14	—	10	19	33	59	82	138	198	310	418	542	860	1233

(continued)

Table C.10 Maximum Number of Conductors or Fixture Wires in Rigid PVC Conduit, Schedule 80 (Based on Chapter 9: Table 1, Table 4, and Table 5) ***Continued***

Type	Conductor Size (AWG/kcmil)	Trade Size (Metric Designator)												
		⅜ (12)	½ (16)	¾ (21)	1 (27)	1¼ (35)	1½ (41)	2 (53)	2½ (63)	3 (78)	3½ (91)	4 (103)	5 (129)	6 (155)
FIXTURE WIRES														
KF-2, KFF-2	18	—	29	54	91	165	228	383	549	859	1158	1501	2380	3413
	16	—	20	38	64	115	159	267	383	599	808	1047	1661	2381
	14	—	13	25	43	77	107	179	257	402	543	703	1116	1600
	12	—	9	17	30	53	74	125	179	280	377	489	776	1113
	10	—	6	11	19	35	49	82	118	185	250	324	514	736
KF-1, KFF-1	18	—	33	63	106	190	263	442	633	991	1336	1732	2747	3938
	16	—	23	44	74	133	185	310	445	696	939	1217	1930	2767
	14	—	15	29	50	90	124	209	299	468	632	818	1298	1861
	12	—	10	19	33	59	82	138	198	310	418	542	860	1233
	10	—	7	13	21	39	54	90	129	203	273	354	562	806
XF, XFF	12	—	3	6	10	19	26	44	63	99	133	173	274	394
	10	—	2	5	8	15	20	34	49	77	104	135	214	307

Notes:

1. This table is for concentric stranded conductors only. For compact stranded conductors, Table C.9(A) should be used.

2. Two-hour fire-rated RHH cable has ceramifiable insulation, which has much larger diameters than other RHH wires. Consult manufacturer's conduit fill tables.

*Types RHH, RHW, and RHW-2 without outer covering.

Table C.10(A) Maximum Number of Conductors or Fixture Wires in Rigid PVC Conduit, Schedule 80 (Based on Chapter 9: Table 1, Table 4, and Table 5A)

Type	Conductor Size (AWG/kcmil)	Trade Size (Metric Designator)												
		⅜ (12)	½ (16)	¾ (21)	1 (27)	1¼ (35)	1½ (41)	2 (53)	2½ (63)	3 (78)	3½ (91)	4 (103)	5 (129)	6 (155)
COMPACT CONDUCTORS														
THW, THW-2, THHW	8	—	1	3	5	9	13	22	32	50	68	88	140	200
	6	—	1	2	4	7	10	17	25	39	52	68	108	155
	4	—	1	1	3	5	7	13	18	29	39	51	81	116
	2	—	1	1	1	4	5	9	13	21	29	37	60	85
	1	—	0	1	1	3	4	6	9	15	20	26	42	60
	1/0	—	0	1	1	2	3	6	8	13	17	23	36	52
	2/0	—	0	1	1	1	3	5	7	11	15	19	30	44
	3/0	—	0	0	1	1	2	4	6	9	12	16	26	37
	4/0	—	0	0	1	1	1	3	5	8	10	13	22	31
	250	—	0	0	1	1	1	2	4	6	8	11	17	25
	300	—	0	0	0	1	1	2	3	5	7	9	15	21
	350	—	0	0	0	1	1	1	3	5	6	8	13	19
	400	—	0	0	0	1	1	1	3	4	6	7	12	17
	500	—	0	0	0	1	1	1	2	3	5	6	10	14

(continued)

Table C.10(A) Maximum Number of Conductors or Fixture Wires in Rigid PVC Conduit, Schedule 80 (Based on Chapter 9: Table 1, Table 4, and Table 5A) ***Continued***

Type	Conductor Size (AWG/ kcmil)	Trade Size (Metric Designator)												
		⅜ (12)	½ (16)	¾ (21)	1 (27)	1¼ (35)	1½ (41)	2 (53)	2½ (63)	3 (78)	3½ (91)	4 (103)	5 (129)	6 (155)
COMPACT CONDUCTORS														
	600	—	0	0	0	0	1	1	1	3	4	5	8	12
	700	—	0	0	0	0	1	1	1	2	3	4	7	10
	750	—	0	0	0	0	1	1	1	2	3	4	7	10
	900	—	0	0	0	0	0	1	1	1	3	4	6	8
	1000	—	0	0	0	0	0	1	1	1	2	3	5	8
THHN, THWN, THWN-2	8	—	—	—	—	—	—	—	—	—	—	—	—	—
	6	—	1	3	6	11	15	25	36	57	77	99	158	226
	4	—	1	1	3	6	9	15	22	35	47	61	98	140
	2	—	1	1	2	5	6	11	16	25	34	44	70	100
	1	—	1	1	1	3	5	8	12	19	25	33	53	75
	1/0	—	0	1	1	3	4	7	10	16	22	28	45	64
	2/0	—	0	1	1	2	3	6	8	13	18	23	37	53
	3/0	—	0	1	1	1	3	5	7	11	15	19	31	44
	4/0	—	0	0	1	1	2	4	6	9	12	16	25	37

	250	—	0	0	1	1	1	3	4	7	10	12	20	29
	300	—	0	0	1	1	1	3	4	6	8	11	17	25
	350	—	0	0	0	1	1	2	3	5	7	9	15	22
	400	—	0	0	0	1	1	1	3	5	6	8	13	19
	500	—	0	0	0	1	1	1	2	4	5	7	11	16
	600	—	0	0	0	1	1	1	1	3	4	6	9	13
	700	—	0	0	0	0	1	1	1	3	4	5	8	12
	750	—	0	0	0	0	1	1	1	3	4	5	8	11
	900	—	0	0	0	0	1	1	1	1	3	4	6	9
	1000	—	0	0	0	0	0	1	1	1	3	3	5	8
XHHW, XHHW-2	8	—	1	4	7	12	17	29	42	65	88	114	181	260
	6	—	1	3	5	9	13	21	31	48	65	85	134	193
	4	—	1	1	3	6	9	15	22	35	47	61	98	140
	2	—	1	1	2	5	6	11	16	25	34	44	70	100
	1	—	1	1	1	3	5	8	12	19	25	33	53	75
	1/0	—	0	1	1	3	4	7	10	16	22	28	45	64
	2/0	—	0	1	1	2	3	6	8	13	18	24	38	54
	3/0	—	0	1	1	1	3	5	7	11	15	19	31	44
	4/0	—	0	0	1	1	2	4	6	9	12	16	26	37

(continued)

Table C.10(A) Maximum Number of Conductors or Fixture Wires in Rigid PVC Conduit, Schedule 80 (Based on Chapter 9: Table 1, Table 4, and Table 5A) ***Continued***

Type	Conductor Size (AWG/ kcmil)	Trade Size (Metric Designator)												
		⅜ (12)	½ (16)	¾ (21)	1 (27)	1¼ (35)	1½ (41)	2 (53)	2½ (63)	3 (78)	3½ (91)	4 (103)	5 (129)	6 (155)
COMPACT CONDUCTORS														
	250	—	0	0	1	1	1	3	5	7	10	13	21	30
	300	—	0	0	1	1	1	3	4	6	8	11	17	25
	350	—	0	0	1	1	1	2	3	5	7	10	15	22
	400	—	0	0	0	1	1	1	3	5	7	9	14	20
	500	—	0	0	0	1	1	1	2	4	5	7	11	17
	600	—	0	0	0	1	1	1	1	3	4	6	9	13
	700	—	0	0	0	0	1	1	1	3	4	5	8	12
	750	—	0	0	0	0	1	1	1	2	3	5	7	11
	900	—	0	0	0	0	1	1	1	2	3	4	6	9
	1000	—	0	0	0	0	0	1	1	1	3	3	6	8

Definition: *Compact stranding* is the result of a manufacturing process where the stranded conductor is compressed to the extent that the interstices (voids between strand wires) are virtually eliminated.

Table C.11 Maximum Number of Conductors or Fixture Wires in Rigid PVC Conduit, Schedule 40 and HDPE Conduit (Based on Chapter 9: Table 1, Table 4, and Table 5)

Type	Conductor Size (AWG/kcmil)	Trade Size (Metric Designator)												
		⅜ (12)	½ (16)	¾ (21)	1 (27)	1¼ (35)	1½ (41)	2 (53)	2½ (63)	3 (78)	3½ (91)	4 (103)	5 (129)	6 (155)
CONDUCTORS														
RHH, RHW, RHW-2	14	—	4	7	11	20	27	45	64	99	133	171	269	390
	12	—	3	5	9	16	22	37	53	82	110	142	224	323
	10	—	2	4	7	13	18	30	43	66	89	115	181	261
	8	—	1	2	4	7	9	15	22	35	46	60	94	137
	6	—	1	1	3	5	7	12	18	28	37	48	76	109
	4	—	1	1	2	4	6	10	14	22	29	37	59	85
	3	—	1	1	1	4	5	8	12	19	25	33	52	75
	2	—	1	1	1	3	4	7	10	16	22	28	45	65
	1	—	0	1	1	1	3	5	7	11	14	19	29	43
	1/0	—	0	1	1	1	2	4	6	9	13	16	26	37
	2/0	—	0	0	1	1	1	3	5	8	11	14	22	32

(continued)

Table C.11 Maximum Number of Conductors or Fixture Wires in Rigid PVC Conduit, Schedule 40 and HDPE Conduit (Based on Chapter 9: Table 1, Table 4, and Table 5) ***Continued***

Type	Conductor Size (AWG/kcmil)	Trade Size (Metric Designator)												
		⅜ (12)	½ (16)	¾ (21)	1 (27)	1¼ (35)	1½ (41)	2 (53)	2½ (63)	3 (78)	3½ (91)	4 (103)	5 (129)	6 (155)
CONDUCTORS														
	3/0	—	0	0	1	1	1	3	4	7	9	12	19	28
	4/0	—	0	0	1	1	1	2	4	6	8	10	16	24
	250	—	0	0	0	1	1	1	3	4	6	8	12	18
	300	—	0	0	0	1	1	1	2	4	5	7	11	16
	350	—	0	0	0	1	1	1	2	3	5	6	10	14
	400	—	0	0	0	1	1	1	1	3	4	6	9	13
	500	—	0	0	0	0	1	1	1	3	4	5	8	11
	600	—	0	0	0	0	1	1	1	2	3	4	6	9
	700	—	0	0	0	0	0	1	1	1	3	3	6	8
	750	—	0	0	0	0	0	1	1	1	2	3	5	8
	800	—	0	0	0	0	0	1	1	1	2	3	5	7
	900	—	0	0	0	0	0	1	1	1	2	3	5	7

	1000	—	0	0	0	0	0	1	1	1	1	3	4	6
	1250	—	0	0	0	0	0	0	1	1	1	1	3	5
	1500	—	0	0	0	0	0	0	1	1	1	1	3	4
	1750	—	0	0	0	0	0	0	1	1	1	1	2	3
	2000	—	0	0	0	0	0	0	0	1	1	1	2	3
TW, THHW, THW, THW-2	14	—	8	14	24	42	57	94	135	209	280	361	568	822
	12	—	6	11	18	32	44	72	103	160	215	277	436	631
	10	—	4	8	13	24	32	54	77	119	160	206	325	470
	8	—	2	4	7	13	18	30	43	66	89	115	181	261
RHH*, RHW*, RHW-2*	14	—	5	9	16	28	38	63	90	139	186	240	378	546
	12	—	4	8	12	22	30	50	72	112	150	193	304	439
	10	—	3	6	10	17	24	39	56	87	117	150	237	343
	8	—	1	3	6	10	14	23	33	52	70	90	142	205
TW, THW, THHW, THW-2, RHH*, RHW*, RHW-2*	6	—	1	2	4	8	11	18	26	40	53	69	109	157
	4	—	1	1	3	6	8	13	19	30	40	51	81	117
	3	—	1	1	3	5	7	11	16	25	34	44	69	100
	2	—	1	1	2	4	6	10	14	22	29	37	59	85
	1	—	0	1	1	3	4	7	10	15	20	26	41	60

(continued)

Table C.11 Maximum Number of Conductors or Fixture Wires in Rigid PVC Conduit, Schedule 40 and HDPE Conduit (Based on Chapter 9: Table 1, Table 4, and Table 5) ***Continued***

Type	Conductor Size (AWG/kcmil)	Trade Size (Metric Designator)												
		3/8 (12)	1/2 (16)	3/4 (21)	1 (27)	1¼ (35)	1½ (41)	2 (53)	2½ (63)	3 (78)	3½ (91)	4 (103)	5 (129)	6 (155)
		CONDUCTORS												
	1/0	—	0	1	1	2	3	6	8	13	17	22	35	51
	2/0	—	0	1	1	1	3	5	7	11	15	19	30	43
	3/0	—	0	1	1	1	2	4	6	9	12	16	25	36
	4/0	—	0	0	1	1	1	3	5	8	10	13	21	30
	250	—	0	0	1	1	1	3	4	6	8	11	17	25
	300	—	0	0	1	1	1	2	3	5	7	9	15	21
	350	—	0	0	0	1	1	1	3	5	6	8	13	19
	400	—	0	0	0	1	1	1	3	4	6	7	12	17
	500	—	0	0	0	1	1	1	2	3	5	6	10	14
	600	—	0	0	0	0	1	1	1	3	4	5	8	11
	700	—	0	0	0	0	1	1	1	2	3	4	7	10
	750	—	0	0	0	0	1	1	1	2	3	4	6	10
	800	—	0	0	0	0	1	1	1	2	3	4	6	9
	900	—	0	0	0	0	0	1	1	1	3	3	6	8

	1000	—	0	0	0	0	0	1	1	1	2	3	5	7
	1250	—	0	0	0	0	0	1	1	1	1	2	4	6
	1500	—	0	0	0	0	0	1	1	1	1	1	3	5
	1750	—	0	0	0	0	0	0	1	1	1	1	3	4
	2000	—	0	0	0	0	0	0	1	1	1	1	3	4
THHN, THWN, THWN-2	14	—	11	21	34	60	82	135	193	299	401	517	815	1178
	12	—	8	15	25	43	59	99	141	218	293	377	594	859
	10	—	5	9	15	27	37	62	89	137	184	238	374	541
	8	—	3	5	9	16	21	36	51	79	106	137	216	312
	6	—	1	4	6	11	15	26	37	57	77	99	156	225
	4	—	1	2	4	7	9	16	22	35	47	61	96	138
	3	—	1	1	3	6	8	13	19	30	40	51	81	117
	2	—	1	1	3	5	7	11	16	25	33	43	68	98
	1	—	1	1	1	3	5	8	12	18	25	32	50	73
	1/0	—	1	1	1	3	4	7	10	15	21	27	42	61
	2/0	—	0	1	1	2	3	6	8	13	17	22	35	51
	3/0	—	0	1	1	1	3	5	7	11	14	18	29	42
	4/0	—	0	1	1	1	2	4	6	9	12	15	24	35

(continued)

Table C.11 Maximum Number of Conductors or Fixture Wires in Rigid PVC Conduit, Schedule 40 and HDPE Conduit (Based on Chapter 9: Table 1, Table 4, and Table 5) ***Continued***

Type	Conductor Size (AWG/kcmil)	Trade Size (Metric Designator)												
		⅜ (12)	½ (16)	¾ (21)	1 (27)	1¼ (35)	1½ (41)	2 (53)	2½ (63)	3 (78)	3½ (91)	4 (103)	5 (129)	6 (155)
CONDUCTORS														
	250	—	0	0	1	1	1	3	4	7	10	12	20	28
	300	—	0	0	1	1	1	3	4	6	8	11	17	24
	350	—	0	0	1	1	1	2	3	5	7	9	15	21
	400	—	0	0	0	1	1	1	3	5	6	8	13	19
	500	—	0	0	0	1	1	1	2	4	5	7	11	16
	600	—	0	0	0	1	1	1	1	3	4	5	9	13
	700	—	0	0	0	0	1	1	1	3	4	5	8	11
	750	—	0	0	0	0	1	1	1	2	3	4	7	11
	800	—	0	0	0	0	1	1	1	2	3	4	7	10
	900	—	0	0	0	0	1	1	1	2	3	4	6	9
	1000	—	0	0	0	0	0	1	1	1	3	3	6	8

FEP, FEPB, PFA, PFAH, TFE	14	—	11	20	33	58	79	131	188	290	389	502	790	1142
	12	—	8	15	24	42	58	96	137	212	284	366	577	834
	10	—	6	10	17	30	41	69	98	152	204	263	414	598
	8	—	3	6	10	17	24	39	56	87	117	150	237	343
	6	—	2	4	7	12	17	28	40	62	83	107	169	244
	4	—	1	3	5	8	12	19	28	43	58	75	118	170
	3	—	1	2	4	7	10	16	23	36	48	62	98	142
	2	—	1	1	3	6	8	13	19	30	40	51	81	117
PFA, PFAH, TFE	1	—	1	1	2	4	5	9	13	20	28	36	56	81
PFA, PFAH, TFE, Z	1/0	—	1	1	1	3	4	8	11	17	23	30	47	68
	2/0	—	0	1	1	3	4	6	9	14	19	24	39	56
	3/0	—	0	1	1	2	3	5	7	12	16	20	32	46
	4/0	—	0	1	1	1	2	4	6	9	13	16	26	38
Z	14	—	13	24	40	70	95	158	226	350	469	605	952	1376
	12	—	9	17	28	49	68	112	160	248	333	429	675	976
	10	—	6	10	17	30	41	69	98	152	204	263	414	598
	8	—	3	6	11	19	26	43	62	96	129	166	261	378
	6	—	2	4	7	13	18	30	43	67	90	116	184	265

(continued)

Table C.11 Maximum Number of Conductors or Fixture Wires in Rigid PVC Conduit, Schedule 40 and HDPE Conduit (Based on Chapter 9: Table 1, Table 4, and Table 5) ***Continued***

Type	Conductor Size (AWG/kcmil)	Trade Size (Metric Designator)												
		⅜ (12)	½ (16)	¾ (21)	1 (27)	1¼ (35)	1½ (41)	2 (53)	2½ (63)	3 (78)	3½ (91)	4 (103)	5 (129)	6 (155)
CONDUCTORS														
	4	—	1	3	5	9	12	21	30	46	62	80	126	183
	3	—	1	2	4	6	9	15	22	34	45	58	92	133
	2	—	1	1	3	5	7	12	18	28	38	49	77	111
	1	—	1	1	2	4	6	10	14	23	30	39	62	90
XHHW, ZW, XHHW-2, XHH	14	—	8	14	24	42	57	94	135	209	280	361	568	822
	12	—	6	11	18	32	44	72	103	160	215	277	436	631
	10	—	4	8	13	24	32	54	77	119	160	206	325	470
	8	—	2	4	7	13	18	30	43	66	89	115	181	261
	6	—	1	3	5	10	13	22	32	49	66	85	134	193
	4	—	1	2	4	7	9	16	23	35	48	61	97	140
	3	—	1	1	3	6	8	13	19	30	40	52	82	118
	2	—	1	1	3	5	7	11	16	25	34	44	69	99

XHHW, XHHW-2, XHH	1	—	1	1	1	3	5	8	12	19	25	32	51	74
	1/0	—	1	1	1	3	4	7	10	16	21	27	43	62
	2/0	—	0	1	1	2	3	6	8	13	17	23	36	52
	3/0	—	0	1	1	1	3	5	7	11	14	19	30	43
	4/0	—	0	1	1	1	2	4	6	9	12	15	24	35
	250	—	0	0	1	1	1	3	5	7	10	13	20	29
	300	—	0	0	1	1	1	3	4	6	8	11	17	25
	350	—	0	0	1	1	1	2	3	5	7	9	15	22
	400	—	0	0	0	1	1	1	3	5	6	8	13	19
	500	—	0	0	0	1	1	1	2	4	5	7	11	16
	600	—	0	0	0	1	1	1	1	3	4	5	9	13
	700	—	0	0	0	0	1	1	1	3	4	5	8	11
	750	—	0	0	0	0	1	1	1	2	3	4	7	11
	800	—	0	0	0	0	1	1	1	2	3	4	7	10
	900	—	0	0	0	0	1	1	1	2	3	4	6	9
	1000	—	0	0	0	0	0	1	1	1	3	3	6	8
	1250	—	0	0	0	0	0	1	1	1	1	3	4	6
	1500	—	0	0	0	0	0	1	1	1	1	2	4	5
	1750	—	0	0	0	0	0	0	1	1	1	1	3	5
	2000	—	0	0	0	0	0	0	1	1	1	1	3	4

(continued)

Table C.11 Maximum Number of Conductors or Fixture Wires in Rigid PVC Conduit, Schedule 40 and HDPE Conduit (Based on Chapter 9: Table 1, Table 4, and Table 5) ***Continued***

Type	Conductor Size (AWG/kcmil)	Trade Size (Metric Designator)												
		⅜ (12)	½ (16)	¾ (21)	1 (27)	1¼ (35)	1½ (41)	2 (53)	2½ (63)	3 (78)	3½ (91)	4 (103)	5 (129)	6 (155)
FIXTURE WIRES														
RFH-2, FFH-2, RFHH-2	18	—	8	14	23	40	54	90	129	200	268	346	545	788
	16	—	6	12	19	33	46	76	109	169	226	292	459	664
SF-2, SFF-2	18	—	10	17	29	50	69	114	163	253	338	436	687	993
	16	—	8	14	24	42	57	94	135	209	280	361	568	822
	14	—	6	12	19	33	46	76	109	169	226	292	459	664
SF-1, SFF-1	18	—	17	31	51	89	122	202	289	447	599	772	1216	1758
RFH-1, TF, TFF, XF, XFF	18	—	13	23	38	66	90	149	213	330	442	570	898	1298
	16	—	10	18	30	53	73	120	172	266	357	460	725	1048
XF, XFF	14	—	8	14	24	42	57	94	135	209	280	361	568	822
TFN, TFFN	18	—	20	37	60	105	144	239	341	528	708	913	1437	2077
	16	—	16	28	46	80	110	183	261	403	541	697	1098	1587

PF, PFF, PGF, PGFF, PAF, PTF, PTFF, PAFF	18	—	19	35	57	100	137	227	323	501	671	865	1363	1970
	16	—	15	27	44	77	106	175	250	387	519	669	1054	1523
	14	—	11	20	33	58	79	131	188	290	389	502	790	1142
ZF, ZFF, ZHF	18	—	25	45	74	129	176	292	417	646	865	1116	1756	2539
	16	—	18	33	54	95	130	216	308	476	638	823	1296	1873
	14	—	13	24	40	70	95	158	226	350	469	605	952	1376
KF-2, KFF-2	18	—	38	67	111	193	265	439	626	969	1298	1674	2634	3809
	16	—	26	47	77	135	184	306	436	676	905	1168	1838	2657
	14	—	18	31	52	91	124	205	293	454	608	784	1235	1785
	12	—	12	22	36	63	86	143	204	316	423	546	859	1242
	10	—	8	14	24	42	57	94	135	209	280	361	568	822
KF-1, KFF-1	18	—	44	78	128	223	305	506	722	1118	1498	1931	3040	4395
	16	—	31	55	90	157	214	355	507	785	1052	1357	2136	3088
	14	—	20	37	60	105	144	239	341	528	708	913	1437	2077
	12	—	13	24	40	70	95	158	226	350	469	605	952	1376
	10	—	9	16	26	45	62	103	148	229	306	395	622	899
XF, XFF	12	—	4	8	13	22	30	50	72	112	150	193	304	439
	10	—	3	6	10	17	24	39	56	87	117	150	237	343

Notes:

1. This table is for concentric stranded conductors only. For compact stranded conductors, Table C.10(A) should be used.
2. Two-hour fire-rated RHH cable has ceramifiable insulation, which has much larger diameters than other RHH wires. Consult manufacturer's conduit fill tables.

*Types RHH, RHW, and RHW-2 without outer covering.

Table C.11(A) Maximum Number of Conductors or Fixture Wires in Rigid PVC Conduit, Schedule 40 and HDPE Conduit (Based on Chapter 9: Table 1, Table 4, and Table 5A)

Type	Conductor Size (AWG/ kcmil)	Trade Size (Metric Designator)												
		⅜ (12)	½ (16)	¾ (21)	1 (27)	1¼ (35)	1½ (41)	2 (53)	2½ (63)	3 (78)	3½ (91)	4 (103)	5 (129)	6 (155)
COMPACT CONDUCTORS														
THW, THW-2, THHW	8	—	1	4	6	11	15	26	37	57	76	98	155	224
	6	—	1	3	5	9	12	20	28	44	59	76	119	173
	4	—	1	1	3	6	9	15	21	33	44	57	89	129
	2	—	1	1	2	5	6	11	15	24	32	42	66	95
	1	—	1	1	1	3	4	7	11	17	23	29	46	67
	1/0	—	0	1	1	3	4	6	9	15	20	25	40	58
	2/0	—	0	1	1	2	3	5	8	12	16	21	34	49
	3/0	—	0	1	1	1	3	5	7	10	14	18	29	42
	4/0	—	0	1	1	1	2	4	5	9	12	15	24	35
	250	—	0	0	1	1	1	3	4	7	9	12	19	27
	300	—	0	0	1	1	1	2	4	6	8	10	16	24
	350	—	0	0	1	1	1	2	3	5	7	9	15	21
	400	—	0	0	0	1	1	1	3	5	6	8	13	19
	500	—	0	0	0	1	1	1	2	4	5	7	11	16

	600	—	0	0	0	1	1	1	1	3	4	5	9	13
	700	—	0	0	0	0	1	1	1	3	4	5	8	12
	750	—	0	0	0	0	1	1	1	2	3	5	7	11
	900	—	0	0	0	0	1	1	1	2	3	4	6	9
	1000	—	0	0	0	0	1	1	1	1	3	4	6	9
THHN, THWN, THWN-2	8	—	—	—	—	—	—	—	—	—	—	—	—	—
	6	—	2	4	7	13	17	29	41	64	86	111	175	253
	4	—	1	2	4	8	11	18	25	40	53	68	108	156
	2	—	1	1	3	5	8	13	18	28	38	49	77	112
	1	—	1	1	2	4	6	9	14	21	29	37	58	84
	1/0	—	1	1	1	3	5	8	12	18	24	31	49	72
	2/0	—	0	1	1	3	4	7	9	15	20	26	41	59
	3/0	—	0	1	1	2	3	5	8	12	17	22	34	50
	4/0	—	0	1	1	1	3	4	6	10	14	18	28	41
	250	—	0	0	1	1	1	3	5	8	11	14	22	32
	300	—	0	0	1	1	1	3	4	7	9	12	19	28
	350	—	0	0	1	1	1	3	4	6	8	10	17	24
	400	—	0	0	1	1	1	2	3	5	7	9	15	22
	500	—	0	0	0	1	1	1	3	4	6	8	13	18

(continued)

Table C.11(A) Maximum Number of Conductors or Fixture Wires in Rigid PVC Conduit, Schedule 40 and HDPE Conduit (Based on Chapter 9: Table 1, Table 4, and Table 5A) ***Continued***

Type	Conductor Size (AWG/kcmil)	Trade Size (Metric Designator)												
		⅜ (12)	½ (16)	¾ (21)	1 (27)	1¼ (35)	1½ (41)	2 (53)	2½ (63)	3 (78)	3½ (91)	4 (103)	5 (129)	6 (155)
COMPACT CONDUCTORS														
	600	—	0	0	0	1	1	1	2	4	5	6	10	15
	700	—	0	0	0	1	1	1	1	3	4	5	9	13
	750	—	0	0	0	1	1	1	1	3	4	5	8	12
	900	—	0	0	0	0	1	1	1	2	3	4	7	10
	1000	—	0	0	0	0	1	1	1	2	3	4	6	9
XHHW, XHHW-2	8	—	3	5	8	14	20	33	47	73	99	127	200	290
	6	—	1	4	6	11	15	25	35	55	73	94	149	215
	4	—	1	2	4	8	11	18	25	40	53	68	108	156
	2	—	1	1	3	5	8	13	18	28	38	49	77	112
	1	—	1	1	2	4	6	9	14	21	29	37	58	84
	1/0	—	1	1	1	3	5	8	12	18	24	31	49	72
	2/0	—	1	1	1	3	4	7	10	15	20	26	42	60
	3/0	—	0	1	1	2	3	5	8	12	17	22	34	50
	4/0	—	0	1	1	1	3	5	7	10	14	18	29	42

	250	—	0	0	1	1	1	4	5	8	11	14	23	33
	300	—	0	0	1	1	1	3	4	7	9	12	19	28
	350	—	0	0	1	1	1	3	4	6	8	11	17	25
	400	—	0	0	1	1	1	2	3	5	7	10	15	22
	500	—	0	0	0	1	1	1	3	4	6	8	13	18
	600	—	0	0	0	1	1	1	2	4	5	6	10	15
	700	—	0	0	0	1	1	1	1	3	4	5	9	13
	750	—	0	0	0	1	1	1	1	3	4	5	8	12
	900	—	0	0	0	0	1	1	1	2	3	4	7	10
	1000	—	0	0	0	0	1	1	1	2	3	4	6	9

Definition: *Compact stranding* is the result of a manufacturing process where the stranded conductor is compressed to the extent that the interstices (voids between strand wires) are virtually eliminated.

INDEX